W9-BWK-468

TJ
930
E83
2006

LIBRARY
NSCC, STRAIT AREA CAMPUS
226 REEVES ST.
PORT HAWKESBURY, NS B9A 2A2 CANADA

Piping and Pipeline Assessment Guide

Piping and Pipeline Assessment Guide

A. Keith Escoe

LIBRARY
NSCC, STRAIT AREA CAMPUS
226 REEVES ST.
PORT HAWKESBURY, NS B9A 2A2 CANADA

AMSTERDAM • BOSTON • HEIDELBERG • LONDON
NEW YORK • OXFORD • PARIS • SAN DIEGO
SAN FRANCISCO • SINGAPORE • SYDNEY • TOKYO

ELSEVIER

Gulf Professional Publishing is an imprint of Elsevier

Gulf Professional Publishing is an imprint of Elsevier
30 Corporate Drive, Suite 400, Burlington, MA 01803, USA
Linacre House, Jordan Hill, Oxford OX2 8DP, UK

Copyright © 2006, Elsevier Inc. All rights reserved.

No part of this publication may be reproduced, stored in a retrieval system, or transmitted in
any form or by any means, electronic, mechanical, photocopying, recording, or otherwise,
without the prior written permission of the publisher.

Permissions may be sought directly from Elsevier's Science & Technology Rights
Department in Oxford, UK: phone: (+44) 1865 843830, fax: (+44) 1865 853333,
E-mail: permissions@elsevier.com. You may also complete your request on-line
via the Elsevier homepage (http://elsevier.com), by selecting "Support & Contact"
then "Copyright and Permission" and then "Obtaining Permissions."

 Recognizing the importance of preserving what has been written, Elsevier prints its books on
acid-free paper whenever possible.

Library of Congress Cataloging-in-Publication Data
Escoe, A. Keith.
 Piping and pipeline assessment guide/Keith Escoe.—1st ed.
  p. cm.—(Stationary equipment assessment series; v. 1)
 Includes bibliographical references and index.
 ISBN-13: 978-0-7506-7880-3 (casebound : alk. paper)
 ISBN-10: 0-7506-7880-1 (casebound : alk. paper)
 1. Piping—Maintenance and repair. 2. Pipelines—Maintenance and repair.
 3. Service life (Engineering) I. Title. II. Series.
 TJ930.E83 2006
 621.8'672—dc22

                                                                    2005027032

British Library Cataloguing-in-Publication Data
A catalogue record for this book is available from the British Library.

ISBN 13: 978-0-7506-7880-3
ISBN 10: 0-7506-7880-1

For information on all Gulf Professional Publishing publications
visit our Web site at www.books.elsevier.com

06 07 08 09 10   10 9 8 7 6 5 4 3 2 1

Printed in the United States of America

Working together to grow
libraries in developing countries

www.elsevier.com | www.bookaid.org | www.sabre.org

ELSEVIER   **BOOK AID** International   **Sabre Foundation**

This book is dedicated to my beloved wife, Emma. We have lived and seen so much together!

Contents

2 An Introduction to Engineering Mechanics of Piping   50

3 Fitness-for-Service Topics of Local Thin Areas, Plain Dents, Dents-Gouges, and Cracks for Piping   100

5 Piping Support Systems for Process Plants   237

6 Piping Maintenance and Repairs   323

Preface

This book is written to be an assessment guide from the plant engineering, pipeline engineering and operations perspective. It is intended to serve as a guide for the practicing plant and pipeline engineer, operations personnel, and central engineering groups in operating companies. It will serve as a helpful guide for those in the engineering and construction companies to provide insight to plant and pipeline operations from their client's eyes and to writing specifications and procedures. It also will offer engineering students a perspective about plant and pipeline operations for a more productive career. Also the book will be a helpful guide for plant and pipeline inspectors who are so critical to the satisfactory operation of plant and pipeline facilities. The role and function of inspectors cannot be over emphasized.

The book is a fitness-for-service guide with emphasis on remediation of piping and pipelines containing flaws. The book is divided into eight chapters. Chapter 1 is about the basic concepts of fitness-for-service based on the work of the great pioneer Dr. John F. Kiefner and others who developed the field in the 1960s. The field of fracture mechanics was in its early stages of development, but the work by Kiefner, et al., served to translate the theory into practical use in pipelines. Chapter 2 is about the ASME piping and pipeline codes and the basic equations. Chapter 3 is fitness-for-service based on the API RP 579 with emphasis on local thin areas, plain dents, dents-gouges, grooves, and crack-like flaws for piping. The methodology of the API 579 is reorganized into methodology that simplifies the assessment for the practitioner. In Chapter 3, there is an extensive discussion about mechanical damage mechanisms. Chapter 4 is about the concerns of brittle fracture and how to assess it. After the basic fitness-for-service for piping is presented,

Chapter 5 concerns piping support mechanisms and the vital role they play in plant operations. This chapter discusses the maintenance function of plants and how various supports affect piping loads must be considered in fitness-for-service assessments. Chapter 6 is about piping maintenance and repairs with the emphasis on remediation of piping with flaws. This material is based on years of operating experience and combines into one chapter remediation techniques to solve maintenance and repair problems. Chapter 7 is about hot tapping and freezing. These techniques are invaluable in plant and pipeline operations to maintain operability of existing piping and pipelines. Finally, Chapter 8 is exclusively about pipelines with an insight of how the methodology of the API RP 579 can be used with pipelines. Currently the API RP 579 does not cover pipelines, but the methodology presented will help pipeline engineers and operators with methods to assess pipelines. Cathodic protection is briefly covered with a discussion about pigging technology and the various types of pigs and how they are used to detect mechanical flaws. Next remediation is discussed with presentations of various repair techniques in pipelines with a summary table from the upcoming ASME B31.4 classifying repair techniques and their limitations. Finally, there is a discussion concerning buried pipelines, the thermal expansion and consequent bowing of pipelines and practical solutions.

All chapters contain examples based on actual field problems. The author has tried to give examples in both the American Engineering System (AES—English or Imperial) of units and the metric SI unit system. This book is intended for world-wide use, so it is proper to present both unit systems. Also the metric SI unit system is now the preferred system in the ASME codes. However, the book is slanted toward the English system of units, but there are discussions about proper conversions between the systems. There are examples in the metric SI unit system. This should help U.S. engineers to become better acquainted with the metric SI system. It is expected over time that the metric SI system will become standard use everywhere; acknowledging that there are those, for obvious reasons, emotionally attached to one particular system of units.

For many years there were design codes for new equipment. Standards and recommended practices for assessing existing equipment were slower in development. Like the reasons for developing the ASME design codes for new equipment, operational problems in the plants and pipelines and explosions dictated the need for fitness-for-service.

When writing this book, the author thought of the many times he was called out to the plant in the middle of the night to face an operational problem. The specialists and support personnel were thousands of miles away and were not available for the situation. Many successful engineering solutions are performed in the far-away jungles or deserts of the world. It is against that background that this book has been developed. One classic example is having a contractor undersize several spring supports, and the engineer being faced with the hazard of a very hot pipe that contains a highly explosive and toxic process fluid trying to thermally expand and cannot because the springs have bottomed out and have locked-up to become rigid hangers. Spring hangers can't be delivered for weeks, so improvising is a must. Another situation is facing the leakage of a toxic substance because of a crack, and faced with placing a clamp in service in a hostile environment. These events have happened, do happen, and will continue to happen.

One purpose of this book is to assess such failures and prevent them from happening. The other is to show tools available to correct the problem when it occurs.

This book is the first volume in a series called the *Stationary Equipment Assessment Series*, or SEAS. The following volumes in this series will include assessment guides for pressure vessels and tubular exchangers and various other types of stationary equipment. The SEAS series is based on authentic actual field problems at facilities throughout the world. This series is written to provide helpful guides for mechanical engineers, plant operators, pipeline operators, maintenance engineers, plant engineers and inspectors, and pipeline engineers and inspectors, materials specialists, consultants, contractors, and academicians.

A. Keith Escoe

Chapter One
An Introduction to In-plant Piping and Pipeline Fitness-for-Service

Introduction

The field of fitness-for-service is a multidiscipline task composed of three technology areas—inspection, materials, and mechanical. This technology triad is shown in **Figure 1-1**, with various functions under each area. All three areas must be a part of the triangle, or the fitness-for-service concept doesn't work. The inspectors are the eyes and ears of every operating unit. They survey, look, and gather data critical to the assessment process. Many times they may assess the problem, and they usually write the repair procedures as their routine. The material specialists are the ones who are knowledgeable about the materials being applied, their properties, their limitations, and the mechanisms that attack the materials. The mechanical function, which is what this book is mainly about, is the analytical assessment of the piping or pipeline, using various techniques (e.g., finite element, fracture mechanics). Most mechanical specialists have stress engineering backgrounds. Pipe stress, or piping flexibility analysis, is rudimentary to this function.

The need for a more accurate assessment is the impetus of the development of this methodology. Economic conditions led to many process plants having restricted budgets; consequently, replacing piping as a method of solving flawed piping problems is sometimes economically prohibitive. Also, the ability to shut down a process stream to replace a pipe is not always an available option. This effort is a consequence of these concerns.

The need for fitness-for-service comes from flaws developed in piping. Such flaws as cracks, pitting, local thin areas (LTAs), and blisters are

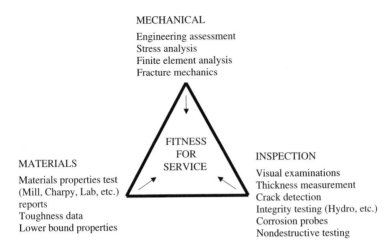

MECHANICAL

Engineering assessment
Stress analysis
Finite element analysis
Fracture mechanics

FITNESS
FOR
SERVICE

MATERIALS

Materials properties test
(Mill, Charpy, Lab, etc.)
reports
Toughness data
Lower bound properties

INSPECTION

Visual examinations
Thickness measurement
Crack detection
Integrity testing (Hydro, etc.)
Corrosion probes
Nondestructive testing

Figure 1-1. Technology triad required for fitness-for-service assessment.

generic terms that classify flaws. These flaws are caused by corrosion, erosion, and environmental exposure to various substances. Cracks are self-explanatory; we have all seen them in grandma's favorite porcelain platter and in house structures that have settled. Pitting is exactly what it says; it is localized areas of metal loss forming pits in pipe walls. These can be formed by CUI (corrosion under insulation) for the outside surface or corrosion or erosion on the inside surface of the pipe. LTAs are formed by CUI on the outside or corrosion or erosion on the inside surface. One classic example is seawater in carbon steel pipelines. Seawater, which is one of the most corrosive substances, attacks carbon steel and eats away portions on the inside pipe wall, forming LTAs. Some people call LTAs "lakes" because they look like a lake as seen on a map or from an airplane. We shall prefer the more accurate term, LTA. Blisters are formed by hydrogen attack and basically cause the pipe wall to delaminate. There are various forms of delamination—gross delamination occurs when a pocket forms within the pipe wall, but clusters of these pockets can also be scattered throughout the pipe wall. Blisters often form when carbon steel is exposed to hydrogen sulfide, but these are most common in refineries.

Carbon steel is the most common construction material in the petrochemical, refinery, and pipeline industries. The reason for this is simple economics: carbon steel is widely available, inexpensive, and maintainable (easy to work on). Certainly, many other metals are used. Austenitic stainless steels (normally 18% Cr-8% Ni, "eighteen-eight") are used for elevated and very low process temperatures. Chrome nickel alloys are used in various alloys for corrosion resistance at elevated temperatures. Chromium-molybdenum (called "chrome-moly") is used in the temperature range where carbon steel is limited (750–800°F, 400–426°C) and on up to around approximately 1000°F (538°C). Nickel works well for sodium hydroxide (NaOH, caustic soda, or "caustic") at high temperatures. At ambient and slightly higher temperatures, carbon steel can be used for caustic service. Titanium fans believe that this metal is the panacea for all material problems, but it is very expensive, hard to weld, and extremely difficult to decontaminate when it is exposed to caustic and has to be welded. Certainly titanium has its place, especially in highly corrosive environments; however, when it comes to selecting materials, it is always best to start with the simplest materials available and work up to the more noble materials and alloys. There is a common misconception that substituting a carbon steel component (e.g., a valve) for austenitic steel is an "upgrade." The consequences have been disastrous when the process contains chlorides because the chlorides attack the austenitic stainless, resulting in chloride stress corrosion cracking. On the other hand, for temperatures above the limit for carbon steel and Cr-Mo steels (such as 1Cr-Mo and $1\frac{1}{4}$Cr-$\frac{1}{2}$Mo), austenitic stainless may be the right choice. Here we are concerned mainly with carbon steels (there are many types!), but the methodology can be used for other materials. We will say when we are dealing with other materials.

This book is not about materials, but to discuss fitness-for-service intelligently, we need to discuss materials. Later in this chapter we will discuss various functions of the other part of the fitness-for-service triad—inspection. This book also is not about inspection, but we will address NDE (nondestructive examination) and other inspection topics as necessary.

Numerous pipe failures in the past have led to the need for assessing flawed pipe. The occurrence of several dramatic explosions resulted in extensive property damage and, in some cases, loss of human lives. In the early 1970s a passing automobile's spark plug initiated an explosion in eastern Texas that killed five people. In 1989 the Garden City line at Baton Rouge erupted and resulted in damages totaling over one hundred million dollars. In 1992 a Tenneco pipeline in White Bluff, Tennessee, exploded and resulted in considerable property damage. Thus, a piping fitness-for-service program is justified. Explosions such as those mentioned could conceivably destroy an entire process facility and kill many people. There have been more explosions related to piping and pipelines than to any other type of equipment.

The first fitness-for-service work was in the area of pipelines because it was first recognized that safety was of utmost importance in this industry. The methodology was later developed for process plants and then recorded in API 579, Fitness-For-Service [Reference 1]. Before the publication of API 579, this methodology was used with plant piping. Its simplicity and ease of use makes it a valuable tool for plant piping today.

The development of a suitable methodology was started in the 1960s and culminated in a classic work by John F. Kiefner and others in 1969 for the Battelle Institute. The basis of this technology was the many burst tests performed at the Battelle Memorial Institute. Development of closed-form solutions to fit the burst test data resulted in industry standards. Then API 579, Fitness-for-Service Recommended Practice (RP), did not exist. Burst tests were made to find the strength of pipelines with metal loss, either in the form of pits or corroded regions. Since that classic work, Kiefner and others have refined the methodology for assessing flawed pipelines. It was through Kiefner's work that the ASME ANSI B31G was developed. "Manual for Determining the Remaining Strength of Corroded Pipelines" came into being in 1984. When this method was developed, a simplified expression was needed; this methodology had to be easily applied in industry but also be conservative. The algorithms developed predicted the nominal hoop stress level that would cause failure of a corroded pipe with a specified wall thickness, pipe wall thickness, and SMYS (specified minimum yield strength). The method was inherently conservative, because it is assumed that the flow stress is equal to 1.1 times the yield strength, and that the corrosion flaw has a parabolic shape. This parabolic shape results in the "$\frac{2}{3}$ area factor." The method was widely accepted and used; nevertheless, it was criticized for being excessively conservative.

With the advent and use of personal computers, the assessment criterion was revisited in the late 1980s in an attempt to reduce the simplifying assumptions and associated conservatism. Additional burst tests were performed on corroded pipe and pipe with machined flaws, and algorithms were fit to the burst tests data. In the criterion, the flow stress was given by the less conservative definition of 10,000 plus the yield strength. Instead of using the "$\frac{2}{3}$ area factor," this criterion used an empirical factor of 0.85, as well as a more accurate three-term expression of the Folias bulging factor (see Chapter 3). The method resulted in a less conservative and more reliable estimate of the failure pressure than the earlier B31G. The result was the closed-form Modified B31G criterion published in 1991 along with the software RSTRENG (which means Remaining Strength). The B31G-1991 [Reference 2] is the latest edition. The Modified B31G method attempted to approximate a more exact effective area method (this iterative computation is described later), but it was unsuccessful, yielding more conservative results because of its approximation of the 0.85 dL. Pipeline operators and inspectors during the mid 1980s wished for a simpler method than the effective area method, which requires iterations. This was before the advent of the personal computer when electronic calculators were in widespread use. Unfortunately, some confuse the Modified B31G criterion with the term RSTRENG. The term RSTRENG in its more exact computation is the effective area method. The Modified B31G method passes the Einstein criterion, "Everything should be made as simple as possible, but not simpler." The Modified B31G criterion was too simple for accuracy.

Since the publication of the ASME/ANSI B31G-1991, further refinement of less conservative techniques appeared with data from additional burst tests. The new criterion employed the less conservative Folias bulging factor used in the Modified B31G, but incorporated the "equivalent axial" profile, which can be made by plotting points along the deepest path of the corroded area. The new version of RSTRENG, which is known as the *effective area method*, evaluated each different depth over a corroded area, assessing each area separately and as a whole. It then used the lowest failure pressure computed in the iterative analysis. As of this writing, the iterative approach will be the preferred method in the upcoming ASME B31.4 code, "Pipeline Transportation Systems for Liquid Hydrocarbons and Other Liquids." It is described in Chapter 8.

Figure 1-2 illustrates the development of the piping/pipeline fitness-for-service. The various tools used in piping fitness-for-service are shown in

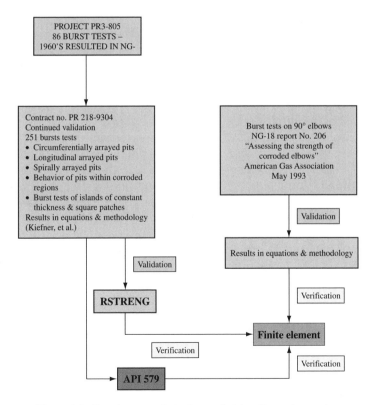

Figure 1-2. Development of pipeline and piping fitness-for-service.

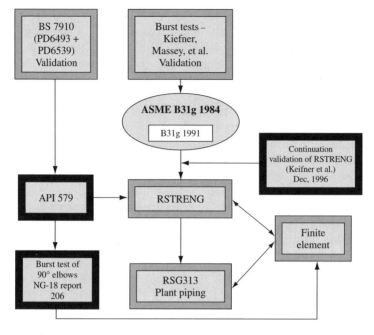

Figure 1-3. Tools used to perform piping and pipeline fitness-for-service (Newer version of RSG313 is called RGB313).

Figure 1-3. The history of the development is very interesting, but is not the subject of this discussion.

The United States was not the only country where burst tests were made. The Det Norske Veritas (DNV) produced an alternative method, known as RP-F101. Additionally, company standards, namely Shell-92, were produced independently of these efforts. Other methods are the PCORRC method and the LPC methods. PCORRC is predicting the remaining strength of corrosion defects in moderate- to high-toughness steels that fail by the mechanism of plastic collapse. LPC is the Line Pipe Corrosion Equation. Even though it is not practical to apply every standard to these problems, we selected the RSTRENG, based on the modified data, for this use because it is based on the results of pipeline steels that have SMYS up to 65,000 psi. For steels with higher yield strengths (e.g., X70 and X80 steel), the RSTRENG equation is not suitable. The Shell-92, PCORRC and LPC methods' equations are more suitable for these high strength steels, but are not applicable to low toughness pipes and pipes with high transition temperatures. The RSTRENG equation gives the least scatter and most consistent failure predictions. Most pipelines throughout the world are made of lower strength steel and not the X70 and X80 steel. For the relationship of RSTRENG to the API 579, refer to the WRC Bulletin 465 [Reference 3].

In Chapters 3 and 8, we will discuss applying the finite element method to these problems. We will also introduce the API 579, "Fitness-for-Service." The use of RSTRENG here is augmented by the fact that it was used in the development of the API 579.

RSTRENG has seen widespread use in the pipeline industry. It was first produced as a DOS-based program, and later an updated Windows version, known as RSTRENG2 [Reference 4] was developed. The two basically provide the same assessment, even though RSTRENG2 is based on more recent research. Kiefner and others have produced a set of Excel spreadsheets that are used for the same purpose, KAPA (Kiefner and Associates, Inc.'s Pipe Assessment) [Reference 5]. Additionally, RSTRENG3, a newer Windows-based application, has recently been released.

Our application here extends from applications to plant piping. The same software is utilized, except that the ASME B31.3 piping code rules are used. The most significant difference is that the pipeline codes ASME B31.4/B31.8 have allowable stresses of $F*$ (SMYS), with design factors (F) varying from 0.5 to 0.72 applied in B31.8 (B31.4 uses 0.72); whereas in B31.3 the allowable stress for most carbon steels is one third of the UT (ultimate tensile strength) of the material. For example, API 5L GR B has a specified tensile strength of 60,000 psi at ambient temperature. The allowable stress at ambient is 20,000 psi. API 5L X52 has an ultimate strength of 66,000 psi and an allowable stress of 22,000 psi (SMYS = 52,000 psi, hence X52). The reader is cautioned that one third of the UT as an allowable stress is not valid for all steels. The burst tests made by Kiefner and others initially were only on low-grade carbon steels with a wall thickness of one inch or less; later higher grades of steel were tested, and the results were published in the report "Database of Corroded Pipe Tests, Contract PR-218-9206" [Reference 6], or in "Continued Validation of RSTRENG, Contract No. PR-218-9304" [Reference 7]. This methodology was not tested for other materials (e.g., austenitic stainless steels) because the vast majority of pipelines are constructed of carbon steels. For austenitic stainless steels, two thirds of the SMYS is used as the allowable stress in the ASME B31.3. Thus, use of this methodology is limited to carbon steel pipe of a wall thickness of one inch or less. For other materials, other methodology (e.g., API 579 or finite element) should be used.

In process plants, pipe elbows and branch connections are far more numerous than pipelines. Battelle Memorial Institute performed burst tests on elbows ("Assessing the Strength of Corroded Elbows" for the Line Pipe Research Supervisory Committee of the American Gas Association, May 1993) [Reference 8]. The basic methodology for straight piping is based on that developed by Kiefner and others, but it is tailored to ASME B31.3 requirements. For elbows we use finite element studies, which have been verified with the work by Battelle Memorial Institute.

The methodology developed by these researchers is valid for piping with corroded areas. The methods presented are too cumbersome to accomplish by hand; consequently, RSTRENG is available to do this. This software package will be discussed in more detail later in this chapter. It must be emphasized that the burst tests performed by Kiefner were based on the following:

1. The burst tests were made on carbon steel piping.
2. The burst tests were made on piping 0.593 inch (15 mm) in thickness and less.
3. The maximum SMYS was X65.

Use of RSTRENG is valid only if the preceding three conditions are met. Oftentimes RSTRENG is used for pipelines with wall thickness up to one inch, but such cases where the nominal wall exceeds 15 mm should be checked with another method. This fact is based on the latest RSTRENG validation; see [Reference 7]. We will discuss these methods in Chapter 3 for in-plant piping and Chapter 8 for pipelines. The critical value of Kiefner's work is that it is validated by burst tests. The results being validated are extremely valuable; however, using the analysis for another material is risky. Using the method for piping over an inch thick may lead to error. Also the testing was mainly concerned with pitting or local thin areas (see Chapter 3) along the longitudinal axis, although Kiefner provides rules for corrosion in the circumferential direction. For large LTAs and conditions that do not meet the preceding three criteria, finite element and/or the API 579 is recommended, as discussed in Chapter 3. Here we will first concentrate on the RSTRENG method because of its simplicity, ease of use, and ubiquitousness in use.

What Is Piping?

This question seems very basic, but it is seemingly confusing to some. The prime function of piping is to transport fluids from one location to another. Pressure vessels, on the other hand, basically store and process fluids. Can piping be used as pressure vessels? Yes, it's done all the time; however, when piping is used as stamped ASME vessels, it no longer falls under the piping codes, but the vessel codes (e.g., ASME Section 8 Div. 1).

The allowable stresses in piping are categorized differently than those for vessels. In piping, for example, sustained stress and expansion stress are used versus primary and secondary stresses for pressure vessels.

The word "piping" generally refers to in-plant piping—process piping, utility piping, etc.—inside a plant facility. The word "pipeline" refers to a long pipe running over distances transporting liquids or gases. Pipelines do often extend into process facilities (e.g., process plants and refineries). There is a detailed discussion of pipelines in Chapter 8. *Do not confuse piping with pipelines; they have different design codes and different functions.* Each one has unique problems that do not exist in the other.

Areas Where Corrosion Attacks Piping

The plant maintenance engineers and inspectors have categorized locations where corrosion on piping systems is most likely. Approximately *80–85%* of corrosion occurs at pipe supports on straight pipe runs; *15–20%* of piping corrosion occurs at elbows. Corrosion at branch connections (e.g., tees) is rare.

Described herein is an improved method for assessing the remaining strength of corroded pipe. A new and improved method was desired because of the known excess conservatism in the B31G method. As previously noted, one reason for the conservatism is the use of the parabolic area method. The iterative calculations in the RSTRENG software yield more accurate assessments. Even though the B31G method has helped avoid unnecessary repairs and replacements, the excess conservatism continues to cause some unnecessary repairs. The methodology described herein was devised not only to ensure adequate piping integrity but also to eliminate as much as possible the excess conservatism embodied in the B31G method.

The Maximum Allowable Operating Pressure (MAOP)

The pipeline codes use the term Maximum Allowable Operating Pressure (MAOP); whereas, the API 579 uses the term Maximum Allowable Working Pressure (MAWP). Even though the ASME B31.3 uses design pressure, we use MAWP for assessment purposes (API 579). The ASME B31.3 is for new piping, not assessment. The design pressure (or current operating pressure) of the pipeline must be compared to a calculated "safe" pressure for both axial and circumferential extent of corrosion. This process can occur only after the parameters L, d (or $t - d$), and C are determined. Whenever the evaluation of a corroded region indicates that the safe operating pressure exceeds the maximum operating or design pressure, the piping under evaluation may be returned to service. If this is not the case, then further assessment, retirement, or reduction of pressure is necessary.

The ASME B31G, the ASME B31G modified, and the effective area method (RSTRENG) were developed exclusively for pipelines, but they can be used for in-plant piping with modifications, as shown in

Flow chart for evaluating metal loss areas

NOTE: SMTST = Lower of SMTS at ambient or operating temperature

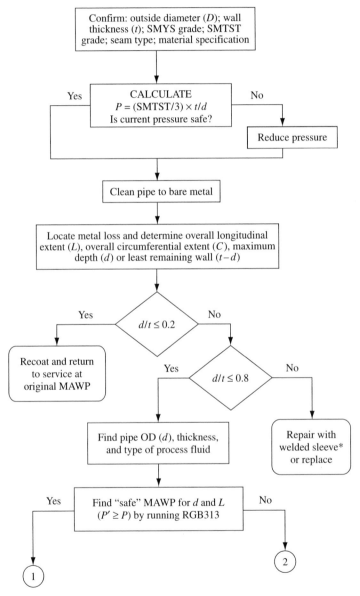

Figure 1-4. Flowchart for evaluating metal loss areas. *See chapter 6. Adapted from O'Grady et al. [Reference 9].

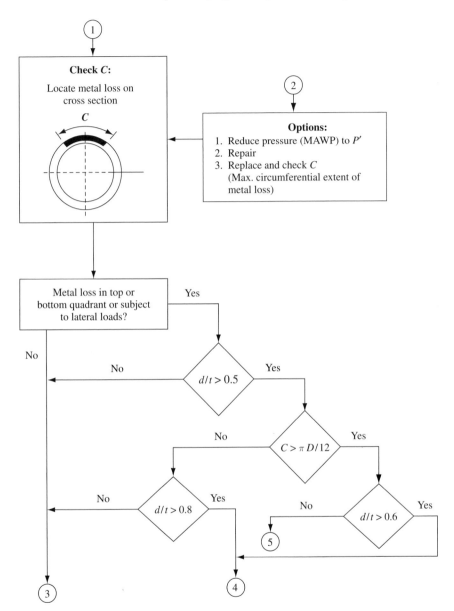

Figure 1-4. cont'd.

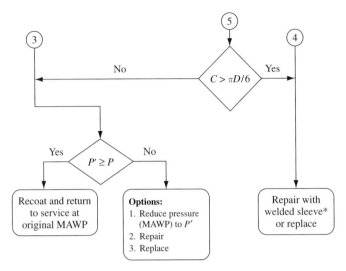

Figure 1-4. cont'd. *See chapter 6.

Figure 1-4. This methodology was used successfully for many years before the publication of the API RP 579, which was first published in 2000. To use **Figure 1-4** for in-plant piping, namely ASME B31.3, one must make modifications. The calculated pressure becomes

$$P = \frac{2\sigma_a E t}{2(\sigma_a E W + 0.4t)}$$ Eq. 1-1

where σ_a = ASME B31.3 allowable stress at temperature, psi (MPa)
E = weld joint efficiency factor
t = nominal wall thickness, in. (mm)
W = weld strength reduction factor, for temperatures below 950°F (510°C), $W = 1$

In the ASME B31.3, $Y = 0.4$ for temperatures below 900°F (482°C).
For the grade of carbon steels used in the burst tests developing the ASME B31G methods and the effective area method, namely API 5L Gr B, X42, X46, X52, X60, and X65, the allowable stress is

$$\sigma_a = \frac{SMTS}{3}$$ Eq. 1-2

where $SMTS$ = lower of specified minimum tensile strength at ambient or operating temperature, psi

For this reason, when using RSTRENG the maximum allowable pressure is

$$P_a = \frac{P_f}{3}$$
                                                          Eq. 1-3

where P_f is the minimum failure pressure determined by RSTRENG, ASME B31G, or ASME B31G modified method.

One can run the effective area method using software (e.g., RSTRENG) and dividing the minimum failure pressure by 3.0 to obtain the maximum allowable pressure. In in-plant piping, most companies do not use sleeves, such as those used in pipelines. Sleeves can be used for in-plant piping, but the thermal stresses resulting from the differential thermal expansion between the sleeve and process pipe must be considered. Refer to Chapter 6 for details and examples. Other methods of repair are possible, and these are also covered in Chapter 6. Example 1-2 provides an example of in-plant piping. In using the methodology of burst test algorithms, external loads (e.g., bending moments and torsion) are not considered. To consider these loadings, one must use the API RP 579 (see Chapter 3) or finite element.

In **Figure 1-4**, the use of a welded sleeve is mentioned. There are two types of welded sleeves—pressure containing and nonpressure containing. Chapter 8 offers more details about these sleeves and also discusses nonmetallic composite sleeves. These types of sleeves are creeping into use for in-plant piping where the operating temperature does not exceed the temperature limitations of the composite manufacturer's recommendations.

Four levels of assessment of corroded pipe are available. These levels should not be confused with the API 579 Levels 1, 2, and 3. The first method involves finding the safe operating pressure on the basis of the overall longitudinal length and maximum depth of the corroded area. This level is the ASME B31G analysis.

The second level of assessment involves the more rigorous iterative method to find the safe operating pressure on the basis of a detailed map of pit depths or remaining wall thickness. This second level is used only when the piping under consideration fails the first level and replacement or repair is necessary or when there has to be a reduction in operating pressure.

The third level is generally referred to as grandfathering (i.e., the piping is considered satisfactory by using the accepted criteria, or is compatible to a piping system that has been determined to be acceptable).

The fourth level involves a more refined analysis to assess the corrosion damage. This may involve application of the API 579 or the finite element method, discussed in later chapters.

The detailed analysis should be based on pit-depth measurements taken at a maximum distance of one inch generally along the longitudinal axis of the pipe for a minimum distance of D (diameter) on both sides of the deepest pit. This establishes the profile for that corroded region that is used in the assessment.

Several profiles of the corroded regions may be necessary to obtain the optimum for evaluation. For example, it may be preferable to select profiles that are slightly at an angle to the axis of the pipe to reduce the number of profiles considered.

The maximum deviation from the pipe longitudinal axis should not exceed $\pm 5°$. This is to keep the profile from being significantly affected by the curvature of the pipe. The main objective is to select the profiles that have the deepest pits.

Assessment Procedure

The evaluator needs to confirm the pipe outside diameter (D), wall thickness (t), grade of pipe material, type of weld seam, design pressure, and process fluid for the pipe segment under consideration. Next the evaluator needs to find the maximum depth of corrosion pitting (d) or the least remaining wall thickness ($t - d$) of each separate corroded area. The evaluator then records the overall axial and circumferential extent of the corrosion and the deepest penetration or least remaining wall thickness. This evaluation needs to be made after examining all the metal loss and identifying with certainty the deepest penetration or minimum remaining wall thickness. Next, the evaluator needs to calculate the maximum allowable operating pressure (MAOP) that corresponds to the applicable design code.

The following criteria of piping assessment are recommended by Kiefner and others:

1. If the deepest pit (d) is found to be 20% of the specified (nominal) wall thickness (t) and the least remaining wall thickness ($t - d$) is at least 80% of the original (nominal) wall thickness, the segment containing the corroded region is acceptable for continued service.
2. If the deepest penetration (d) is determined to be greater than 80% of the specified (nominal) wall thickness or the least remaining wall thickness is less than 20% of the specified (nominal) wall thickness, the segment containing the corrosion should be replaced or repaired.
3. If the failure pressure equals or exceeds the original MAOP of the pipe, the segment containing the corroded region is acceptable for continued service.

4. If the safe pressure is less than the original MAOP but greater than or equal to the actual MAOP, a new safe working pressure is required, or an appropriate repair needs to be made.
5. The need to establish safe conditions is necessary if the safe pressure is less than the actual MAOP pressure.

Classification of Corroded Regions

To make the evaluation and measurement of corrosion damage in piping systems easier, it is necessary to apply several criteria. The parameters and criteria used to classify the effects of metal loss are discussed next.

External Versus Internal Corrosion

The effect of metal loss on the inside and outside is the same. What is not the same are various opinions on how to evaluate the damage. With external corrosion, measurements can be made with simple tools (e.g., pit gauges, rulers). Once the surface of the pipe has been cleaned, the extent of the damage is obvious.

Determining the extent of damage for internal corrosion requires the application of an ultrasonic device. The use of a corrosion allowance for internal corrosion is a consideration not used for external corrosion. For external corrosion, once the damaged surface has been cleaned, evaluated, and recoated, it may be assumed that the corrosion will not continue. This is not the case with internal corrosion, where a specified amount (corrosion allowance) may be added to the measured depth of the corrosion to render the evaluation valid for a given time span. The amount of corrosion allowance applied should be consistent with the corrosion rate and time anticipated before the next inspection.

Localized Versus General Corrosion

Localized pitting consists of well-defined, relatively isolated regions of metal loss. The longitudinal and circumferential extents of such regions are easily determined by measuring the depths of the pits. A pit gauge will often be sufficient to measure pit depths because the pipe's original surface can usually be used as a plane of reference.

Unlike localized corrosion, general or widespread corrosion can make measurement difficult. When corrosion exists, small islands of original pipe can be linked with a straight edge and then a scale can be used to measure the pit depths along the straight edge. When such islands do not exist, or if the piping is not straight, the remaining wall thickness can be determined from an ultrasonic device. The application of such devices often requires grinding flat spots on the pipe. Remember that removing any metal potentially lowers the failure pressure of the pipe. Consequently, it is best to start such grinding at high spots to obtain an indication of the remaining wall thickness.

If none of the pits within an area of general corrosion exceeds 20% of the wall thickness, no repair is required regardless of the length of the corrosion. Ultrasonic measurements should be used to determine the remaining wall thickness is at least 80% of the minimum wall thickness for the design stress level.

Interaction of Closely Spaced Areas of Corrosion

The consideration of regional interaction is necessary if the regions are near enough to each other to have a combined effect on the strength of the pipe. **Figure 1-5** shows three types of interaction.

Type 1 regions are separated in the circumferential direction but overlap in the longitudinal direction. When the circumferential separation between the outside boundaries of the individual pits is less than six times the pipe wall thickness (t), the pits should be considered to act together. In **Figure 1-5**, the length should be taken as the overall length (L). When the

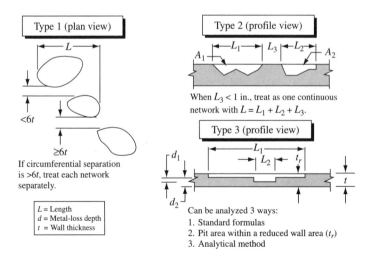

Figure 1-5. Types of metal loss interactions. Courtesy of Dr. John F. Kiefner, PRCI, and the *Oil and Gas Journal* [Reference 8].

Table 1-1
Methods for Assessing Type 3 Regions

Method	Defect Length	Defect Depth
1. Use formula (see section on Methodology)	L_1	D_2
2. Consider as short flaw in pipe with reduced wall thickness	L_2	$d_2 - d_1$, then multiply safe pressure by $(1 - d_1/t)$
3. Iterative analysis	Effective area	Effective area

corroded pits are separated by $6t$ or greater, the pits should be considered separately.

Type 2 regions lie on the same longitudinal line but are separated by an island of full pipe wall thickness. When this island of full wall thickness is less than one inch long, the pits should be considered as a continuous flaw that has an overall length of $L_1 + L_2 + L_3$. When the island of full wall thickness is greater than one inch, the pits can be considered separately. In this case L_1, L_2, and L_3 represent then individual lengths of the respective pits.

Type 3 regions consist of one or more deep pits within long, shallow corroded regions. These types of regions can be considered in three ways: by using standard formulas; by considering the region as a shorter, deeper flaw in a pipe of reduced wall thickness (t_r); or by iterative analytical methods (see **Table 1-1**).

Circumferential Extent of Damage

In assessing a corroded region where the depth of the corrosion is greater than 50% of the original pipe wall thickness and the circumferential extent is greater than 1/12 (8.33%) of the circumference, the circumferential extent should be measured and recorded.

Welds, Elbows, and Branch Connections

For the various types of welds, the following rules of thumb may be applied:

Submerged-Arc Seam Welds. Corrosion in a submerged-arc seam weld should be assessed as though it were in the parent pipe.

Electric-Resistance Weld (ERW) Seam Welds. In these types of welds, the corrosion is sometimes located on the centerline area of an ERW. Often this type of corrosion is quite severe and cannot be evaluated with the lower levels of assessment. Corrosion in an ERW requires a Level 4 type of assessment. See the section on Methodology.

Girth Welds. These types of welds are vulnerable to preferential internal erosion-corrosion, which causes circumferential groove-like metal loss. This phenomenon is common where tapewrap is used and water migrates into the trap.

Corrosion in Elbows. See section on Methodology.

Branch Connections. See section on Methodology.

Corroded Pit Region Interaction Parameters

With one corroded pit region, one needs only to consider the length (L) of the region and its maximum depth (d) to assess its effect on the integrity of the pipe. If two or more such corroded regions are in close proximity, then their combined effects must be considered.

Burst tests data show that the origin of failure is in the thinnest regions, which one might logically assume. Thus, combining the profiles by considering the composite of all the profiles within a metal loss region is required. There are limits beyond which widely separated corroded regions need not be combined for assessment. Shown in **Figure 1-6** are the parameters used in these criteria. The actual criteria for determining whether corroded regions are in proximity are shown in **Table 1-2**.

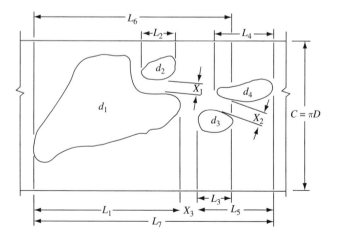

Figure 1-6. Several metal loss profiles. Courtesy of Dr. John F. Kiefner, PRCI, and the *Oil and Gas Journal* [Reference 8].

Table 1-2
Corroded Pit Interaction Assessment Criteria

Overall Flaw Length	Separate or Interfacing	Conditions*	Maximum Depth	Conditions
L_1	Separate	If $X_3 > 1$ in.	d_1	If $d_1 > d_2$
			d_2	If $d_2 > d_1$ and $x_1 > 6t$
L_2	Separate	If $X_1 > 6t$	d_2	
L_3	Separate	If $X_3 > 1$ in. and $X_2 > 6t$	d_3	
L_4	Separate	If $X_3 > 1$ in. and $X_2 > 6t$	d_4	
L_5	Interacting	If $X_3 > 1$ in. and $X_2 < 6t$	d_3	If $d_3 > d_4$
			d_4	If $d_4 > d_3$
L_6	Interacting	If $X_3 < 1$ in. and $X_2 > 6t$	d_1	If $d_1 > d_2, d_3$
			d_2	If $d_2 > d_1, d_3$
			d_3	If $d_3 > d_1, d_2$
L_7	Interacting	If $X_3 < 1$ in. and $X_2 < 6t$	d_1	If $d_1 > d_2, d_3, d_4$
			d_2	If $d_2 > d_1, d_2, d_4$
			d_3	If $d_3 > d_1, d_2, d_4$
				If $d_4 > d_1, d_2, d_3$

*X = Distance of full wall thickness between metal loss areas (corroded regions).

The combining of corroded regions requires some experience and judgment. When experienced personnel are unavailable, then the most conservative estimates should be made.

Methodology

The ASME/ANSI B31G criterion is based on a semiempirical fracture mechanics relationship referred to as the NG-18 surface flaw equation. This name comes from the work sponsored by the NG-18 Line Pipe Research Committee of the American Gas Association (AGA). Before this criterion is described any further, it is appropriate to discuss the basic parameters used in this criterion.

The concept of flow stress ($\overline{S}$) is a concept used to account for strain-hardening in terms of an equivalent elastic-plastic material having a yield strength ($\overline{S}$). This allows one to express the plastic flow in a real material as a single parameter. The flow stress magnitudes for piping materials have been found to average approximately 10,000 psi greater than the yield strength (σ_y) of the material. It is measured by means of a transverse flattened tensile specimen.

The Folias factor (M_T) is a term that describes the "bulging" effect of a shell surface that is thinner in wall thickness than the surrounding shell. This phenomenon exists for internal pressure and is more pronounced for smaller diameter shells. The Folias factor is a function of the flaw length divided by the shell's diameter and thickness. As the shell diameter increases, the Folias factor approaches 1.0. Thus, the parameter generally is more important to piping than vessels. It was first analytically derived by Efthymios S. Folias in 1964.

SMYS and yield strength (sometimes referred to as yield stress) are used differently in the piping assessment literature. The SMYS is the absolute minimum yield strength for a particular grade of material and is specified by ASTM. This yield strength must be obtained by the mill in the tensile tests. Yield strength is the actual yield strength of the material indicated in the mill reports. If the yield strength is not known, the SMYS value should be used in the calculations. If the yield strength is known, then it should be used to obtain a more realistic assessment.

The failure stress level (S_f) represents the predicted burst pressure of the pipe. The failure pressure S_f should never be less than the value of the SMYS.

The NG-18 surface flaw equation is

$$s = \bar{S} \left[\frac{1 - \dfrac{A}{A_o}}{1 - \left(\dfrac{A}{A_o}\right) M_T^{-1}} \right] \qquad \text{Eq. 1-4}$$

where M_T = the Folias factor, a function of L, D, and t
S = the hoop stress level at failure, psi
$\bar{S}$ = the flow stress of the material, a material property related to its yield strength, psi
A = area of crack or defect in the longitudinal plane through the wall thickness, in.^{2}
$A_o = Lt$, in.^{2}
L = the axial extent of the defect, in.
t = the wall thickness of the pipe, in.
D = the pipe diameter, in.

This equation is used to calculate the failure stress level of a pipe loaded with internal pressure containing an axially oriented (as mentioned previously, $\pm 5°$ deviation from axial direction is maximum permitted) crack or defect. It is also used to determine the remaining strength of corroded pipe where the parameters of the metal loss are found as shown in **Figure 1-7**.

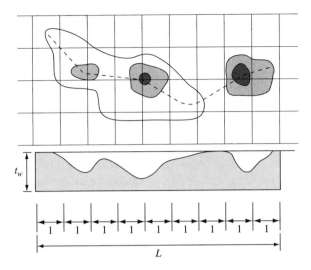

Figure 1-7. Projection of metal loss onto longitudinal axis.

The use of the parameters L and A as defined in Eq. 1-4 for an array of pits generally yields a conservative assessment of the remaining strength for several reasons. The first reason is that Eq. 1-4 assumes that the corroded pits are perfectly lined up in the axial direction of the pipe. When they are not so lined up, Eq. 1-4 will underestimate the remaining strength of the pipe. As shown in **Figure 1-7**, the line connecting the corroded pits is projected onto the longitudinal axis of the pipe. The length of the corroded region is the length of the projected line on the longitudinal axis. The longitudinal axis is used for the projection because the hoop stress is the maximum stress for internal pressure.

The second reason for the conservative assessment of Eq. 1-4 is that corrosion pits are blunt defects compared to cracks and many other types of flaws found in piping systems. It has been shown that blunt surface flaws have significantly higher failure stress levels than sharp surface flaws. For this reason, Eq. 1-4 was developed on the data of burst tests of pipe having relatively sharp flaws. Thus, Eq. 1-4 will give conservative predictions for the effects of blunt flaws. This conservatism has not been changed in this modified approach.

The B31G criterion is based on the original Battelle work, which represents a simplified algorithm for the determination of the effect of a corroded defect on the hoop stress of a pipe. It is not the purpose of this document to duplicate the development or application of the B31G criterion. Rather, we will address the practical application of the modified criteria.

The conservatism of the original B31G criterion leads to excessive amounts of piping either being unnecessarily repaired or replaced. The sources for the conservatism follow:

1. The expression for the flow stress.
2. The approximation used for the Folias factor.
3. The parabolic representation of the metal loss (as used within the B31G limits), primarily the limitation when applied to long areas of corrosion.
4. The inability to consider the strengthening effect of islands of full pipe wall thickness or near-full thickness at the ends of or between arrays of corrosion pits.

It was known even when the original B31G criterion was developed that 110% of the SMYS substantially underestimates the flow stress of the piping material. For the Kiefner modified criterion, the value of the flow stress is taken as SMYS + 10,000 psi. As will be shown later, this value can be adjusted with a factor of safety.

The two-term approximation for the Folias factor used in the original B31G criterion has been replaced by more exact and less conservative approximations. These approximations follow:

For values of $\dfrac{L^2}{Dt} \leq 50$,

$$M_T = \sqrt{1 + 0.6275\left(\frac{L^2}{Dt}\right) - 0.003375\left(\frac{L^2}{Dt}\right)^2} \qquad \text{Eq. 1-5}$$

For values of $\dfrac{L^2}{Dt} \geq 50$,

$$M_T = 0.032\left(\frac{L^2}{Dt}\right) + 3.3 \qquad \text{Eq. 1-6}$$

The parabolic representation of the area (A), where $A = \frac{2}{3}Ld$, was developed after Kiefner performed 47 burst tests. In reality, the parabolic representation of the area (A) has significant limitations. If the corroded region is very long, the effect of the metal loss is underestimated, and the remaining strength is overestimated. To prevent the misuse of the criterion for long, deep corroded regions, the method was limited to defects for which $L^2/Dt \leq 20$.

A second method of analyzing the remaining strength on the basis of the profile, which is more accurate than the parabolic method, is called the equivalent length method. In this method, the metal loss area (A) is defined by

$$A = l\left(\frac{d_0 + d_1}{2}\right) + l\left(\frac{d_1 + d_2}{2}\right) + \cdots + l\left(\frac{d_{n-1} + d_n}{2}\right)$$

$$= l\left(\frac{d_0 + d_n}{2} + \sum_{i=1}^{n-1} d_i\right) = x_i \sum_{i=1}^{n-1} d_i = L_{TOTAL} d_{avg} \qquad \text{Eq. 1-7}$$

from which

$$A = L_{eq} d_{avg} \qquad \text{Eq. 1-8}$$

In the equivalent length method, the total length is replaced by the equivalent length, and the defect is represented by a rectangle $L_{eq}d$. The parameter L_{eq} is used in Eq. 1-8 for M_T to calculate the failure stress level (S_f). In this method, the term A/A_o in Eq. 1-4 becomes d/t. An arbitrary constant of 0.85 was chosen to calculate the corroded area by 0.85 dL. This is known as the Modified B31G method. A comparison of the parabolic area to the modified B31G method is shown in **Figure 1-8**. The parabolic method was developed in the early to mid 1980s before personal computers were common. Its development was in response to pipeline operators who thought that the iterations in the effective area method were too cumbersome for hand calculations. With the advent of personal computers, this method has been dropped in favor of the effective area method.

Various tests and finite element studies have shown that for irregular defects such as corrosion the calculation of the remaining strength on the

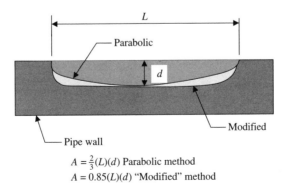

$$A = \tfrac{2}{3}(L)(d) \text{ Parabolic method}$$
$$A = 0.85(L)(d) \text{ "Modified" method}$$

Figure 1-8. Comparison of parabolic method with modified method.

basis of the total area and total length of the defect does not always lead to the minimum value of remaining strength.

A third more accurate method to predict the remaining strength involves calculations based on the various subsections of the total metal loss. One could calculate as many predicted failure pressures based on the number of pits that existed. Each calculation involves a length L_i with i being the interval between each pit. The area of each flaw is then calculated as the sum of the areas of the trapezoids made up by the discrete depth points within L_i. This methodology generally results in a minimum predicted failure stress that is less than the values determined from the exact area or total length methods. This method is called the effective area method because it is based on the effective length and area of the defect. It is an iterative analysis, solving for the minimum failure pressure using Eqs. 1-1, 1-2, and/or 1-3 and running an iteration of each corroded pit. If one inputs ten pits, then ten iterations are performed to solve for the minimum failure pressure of each, using the correct Folias factor in each case.

In the effective area method, each pit is assessed with a combination of other corroded regions in an iterative procedure. The number of calculations performed follows: Where n = number of pits, the number of calculations is given by an arrangement of n pit readings taken into a sequence of two terms. Mathematically this is written as

$$_2C_n = \frac{n!}{2!(n-2)!} = \binom{n}{2} \qquad \text{Eq. 1-9}$$

Thus, for 24 pit readings, the number of iterations is 276. This iterative approach is considered very accurate and has been validated by hundreds of burst tests. When a line is drawn between the pits, these pits are projected onto the longitudinal axis. The length of the projected line on the longitudinal is the length of the flaw.

For example, if one takes seven pit readings and runs RSTRENG, the program will make the following number of iterations:

$$_2C_7 = \frac{7!}{2!(7-2)!} = 21$$

This is illustrated in **Figure 1-9**.

In-plant piping that falls under ASME B31.3, making the maximum allowable pressure is one third of failure pressure (i.e., one third of ultimate strength in the calculations). This new modified version is called RGB313. In RGB313, the failure stress for 100% and 33.3% of ultimate are printed out. In fact, one can run RSTRENG, divide the minimum failure pressure

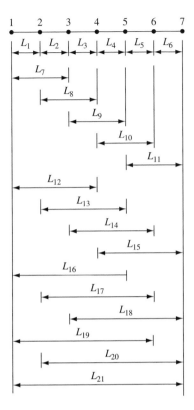

Figure 1-9. For seven pit readings, RSTRENG performs twenty-one computations. The lowest value is the minimum failure pressure.

by 3, and get the same results (see following discussion). This practice was used before API RP 579 with success—it works. One must be cognizant of the limitations of RSTRENG when using it, as previously stated. However, this methodology is valid only for internal pressure loads. If there are external bending, torsion, or shear loads, RSTRENG does not consider this. For these loads, refer to Chapter 3. For most carbon steels and steels in which CUI (corrosion under insulation) and other corrosion modes are a problem, the allowable stress is one third of ultimate strength. For austenitic stainless steels, the allowable is two thirds of the specified minimum yield strength at temperature. Normally these materials are not our prime focus.

Note that RSTRENG is the name of the original DOS-based software developed by Kiefner et al., based on burst tests of flawed piping. Occasionally in this book we use the term RGB313. RGB313 is the exact same software except the failure pressure is divided by 3 per ASME B31.3 for carbon steels. Some versions of RSTRENG multiply the failure pressure by 0.72 per ASME B31.4 or ASME B31.8. When the term RGB313 is used, it means RSTRENG applied to B31.3. In some of the examples, we use the term RGB313.

Determining the Allowable Length of Corrosion

It is desirable to find the allowable length of a defect. To find this parameter, some mathematical manipulation is required. First, let the failure stress $S_f = SMYS$ and $\bar{S} = SMYS + 10,000$. We define stress flow ratio (q) using

$$q = \frac{SMYS + 10,000}{SMYS} \qquad \text{Eq. 1-10}$$

Solving for M_T using Eqs. 1-5 and 1-10, we have

$$M_T = \frac{(-q)\left(\dfrac{A}{A_o}\right)}{1 - \dfrac{A}{A_o} - \dfrac{1}{q}} \qquad \text{Eq. 1-11}$$

Setting Eq. 1-5 equal to the right-hand side of Eq. 1-11, we solve for length L. The fourth-order equation has a solution with four roots, of which three are trivial and one is valid for L. These solutions are for $L \leq \sqrt{50Dt}$

$$L = \sqrt{Dt}\left[92.963 - \left[92.963^2 + 296.296\left[1 - \frac{\left(\dfrac{-1}{q}\right)^2\left(\dfrac{A}{A_o}\right)}{\left(1 - \dfrac{A}{A_o} - \dfrac{1}{q_o}\right)}\right]\right]^{0.5}\right]^{0.5} \qquad \text{Eq. 1-12}$$

for $L > \sqrt{50Dt}$

$$L = \sqrt{Dt}\left[\frac{31.25\left(\dfrac{-1}{q}\right)\left(\dfrac{A}{A_o}\right)}{1 - \dfrac{A}{A_o} - \dfrac{1}{q}} - 103.125\right]^{0.5} \qquad \text{Eq. 1-13}$$

The original MAOP of the pipe is found from the following:

$$P = \frac{2(SMYS)(t)F}{D}$$ Eq. 1-14

Under the new criteria, the reduced operating pressure (P') is found as follows:

for $L \leq \sqrt{50Dt}$

$$P' = \frac{\dfrac{2Ft(SMYS + 10000)}{D}\left(1 - \dfrac{A}{A_o}\right)}{1 - \dfrac{A}{A_o}\left[1 + 0.6275\left(\dfrac{L^2}{Dt}\right) - 0.003375\left(\dfrac{L^2}{Dt}\right)^2\right]^{-0.5}}$$

Eq. 1-15

for $L > \sqrt{50Dt}$

$$P' = \frac{\dfrac{2Ft(SMYS + 10000)}{D}\left(1 - \dfrac{A}{A_o}\right)}{1 - \left(\dfrac{A}{A_o}\right)\left[0.032\left(\dfrac{L^2}{Dt}\right) + 3.3\right]^{-1}}$$ Eq. 1-16

If the safe operating pressure (P') is used for values of L that are too large to satisfy Eq. 1-15, then Eq. 1-16 is used. In using Eqs. 1-15 and 1-16, P' must be less than or equal to P.

Corrosion Allowance

To account for future corrosion, Kiefner et al. chose a method that is neither the most conservative nor the least. This approach adds to the calculations the desired allowance to the measured depth of existing corrosion. This approach yields a new MAOP based on additional corrosion in the same location as the existing damage. We define a value for the MAOP of the original pipe after the corrosion allowance (P_o) has been taken into account. This value is defined as

$$P_o = \frac{2F(SMYS)(t - CA)}{D}$$ Eq. 1-17

where

$$\frac{A}{A_o} = \frac{0.85(d + CA)L}{tL} = \frac{0.85(d + CA)}{t}$$ Eq. 1-18

For $L \leq \sqrt{Dt}$

$$P' = \frac{\dfrac{2Ft(SMYS + 10000)}{D}\left(1 - \dfrac{0.85(d + CA)}{t}\right)}{1 - \left(\dfrac{0.85(d + CA)}{t}\right)\left(1 + 0.6275\dfrac{L^2}{Dt} - 0.003375\left(\dfrac{L^2}{Dt}\right)^2\right)^{-0.5}}$$

Eq. 1-19

For $L > \sqrt{Dt}$

$$P' = \frac{\dfrac{2Ft(SMYS + 10000)}{D}\left(1 - \dfrac{0.85(d + CA)}{t}\right)}{1 - \left(\dfrac{0.85(d + CA)}{t}\right)\left(3.3 + 0.032\left(\dfrac{L^2}{Dt}\right)\right)^{-1}}$$ Eq. 1-20

The values of the safe operating pressure (P') for values of L that are too large to satisfy (Eq. 1-18) are calculated by Eq. 1-19. Once again, P' must be $\leq P_o$.

Assessing Type 3 Flaws

The three types of flaws were described previously and are shown in **Figure 1-5**. First, one can always use RSTRENG and input the detailed profiles of pit depths or remaining wall thickness to obtain the minimum failure pressure. Second, one can always use the overall length (L) and the maximum depth (d) to calculate a safe working pressure using the preceding equations. However, using these equations instead of RSTRENG will yield an overly conservative safe operating pressure.

If RSTRENG is not available, the type 3 flaw can be assessed by the following method. Refer to **Figure 1-9** to see the type 3 flaws. In this method, the B31G criterion is applied for a short defect by assuming that the remaining wall thickness in the long shallow defect is an uncorroded pipe of lesser wall thickness. This use of net remaining wall shows the application of the more liberal short-defect equations in B31G.

To apply this method, follow these four steps:

1. Determine the lengths (L_1, L_2, etc.) and depths (d_1, d_2, etc.) of all the shorter, deeper flaws.
2. The inspector should examine the pipe for a distance $L/2$ on both sides of the flaws to assess the remaining thicknesses (t_{r1}, t_{r2}, etc.) that represent the deepest portions of the longer, shallower flaw.
3. The inspector next determines t_{net1}, t_{net2}, etc., the actual net thickness below each of the shorter, deeper flaws.
4. Substitute values for d_1, d_2, etc., and L_1, L_2, etc., and t_{r1}, t_{r2}, etc., in place of t into Eqs. 1-15 or 1-16. It is critical to use the values of t_{r1}, t_{r2}, etc., in place of t.

The conditions for an unlimited corroded region length to exist follow:

Acceptable Conditions for an Unlimited Corroded Region

1. If $t_a \geq t_{nom}$ (as required by design) and $t_{net} \geq 0.8t_{nom}$, where $d =$ measured depth with reference to the uncorroded surface, d may be $>0.20t_{nom}$ (see Figure 1-9a)
2. If $t_a < t_{nom}$ and $t_{net} > 0.8t_{nom}$ and $d < 0.20t_{nom}$

Note 1: The pit depth d is *not* compared to t_a, the actual measured thickness in application of the 20% rule. D must *always* be compared to t_{nom}.

Note 2: The condition $t_{net} > 0.8t_{nom}$ must *always* be met for L to be unlimited.

Note 3: The 20% rule does not allow portions of corroded areas to be neglected in determining L. If $t_{net} < 0.8t_{nom}$, where t_{net} is not $<0.8t_{nom}$, this rule must not be applied to those portions where $t_{net} < 0.8t_{nom}$.

Note 4: All of the corroded area is to be included in determining L. Only if the effective length as determined by RSTRENG is less than L is one permitted to consider a length less than L.

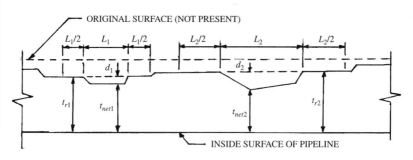

Figure 1-9a. Adapted from O'Grady *et al.*

Burst Tests Validation

Reference 9 rigorously discusses the continued validation of RSTRENG. Included are 129 new tests to further validate the methodology that has been successfully used for many years. A pipe test specimen before testing is shown in **Figure 1-10**. Note that pits exist on the outside surface of the pipe. Such pitting is quite common with CUI and soil contamination on pipelines. **Figure 1-11** shows the pipe specimen after the burst test with the origin of

Figure 1-10. Test specimen before burst test. Courtesy of Dr. John F. Kiefner, PRCI, and the *Oil and Gas Journal.*

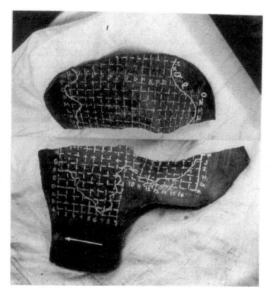

Figure 1-11. Test specimen after burst test. Courtesy of Dr. John F. Kiefner, PRCI, and the *Oil and Gas Journal.*

rupture. Such testing is the basis of this methodology and is invaluable in developing a fitness-for-service approach to piping and pipelines.

Circumferential Corrosion

It has been said that with pressurized pipelines one will need to assess corrosion in the circumferential direction only rarely. This is not true with pipelines containing seawater. The author has worked with seawater injection pipelines where seawater is injected into underground oil reservoirs at pressures of 3000 psig (20,689.7 KPa$_g$). As the oil is pumped out of the reservoir, seawater is injected into the reservoir to prevent it from collapsing. In these systems, corrosion along the circumferential direction is quite common. We will discuss the methodology of addressing this subject in Chapter 3. In plant piping, circumferential corrosion can also occur more often. There are situations where this condition occurs and the situation must be assessed.

Criteria for Circumferential Metal Loss

The key dimensions related to the effect of circumferential metal loss are shown in **Figure 1-12**. It must be emphasized that the longitudinal extent of corrosion is always of greatest importance and should be considered first.

The analysis of circumferential corrosion is based on assuming maximum longitudinal stresses for the in-plane direction (maximum longitudinal stresses at the top and bottom). For maximum longitudinal stresses in the out-of-plane direction, which can be caused by high lateral loads, the analysis is the same as that for in-plane bending.

As shown in **Figure 1-12**, if $d > 0.5t$ and the circumferential extent of the corrosion is greater than 1/12 (8.33%) of the circumference, the maximum circumferential extent (S) of the damage should be determined and indicated on a cross-sectional drawing.

The regions of metal loss are laid out on the drawing around the circumference, using the top of the pipe as the reference. The parameters are the maximum circumferential extent (S) and the maximum depth (d).

In Kiefner's method, if the corroded regions extend beyond the top or bottom quadrants, the part that extends beyond the quadrant is not considered in the analysis. Only if the location of the corrosion is near an expansion loop, or otherwise has high lateral loading, are these extended corroded areas relevant to the analysis. In plant piping, compressive loads are not necessarily limited to the bottom and top quadrants. Large bending moments

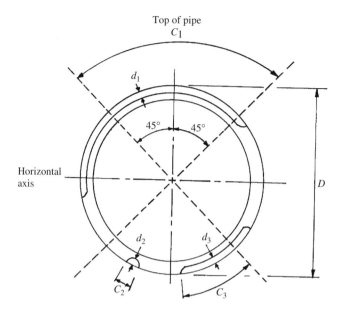

Figure 1-12. Circumferential metal loss.

can exist at any plane, creating tensile loads on one side and compressive loads on the opposite side. In **Figure 1-4**, only the acceptance criteria limits the extent of circumferential corrosion to the top or bottom coordinates. For example, $C > \pi D/12$ (30°) and $C > \pi D/6$ (60°) are the criteria for further action. The parameters S and d may not necessarily occur at the same cross section. Once d is located, the inspector needs to examine the pipe for a distance (D) on both sides of the location of pipe containing d. The maximum combination of S within D of the section of pipe containing d should be used to define the effect of the circumferential corrosion.

Metal loss in the top and bottom quadrants should be considered separately. Interaction between the two regions would not be expected; hence, the maximum metal loss may exist simultaneously in both quadrants of a given segment.

1. The circumferential extent of corrosion is significantly larger than the longitudinal extent of corrosion. Also, the corrosion in the longitudinal direction is too small to result in the piping having to be repaired, or such repair will not alleviate the weakness in the circumferential direction.
2. The piping is subjected to high longitudinal tensile stress.
3. The location of corrosion is in an area of compressive stress and any metal loss could result in local buckling.

The allowable circumferential metal loss may be as much as half the original pipe wall thickness with no restrictions on length. If it is found in the analysis that the remaining hoop strength of the pipe (longitudinal defects) is not altered, one may use the following criteria to determine the circumferential corrosion:

When the depth of corrosion exceeds $0.5t$, the circumferential extent shall be limited as follows:

$d < 0.5t$, no limit on S
$0.5t < d < 0.6t$, $S \leq \pi D/6$
$0.6t < d < 0.8t$, $S \leq \pi D/12$
$d > 0.8t$, remove or repair

Another guideline to remember is that mentioned by API 579, Paragraph 4.4.3.3.c.3, which says that if the metal loss in the circumferential direction is significant enough to alter the section modulus of the pipe, then the new section modulus is computed and entered into the pipe stress analysis for reassessment. This would be done if piping loads, other than pressure, were of such magnitude to affect the analysis. Judgment should be used; for example, corroded regions with high external loads should be assessed with numerical methods (e.g., finite element).

Methodology of Circumferential Metal Loss

Kiefner et al. recommended the Wilkowski-Eiber method. This method is conservative, as the depth of the flaw is considered uniform over a circumferential direction S (see **Figure 1-12**).

In the analysis of circumferential defects, the parameter S_L is the longitudinal stress instead of the hoop stress as in the situation with longitudinally oriented flaws. A circumferential flaw is affected by longitudinal stress in the pipe and not by the hoop stress. Thus, knowledge of the longitudinal stress is essential to the analysis of such flaws.

An empirical stress-intensification factor (M_c) analogous to the Folias factor was derived from test data. This factor, which is greater than unity, is a function of the through-wall flaw's arc length. It is divided into the flow stress of the material ($\overline{S}$) to obtain the predicted longitudinal bending stress that will produce failure of a through-wall flaw.

The Wilkowski-Eiber method was intended for use in relatively narrow defects' repair grooves in pipes subjected only to bending stress (no uniform axial stress). To apply this method, one needs only the flow stress ($\overline{S}$) and the flaw and pipe geometries to calculate the failure stress in the following equations:

$$M_C = \left[1 + 0.26\left(\frac{\lambda}{\pi}\right) + 47\left(\frac{\lambda}{\pi}\right)^2 - 59\left(\frac{\lambda}{\pi}\right)^3\right]^{\frac{1}{2}} \qquad \text{Eq. 1-21}$$

where λ = one half the angle of the arc, α, where α is C in **Figure 1-13**.

$$S_L = \overline{S}\left[\frac{1 - \dfrac{d}{t}}{1 - \left(\dfrac{d}{t}\right)M_C^{-1}}\right] \qquad \text{Eq. 1-22}$$

These equations only hold when $0.5 \leq d/t \leq 0.8$. If this criterion is not met, then a higher level study is required. In this case, use API 579, Paragraph 5.4.3.3.d, "Alternate Assessment Procedures." It states that if the circumferential plane cannot be approximated by a single area, then a numerical procedure (finite element) be used to compute the section properties and the membrane and bending stresses resulting from pressure and supplemental loads. The acceptance criteria for the stress results shall be per Step 6 in Paragraph 5.4.3.3.c.

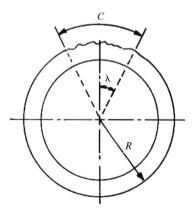

Figure 1-13. Computing the Folias factor (M_C).

Corrosion in Pipe Bends

The internal pressure stress distribution in a 90° elbow varies in the circumferential direction. The ASME B31.3 gives formulations for these stresses in Paragraph 304.2.1 in equations 3c, 3d, and 3e. Equation 1-19 is the general formula for the required wall thickness for internal pressure, which reads as follows:

$$t = \frac{PD}{2\left[\left(\dfrac{SE}{I}\right) + PY\right]} \qquad \text{Eq. 1-23}$$

The term I used in Eq. 1-23 is defined by the code in Eq. 1-20 for the intrados as follows:

$$I = \frac{4\left(\dfrac{R_1}{D}\right) - 1}{4\left(\dfrac{R_1}{D}\right) - 2} \qquad \text{Eq. 1-24}$$

The equation for I at the extrados, Eq. 1-21, is as follows:

$$I = \frac{4\left(\dfrac{R_1}{D}\right) + 1}{4\left(\dfrac{R_1}{D}\right) + 2} \qquad \text{Eq. 1-25}$$

where R_1 = bend radius, in.
     D = outside diameter of the pipe, in.

These parameters are shown in the code in the ASME B31.3 Figure 304.2.1 (not shown). The pressure stress profile, shown in **Figure 1-14**, illustrates how the hoop stress varies.

The assessment of corroded elbows goes beyond Eqs. 1-23, 1-24, and 1-25, just like that for piping with pits and LTAs. This was accomplished in the research study "Assessing the Strength of Corroded Elbows, NG-18 Report No. 206," by T. A. Bubenik and M. J. Rosenfeld for the Line Pipe Research Supervisory Committee of the Line Pipe Research Committee of the American Gas Association [Reference 8]. In this study, burst tests were made on 90° elbows. This report forms the basis of assessment for corroded elbows. It is a very important

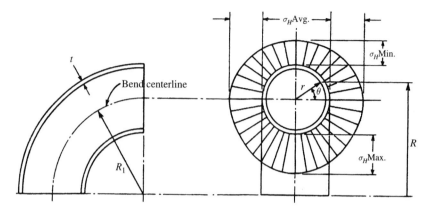

Figure 1-14. Internal pressure stress in a 90° elbow.

work because the burst tests provide validation for the assessment procedures.

In the NG-18 Report No. 206, equations for theoretical elastic stress distributions are presented as the Lorenz factor, which accounts for the uniform stress distribution around the circumference. This theoretical stress distribution is a function of the Lorenz factor, which is defined as

$$LF = \left[\frac{\dfrac{R_b}{R_m} + \dfrac{\sin\alpha}{2}}{\dfrac{R_b}{R_m} + \sin\alpha} \right]$$

Eq. 1-26

where LF = the Lorenz factor
  R_b = the bend radius
  R_m = the pipe radius
  α = the circumferential angle measured from the crown of the elbow ($+90°$ at the extrados, $-90°$ at the intrados)

The Lorenz factor is an indicator of the increase or decrease in the nominal stresses in an elbow relative to a straight pipe (e.g., the I factor used in B31.3). For a long radius elbow, where $R_b/R_m = 3$, the Lorenz factor reduces to the following:

$$LF = \left[\frac{\dfrac{R_b}{R_m} - 0.5}{\dfrac{R_b}{R_m} - 1.0} \right] = 1.25$$

$$LF = \left[\frac{\dfrac{R_b}{R_m} + 0.5}{\dfrac{R_b}{R_m} + 1.0} \right] = 0.875$$

For bends where the bend radius $\geq 1.5D$ (D = OD, outside diameter of pipe), these can be assessed by the methods used for straight pipe. The assessment of 90° elbows involves the following discussion.

Example 3 addresses an LTA on a 90° elbow. The program RGB313 (as well RSTRENG) can be used on the extrados of the elbow only. For the area at or near the intrados, either finite element of API 579 can be used. The finite element results were benchmarked with the burst tests in NG-18 Report No. 206 and API 579.

Branch Connections and Fittings

As mentioned previously, plant field inspectors report that the incidence of corrosion at branch connections is rare. RGB313 (and RSTRENG) cannot assess corroded regions at branch connections. Here API 579 methodology and/or finite element techniques must be applied.

Determining a Maximum Allowable Operating Pressure

When one finds the minimum failure pressure using the effective area method (RSTRENG, KAPA), the ASME B31G modified method, or the ASME B31G method, it is desirable to find the MAOP. The KAPA software automatically computes the MAOP, but the RSTRENG DOS version does not. It is important to know how the MAOP is computed from the failure pressure, P_f. The MAOP is computed from the failure pressure as follows:

$$MAOP = \left(\frac{YS}{\sigma_f} \right)(F)P_f \qquad \text{Eq. 1-27}$$

where YS = actual yield strength of heat of steel, psi
F = design factor of 0.72 per ASME B31.4 or value given in B31.8
P_f = failure pressure computed by RSTRENG, psi
σ_f = failure stress computed by RSTRENG, psi

If the yield strength (*YS*) is greater than the failure pressure computed by RSTRENG (σ_f) then the ratio of YS/σ_f is taken to be 1.0. This situation, which has happened in several burst tests, will be discussed further in Chapter 8.

In the case of liquid hydrocarbons or liquids that fall under ASME B31.4, *F* = 0.72. Some companies elect to use a lower design factor. If this is the case, then the design factor is adjusted accordingly.

Flaws in Heat-Affected Zones of Welds

RSTRENG does not consider the effects of residual stress or the toughness of a material. This is necessary in assessing flaws in the heat-affected zone (HAZ) of welds. One tool commonly used for this purpose is BS 7910, "Guide on Methods for Assessing the Acceptability of Flaws in Fusion Welded Structures" by the British Standard Institute. This standard, like RSTRENG, is validated by actual testing. The API 579 Paragraph 9.4.4.1e recognizes the BS 7910 as an acceptable method for assessing cracks. The discussion here is about metal loss and pitting. RSTRENG can and is used for cracks, but our discussion pertains to metal loss applications in this book.

A pit can be assessed like a crack if it is assumed that it is a blunt crack. A blunt crack is approximated by a crack located in a high toughness material. The surrounding high toughness makes the crack blunt because it limits its ability to propagate. The API 579 provides upper shelf values for what a high toughness material is. In Appendix F, Paragraph 4.4.1e, high toughness is defined as 100 ksi $\sqrt{\text{in.}}$ (110 MPa $\sqrt{\text{m}}$). Using this value, we can predict the behavior of pits in the HAZ of a weld.

Residual stresses are present after fusion welding. They can be mechanically relieved through the application of loads, such as hydro test. A pipe may be loaded the first time to the maximum combination of loads it is expected to experience. Then local yielding may occur in areas of high secondary and peak stresses, especially if high residual stresses exist in these areas prior to loading. Once this local yielding has occurred, the pipe may be reloaded to this level without further yielding. This is called elastic shakedown or shakedown to elastic conditions. After shakedown has occurred, it is reasonable to assume that the sum of the operating stress plus the residual stress will not exceed yield except in areas with high local stress concentrations. The minimum value of mechanically reduced residual stress that may be used in a fracture assessment is 15% of the yield strength. For more detail on residual stress, refer to API 579 Section 9 and BS 7910, Paragraph 7.2.4.4. In the BS 7910, Paragraph 7.2.4.4, credit on residual stress reduction from elastic shakedown can only be done if the hydro test was made before the flaw occurred. Another rule of thumb is that when post-weld heat treating is applied, a maximum reduction of 80% of the yield strength is a reasonable assumption for a *longitudinal flaw*. This is from API 579, Paragraph E.4.4.1c. If the orientation is a *circumferential flaw*, a maximum reduction of 70% is allowed per Paragraph E.4.4.2c in API 579. Crack-like flaws will be discussed in Chapter 3.

Example 1-1: This example demonstrates how the equations are applied to a practical application. A single iteration that demonstrates the use of the equations is required in this problem. In most applications, depending on the number of different pit depths, several or many iterations are necessary. For example, for Case 1, do not conclude that a single calculation is sufficient. The pipeline is designed per ASME B31.4 and has the following properties:

Diameter (OD) = D = 24.00 in.
Nominal thickness = t = 0.365 in.
Yield strength = 41,800 psi
$SMYS$ = 35,000 psi
Length of corrosion = L = 2.75 in.
Increment of corrosion readings = 0.25 in.

The pit measurements are given in mils, where 1 mil = 1/1000 inch. The pit readings in mils follow:

0; 136; 188; 261; 219; 188; 157; 178; 178; 157; 136; 0.

$$d_{avg} = \frac{136 + 188 + 261 + 219 + 188 + 157 + 178 + 178 + 157 + 136}{10}$$

d_{avg} = 179.8 mils, or approx. 0.180 in.

Case 1: Effective Area Method

First, we calculate the failure pressure with YS = 41,800 psi

$$AR = A/A_O = (10)(0.25)(0.180)/((2.75)(0.365)) = 0.45$$

Then, we compute the Folias factor as

$$M_{T1} = \left[1 + \frac{1.255L^2}{2Dt}\right]$$

The failure pressure is calculated as

$$P_{FAIL} = \left(2\frac{t}{D}\right)(YS + 10,000)\left(\frac{1 - AR}{1 - \dfrac{AR}{M_{T1}}}\right)$$

$$P_{FAIL} = 1360 \text{ psi}$$

Case 2: Modified B31G Method

$$A = 0.85 \text{ dL}$$
$$A = 0.61$$
$$AR = 0.85(d)/t$$
$$AR = 0.608$$

$$\frac{L^2}{(Dt)} = 0.863$$

$$q = \frac{SMYS + 10,000}{SMYS}$$

$$q = 1.286$$

$$L = \sqrt{Dt}\left[92.963 - \left[92.963^2 + 296.296\left[1 - \frac{\left(\dfrac{-1}{q}\right)^2 AR^2}{\left(1 - AR - \dfrac{1}{q}\right)^2}\right]\right]^{0.5}\right]^{0.5}$$

$$L = 2.656 \text{ in.}$$

$$M_T = \left[1 + 0.6275\left(\frac{L^2}{Dt}\right) - 0.003375\left[\frac{L^2}{(Dt)^2}\right]^2\right]^{0.50}$$

$$M_T = 1.227$$
$$F = 1.0$$

$$P' = \frac{\dfrac{2Ft(SMYS + 10,000)}{D}(1 - AR)}{1 - AR\left[1 + 0.6275\left(\dfrac{L^2}{Dt}\right) - 0.003375\left(\dfrac{L^4}{D^2t^2}\right)\right]^{-0.5}}$$

Library, Nova Scotia Community College

$P' = 1065$ psi

$$MAOP = \frac{2SMYS(t)}{D}F$$

$MAOP = 1065$ psi

Case 3: Parabolic Method B31G Method

$A_O = Lt$

$$A_2 = \left(\frac{2}{3}\right)(d)L$$

$$AR = \frac{A_2}{A_o}$$

$d = 0.261$

$SMYS = 35,000$ psi

$L = 2.75$ in.

$$M_T = \sqrt{1 + \frac{0.805(L)^2}{Dt}}$$

$M_T = 1.302$

$$P_{FAIL} = \left(2\left(\frac{t}{D}\right)\right)(1.1SMYS)\frac{1 - AR}{\left(1 - \frac{AR}{M_T}\right)}$$

$P_{FAIL} = 966.8$ psi

According to B31G-1991, Part 2, the maximum allowable longitudinal extent of corrosion is

$$B = \sqrt{\left[\frac{\dfrac{d}{t}}{1.1\left(\dfrac{d}{t}\right) - 0.15}\right]^2 - 1}$$

$B = 0.512$

$L = 1.12B\sqrt{Dt}$

$L = 1.696$ in.

These results are shown in **Figure 1-13**. This version of RSTRENG is the original DOS version and should not to be confused with RSTRENG2 and RSTRENG3 Windows versions.

The RSTRENG software results follow:

EXAMPLE 1-1 IN SEASBK CHAPTER 1

```
FILENAME: C:\RSTRENG\EXAMPLE1.RST

Specimen = EXAMPLE 1-1          IN Date = 01-26-2004
Diameter = 24.00 in.           Thickness = 0.365 in.
Yield Str. = 41,800 psi.       Comment =
!-------------------------------------!
012
0.00*------------------------------------------*
0.04-
0.08-
0.12-**
0.16-****
0.20-***
0.24-
0.28-*
0.32-
0.36-
0.40-

CASE 1 MINIMUM         CASE 1 MINIMUM         72% MINIMUM
Failure Stress         Failure Pressure       FAILURE PRESSURE
psi.                   psi.                   psi.
44,759                 1,361                  980

(*) NOTE: NO SAFETY FACTOR APPLIED TO CASE 1, 2, or 3
SMYS = 35,000 psi. (NOTE: No Safety Factors Applied to
CASE 1, 2 or 3)
100% 72%

CASE 2 MODIFIED METHOD USING AREA = 0.85 dL : 1,052 psi.
758 psi.

CASE 3 EXISTING B31G METHOD USING AREA = 2/3 dL : 968 psi.
697 psi.

---PIT DEPTH MEASUREMENTS (MILS)---(MAX. Pit Depth =
0.261 inch)---

0.00 0
0.25 136
0.50 188
0.75 261
1.00 219
1.25 188
1.50 157
```

```
1.75 178
2.00 178
2.25 157
2.50 136
2.75 0
```

Using Eq. 1-27, we calculate the MAOP as follows:

$$MAOP = \left(\frac{41,800}{44,759}\right)(0.72)(1361) = 915.1 \text{ psig}$$

Note that in the code for RSTRENG (DOS version), there is a scale, 0.00, 0.04, 0.08, 0.12, 0.16, 0.20, 0.24, 0.28, 0.32, 0.36, 0.40. The program divides the pipe (or pipeline) wall thickness into tenths. Thus, the 0.365 in. wall is rounded to 0.40 in., and the wall is scaled into tenths of 0.40 in.

Example 1-2: A 48 in. duct connects to a vessel made of SA-516-70 and has a nominal wall of 0.75 in. The pipe is fabricated from rolled plate and has a weld seam with a joint efficiency of 0.85. The uniform metal loss (UML) is zero, but the pipe has a future corrosion allowance (FCA) of 0.10 in. The design conditions are 300 psig at 350°F. Inspection has found an external corroded area shown in **Figure 1-15**. The pipe has a

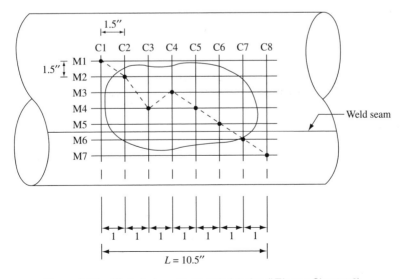

Figure 1-15. 48″ duct pipe with corroded region (LTA; see Chapter 3).

design pressure of 300 psig at 350°F. Perform a fitness-for-service assessment for the pipe.

We can expect $FCA = 0.10$ in. to occur within the next operating period. Thus,

Wall thickness $= 0.75 - 0.10 = 0.65$ in.

At 350°F, from the ASME Section II, Part D, $SMYS = 33,000$ psi. Because we have no indication in this example that a specimen was extracted from the cylinder and a test made for finding the yield strength, we assume the yield strength is equal to the SMYS. For carbon steels and most ferrous metals, the allowable strength is the tensile strength divided by 3 according to ASME B31.3. In the example in API 579, the allowable stress is modified by multiplying it by the joint factor (E) of 0.85. For RSTRENG we adjust the SMYS by multiplying it by E, giving

$SMYS = 33,000(0.85) = 28,050$ psi

The flaw length is 10.50 in. with 1.50 increments between point readings. From **Table 1-3** we connect a line between the points that represent the deepest pits. The points that represent the deepest pits along the circumferential and longitudinal axes are known as the critical thickness profiles (CTP). This line is projected onto the longitudinal axis as shown in **Figure 1-15**. The data on the line representing the longitudinal axis are entered into RGB313 (RSTRENG modified for ASME B31.3).

Table 1-3
Inspection Data for LTA for 48″ Duct Pipe

Long. Inspect. Planes	C1	C2	C3	C4	C5	C6	C7	C8	Circum. CTP
M1	0.75	0.75	0.75	0.75	0.75	0.75	0.75	0.75	0.75
M2	0.75	0.48	0.52	0.57	0.56	0.58	0.60	0.75	0.48
M3	0.75	0.57	0.59	0.55	0.59	0.60	0.66	0.75	0.55
M4	0.75	0.61	0.47	0.58	0.36	0.58	0.64	0.75	0.36
M5	0.75	0.62	0.59	0.58	0.57	0.48	0.62	0.75	0.48
M6	0.75	0.57	0.59	0.61	0.57	0.56	0.49	0.75	0.49
M7	0.75	0.75	0.75	0.75	0.75	0.75	0.75	0.75	0.75
Long. CTP	0.75	0.48	0.47	0.55	0.36	0.48	0.49	0.75	

The computer run is as follows:

EXAMPLE 1-2

```
FILENAME: D:\RSTRENG\EXAMPLE1.PHR

Specimen = EXAMPLE 1-2      IN Date = 06-08-2004
Diameter = 48.00 in.        Thickness = 0.650 in.
Yield Str. = 28,050 psi.    Comment = USING ASME B31.
!------------------------------------!
0 7.5
0.00*------------------------------------*
0.04-
0.08-
0.12-****
0.16-*
0.20-*
0.24-
0.28-
0.32-
0.36-
0.40-


CASE 1 MINIMUM         CASE 1 MINIMUM        B31.3 MAXIMUM
Failure Stress         Failure Pressure      ALLOW. PRESSURE
psi.                   psi.                  psi.
30,332                 821                   274
```

(*) NOTE: NO SAFETY FACTOR APPLIED TO CASE 1, 2, or 3
SMYS = 28,050 psi. (NOTE: No Safety Factors Applied to
CASE 1, 2 or 3)
100% B31.3

CASE 2 MODIFIED METHOD USING AREA = 0.85 dL : 707 psi.
236 psi.

CASE 3 EXISTING B31G METHOD USING AREA = 2/3 dL : 630 psi.
210 psi.

--- PIT DEPTH MEASUREMENTS (MILS) --- (MAX. Pit Depth =
0.390 inch) ---

```
 0.00  0
 1.50  270
 3.00  280
 4.50  200
 6.00  390
 7.50  270
 9.00  260
10.50  0
```

The maximum allowable pressure of 274 psig is less than the design pressure of 300 psig, so the cylinder is not satisfactory.

Checking for the Circumferential Direction Criteria

RSTRENG does not have an algorithm for circumferential corrosion. However, as mentioned previously, Wilkowski and Eiber developed a method that allows one to assess corrosion in the circumferential direction. The method was developed for predicting the critical size of a circumferential repair groove that could be made at a girth weld in the pipeline as it was being laid off a pipe-laying barge into the water.

An empirical Folias factor was derived from test data. The factor was derived as a stress intensification factor that is a function of the through-the-wall flaw's arc length and divided into the flow stress of the material to obtain the predicted longitudinal bending stress that will produce failure in a through-the-wall flaw.

The failure stress levels for surface flaws can then be calculated on the basis of the empirical stress intensification factor (M_c). To use the criteria, we must follow the flow chart in **Figure 1-4**.

Now the maximum depth of the flaw (d) is

$$0.75 - 0.10 - (0.36 - 0.10) = 0.39 \text{ in.}$$

Thus, $d/t = 0.39/0.65 = 0.6$, which means that $d/t > 0.5$. Now $R = 48.0/2 = 24.0$ in. There are six intervals along the circumferential length of the LTA, giving a chord length of $6(1.5) = 9.0$ in.

Referring to **Figure 1-13**, we see that

$$\alpha/2 = \lambda = \text{ARCSIN} \left(\frac{9.0(0.5)}{24.0} \right) = 10.81 \text{ deg} = 0.189 \text{ rad}$$

Now,

$$\frac{\pi D}{12} = \frac{\pi (48.0)}{12} = 12.57$$

$$C = R\alpha = (24.0)(0.189)(2) = 9.07 \text{ in.}$$

Thus,

$C < \pi D/12$ and $d/t = 0.6$ but not greater than 0.6

Referring to **Figure 1-4**, we have

$C < \pi D/6 = \pi(49.30)/6 = 25.81$ in.

We now must calculate the pressure (P') and the resulting stress (S_L). We calculated P' using RGB313 as 274 psig.

Computing the factor M_C, we have $\lambda = 0.189$ rad, so

$$M_C = \left[1 + 0.26\left(\frac{\lambda}{\pi}\right) + 47\left(\frac{\lambda}{\pi}\right)^2 - 59\left(\frac{\lambda}{\pi}\right)^3 \right]^{1/2}$$

$$M_C = 1.083$$

With $SMYS = 28,050$ psi, then $\overline{S} = 38,050$ psi

$$s = \overline{S} \left[\frac{1 - d/t}{1 - \left(\frac{d}{t}\right)M_C^{-1}} \right]$$

$$S = (38,050) \left[\frac{1 - 0.6}{1 - [(0.6)(1.083)^{-1}]} \right] = 34,125.6 \text{ psi}$$

With $S = 34,125.6$ psi $> 20,000$ psi, the allowable stress, the circumferential criteria is not satisfactory, and the cylinder fails to pass both the longitudinal and circumferential criteria of RSTRENG.

The pipe must be either replaced or rerated to 274 psig.

Example 1-3: In this example we use RGB313 to approximate that in Example 5.11.8 in API 579. This example consists of a seamless 12 in. Sch 40 LR ell made of ASTM A234 Gr WPB. The nominal wall thickness is 0.406 in. The design conditions are 700°F at 600 psig. The spacing to the nearest discontinuity is 32 in. The UT readings indicate that the LTA is located in the middle one-third section of the elbow on the extrados. The critical thickness profiles in the longitudinal and circumferential directions are 6.5 and 3.0 in., respectively. A visual inspection in conjunction with the UT readings indicates that the metal loss can be assumed to be uniform with a thickness reading of $t_{mm} = 0.18$ in. The FCA is 0.05 inches.

Data: For A234 Gr WPB at 700°F, $SMYS = 25,100$ psi (from ASME Section 2, Part D). The pit depth ($PIT = 0.406 - 0.180 = 0.176$ in.) is constant throughout the region.

We know from ASME B31.3, Paragraph 304.2,1, that the minimum required thickness for a pipe bend is determined from Eq. 1-23. In this equation the allowable stress (S) is adjusted by the Lorenz factor (LF), denoted by I. This compensates the stress at the location of interest. The stress level at the extrados is lower than that of the intrados. To compute this, the code gives Eqs. 1-24 and 1-25. Equation 1-25 is for the extrados. Now using Eq. 1-25, we determine the relative parameters as follows:

$R_i = 1.5(12) = 18$ in.; $D = 12$ in.

$$I = \frac{4\left(\dfrac{R_1}{D}\right) + 1}{4\left(\dfrac{R_1}{D}\right) + 2} \qquad D = 12.75 \text{ in.}, R_1 = 1.5(12) = 18$$

$I = [4(18/12.75) + 1]/ [4(18/12.75) + 2] = 0.869$

Adjusting the allowable stress, which we do by altering the SMYS in RGB313, we have

$SMYS = 25,100$ psi/$0.869 = 28,884$ psi

Based on the UT readings, the remaining wall thickness in the LTA is 0.18 in. So the pit depth is calculated as follows:

$PIT = (0.406 -$ Uniform metal loss $-$ FCA$) -$ remaining wall
$= (0.406 - 0 - 0.05) - 0.18 = 0.176$ in.

Thus, we enter a pit depth of 176 mil (1 mil $= 0.001$ in.) RGB313, and we find a B31.3 MAOP of 462 psi.

Example 5.11.8 in API 579 came up with 478 psig as the MAOP. Thus, the error is

ERROR $= (478 - 461)(100)/478 = 3.6\%$ error

References

1. API Recommended Practice 579, *Fitness-for-Service*, 1st edition, January 2000, American Petroleum Institute.
2. ASME B31G-1991, *Manual for Determining the Remaining Strength of Corroded Pipelines, A Supplement to ASME B31 Code for Pressure Piping*, 1991, American Society of Mechanical Engineers.
3. D. A. Osage et al., WRC 465, *Technologies for the Evaluation of Non-Crack-Like Flaws in Pressurized Components-Erosion/Corrosion*,

Pitting, Blisters, Shell Out-of-Roundness, Weld Misalignment, Bulges and Dents, 2001, Welding Research Council.

4. *RSTRENG2 User's Manual PR-218-9205*, March 31, 1993, Pipeline Research Council International.

5. KAPA, Kiefner and Associates, Inc.'s Pipe Assessment, 2005, http://www.kiefner.com.

6. P. H. Vieth and J. F. Kiefner, *Database of Corroded Pipe Tests, Contract PR-218-9206*, January 1993, PRCI.

7. J. F. Kiefner, P. H. Vieth, and I. Roytman, *Continued Validation of RSTRENG, Contract No. PR-218-9304*, December 1996, Pipeline Research Council International.

8. T. A. Bubenik and M. J. Rosenfield, "Assessing the Strength of Corroded Elbows," May 1993, American Gas Association.

9. Thomas J. O'Grady, Daniel T. Hisey, and John F. Kiefner, "Method for Evaluating Corroded Pipe Addresses Variety of Parameters," *Oil and Gas Journal*, October 12, 1992.

Chapter Two
An Introduction to Engineering Mechanics of Piping

Before we go further down the road of piping fitness-for-service, we must take a detour and consider the engineering mechanics of piping. We use the term "engineering mechanics of piping" instead of "piping flexibility analysis," which is used by the ASME B31.3 code. This term is appropriate, but it is not comprehensive enough for our purposes; adding flexibility to a piping problem is just a portion of what is required to solve piping problems. We do not wish to argue semantics, but we do wish to describe what we are doing. Probably most people refer to this subject as pipe stress, and that term is more suitable than piping flexibility. Frequently, limited space prohibits the addition of pipe loops to make the piping more flexible (e.g., ships, offshore platforms, and skid mounted units). In these applications, the pipe support scheme and stiffness of the pipe supports play as much a role in piping behavior and response as flexibility. Certainly in cyclic systems additional flexibility can work against the engineer who is trying to limit dynamic response. The term "flexibility" began to be used in the days before computers when one had unlimited freedom to add pipe loops to a piping system to lower the stress and reaction loads in the piping system.

Using the term "flexibility" to describe piping design is appropriate because every system must have a certain amount of flexibility to be satisfactory. The pipe stress software packages available today are designed to perform this flexibility analysis. As of this writing, we cannot use this software to perform detailed component design for bolted clamps or equipment attached to piping. Such design is performed with separate software designed specifically for this purpose, finite element software, or software designed for pressure vessels. In this chapter, we refer to the "stiffness" approach utilizing the stiffness of structural components in piping flexibility software. The classical flexibility approach to piping design is also covered.

Frequently, a piping system can be well under the code allowable stress but have excessive nozzle loads on equipment, particularly machinery. In fact, the piping stress engineer, more often than not, is trying to decrease the loads on machinery nozzles.

In this chapter, we will look at the thermal movements that will exert loadings—forces and moments—on the piping and attached equipment when the piping is heated or cooled. Such loadings can affect the assessment of flaws—cracks, LTAs (local thin areas), and pitting. In Chapter 1 we were primarily interested in internal pressure alone. Now we will discuss the other loading that piping exhibits because of thermal expansion or contraction. To do this we must understand the mechanics, knowing what forces and moments are acting on the pipe and, consequently, the corroded region or defect.

Piping Criteria

In analyzing piping mechanics, we need to consider the following parameters:

1. The appropriate code that applies to the system.
2. The design pressure and temperature.
3. The type of material. This includes protecting the material from critical temperatures, either on the high or low end.
4. The pipe size and wall thickness.
5. The piping geometry.
6. The movements of anchors and restraints.
7. The allowable stresses for the design conditions set by the appropriate code.
8. The limitations of forces and moments on equipment nozzles set by API (American Petroleum Institute), NEMA (National Electrical Manufacturers Association), or the equipment manufacturers. Please refer to Chapter 5 for more detail of these standards.

It is not a code requirement to assess the piping system at the "design" temperature. Although it is permitted, it may not represent the worst (high or lowest) temperature. Consider, for example, a low pressure steam out where the temperature may exceed the design temperature. Some engineering companies use both a design temperature and a separate "flux" temperature that represents upsets or excursions. These conditions occur routinely enough to not qualify for allowances for pressure and temperature variations (see ASME B31.3, Paragraph 302.2.4). However, it is the responsibility of the process engineers and the specification engineers to define the worst conditions adequately.

After the first six criteria are met, the next and foremost factor to consider in assessment is the allowable stress of the pipe at the given temperature. What is important to remember here is that we are discussing the fitness-for-service of operating piping systems, *not* new pipe. The piping codes give us rules for new pipe that we will use; however, once a piping system is in operation, the fitness-for-service rules take over—API 570, API 579, and general engineering practice.

Because we are concerned in this book with assessment rather than design and consequently are focused on computer-based solutions (e.g., numerical methods like finite element), we will not delve into manual methods of pipe stress. All our approaches will have computer-based solutions; manual methods fall short of accurately predicting forces and moments acting in a particular region of a pipe system, especially in a complex system.

The focus here will be U.S. codes, as it would be quite lengthy to discuss all the codes of the world in the discussion in this chapter. The United States uses a form of the English, or Imperial, system of units. This form of units is called the American Engineering System (AES) system of units and is described in detail in Chapter 8. We will try to insert the SI system whenever we can for readers outside the United States, providing conversions for the readers as the discussion progresses.

In the United States, the piping codes used for new pipe are as follows:

ASME B31.1, *Power Piping*—governs piping in the power industries (e.g., power plants).
ASME B31.2, *Fuel Gas Piping*.
ASME B31.3, *Chemical Plant and Petroleum Refinery Piping*—governs piping systems used in the chemical and petroleum industry.
ASME B31.4, *Liquid Petroleum Transportation Piping Systems*—governs liquid hydrocarbons and other liquids in pipeline systems.
ASME B31.5, *Refrigeration Piping and Heat Exchanger Components*.
ASME B31.8, *Gas Transmission and Distribution Systems*—governs gas pipelines.
ASME B31.9, *Building Services Piping*.
ASME Section III, *Nuclear Piping*.

The ASME codes are based on the maximum shear stress theory of yield—the Tresca stress. Several European codes are based on the more accurate von Mises theory of yield. The maximum shear stress provides simpler formulation, but as we get into fitness-for-service in later chapters we will utilize both Tresca and von Mises stresses. Also the ASME codes are design by rule. This is an important concept in that it stipulates that the designer follows the rules even though it does not tell the designer how to accomplish the assessment. The step-by-step or "cookbook" approach falls down when the methodology stipulated becomes obsolete. As opposed to the ASME piping and boiler and pressure vessel codes, the API standards, which are referred to as recommended practices, do not follow the design-by-rule approach.

Stress Categories

The ASME piping codes classify stresses as sustained, expansion, or occasional. The concepts of primary and secondary stresses will be discussed as we learn about the basic mechanics. Various European codes classify stress into primary and secondary, but our discussion here requires a deeper understanding. *Secondary stresses* are called self-limiting because, as they increase in magnitude, local yielding causes local deformation, which in turn reduces the stresses. Thermal stresses that result when pipe expands are reduced when the pipe deforms. Thus, thermal stresses are self-limiting.

Primary stresses are not self-limiting because, as they increase, local yielding does not reduce them. Internal pressure is one example of primary stress. Under sufficient pressure, a pipe will undergo local yielding and deform, but the stress will not diminish. As the internal pressure increases further, the pipe wall deformations will increase excessively to the point of rupture. For this reason, it is necessary to assign lower allowable stress limits to primary stresses rather than to secondary stresses. The piping code allowable stresses are based on primary stresses. The allowable stress for secondary stresses is determined by consideration of the self-limiting, or self-relieving, property. This self-relieving characteristic is discussed in the following section.

Allowable Stress Range for Secondary Stresses

The most important secondary stresses are those induced by thermal expansion (or contraction) and surface discontinuities. The ubiquitous practice is to keep the induced stresses in the elastic range. In the case of ductile materials, the elastic range is well defined by the minimum yield point. This value is a variable for all materials, even those that have the same chemistry. To talk of a standard stress-strain curve for a particular material is meaningless because it will vary with each heat from the mill. Taking a specimen from the longitudinal direction rather than the transverse direction will also make a difference in the stress-strain curve. So the code bodies have taken many tests of each category of metal and have found that the yield strength of the metal specimens gives a scatter of data points. Thus, the minimum value in the scatter of points becomes the specified minimum yield strength for the given American Society for Testing of Materials (ASTM) specification. All specimens stamped with each ASTM category must meet the SMYS for that category over a given temperature range because the yield strength will reduce with an increase in temperature. As will be seen in later chapters, the component being assessed may have a yield strength greater than the SMYS.

In pressure-containing equipment, ductile materials are preferred because the elastic range is well defined. Where secondary stresses are to be encountered, ductile materials offer an advantage. Materials that do not have a well-defined minimum yield point are designed on the basis of ultimate strength. The ultimate, or tensile, strength of the material is the point of rupture. The region between the minimum yield point and the ultimate strength is known as the plastic region. Materials that do not have a well-defined minimum yield point are generally not used in piping systems subjected to thermal stresses and internal pressures of significant magnitude. This discussion applies to materials with well-defined minimum yield points.

Consider the stress-strain curve shown in **Figure 2-1**. Here the metal specimen is loaded to point A and then is unloaded. Because point A is the minimum yield point, no deformation occurs because the material is still in the elastic range. Now consider **Figure 2-2** where the material is loaded beyond point A. Because the minimum yield point is exceeded, plastic

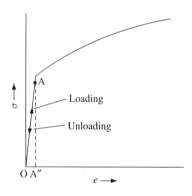

Figure 2-1. Stress-strain curve showing loading and unloading within the elastic range.

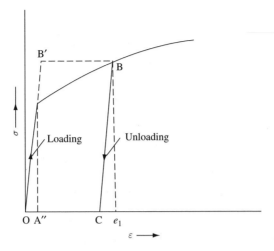

Figure 2-2. Stress-strain curve showing loading and unloading beyond the yield strength to induce strain hardening and elastic shakedown.

deformation sets in and deforms the material to point B. When the specimen is unloaded, e_1 is the amount of permanent deformation, denoted by point C. Point B′ is the theoretical stress point if the material had not been deformed to point B. **Figure 2-3** shows a case where a specimen is loaded into the plastic region. For complete plastic deformation to occur, the entire area of the pipe wall must exceed the minimum yield point; that is, yielding

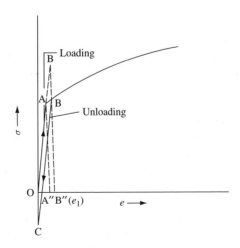

Figure 2-3. Stress-strain curve showing elastic shakedown with repetitive loading beyond the yield point.

must occur throughout the entire thickness of the pipe. This is not acceptable in practice because permanent deformation throughout the entire pipe wall would lead to rupture, ignoring the effects of strain hardening. Also with enough strain hardening, the component will fail at the yield point.

There are acceptable cases where loads will fall between **Figures 2-1** and **2-2**. This condition is shown in **Figure 2-3**, where part of the pipe wall is in the elastic range and the other part is in the plastic range. For cases where the portion in the plastic range is small compared to the portion in the elastic range, the amount of permanent plastic deformation is imperceptible. For this reason, the distance between points A and B in **Figure 2-3** is small compared to **Figure 2-2** because the portion of material in the elastic range limits the amount of permanent deformation. Thus, when the specimen is unloaded, residual stresses that cause reverse yielding when the material exceeds the compressive yield point are developed. This is shown graphically in **Figure 2-4**. The specimen is loaded to point A, and an excessive load deforms it to point B. At point B, part of the material is in the plastic range and the other portion is in the elastic range. When the specimen is unloaded, the stresses in the material go into compression shown at point C. Residual stresses caused by the combination of material in the elastic and plastic regions make part of the material exceed the compressive yield point and the specimen deforms from point C to point D. Upon application of the same initial tensile load, the material is loaded to point E. Point E is larger in value and, thus, to the right of point A because the initial loading of part of the specimen into the plastic range causes strain hardening and thus increases the minimum yield point of the material. As excessive loads are applied, the minimum yield point E is exceeded, and the material deforms to point F. As the material is unloaded again, the initial process repeats itself and the stresses in the material move to point G and then to point H as the compressive yield point is exceeded. Note that, for most ductile materials, the minimum yield strength in compression is equal to the minimum yield strength in tension.

Point Q represents the stress in the loaded condition after several loading cycles, and point P represents the stress in the unloaded condition. It is possible that no significant plastic deformation will occur after many load cycles. However, should stress values of Q or P exceed the fatigue limit of the material, small cracks will propagate throughout the strain-hardened material. After significant cracks appear, further cyclic loading will result in fracture. The stress magnitude P results from the specimen being unloaded when the load condition, point Q, is reached. Thus, because Q is the tensile stress opposite to the compressive stress P in the parallelogram OB'QP, the sides OB' and OP are equal in length. Therefore, Q = 0.5B'. Fracture by strain hardening will not occur if the theoretical tensile stress B' does not exceed twice the minimum yield

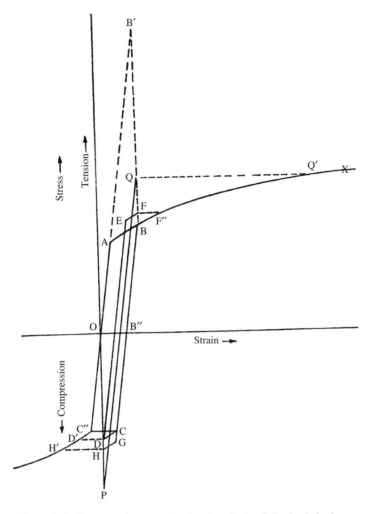

Figure 2-4. Stress-strain curve showing the effects of elastic shakedown.

stress of point A, and the magnitude of Q does not exceed the ultimate strength of point A.

When a *ductile material*, a material with a defined minimum yield point, is subjected to repeated loading, a certain behavior occurs. When a component such as a branch connection in a header pipe is repeatedly loaded and unloaded, the strain hardening makes the material stronger from one load cycle to the next load cycle. As the material becomes harder, it is better able to resist yield, and thus deformation. However, the maximum point at which this repeated loading cycle can occur is $2\sigma_{yp}$. This stress $s = 2\sigma_{yp}$ is the limit of the maximum stress range. This process is called *elastic shakedown* because the material shakes down to an elastic response and undergoes deformations or strains induced by loads beyond the minimum yield point of the material.

Note that at elevated temperatures the value of $2\sigma_{yp}$ can be altered by hydrogen attack. One form of hydrogen attack is hydrogen embrittlement. Carbon steel exposed to hydrogen at elevated temperatures can fail during elastic shakedown because the hydrogen combines with the carbon causing the embrittlement.

The relationship between the maximum stress range and the initial yield point can be expressed as

$$\sigma_{MR} < 2\sigma_{yp} \qquad\qquad \text{Eq. 2-1}$$

where $_{MR}$ = maximum local stress range not producing fatigue failure, psi
$_{yp}$ = initial yield point of the material at the operating temperature, psi

The material's ability to revert into compression and limit itself to the amount of permanent plastic deformation is called shakedown. When a material shakes down, it limits the amount of deformation and, thus, creates an elastic response.

From this discussion, we see that there is a range of allowable stresses available. Direct membrane stresses are limited by $1.5\sigma_y$, and a limited, one-time permanent deformation from A to B occurring from secondary stresses is limited by $2\sigma_y$.

The discussion in the next section will deal with allowable stresses used in piping codes. Elastic shakedown is not found exclusively in metals. In oil exploration, clays that drill bits protrude through also exhibit elastic shakedown properties.

Stresses Acting on Piping Elements

An element of pipe wall subjected to four stresses is shown in **Figure 2-5**. The pipe is under internal pressure and the four stresses are as follows:

σ_L = longitudinal stress, psi (MPa)
σ_C = circumferential or hoop stress, psi (MPa)
σ_R = radial stress, psi (MPa)
σ_T = shear or torsional stress, psi (MPa)

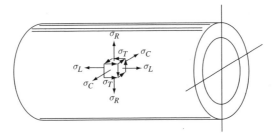

Figure 2-5. A pipe with stress components of internal pressure.

The longitudinal stress is the sum of the following three components:

1. Bending stress induced by thermal expansion (or contraction). For straight pipe:

$$\sigma_B = \frac{M}{Z_M} \qquad\qquad \text{Eq. 2-2}$$

   For curved pipe:

$$\sigma_B = \frac{M}{Z_m} i \qquad\qquad \text{Eq. 2-3}$$

2. Bending stress induced by the weight of the pipe. (This stress should be minimal if the pipe is properly supported.)
3. Longitudinal stress induced by internal pressure.

$$\sigma_P = P \frac{A_i}{A_m} \qquad\qquad \text{Eq. 2-4}$$

Because both longitudinal stress caused by internal pressure and bending stress act in the same direction,

$$\sigma_L = \sigma_{BL} + \sigma_P \qquad\qquad \text{Eq. 2-5}$$

The circumferential or hoop stress is caused primarily by internal pressure. Thus,

$$\sigma_C = \frac{P(D - 2Py)}{2tE} \qquad\qquad \text{Eq. 2-6}$$

For thin-walled cylinders, σ_R is negligible. However, for thick-walled pipe, the following relationship may be used to determine the radial stress:

$$\sigma_R = \frac{r_i^2 P_i - r_o^2 P_o}{r_o^2 - r_i^2} - \frac{r_o^2 r_i^2 (P_i - P_o)}{(r_o^2 - r_i^2)r} \qquad \text{Eq. 2-7}$$

For external pressure, $P_o = 0$, we have

$$\sigma_R = \frac{r_i^2 P_i}{r_o^2 - r_i^2} - \frac{r_o^2 r_i^2 P_i}{(r_o^2 - 4r_i^2)r} \qquad \text{Eq. 2-8}$$

Torsional or shear stress is

$$\sigma_T = \frac{T}{Z_m} \qquad \text{Eq. 2-9}$$

where torsion is generated in a multiplane system.

Direct shear stress is negligible and is not considered when it is caused by the piping temperature because local yielding or cold forming, known as ratcheting, reduces the stress at the piping components. Local strain hardening restricts the local yielding and prevents the material from rupturing. When local yielding reduces stress it is called self-springing, and has the same or similar effect as cold or hot springing. The operating stress ("operating" is used because it can be either hot or cold) diminishes with time. This change in stress is compensated for by the allowable stress range, which is the sum of the operating and down condition stresses and remains practically constant for one cycle. The sum is obtained as follows:

$$\sigma_A = f(1.25\sigma_C + 0.25\sigma_O) \qquad \text{Eq. 2-10}$$

where f = stress range reduction factor for cyclic condition
   Total number of full temperature cycles over expected life

   $1.0 \le 7,000$
   $0.9 \le 14,000$
   $0.8 \le 22,000$
   $0.7 \le 45,000$
   $0.6 \le 100,000$

Expansion stress, caused by thermal expansion, must not exceed the allowable stress range (σ_E) and is defined as

$$\sigma_E = [(\sigma_B)^2 + 4(\sigma_T)^2] \qquad \text{Eq. 2-11}$$

The piping codes further state that the sum of the longitudinal stresses caused by pressure, weight, and other sustained loadings shall not exceed *so* (referred to as *sh* in the code). This also includes the longitudinal stress caused by internal pressure (σ_P), defined earlier.

When torsional stress becomes significant, as it does in many multiplane systems, the resultant fiber stress, or combined stress, is determined by the following:

$$\sigma = 0.5[\sigma_L + \sigma_O + [4(\sigma_T)^2 + (\sigma_L - \sigma_O)^2]^{0.5}] \qquad \text{Eq. 2-12}$$

According to the ASME B31.3 guideline, a thin wall pipe is when $D/t < 0.6$, where D is the pipe outside diameter and t is the nominal wall thickness. The vast majority of standard schedules of piping are thin walled. The only exceptions are found in small diameter piping or specially made thick wall piping. Thus, thin wall rules generally always apply to piping. This fact will have significance later when we discuss fitness-for-service methodologies.

Stress Calculations

Stress computations are based on the formulations defined by the applicable code. The general hoop stress formula follows:

$$\sigma_H = \frac{P[R_o - 0.4(t - CA)]}{Et} \qquad \text{Eq. 2-13}$$

where CA = corrosion allowance, in. (mm)
R_o = outside radius of pipe, in. (mm)
P = internal design pressure of pipe, psi (MPa)
t = nominal wall thickness of pipe, in. (mm)
E = weld joint efficiency

The ASME B31.3 code committee has modified Eq. 2-13 to include the weld joint factor (W). The new revised equation reads as follows:

$$t_{\min} = \frac{PD}{2(SEW + PY)} \qquad \text{Eq. 2-13a}$$

The weld joint reduction factor (W) applies at temperatures above 950°F (510°C) and is based on the effects of creep. The new code edition will contain charts for determining W. For our purposes here, we will assume $W = 1$ unless otherwise indicated.

This formula may not be conservative for cast iron pipe.
Longitudinal stress:

$$\sigma_L = \frac{[(i_iM_i)^2 + (i_oM_o)^2]^{0.5}}{Z} + \frac{F_a + P_a}{A_C} \qquad \text{Eq. 2-14}$$

where A_C = corroded cross-sectional area, in.^{2} (mm^2)

$$A_C = \frac{\pi}{4}[D_o^2 - [D_o - 2(t - CA)]^2]$$

F_a = axial force, lb$_f$ (N)
i_i = in-plane stress intensification factor
i_o = out-of-plane stress intensification factor
M_i = in-plane moment, in.-lb$_f$ (N-mm)
M_o = out-of-plane moment, in.-lb$_f$ (N-mm)
P_a = Axial force induced from internal pressure thrust, lb$_f$ (N)

$$P_a = \frac{\pi}{4}[P[D_o - 2(t - CA)]^2]$$

Z = section modulus of pipe, in.^{3} (mm^3)

Secondary shear stress:

$$\sigma_{SH}\sqrt{\left[\left(\frac{\sigma_L - \sigma_H}{2}\right)^2 + \left(\frac{M_a}{2Z}\right)\right]} \qquad \text{Eq. 2-15}$$

where M_a = torsional moment, in.-lb$_f$ (N-mm)
Principal stress:

$$\sigma_P = \frac{\sigma_L + \sigma_H}{2} + \sigma_{SH} \qquad \text{Eq. 2-16}$$

Maximum shear stress:

$$\sigma_S = \frac{\sigma_P}{2} \qquad \text{Eq. 2-17}$$

Axial stress:

$$\sigma_A = \frac{F_a + P_a}{A_C} \qquad \text{Eq. 2-18}$$

Bending stress:

$$\sigma_B = \frac{[(i_iM_i)^2 + (i_oM_o)^2]^{\frac{1}{2}}}{Z}$$

Eq. 2-19

Torsional stress:

$$\sigma_T = \frac{M_a}{2Z}$$

Eq. 2-20

ASME B31.1 Code Stress

An *occasional load* is defined as acting less than 1% of the time in any 24 hour operating period. Seismic and wind loads are examples of occasional loads. Refer to the code [Reference 1, Paragraph 104.8].
Sustained stress:

$$\sigma_L = \frac{PD_o}{4t_n} + \frac{0.75iM_A}{Z}$$

Eq. 2-21

Substained and occasional stress:

$$\sigma_O = \frac{PD_o}{4t_n} + \frac{0.75iM_A}{Z} + \frac{0.75iM_B}{Z}$$

Eq. 2-22

Expansion stress:

$$\sigma_E = \frac{iM_C}{Z}$$

Eq. 2-23

where i = stress intensification factor. The product $0.75i$ shall never be taken less than 1.0.

ASME B31.3 Code Stress

Refer to [Reference 2]. In this section, F = axial force, lb_f (N).
Longitudinal sustained stress:

$$\sigma_L = \frac{PD_o}{4t_n} + \frac{F}{A} + \frac{M_B}{Z}$$

Eq. 2-24

Circumferential sustained stress:

$$\sigma_H = \frac{PD_o}{2Et_n}$$

Eq. 2-25

Occasional stress:

$$\sigma_O = \frac{F}{A} + \frac{M_b}{Z}$$

Eq. 2-26

Expansion stress (refer to Reference 2, Paragraph 319.4.4):

$$\sigma_E = \sqrt{\sigma_b^2 + 4\sigma_t^2}$$

Eq. 2-27

where

$$\sigma_b = \frac{[(i_iM_i)^2 + (i_oM_o)^2]^{\frac{1}{2}}}{Z}$$

Eq. 2-28

$$\sigma_t = \frac{M_t}{2Z}$$

Eq. 2-29

Equation 2-27 has been revised in the Alternative Rules in Appendix P in the new revised B31.3 to account for axial loads. The new rules are more comprehensive because they were designed around computer flexibility software. To compute the stress range, the difference in stress states, taking all loads into account, is considered. The equation for calculating stress is revised as follows:

$$\sigma = \sqrt{(|\sigma_a| + \sigma_b)^2 + 4\sigma_t^2}$$

Eq. 2-27a

where σ_a = stress due to axial force = i_aF_a/A_p
$\quad\quad F_a$ = axial force, including that due to internal pressure
$\quad\quad i_a$ = axial force stress intensification factor. In the absence of more applicable data, $i_a = 1.0$ for elbows and $i_a = i_o$ from Appendix D for other components.
$\quad\quad A_p$ = cross-sectional area of pipe
$\quad\quad \sigma = \sigma_E$ or σ_{om}, where σ_{om} is the maximum operating stress

For a detailed and comprehensive discussion of the ASME B31.3 code revisions and a complete guide to the code, the reader is referred to Becht [Reference 3].

The Pipeline Codes—ASME B31.4 and B31.8

The ASME pipeline codes work with specified minimum yield strength rather than a given allowable stress. The allowable stress in these codes becomes the SMYS multiplied by a safety design factor. The design factor is partly dependent on the location of the pipeline. Portions of the pipeline running through populated areas or under roads have a lower design factor than for portions in the open country. Generally, pipelines operate at much higher pressures than plant piping and extend over much longer distances. However, pipelines do not have large external forces and moments operating at very high or low temperatures. The metallurgy of pipelines is generally fine grain carbon steels.

ASME B31.4—Liquid Transportation Pipelines Code

The ASME B31.4 code for liquid transportation specifies different limits on allowable longitudinal expansion stresses for restrained and unrestrained piping. Buried or similarly restrained portions are exposed to substantial axial restraint, which are quite different from unrestrained or above ground portions.

Restrained piping is defined in ASME B31.4, Paragraph 419.6.4 [Reference 4]:

$$\sigma_L = E\alpha(T_i - T_o) - \mu\sigma_h \qquad \text{Eq. 2-30}$$

$$\sigma_A = 0.90(SMYS) \qquad \text{Eq. 2-31}$$

where σ_L = longitudinal expansion stress, psi (MPa)
E = modulus of elasticity, psi (MPa)
α = linear coefficient of thermal expansion, in./in.-°F (mm/mm-°C)
T_i = temperature at time of installation, °F (°C)
T_o = maximum or minimum operating temperature, °F (°C)
μ = Poisson ratio
σ_A = expansion allowable stress, psi (MPa)
σ_h = hoop stress induced by internal pressure, psi (MPa)
$\sigma_h = \dfrac{PD}{2t}$
$SMYS$ = specified minimum yield strength of pipeline metal, psi (MPa)

Sustained stress:
Longitudinal stress (ASME B31.4, Paragraph 402.3.2):

$$\sigma_L = \frac{PD_o}{4t_n} + \frac{F}{A} + \frac{M_b}{Z} \qquad \text{Eq. 2-32}$$

Occasional stress (Paragraph 402.3.3)

$$\sigma_o = \frac{F}{A} + \frac{M_b}{Z}$$

Eq. 2-33

Expansion stress (ASME B31.4, Paragraph 419.6.4):

$$\sigma_E = [\sigma_b^2 + 4\sigma_t^2]^{\frac{1}{2}}$$

Eq. 2-34

where

$$\sigma_b = \frac{[(i_iM_i)^2 + (i_oM_o)^2]^{\frac{1}{2}}}{Z}$$

Eq. 2-35

$$\sigma_t = \frac{M_t}{2Z}$$

Eq. 2-36

Allowable stress calculation:

  Expansion stress $= 0.72(SMYS)$
  Sustained stress $= 0.75(0.72)(SMYS) = 0.54(SMYS)$
  Occasional stress $= 0.80(SMYS)$

ASME B31.8—Gas Transmission and Distribution Pipeline Code

The B31.8 stress criterion, or allowable stress, is a function of the design factor and temperature derating factors multiplied by the SMYS. The stress formulations follow:
Sustained stress:
Circumferential or hoop stress:

$$\sigma_h = \frac{PD_o}{2tFET}$$

Eq. 2-37

Longitudinal stress:

$$\sigma_h = \frac{PD_o}{2t_n} + \frac{F_A}{A} + \frac{M_b}{Z}$$

Eq. 2-38

Occasional stress (ASME B31.8, Paragraph 833.4):

$$\sigma_o = \frac{F_A}{A} + \frac{M_b}{Z}$$

Eq. 2-39

Expansion stress (ASME B31.8, Paragraph 833.3):

$$\sigma_E = [\sigma_b^2 + 4\sigma_t^2]^{\frac{1}{2}}$$ Eq. 2-40

where

$$\sigma_b = \frac{[(i_iM_i)^2 + (i_oM_o)^2]^{\frac{1}{2}}}{Z}$$

$$\sigma_t = \frac{M_t}{2Z}$$

The allowable stress calculation follows:

Expansion stress = $0.72(SMYS)$
Sustained stress = $0.75(SMYS)FT$
Sustained + Occasional = $0.75(SMYS)FT$
Sustained + Expansion + Occasional = $SMYS$

where F = design factor, usually 0.72 (refer to ASME B31.8, Table 841.111 [Reference 5])
F_A = axial force, lb_f (N)
P = internal pressure, psig (MPa)
M_b = bending moment, in.-lb_f (N-mm)
T = temperature derating factor (ASME B31.8, Table 841.113 [Reference 5])
T = 1.000 at 250°F (121.1°C)
T = 0.967 at 300°F (148.9°C)
T = 0.933 at 350°F (176.7°C)
T = 0.900 at 400°F (204.4°C)
T = 0.867 at 450°F (232.2°C)

Allowable stresses:
Appendix A of the ASME B31.1 and B31.3 piping codes gives allowable stress value tables for metallic piping. The allowable stresses for different load categories follow:

Sustained stress = σ_h Eq. 2-41

Sustained + Occasional = $K\sigma_h$ Eq. 2-42

Expansion = $f(0.25\sigma_h + 1.25\sigma_c)$ or $f(1.25\sigma_h + 1.25\sigma_c) - \sigma_L$ Eq. 2-43

Expansion + Sustained = $f(1.25\sigma_h + 1.25\sigma_c)$ Eq. 2-44

The last formulation is not an official interpretation of the code, but it is used widely in practice. If the conditions of this formula, along with the preceding ones, are not met, then you are not in compliance with the code. The conditions set out in all formulas must be met to be in code compliance. where σ_h = Hot (or for cold service, the cold operating), psi (MPa)

   σ_c = cold, or ambient allowable stress, psi (MPa)

   K = occasional load factor, psi (MPa)

   f = stress range reduction factor for cyclic service

   σ_L = longitudinal sustained stress, the sum of longitudinal stresses in any component due to pressure, weight, and other sustained loadings

Flexibility and Stiffness of Piping

The concepts of flexibility and stiffness are two very important concepts in engineering mechanics. The two are mathematically the inverse of the other, but in application both must be understood. As previously mentioned, the piping code refers to the subject of analysis of loading in piping systems as flexibility analysis. Flexibility is an easy concept for most, but stiffness is just as important a concept.

In practical terms, flexibility refers to the piping configuration being able to absorb more thermal movements by using loops that allow the pipe to displace (either expand or contract) itself, resulting in lower stresses, forces, and moments in the system. Thus, making the piping system more flexible is a useful method of solving piping problems.

Stiffness is the amount of force or moment required to produce unit displacement, either translational or angular (rotational) movement. The simplest concept of stiffness is to imagine using X pounds or Newtons to compress a spring 1 inch or 25.4 mm. Thus, the spring stiffness is X lb$_f$/in. or $(X/25.4)$ N/mm. This simple example illustrates translational stiffness. Similarly, rotational stiffness can be thought of as a spring that resists rotational movement, foot-pounds per unit degree, or Newtons-mm per unit degree, of movement.

A piping element has six degrees of freedom, three in translation and three in rotation, as shown in **Figure 2-6**. The amount of force or moment required to produce unit displacement in each degree of freedom at points all along the piping element is described mathematically as the stiffness matrix K, which is defined as

$P = KU$

where we have an elastic element subjected to a set of n forces and moments

$$P = \{P_1, P_2, \ldots, P_i, \ldots, P_n\} \qquad \text{Eq. 2-45}$$

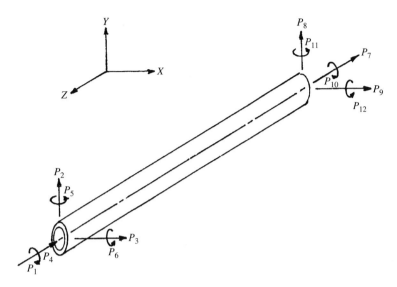

Figure 2-6. Pipe element.

The corresponding displacement of each P_i is described by the matrix

$$U = \{U_1, U_2, \ldots, U_i, \ldots, U_n\}$$ Eq. 2-46

Therefore the stiffness matrix can be expressed as

$$K = \frac{P}{U}$$ Eq. 2-47

which can be expressed in pounds per inch (Newtons per mm) or foot pounds per degree (Newtons-mm per degree). The relationship

$$C = \frac{U}{P}$$ Eq. 2-48

is defined as the compliance or flexibility matrix and can be measured in inches per pounds (mm/per Newtons) or degrees per foot-pounds (degrees per Newtons-mm). Thus, the stiffness matrix K of a system is the inverse of the system compliance or flexibility C, that is, the piping system becomes more flexible, or less stiff, than its initial configuration.

The system stiffness matrix K is made up of elements that are either direct or indirect stiffness components.

$$
P = \begin{vmatrix} K_{11} & . & . & . & . & K_{16} \\ . & & & & & . \\ . & & . & & & . \\ . & & & . & & . \\ . & & & & . & . \\ K_{61} & . & . & . & . & K_{66} \end{vmatrix} \begin{vmatrix} U_1 \\ . \\ . \\ . \\ . \\ U_6 \end{vmatrix} = KU
$$

For a general set of stiffness properties of piping elements, see **Table 2-1**.

To illustrate these concepts applied to piping mechanics, let us consider both a 4 in. schedule 40 pipe and a 10 in. schedule 40 pipe shown in **Figure 2-7**. Here we are considering two pipe spool pieces subjected to a force F as shown. Referring to **Table 2-1**, we see that the translational stiffness for a beam element fixed on one end and pinned on the other end is

$$
K_{ij} = \frac{3EI}{L^3}
$$

For the 4 in. pipe,

$$
K_4 = \frac{3(29 \times 10^6) \dfrac{\text{lb}}{\text{in.}^2} (7.23) \text{ in.}^4}{(48)^3 \text{ in.}^3} = 5687.66 \frac{\text{lb}}{\text{in.}} = 995.34 \frac{\text{N}}{\text{mm}}
$$

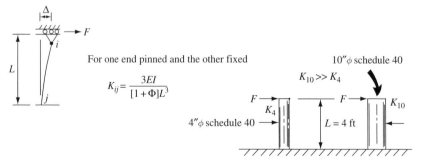

Figure 2-7. Comparative stiffness.

Table 2-1
Stiffness Properties of Piping Elements

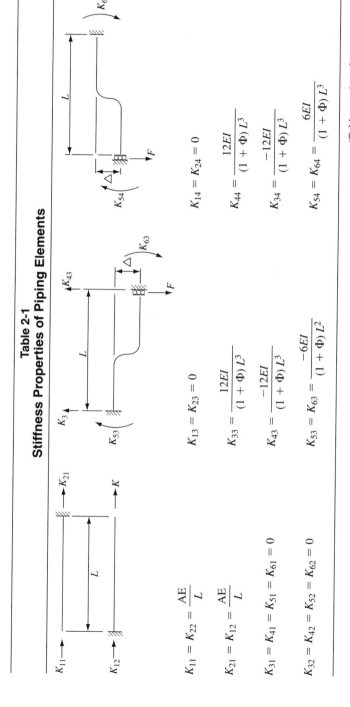

$$K_{11} = K_{22} = \frac{AE}{L}$$

$$K_{21} = K_{12} = \frac{AE}{L}$$

$$K_{31} = K_{41} = K_{51} = K_{61} = 0$$

$$K_{32} = K_{42} = K_{52} = K_{62} = 0$$

$$K_{13} = K_{23} = 0$$

$$K_{33} = \frac{12EI}{(1 + \Phi) L^3}$$

$$K_{43} = \frac{-12EI}{(1 + \Phi) L^3}$$

$$K_{53} = K_{63} = \frac{-6EI}{(1 + \Phi) L^2}$$

$$K_{14} = K_{24} = 0$$

$$K_{44} = \frac{12EI}{(1 + \Phi) L^3}$$

$$K_{34} = \frac{-12EI}{(1 + \Phi) L^3}$$

$$K_{54} = K_{64} = \frac{6EI}{(1 + \Phi) L^3}$$

(Table continued on next page)

Table 2-1—cont'd
Stiffness Properties of Piping Elements

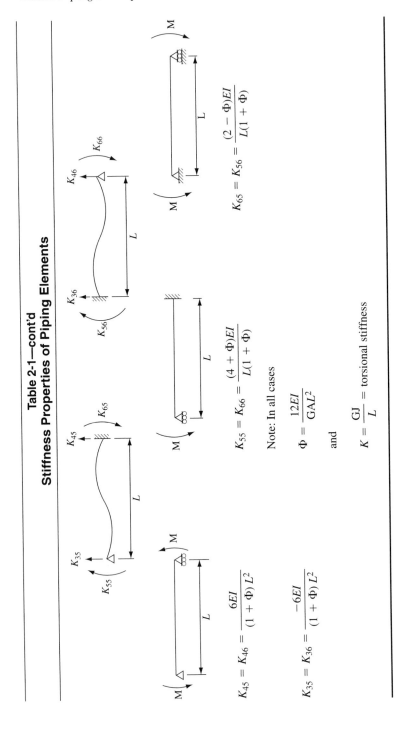

$$K_{45} = K_{46} = \frac{6EI}{(1 + \Phi)\, L^2}$$

$$K_{35} = K_{36} = \frac{-6EI}{(1 + \Phi)\, L^2}$$

$$K_{55} = K_{66} = \frac{(4 + \Phi)EI}{L(1 + \Phi)}$$

$$K_{65} = K_{56} = \frac{(2 - \Phi)EI}{L(1 + \Phi)}$$

Note: In all cases

$$\Phi = \frac{12EI}{GAL^2}$$

and

$$K = \frac{GJ}{L} = \text{torsional stiffness}$$

For the 10 in. pipe,

$$K_4 = \frac{3(29 \times 10^6)\frac{\text{lb}}{\text{in.}^2}(160.8)\,\text{in.}^4}{(48)^3\,\text{in.}^3} = 126{,}497.40\,\frac{\text{lb}}{\text{in.}}$$

$$= 22{,}137.04\,\frac{\text{N}}{\text{mm}}$$

The force required to move the 4 in. pipe $\frac{1}{4}$ in. (6.35 mm) is

$$F = (995.34)\,\frac{\text{N}}{\text{mm}}(6.35)\,\text{mm} = 6320.41\,\text{N} = 1421\,\text{lb}$$

To generate the same amount of force in a 10 in. pipe, the same length would have to move

$$\Delta = \frac{6320.41\,\text{N}}{22{,}137.04\,\frac{\text{N}}{\text{mm}}} = 0.286\,\text{mm} = 0.011\,\text{in.}$$

In other words, if the pipe itself moved because of thermal expansion and there was a restraint of a given stiffness (spring constant) restraining the movement, the 10 in. pipe would only have to move 0.286 mm (0.011 in.) to exert the same force as the 4 in. pipe moving 6.35 mm ($\frac{1}{4}$ in.). Thus, the 10 in. pipe is more than 22 times stiffer than the 4 in. pipe, which is a significant point because it indicates that the larger the piping is, the less it must move to exert excessive forces and moments on nozzle connections and pipe supports. From this example, it is obvious that the larger the piping is, the greater the stiffness is. This basic fact is important in the design of pipe supports, particularly using the concept of stiffness and how it relates to piping.

Carrying the analysis further, consider the two piping configurations shown in **Figure 2-8**. This situation is similar to **Figure 2-7** in that one end is fixed and the other pinned (i.e., both systems have the same boundary conditions). Segment B-C is flexible enough to bend with enough rotational flexibility to consider that end as a pinned joint. If the temperature of the piping is $-100°F$ ($-73.3°C$), segment B-C moves

$$\Delta = (-1.75)\,\frac{\text{in.}}{100\,\text{ft}}(4)\,\text{ft} = -0.070\,\text{in.} = 1.78\,\text{mm}$$

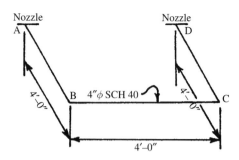

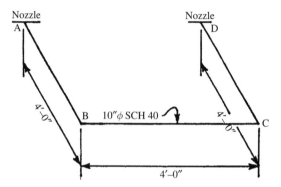

Figure 2-8. Pipe size makes a significant difference in nozzle loadings.

The force required to move a 4 in. schedule 40 pipe 0.070 in. (1.78 mm) is

$$F_4 = (5687.66) \frac{\text{lb}}{\text{in.}} (0.070) \text{ in.} = 398.14 \text{ lb} = 1770.93 \text{ N}$$

The force required to move a 10 in. schedule 40 pipe 0.070 in. (1.78 mm) is

$$F_{10} = (126,497.40) \frac{\text{lb}}{\text{in.}} (0.070) \text{ in.} = 8854.82 \text{ lb} = 39,386.24 \text{ N}$$

resulting in a moment of

$$M_{10} = (8854.82)(4) = 35,419.27 \text{ ft-lb/2} = 17,709.64 \text{ ft-lb}$$
$$= 24,009.85 \text{ N-m}$$

or 24,009.85 Joules at the nozzles A and B.

The 4 in. force of 398.14 lb would produce a moment of

$$M_4 = (398.14) \frac{(4)}{2} = 796.28 \text{ ft-lb} = 1079.70 \text{ N-m}$$

at nozzles A and B.

It is clear that the 10 in. pipe would exert moments well above the allowable moments for most rotating and stationary equipment. To reduce the loading at the nozzle, the engineer is faced with two options: make the piping configuration more flexible or restrain the piping with limited flexibility. To fabricate the piping configuration to within a tolerance of 0.070 in. (1.8 mm) would be well beyond the practical range of any conventional fabricating shop.

Stiffness becomes significant when restraints are placed on flexible structural members. When they are placed on surfaces of very high stiffness (e.g., concrete foundations), the members attaching the pipe to the concrete can become important with high loads. In most pipe stress computer programs, when a restraint is modeled at a particular point in the piping, the program assumes the restraint to be acting at the centerline of the pipe. This can be quite misleading if the restraint is attached some distance from the pipe. To more accurately model the restraint, the structural component must be modeled in to account for the actual stiffness of the restraint. Consider, for example, a 20 in. process pipe attached to a concrete base plate by a 10 in. pipe half a meter long (1.64 ft). With high axial forces in the process pipe, the wall thickness of the 10 in. pipe (hence moment of inertia) makes a difference on the axial displacement of the process pipe at the point of restraint attachment.

A restraint that resists moments by transferring the moments from the pipe to the support location is called a moment restraining support (MRS). Different MRS supports are shown in **Figure 2-9**. These supports are used in cold cryogenic piping systems—their effectiveness is limited in hot systems. They are also used in skid-mounted units where space is limited.

An MRS can vary from a bolted plate connection shown in **Figure 2-9a** to the sophisticated type shown in **Figure 2-9c**. MRS restraints' sophistication is a function of how much rotation is resisted and how much translational movement is allowed. The simplest MRS restraint is the anchor, where the pipe itself or a pipe attachment is welded down to a structural steel or immersed in concrete. In that case, it is resisting three degrees of freedom in translation and three degrees of freedom in rotation. In most applications, the moments at nozzle connections can become excessive, and it is often desirable to resist rotation in one, two, or three axes while allowing translational movement along one axis. Such

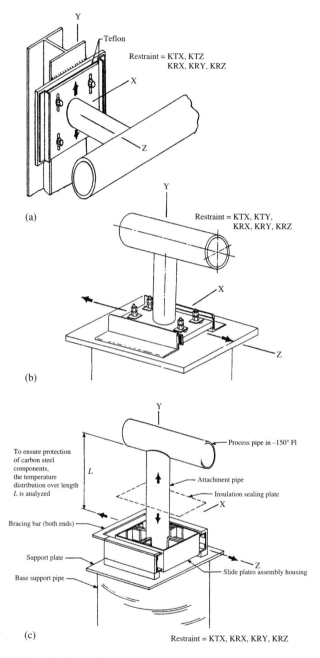

Figure 2-9. Various designs of moment restraint supports (arrows indicate direction of allowed movement).

supports are frequently made of Teflon and other materials with very low coefficients of friction if these materials have been determined to be able to withstand the forces and moments existing. If the material used is not resistant to shear, cold flow will result, leading to uneven force distribution in the pipe and an improperly functioning restraint.

In the engineering of MRS restraints, the principles discussed previously must continuously be applied. No support or restraint can be expected to be infinitely rigid along the degrees of freedom that are being restrained. Placing MRS devices in front of equipment nozzles (in the case of modular structures or ships) will not stop all loading exerted by the piping thermal forces because all restraints have a corresponding stiffness value for each degree of freedom. The engineer must also understand what assumptions in the pipe stress program are being used. Almost all computerized pipe stress packages consider an anchor as six springs, three resisting translational forces of at least 10^9 lb$_f$/in. and three resisting rotational forces resisting 10^9 ft-lb$_f$/deg (in one very popular program, 10^{12} is used instead of 10^9). There is no infinitely rigid anchor in nature, but 10^9 lb$_f$/in. coupled with 10^9 ft-lb$_f$/deg is sufficient to be called an anchor in almost all applications.

In modular plant design of piping with liquefied petroleum gas or cryogenic processes, it is often desirable to enter the actual stiffness of any anchor or restraint to obtain an accurate model of the piping system being analyzed.

Stiffness and Large Piping

Large piping is more difficult to restrain than small piping because of the surface to be restrained. The terms "large" and "small" are quantified in the following discussion. The most common complication of restraining large piping is the phenomenon of shear flow, which occurs longitudinally and circumferentially. As illustrated in **Figure 2-10**, longitudinal shear flow transfers bending moments and shear forces to the equipment nozzle.

In modular construction, longitudinal shear flow does not become a problem until one starts using 10 in. pipe and larger. Shear flow can be resisted to some degree by making the attachment pipe size or structural member size close to that of the pipe, but it is most often impractical (remember that structural wide flange members do not resist torsion very well). What is often desirable is to mount an MRS on opposite ends of the pipe, either on the top or bottom, or off both sides, depending on what space is available. In piping sizes 30 in. and larger, MRS restraints must be attached on four sides for the MRS to be effective. In pipe

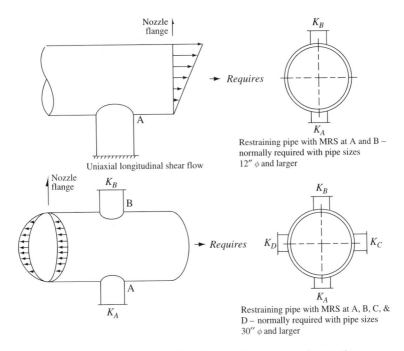

Figure 2-10. Longitudinal shear flow—a phenomenon of large pipe.

diameters 8 in. and smaller, attaching an MRS on one side is sufficient for most applications, especially modular construction.

Circumferential shear flow, on the other hand, is not a factor in most installations because torsion is very effectively transferred to the structural steel by MRS restraints.

Using piping restraints to transfer loads to structural steel or concrete to lower the loads at equipment nozzles is becoming quite popular and more widespread in modular construction. Also, where exotic piping materials are used, the stiffness method can help to reduce the number of elbows used for flexibility and, thus, to reduce the cost of the job because restraints and supports are far cheaper than piping elbows, especially those made of exotic materials.

Flexibility Method of Piping Mechanics

In nonmodular skid construction (block-mounted plants) and areas where there is available space to place equipment, it is often more economical and desirable to design the piping to be flexible enough to reduce loadings on supports and equipment nozzles by making the piping more flexible. This is accomplished by adding elbows to form offsets and loops to allow the piping to expand (or contract) and have the flexibility of the system to "absorb" the thermal movements. This approach is the more conventional approach and what the ASME B31.3 refers to as flexibility analysis. It is by far the most common approach and for most the only approach known. It is fine as long as you have the luxury of space to flex the piping. For pipe racks, long headers, and so on, this method is the only practical approach to solving piping mechanics problems. As previously mentioned, tools used in this method include well-known devices and techniques as the offsets, and loops.

In large flexible piping systems, it is important to control the piping system. If one has a large flexible system with no intermediate anchors or restraints, it is very possible that parts of the system will move in a manner detrimental to equipment. Extraneous motion induced by equipment vibration or wind loads over long periods of time can result in pipe movements not anticipated in the original design. Such movements can be avoided by judicious use of intermediate restraints or anchors.

Pipe Offsets and Loops

The most common types of offsets are "Z" and "L" shapes. Loops always come in "U" shapes. Curves for these shape offset dimensions showing stresses plotted against the offset and loop dimensions are shown in **Figures 2-11**, and their equations are as follows:

$$F_x = A_x B \frac{I_P}{L^2} \text{ lb (N)}, I_P = \text{in.}^4 \text{ (mm}^4\text{)}$$

$$F_Y = A_Y B \frac{I_P}{L^2} \text{ lb (N)}$$

$$\sigma_s = A_b B \frac{D}{L}, \text{ psi (MPa)}; L = \text{ft (m)}; D = \text{in. (mm)}$$

$$B = \frac{\text{Thermal movement (in./100 ft)}}{172,800}$$

$$SIF = 1.0 \text{ (stress identification factor; verified by computer stress analysis)}$$

Offsets and loops such as circle bends, double offsets, and other geometrics involving complicated circular geometry should be avoided unless there are very special process reasons for their use. They are unappealing in appearance and often unnecessary.

In pipe racks, the "U" shape is invariably the most practical shape to use because of its space efficiency. "U" loops are normally spaced together (i.e., lines running together on a pipe rack are looped together

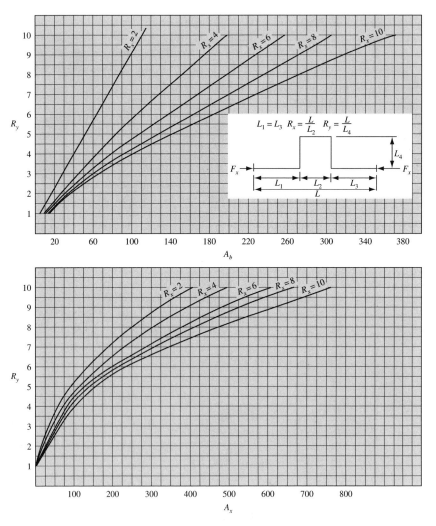

Figure 2-11a. U-loop with equal legs.

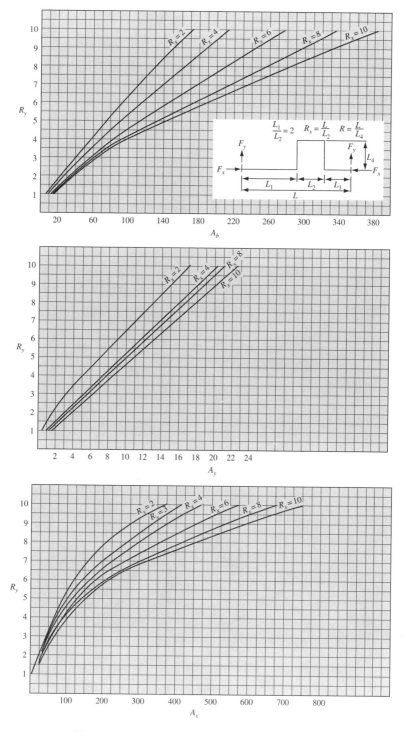

Figure 2-11b. U-loop with one leg twice the other leg.

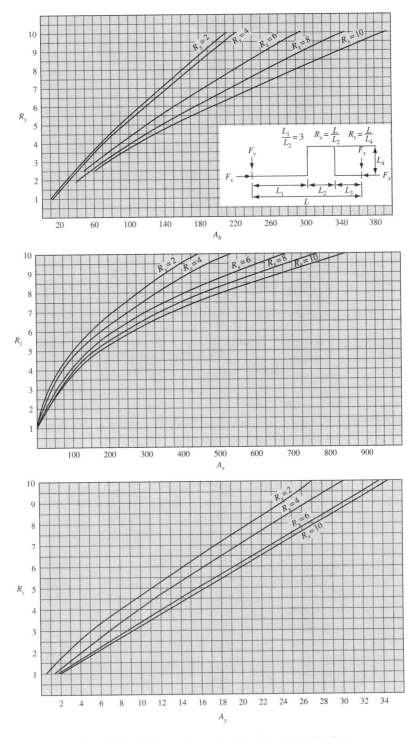

Figure 2-11c. U-loop with one leg three times the other leg.

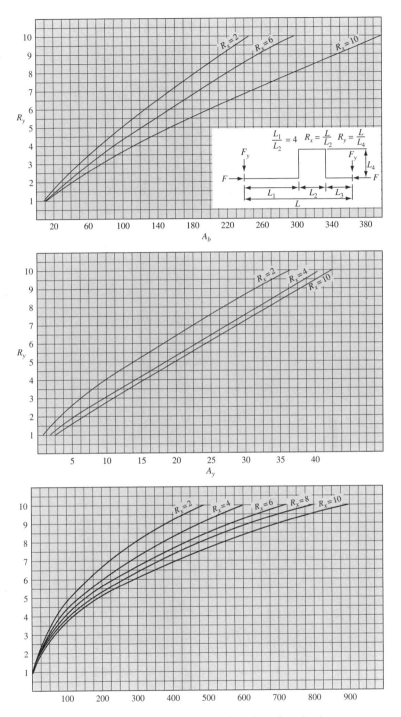

Figure 2-11d. U-loop with one leg four times the other leg.

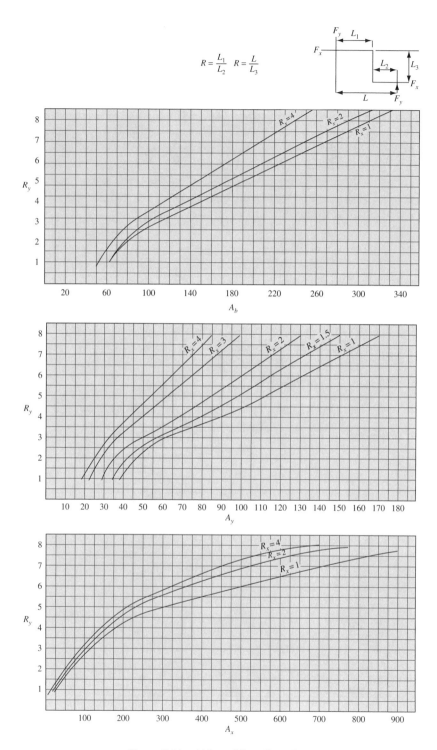

$$R = \frac{L_1}{L_2} \quad R = \frac{L}{L_3}$$

Figure 2-11e. U-loop: "Z" configuration.

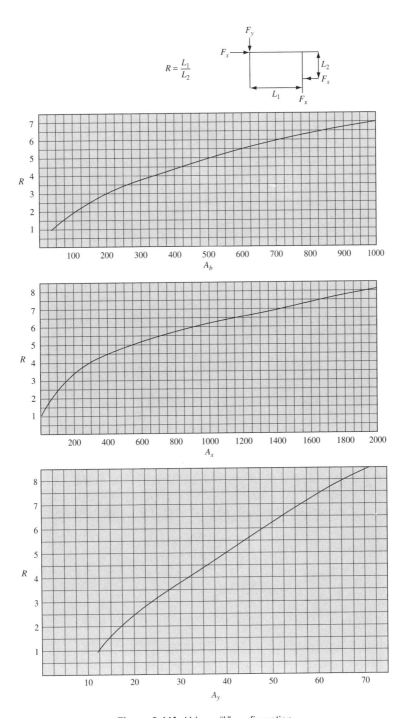

Figure 2-11f. U-loop: "L" configuration.

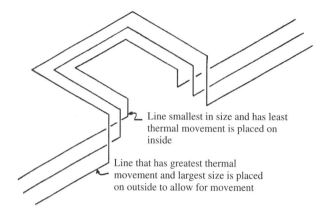

Line smallest in size and has least thermal movement is placed on inside

Line that has greatest thermal movement and largest size is placed on outside to allow for movement

Figure 2-12. Optimum grouping of U-loops.

on a pipe rack as shown in **Figure 2-12**). It is desirable to guide the pipe on each side of the loop and at every other support thereafter as shown in **Figure 2-13**. Make sure the first guide is far enough from the loop to avoid jamming problems. Usually, this distance is twice the bend radius of an elbow of the pipe size being used. If you cannot put piping guides on the pipe coming down from the loop, then put them on the inside of the loop as shown in **Figure 2-12**.

Other configurations, such as "Z" and "L" shapes, are used in the normal routing of piping systems. It must be remembered that when these shapes are anchored on opposite ends, the ratio of the shortest leg to the longest leg should fall in the range of 1.0 to 10.0 to avoid overstressing the pipe. When analyzing the shapes by computer, any ratio can be used, but usually the aforementioned range is valid for most applications.

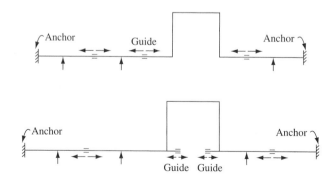

Figure 2-13. Guides are necessary for controlling movement of loops.

Pipe Restraints and Anchors

Pipe restraints are used to counter forces and moments resulting from gravity, thermal displacement, wind, earthquake, vibration, and dynamic pulsations such as water hammer. The most elementary type is the simple support that restrains the force of gravity. A piping restraint can act in one or all degrees of freedom. As discussed previously, no restraints are infinitely rigid—each has its own stiffness or spring constant in translation and rotation. Even a "rigid" restraint has in each degree of freedom a finite translational and rotational stiffness. Some computer programs use 10^9 lb$_f$/in. and 10^9 in.-lb/deg, but others use 10^{12} lb$_f$/in. and 10^{12} in.-lb$_f$/deg. When the stiffness of the restraint has these magnitudes in three degrees of translation and three degrees of rotation, it is termed an *anchor*.

Piping guides are restraints that counter movement in one or several directions but allow total freedom of movement in one or more other directions. Total freedom is defined as a stiffness value of zero. An anchor, by definition, has some value of stiffness in every degree of freedom, even though the anchor itself can have displacement. The displacement occurs while the anchor is still resisting movement at certain stiffness in each degree of freedom. Thus, using the term "sliding anchor" instead of a "pipe guide" is erroneous because guides have a value of zero stiffness in one or two degrees of freedom. An anchor can restrain movement, although it may have displacement. It is important to be cognizant of restraint and anchor terminology to avoid unnecessary confusion.

The stiffness of a support is not only a function of the restraint material but also a function of the structural steel or concrete to which it is attached. Even though concrete is very stiff in compression, it is not infinitely stiff. As shown in **Figure 2-14**, the pipe restraint has a stiffness value of K_R, the concrete has a stiffness value of K_C, and the soil has a value of K_S. Because $K_C >> K_S$, the concrete can sink or move in the soil if the concrete support is designed correctly or in events of subsidence. Movements caused by soil conditions should be the responsibility of the piping engineer as well as the civil/structural engineer. The latter is responsible for limiting such movement as much as possible, and the piping engineer is responsible for entering these movements in the stress software or manual calculations.

As previously mentioned, for a pipe restraint to be considered absolutely rigid, it must restrain one billion (or trillion) pounds per inch of translation and one billion (or trillion) inch-pounds per degree of freedom. However, very few pipe restraints in nature are so rigid. If the actual stiffness of the restraint is modeled into the pipe stress analysis,

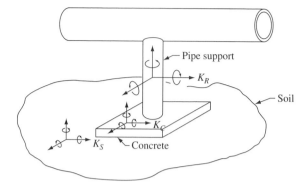

Figure 2-14. Conceptualization of system stiffness. Each component of the system—pipe, pipe supports, concrete, and soil—has translational and rotational values of stiffness (matrices) about each axis. These values can be modeled into the system as springs.

one obtains more realistic reactions in the piping system. In the case of pressure vessels, much work to determine realistic spring constants for nozzles resulted in algorithms. Such algorithms are not intended for machinery nozzles, especially those made of brittle materials such as cast iron. Also, these spring algorithms are used only for ductile materials. Nozzle loadings for machinery should be based on either manufacturer recommendations or applicable standards.

To treat a restraint with elastic end conditions, only rotations are considered significant. Deformations induced by radial force and other translations are ignored because their influence is often insignificant. One classic exception is the bellows expansion joint, which is not intended for this discussion, because we are speaking of piping attached (either welded or flanged) to vessel or piping nozzles.

The basic relationship for rotational deformation of nozzle ends applies the following:

$$K = \frac{P}{U} = \frac{M}{\theta} = \frac{\pi}{180} \left[\frac{EI}{D_N k_f} \right]$$ Eq. 2-49

where $K = KRX$ or KRY, ft-lb/deg (N-mm/deg)
   M = moment, ft-lb (N-mm)
   θ = angle of rotation, deg
   E = modulus of elasticity of vessel metal at ambient temperature, psi
   I = moment of inertia of vessel nozzle, in.[4]
   D_N = diameter of vessel nozzle, in.
   k_f = flexibility factor, referred to in piping codes as "k"

The flexibility factor k_f is a parameter that has had several formulations over the years. One widely used variant was that proposed by the "Oak Ridge ORNL Phase 3 Report-115-3-1966." Since this document was published in 1966, several revisions have been made. The current ASME Section III Division I code gives detailed discussions on the flexibility factor. If one is designing piping for nuclear systems, then that person should consult that code. Outside the nuclear industry, the piping engineer rarely knows all the parameters that are necessary to compute the flexibility factor of Section III. Also, the piping engineer in nonnuclear work rarely knows which vendor will supply the piping components, thereby making many Section III parameters unknown.

WRC Bulletin 329 (December 1987) gives several formulations for flexibility and SIF for unreinforced and various types of reinforced nozzles. For a simple unreinforced pipe on a header, the following algorithms may be applied:

Flexibility factor:

$$\text{Longitudinal} = K_L = C_L \left(\frac{D}{t}\right)^{1.5} \left[\left(\frac{t}{t_B}\right)\left(\frac{D_B}{D}\right)\right]^{0.5} \left(\frac{t_B}{t}\right) \quad \text{Eq. 2-50}$$

$$\text{Circumferential} = K_C = C_C \left(\frac{D}{t}\right)^{1.5} \left[\left(\frac{t}{t_B}\right)\left(\frac{D_B}{D}\right)\right]^{0.5} \left(\frac{t_B}{t}\right)$$

$$\text{Eq. 2-51}$$

where $C_L = 0.1$
$C_C = 0.2$

Rotational spring rate:

$$\text{Longitudinal} = R_L = \frac{EI\pi}{(2,160)D_B K_L} \left(\frac{\text{ft-lb}}{\text{deg}}\right) \quad \text{Eq. 2-52}$$

$$\text{Circumferential} = R_C = \frac{EI\pi}{(2,160)D_B K_C} \left(\frac{\text{ft-lb}}{\text{deg}}\right) \quad \text{Eq. 2-53}$$

Angle of twist:

$$\text{Longitudinal} = \theta_L = \frac{MD_B K_L}{EI} \text{ (radians)} \quad \text{Eq. 2-54}$$

$$\text{Circumferential} = \theta_C = \frac{MD_B K_C}{EI} \text{ (radians)} \quad \text{Eq. 2-55}$$

where C_L = 0.09 for in-plane bending

$\quad$ C_C = 0.27 for out-of-plane bending

$\quad$ D = diameter of vessel or pipe header, in.

$\quad$ D_B = diameter of branch, in.

$\quad$ E = modulus of elasticity, psi

$\quad$ I = moment of inertia of branch, in.[4]

$\quad$ K_L = longitudinal flexibility factor

$\quad$ K_C = circumferential flexibility factor

$\quad$ M = applied moment, in.-lb

$\quad$ θ_L = longitudinal angle of twist, radians

$\quad$ θ_C = circumferential angle of twist, radians

$\quad$ t = wall thickness of vessel or pipe header, in.

$\quad$ t_B = wall thickness of branch, in.

$\quad$ t_n = t_B + Reinforcement—For example, t_n would be the branch (nozzle) hub thickness at the base of a reinforced nozzle; for a nozzle with no nozzle wall reinforcement, $t_n = t_B$.

In-plane bending refers to longitudinal bending in the pipe header or vessel in the plane formed by the intersection of the branch and vessel or pipe header centerlines. Out-of-plane bending refers to circumferential bending in a plane perpendicular to the vessel or pipe header longitudinal axis. These rotational spring rates are necessary when the stiffness of an anchor must be considered in pipe stress analysis.

Criteria for Flexibility Analysis

As per ASME B31.3, Paragraph 319.4.1 (ASME B31.1, Paragraph 119.7.1), a formal flexibility analysis is not required for a piping system that duplicates, or replaces without significant change, a system operating with a successful service record, or that can be readily judged adequate by comparison with previously analyzed systems.

Both the ASME B31.1 Power Piping Code (Paragraph 119.7.1) and the ASME B31.3 Code for Process Piping (Paragraph 319.4) give a criterion when formal flexibility analysis is not required. The term "flexibility analysis" is used throughout the codes and pertains to the amount of flexibility of the piping to absorb thermal movements. Several computer software packages perform flexibility analysis of piping systems. These packages are limited to the flexibility of the system and not designed to compute component design (e.g., a detailed design of a particular component like a pipe support). Component design requires either a closed-form solution or a finite element analysis. A helpful formulation gives a criterion for a two-anchor piping system with no intermediate supports. The formulation follows:

$$\frac{Dy}{(L - U)^2} \leq K_1 \qquad\qquad \text{Eq. 2-56}$$

where D = outside diameter of pipe, in. (mm)
$\quad\ \ y$ = resultant of total displacement strains to be absorbed by the piping system, in. (mm)
$\quad\ \ L$ = developed length of piping between anchors, ft (m)
$\quad\ \ U$ = anchor distance, straight line between anchors, ft (m)
$\quad\ K_1$ = 30 S_A/E_a (in./ft)^2
$\quad\ S_A$ = allowable displacement stress range between anchors, ksi (MPa)
$\quad\ S_A = f(1.25S_c + 0.25S_h)$ $\qquad\qquad\qquad\qquad\qquad$ Eq. 2-43
$\quad\ E_a$ = reference modulus of elasticity at 70°F (21°C), ksi (MPa)

Note: In ASME B31.1, K_1 = 0.03. The K_1 = 30 S_A/E_a is converted to the metric as follows:

$$30.0\left[\left(\frac{25.4 \text{ mm}}{1 \text{ in.}}\right)^2 \left(\frac{3.2808 \text{ ft}}{1 \text{ m}}\right)^2\right] = 208{,}328 \left(\frac{\text{mm}}{\text{m}}\right)^2$$

or K_1 is approximately 208,000 S_A/E_a (mm/m)^2.

In the ASME B31.3, the displacement stress range and reference modulus of elasticity have to be considered. Note that there is no general proof available that Eq. 2-56 will result in accurate or consistently conservative results. It is not applicable to systems under severe cyclic conditions. It should be used with caution in configurations such as unequal leg U-bends ($L/U > 2.5$), near-straight sawtooth runs, large thin walls ($I > 5$), or where extraneous displacements (not in the direction connecting anchor points) constitute a large part of the total displacement. There is no assurance that terminal reactions will be acceptably low, even if a piping system falls within the limitations of Eq. 2-56.

Example Using the Empirical Flexibility Criterion

A 10 in. piping system has the configuration shown in **Figure 2-15**. The pipe is API 5L Grade B. The pipe is filled with a gas at 700°F. The pressure is 50 psig. For our purposes we will consider it without insulation; however, in actual practice such is unthinkable at this temperature.

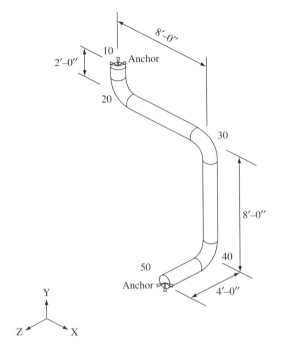

Figure 2-15. Model of piping isometric.

We are just illustrating the use of the flexibility criterion. We have the following:

$D = 10.75$ in.

$L = 2 + 8 + 8 + 4 = 22$ ft

$U = \sqrt{8^2 + (-10)^2 + 4^2} = 13.42$ ft

For low carbon steel pipe at 700°F, the linear coefficient of thermal expansion is 5.63 in per 100 ft of pipe. To compute the parameter y, we have

$$y = 5.63 \text{ in.} \left(\frac{13.42 \text{ ft}}{100 \text{ ft}} \right) = 0.756 \text{ in.}$$

Now,

$S_A = (1.0)[1.25(20,000) + (0.25)(16,500)] = 29,125$ psi

$E_a = 29,500$ ksi

$$K_1 = \left(\frac{29.125}{29,500}\right)(30) = 0.0296$$

$$\frac{Dy}{(L - U)^2} = \frac{(10.75)(0.756)}{(22 - 13.42)^2} = 0.11 > 0.0296$$

Since the flexibility criterion is greater than 0.03, we must modify the piping. The modified configuration is shown in **Figure 2-16**.

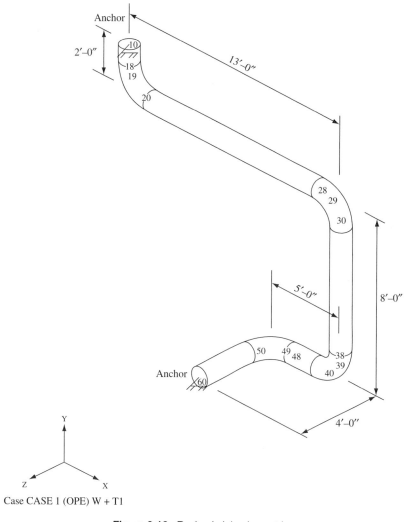

Case CASE 1 (OPE) W + T1

Figure 2-16. Revised piping isometric.

Computing the parameters for the flexibility criterion we have the following:

$$L = 2 + 13 + 8 + 4 + 4 = 30 \text{ ft}$$

$$U = \sqrt{(13 - 4)^2 + (-10)^2 + 4^2} = 14.03$$

$$\frac{Dy}{(L - U)^2} = \frac{(10.75)(0.756)}{(30 - 14.03)^2} = 0.0282 \le 0.0296$$

Thus, the new configuration meets the flexibility criterion.

Looking at the case in **Figure 2-15**, the maximum code stress is at the elbow at node 30. At node 30 for the sustained load (weight plus thermal plus pressure load), the code stress is 1069 psi. For the displacement case (expansion case) the maximum stress is 24,044 psi. The allowable stress for the expansion case is the "unused" stress from the sustained case added to the result in Eq. 2-43, or

$$S_A = 16,500 \text{ psi} - 1069 + (1)[1.25(20,000) + 0.25(16,500)]$$
$$= 44,556 \text{ psi}$$

The Pressure plus Thermal plus Weight run gave a maximum stress of 24,300 at node 30. Since 24,044 psi $<$ 44,556 psi, and 1069 psi $<$ 16,500 psi, the system meets the code allowable stress.

In the case for **Figure 2-16**, the maximum code stress is at the elbow at node 30. At node 30 for the sustained load case, the code stress is 418 psi. The code stress for node 30 for the displacement stress is 13,168 psi. The allowable stress for the sustained case is 16,500 psi. Thus, the allowable stress for the displacement stress at node 30 is

$$S_A = 16,500 \text{ psi} - 418 + (1)[1.25(20,000) + 0.25(16,500)]$$
$$= 45,207 \text{ psi}$$

The code stresses for the two cases are well within the allowable stresses, so the configuration for **Figure 2-16** is satisfactory.

The flexibility criterion is shown to be conservative; however, there is no allowance for insulation or internal fluid weight, which can make a difference in the computed code stresses. Also the reactions at the anchor points are not considered using this criterion. For the configuration in **Figure 2-15**, the reactions at anchor point 10 for the operating case (Weight + Thermal + Pressure) are

$$F_x = -5759 \text{ lb}_f; F_y = 5776 \text{ lb}_f; F_z = -1210 \text{ lb}_f$$

$$M_x = 4952 \text{ ft-lb}_f; M_y = 3084 \text{ ft-lb}_f; M_z = 7115 \text{ ft-lb}_f$$

At the anchor point 50, the reactions are as follows:

$$F_x = 5759 \text{ lb}_f; F_y = -5776 \text{ lb}_f; F_z = 1210 \text{ lb}_f$$
$$M_x = -15,958 \text{ ft-lb}_f; M_y = -16,439 \text{ ft-lb}_f; M_z = -18,491 \text{ ft-lb}_f$$

In **Figure 2-16**, the reactions for the expansion case for anchor node 10 are as follows:

$$F_x = -2081 \text{ lb}_f; F_y = 1595 \text{ lb}_f; F_z = -458 \text{ lb}_f$$
$$M_x = 1222 \text{ ft-lb}_f; M_y = 3237 \text{ ft-lb}_f; M_z = 2729 \text{ ft-lb}_f$$

In **Figure 2-16**, the reactions at the anchor point 60 are as follows:

$$F_x = 2197 \text{ lb}_f; F_y = -2734 \text{ lbf}; F_z = 578 \text{ lb}_f$$
$$M_x = -6275 \text{ ft-lb}_f; M_y = -8053 \text{ ft-lb}_f; M_z = -12,403 \text{ ft-lb}_f$$

Since both systems are below the code allowable stress, the nozzle loads differ considerably. Hence, the major concern of the piping engineer most often is not stresses but nozzle loads. When piping is connected to fragile equipment (e.g., rotating equipment), nozzle loads often govern. It is quite possible for a piping system to be below the code allowable stresses and the nozzles to be overloaded. In the next section, we will discuss recommended situations when a formal computer analysis is required.

The object of this discussion is to understand how to apply the flexibility criterion for situations when only a manual check is necessary in a two-anchor system without intermediate supports.

No matter how many computer runs piping engineers perform all over the world, if the machinery is not correctly maintained, piping is always a central issue. Choosing the incorrect material for machinery components (e.g., bearings and wear rings) can cause a piece of machinery to fail, regardless of the piping loads. In one case, a hot oil pump operating at 700°F (371.1°C) seized up when the wear ring, which was supposed to be 12–13% Cr, was replaced with a wear ring made of austenitic stainless steel. The austenitic stainless steel wear ring had a much higher coefficient of thermal expansion [9.76×10^{-6} in./in.-°F (17.57×10^{-6} mm/mm-°C)] than the 11–13% Cr wear ring [6.60×10^{-6} in./in.-°F (11.88×10^{-6} mm/mm-°C)], and the pump shaft seized upon thermal expansion. The piping had been in place for 20 years and the pumps had operated successfully. The plant was convinced that piping was the cause of the machinery to fail. When the problems were discovered with the wear rings, it was too late—the expensive chromium pipe and fittings had been delivered. This incident is not uncommon. Keyways left out of turbine supports can result in equipment failures; however, it is the responsibility of the piping engineer to meet the requirements for equipment nozzles.

Suggested Criteria for Level of Piping Flexibility Analysis

There is no set of rules that describe when one should use manual versus computer flexibility analysis. With the ubiquitous use of computer piping flexibility software, manual methods have become less frequent. However, much time and expense can be saved by following common guidelines that are proven in practice to be satisfactory.

Formal computer flexibility analysis is performed on all the following piping systems:

1. All process, regeneration, and decoking lines to and from fired heaters and steam generators.
2. All process lines to and from all compressors and all blowers.
3. All main steam lines to and from all turbines.
4. All main lines to and from centrifuges.
5. All main lines to and from all reactors.
6. All lines with design temperatures greater than 371.1°C (700°F).
7. All suction and discharge lines on pumps operating above 121°C (250°F) and below −8°C (0°F).
8. All pumps with discharge nozzles 10 inches or greater.

Engineering analysis by visual inspection and shortcut manual calculations is performed on the following systems not listed in the list for formal computer flexibility analysis:

1. Lines 4 inches and larger to air coolers.
2. All lines 16 inches diameter and larger.
3. Lines to pressure vessels that cannot be disconnected for purging, steam out, and so on.
4. Lines 6 inches and larger at operating temperatures over 260°C (500°F).
5. All relief systems whether closed or relieving to atmosphere with considerations for attached or detached tail pipes. This analysis must include analysis for dynamic load from the worst possible flow

conditions, including, liquid masses accelerated to sonic or subsonic velocities by expanding gases if there is a possibility that this event could occur.

6. Vacuum lines.
7. All nonmetallic piping.
8. All other pumps not covered in the preceding section where computer flexibility analysis is required.

Special consideration should be given to piping systems in the following categories:

1. Lines 3 inches and larger subject to differential settlement of tanks, vessels, equipment, or supports.
2. Category M fluid service—Lines designated as "Class M" per ASME B31.3 are normally identified on the Line List.
3. Lines subjected to mixed phase flow (liquid and vapor) that are normally analyzed for vibration problems.
4. Lines subject to external pressure by reason of vacuum or jacketing.
5. Lines to and from reciprocating pumps.
6. Lines subject to steam purging. Minimum design temperature shall be 149°C (300°F).

A flexibility analysis is made for the most severe operating temperature condition sustained during start-up, normal operation, shutdown, or regeneration. The analysis is for the maximum temperature differential. Empty or nonflowing piping systems are designed for a solar temperature of 66°C (150°F). For stress analysis, an installation temperature of 21°C (70°F) for hot service and 49°C (120°F) for cold service lines in hot climates (e.g., Saudi Arabia) is recommended. For cold services in more moderate climates, the ambient temperature should be the maximum temperature expected during the year. Hence, 120°F (49°C) would be unrealistic in Moscow or St. Petersburg, Russia.

The practice of cold or hot springing is not as common now as in the past. The practice of cold springing involves purposely cutting a pipe short to accommodate thermal expansion. Similarly, a line that is made extra long to accommodate thermal contraction is known as a hot spring. One cannot cold or hot spring a piping system to meet the allowable stresses criteria. Also cold or hot springing should not be used for piping systems connecting to rotating equipment. Engineering contractors generally prohibit the practice of using a cold or hot spring even if it is specified. To use the practice requires very close supervision of the construction process; consequently, most companies design around its use.

Closure

Piping engineering has long suffered respect by other professionals in industry. Much of this is caused by ignorance, mainly lack of knowledge of engineering mechanics. Other factors include rushed schedules and the fact that piping flexibility analysis is seen as a quality control function that "slows down" production. Piping flexibility should be viewed as an integral part of any engineering effort, not just quality control. With today's CAD/CAM drafting design tools, there is less misunderstanding of piping flexibility than in years past; however, all should be cognizant of the Flixborough, England, disaster in 1974. This disaster was caused by a poorly qualified design team that installed temporary piping; their lack of understanding claimed the lives of 28 people and cost $100 billion in material damage. As a result, the United Kingdom passed the Control of Industrial Major Accident Hazard (CIMAH) Act. **Figures 2-17** and **2-18** are photos of the accident caused by inappropriate piping design and a lack of any flexibility analysis.

One only needs to review these photos to realize the importance of piping flexibility analysis. The details of the accident can be found on the

Figure 2-17. A graphic view of the Flixborough accident from the ground.

Figure 2-18. The Flixborough accident from the air.

Internet. Also for those concerned only about schedules, recall the space shuttle *Challenger* Flight 51-L met its launch schedule on January 28, 1986. However, even though it met its schedule, it exploded in mid-air killing seven wonderful and brilliant people.

References

1. ASME B31.1, *Power Piping*, 2004, American Society of Mechanical Engineers.
2. ASME B31.3, *Process Piping*, 2004, American Society of Mechanical Engineers.
3. Charles Becht IV, *Process Piping: The Complete Guide to ASME B31.3*, 2nd edition, 2004, ASME Press, New York.
4. ASME B31.4, *Pipeline Transportation Systems for Liquid Hydrocarbons and Other Liquids*, 2002, American Society of Mechanical Engineers.
5. ASME B31.8, *Gas Transmission and Distribution Systems*, 2003, American Society of Mechanical Engineers.

Chapter Three

Fitness-for-Service Topics of Local Thin Areas, Plain Dents, Dents-Gouges, and Cracks for Piping

In Chapter 1, the concept of fitness-for-service was introduced based on the classic research of Kiefner et al. As pointed out in Chapter 1, others performed similar burst tests to develop algorithms for predicting failure pressures. One difference between the tests was the materials used. Kiefner used ASTM lower strength steels; the British Gas Association used steels of higher strength. Both produced valid algorithms for predicting failure pressures, but we decided to stay with the American Gas Association results produced by Kiefner et al. This will become more apparent when we discuss Level 3 assessments.

Chapter 2 discussed piping mechanics and the various loadings acting on the piping. Here we discussed in-plant piping versus pipeline; albeit, pipeline codes were discussed. In this chapter we seek to integrate the two. The burst tests performed to develop the burst pressure algorithms were primarily based on internal pressure. Here we combine the loadings to develop a complete set of methodology.

The field of fitness-for-service was advanced with the publication of the API Recommended Practice 579 [Reference 1]. This large document is organized into (1) application and limitations of the FFS assessment procedures; (2) data requirements, acceptance techniques, and acceptance criteria; (3) remaining life evaluation; (4) remediation (stopping the damage); (5) in-service monitoring; and (6) documentation.

The application and limitation of FFS procedures involve applying the methodology to piping not designed or constructed to the original design criteria. Limitations of the FFS method involve problems with NDE (nondestructive examination) detection and the limitations of the personnel using these techniques. There are two types of damage mechanisms that are difficult to detect. One can be formed when plate is rolled. Sulfur contained in the metal forms "pancakes," creating pockets when the plate is rolled. These pancakes result in hydrogen-induced cracking, called HICs. HICs are not covered in API 579. They are easily missed by some who perform NDE and can be extremely serious. The ASTM specifications give a wide range of sulfur that can be used in various carbon steels (e.g., ASTM A-516). Some mills in developing countries have been known to use higher concentrations of sulfur. At times this has caused laminations in the steel and eventual failure.

HIC damage is generally thought to proceed slowly, but this is not always true. It normally occurs in the temperature range of 32°F to approximately 130°F (0 to 55°C). Above the 130°F (55°C) threshold, HIC damage proceeds slowly, if at all. HIC damage can result in cooling down a system that normally operates above the 130°F (55°C) threshold. This can be controlled by purging, lowering the H_2S partial pressure, and so on. Such conditions should be regarded as crack-inducing upsets and should be closely monitored.

Stress-oriented hydrogen-induced cracking, called SOHIC, is treated in API 579 like cracks. Even though HICs and SOHICs have similar names, they are two completely different mechanisms. SOHICs occur predominately at welds and can be missed during an NDE. The formation of SOHICs occurs when through-the-wall cracks result from linking up stacked small internal cracks. Normally it is manifested in HAZ (heat-affected zones) of welds that are associated with residual stresses. It is most probable in thick, restrained welds, and sometimes in heavy nozzles, although that is not always true. The mechanism involves two properties, as follows:

1. HIC cracks form and stack in a vertical path. This allows a crack path that is perpendicular to the surface of the shell or plate. These stepwise cracks are usually very short and closely spaced.
2. A through-the-wall crack forms by shearing the ligaments between the HIC cracks. This stepwise cracking, known also as cross-tearing, has been seen mostly in pipelines; it is not as common in vessels, heat exchangers, or tanks. However, SOHIC is seen more in vessels and heat exchangers than in pipelines, even though research indicates that pipeline steels are also vulnerable to this phenomenon. Pipeline steels are usually of higher yield strength than most vessels and heat exchangers, so the residual stress distribution may cause a different behavior in regard to the SOHIC mechanism. It really doesn't have much to do with the strength of the steel; it has to do with the loading conditions and the service

environment. You can SOHIC plate steel just as fast as a pipeline steel, depending on the microstructure and service environment. That is why this mechanism has many people fearful of its consequences. SOHICs form at the interior of the surface, unlike sulfide stress corrosion cracking (SSCC), which forms on the exterior of the pipe when the hydrogen induces embitterment of the steel.

SOHIC damage is generally regarded as proceeding faster than HIC damage; however, there is no cut-off partial pressure for hydrogen for either HICs or SOHICs. In general, NDE personnel need to be trained about damage mechanisms so that effective flaw sizing can be used. They also need to know what the damage looks like to detect and size the flaws.

If material problems exist, the mechanical analyst, assuming a "perfect" material, can perform an analysis and still have a failure. Such can happen while heat-treating thick components that may form internal cracking. This is one of the many reasons that mechanical, materials, and inspection form the FFS integrity triangle mentioned in Chapter 1.

Data requirements are mandatory for a complete FFS program. Files containing design pressure and temperature, fabrication drawings, original design calculations, and inspection records (past and present) form the core of data documentation. Included in this set are maintenance and operational history, including past operational excursions, documentation of changes of service, and records of alterations and repairs.

Assessment techniques and acceptance criteria are made of a triad of assessment levels: The higher levels are less conservative, but they require more detailed analysis and data. These three levels in API 579 are as follows:

- Level 1—Inspector/plant engineer
- Level 2—Plant engineer
- Level 3—Expert engineer

The FFS assessment methods involve one or more of the following:

- Allowable stress
- Remaining strength factor
- Failure assessment diagram

The allowable stress is found from (1) the calculation of stresses from the loads and (2) the superposition of stress results using the classic Hopper Diagram of ASME Section 8 Division 2 Appendix 4, which makes mechanical design easier. The Hopper Diagram shows the stress categories that will be mentioned later. To superimpose stresses requires that the subject component be in the elastic range (i.e., linear). The codes want us to be in the elastic range. However, assessing the true strength of a component requires one to extend beyond the elastic range because a ductile failure normally has to extend into the plastic range. Certainly brittle fractures can occur well within the elastic range, but we are not considering them here. What we are considering is taking a component with flaws and comparing it to an identical specimen without flaws. The ratio of the collapse load of the component with flaws, denoted by "d" (damaged"), to an identical one without flaws is called the *remaining strength factor*, or RSF. The RSF is defined as follows:

$$RSF = \frac{\text{Collapse Load of Damaged Component}}{\text{Collapse Load of Undamaged Component}} \qquad \text{Eq. 3-1}$$

The collapse load of the undamaged component becomes the criterion of determining the value of the RSF. It is the ideal condition to which we are comparing the damaged component. The additional significance of the RSF parameter is that the stress-strain curves "wash out," or become irrelevant because one is taking the ratio of a damaged component to that of an undamaged component, each one with the same stress-strain curve. Note that this concept is *not* valid in an elastic-plastic analysis where the stress-strain curve is used to simulate a stipulated strain in the component, as in a Level 3 assessment. This will be discussed later.

Rewritten, Eq. 3-1 takes the form

$$RSF = \frac{RSF_d}{RSF_a} \qquad \text{Eq. 3-2}$$

The term RSF_a is given in API 579, whereas RSF_d is calculated or determined by a burst or proof test. If RSF_d is calculated, it is done with an elastic-plastic analysis. We will discuss more about that later. The RSF can be rewritten in terms of the maximum allowable working pressure as

$$MAWP_d = MAWP_a\left(\frac{RSF_d}{RSF_a}\right) \text{ when } RSF < RSF_a \qquad \text{Eq. 3-3}$$

where $MAWP_d$ is the MAWP for the damaged component and $MAWP_a$ is the actual MAWP for the undamaged component. If we take the burst tests results that have been already performed, this will give us a further insight into the concept of RSF. As described in Chapter 1, Kiefner burst tested a 10 in. pipe (Index 214 [Reference 2]) with a wall thickness of 0.265 in. The term Index 214 means the burst test case number 214 in

Reference [2]. The SMYS was 49,000 psi. The predicted burst pressure was 1715 psig. Kiefner et al. initially performed burst tests on lower strength steels, with the yield strength being 35,000 psi. In later tests they used higher strength steels. RSF_a is calculated by

$$RSF_a^{RSTRENG} = \frac{1 * \sigma_{yield}}{\sigma_{flow}} = \frac{\sigma_{yield}}{\sigma_{yield} + 10,000} \qquad \text{Eq. 3-4}$$

Using this equation and plotting the yield strength versus values of RSF gives **Figure 3-1**. Note that as the RSF approaches 0.9, the yield strength increases. Some authors have mistakenly referred to the RSTRENG as giving an RSF of 0.778. This is valid only for yield strength of 35,000 psi. As we have seen, we have a locus of points for the higher strength steels, which Kiefner et al. did burst test for. We will see what is practical to use for the RSF_a in the following discussion.

Thus for conservative purposes, the RSF is given as a relative high number, around 0.9. As a point of interest, as of this current writing, the API 579 does not cover pipelines, but a typical RSF for ASME B31.8 would be 0.95. Thus in light of Eq. 3-2, the RSF is

$$RSF = \frac{\text{Collapse Load of Damaged Component}}{\text{Allowable Collapse Load}} \qquad \text{Eq. 3-5}$$

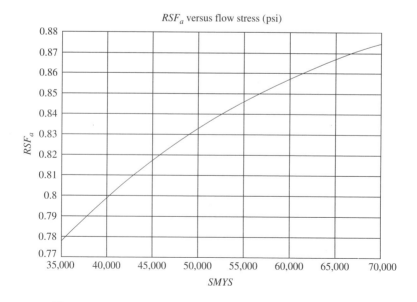

Figure 3-1. RSF_a computed from Eq. 3-4 versus *SMYS* (σ_{yield}) (psi).

Consequently, an accepted RSF has proven to be 0.9. *Note that the RSF cannot be lower than a specified amount. In other words, the collapse load of a damaged component cannot be less than 0.9 of the collapse load of the same component undamaged.* This will vary to the strength of the steel being tested.

The burst tests algorithms (e.g., RSTRENG) are based on burst tests, or by definition, ultimate strength of the pipeline steel. The API 579 is based on the yield strength, thus rendering a more conservative methodology. Thus it is difficult to compare the two methods, being that they have a different basis. As we saw in Eq. 1-27, the MAOP in a pipeline is a function of the failure pressure, not the yield strength. We are concerned here with in-plant piping and piping covered by the ASME B31.1 and 31.3. We will discuss pipelines in more detail in Chapter 8.

Useful RSF Equations Using API 579

Since we are on the topic of RSF, it may be helpful to delve into how this parameter is connected to other problem parameters. The RSF concept is very useful, especially for local thin areas. In FFS, one may know if a component is "acceptable" or "not acceptable," and API 579 is very clear about acceptance criteria. What is not evident is how long it will remain acceptable once it is found to be acceptable. This question is fair and will inevitably be asked by the organization—inspectors and managers alike. To answer this question, we will "jump through some hoops" and get ahead of ourselves since we are talking about RSF and its relevance.

The API 579 mentions maximum allowable working pressure. This term is sanctioned for use with piping by API 570 *Piping Inspection Code*. It is defined in Paragraph 3.21 as "The maximum internal pressure permitted in the piping system for continued operation at the most severe condition of coincident internal or external pressure and temperature (minimum or maximum) expected during service." The term "design pressure" is used in ASME B31.3 and ASME B31.1 and is subject to the same rules relating to allowances for variations of pressure or temperature or both. The pressure design of piping components is based on design conditions. Because piping systems are made of standardized components, often there is significant pressure capacity in the piping system beyond the conditions imposed on the system (with vessels one does not have this "luxury" because the vessel is designed for the specific conditions). This allows for variations in temperature and pressure in the ASME B31.3, Paragraph 302.2.4. Also with pressure-relieving devices, the piping codes refer to the ASME BPVC Section VIII Division 1, with the exception that "design pressure" is substituted for "MAWP." The term

"MAWP" is useful in the discussion of variations, and we must remember that once a piping system is brought into operation, the ASME B31.3 or ASME B31.1 no longer govern; rather, the API 570 and API 579 govern. Recall that the ASME B31.1 and B31.3 are for new piping, not existing piping in operation. This is an important distinction.

The following solves for MAWP in terms of RSF_a and other parameters. The remaining thickness ratio (R_t) is defined in API 579 Eq. (4.2) as follows:

$$R_t \triangleq \left(\frac{t_{mm} - FCA}{t_{min}} \right) \qquad \text{Eq. 3-6}$$

where FCA = future corrosion allowance, in. (mm)
t_{min} = minimum required wall thickness, in. (mm)
t_{mm} = minimum measured thickness, in. (mm)

The RSF_a is the allowable remaining strength factor. Now we are placing a limit on the RSF factor. The parameter, λ, is defined as

$$\lambda = \frac{1.285(s)}{\sqrt{Dt_{min}}} \qquad \text{Eq. 3-7}$$

where s = meridional (axial) dimension of the LTA, in. (mm)
D = ID (inside diameter) of the shell, in. (mm)

API 579 places the value of R_t in Eqs. (5.61), (5.62), and (5.63) as follows:

$$R_t = 0.2 \text{ for } \lambda \leq 0.3475 \qquad \text{Eq. 3-8}$$

$$R_t = \frac{\left(RSF_a - \dfrac{RSF_a}{M_t} \right)}{\left(1.0 - \dfrac{RSF_a}{M_t} \right)} \text{ for } 0.3475 < \lambda < 10.0 \qquad \text{Eq. 3-9}$$

$$R_t = 0.885 \text{ for } \lambda \geq 10.0 \qquad \text{Eq. 3-10}$$

where $M_t = \sqrt{1 + 0.48\lambda^2}$, defined in Eq. (5.12) of API 579 (using the Level 1 value for the Folias factor to be conservative).

The term M_t is called the Folias factor. The API 579, like most standards, sets the procedures and rules, but it is not very didactic. In other words, the Folias factor is a pure abstraction to many readers. It does have physical significance, however. We dealt with it in Chapter 1 with the Keifner et al. algorithms. What it represents is the "bulging effect" of an LTA to internal pressure. Suppose we had a hypothetical pipe containing an LTA that is tissue thin compared to the surrounding cylinder. If we applied pressure inside

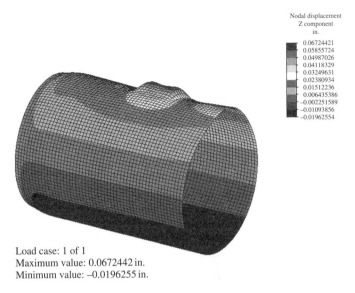

Load case: 1 of 1
Maximum value: 0.0672442 in.
Minimum value: –0.0196255 in.

Figure 3-2. Exaggerated view of LTA displacement relative to thicker pipe.

the cylinder, we would notice the LTA bulging outward as we increased the pressure. This bulging, or balloon, effect causes bending moments at the edge of the LTA junctions with the surrounding shell. For shallow LTAs (i.e., where the LTA, or corroded region, has shallow boundaries or the remaining wall is slightly less than the surrounding wall), these bending moments can be marginal. However, if the remaining wall of the LTA is significantly less than that of the surrounding wall, these bending stresses can become significant. The Folias factor takes this phenomenon into account. Shown in **Figure 3-2** is what the Folias effect means in practical terms.

The basic parameters are defined in Appendix A of API 579. The minimum required wall thickness in the circumferential (hoop) direction and the longitudinal direction (axial) are as follows:

$$t_{\min}^C = \frac{PR_C}{SE - 0.6P} \qquad \text{Eq. 3-11}$$

$$t_{\min}^L = \frac{PR_C}{2SE + 0.4P} \qquad \text{Eq. 3-12}$$

where P = internal pressure, psig (KPa)
   S = allowable stress, psi (MPa)
   E = weld joint efficiency, dimensionless
   R_C = internal radius defined in Paragraph A.3.3, where
   $R_C = R + LOSS + FCA$, in. (mm)

R = ID/2, in. (mm)

$LOSS$ = metal loss in shell prior to the assessment equal to the nominal (or furnished thickness if available) minus the measured minimum thickness at the time of inspection, in. (mm)

FCA = future corrosion allowance as mentioned in Paragraph A.2.7, in. (mm)

The FCA is the expected or anticipated corrosion or erosion that will occur. From Eq. (A.1),

$$t_{req} = t_{min} + FCA \text{ (in., mm)} \qquad\qquad \text{Eq. 3-13}$$

The parameter t_{req} is the required thickness for future operation. Now

$$t_{min} = MAX(t_{min}^C, t_{min}^L) \qquad\qquad \text{Eq. 3-14}$$

Normally, but not always, Eq. 3-11 governs. Assuming this is the case, we proceed.

From Paragraph 4.4.2.1.f.1, we must make a general metal loss assessment (see "General Metal Loss Assessment" later in this chapter); otherwise, skip:

$$t_{am} - FCA \geq t_{min} \qquad\qquad \text{Eq. 3-15}$$

where t_{am} = average measured wall thickness (in., mm)

From Paragraph 4.4.2.1.f.2 — General Metal Loss Assessment (see "General Metal Loss Assessment" later in this chapter), this criterion is used only for General Metal loss; otherwise, skip:

$$t_{mm} - FCA > MAX [0.5t_{min}, 2.5 \text{ mm } (0.10 \text{ in.})]$$

Now we set $R_t = R_{ta}$ and set Eq. 3-6 equal to Eq. 3-9, obtaining

$$R_{ta} = \left(\frac{t_{mm} - FCA}{t_{min}} \right) = \frac{\left(RSF_a - \dfrac{RSF_a}{M_t} \right)}{\left(1.0 - \dfrac{RSF_a}{M_t} \right)} \qquad\qquad \text{Eq. 3-16}$$

From Eq. 3-11 we have

$$t_{min}^C = \frac{P(R + LOSS + FCA)}{SE - 0.6P} \qquad\qquad \text{Eq. 3-17}$$

Now substituting $MAWP = P$, $S_a = S$, and $R_C = R + LOSS + FCA$, we have

$$MAWP = \frac{\left[(t_{mm} - FCA)\left(1.0 - \frac{RSF_a}{M_t}\right)\right]S_a E}{R_C\left(RSF_a - \frac{RSF_a}{M_t}\right) + 0.6\left((t_{mm} - FCA)\left(1.0 - \frac{RSF_a}{M_t}\right)\right)}$$

Eq. 3-18

Now we have obtained a working relationship for the MAWP in terms of the RSF, and we can predict the remaining life. In this equation, we assume that the hoop (circumferential) stress governs. Here we see that the MAWP is inversely proportional to the RSF. This equation results in the lower in value the RSF the higher the value of the MAWP. Shown in **Figure 3-3** is the application of Eq. 3-18 for a specific case, where $R_C = 48.0$ in., $t_{mm} = 0.571$, $M_t = 1.1$, $FCA = 0.01$, $S_a = 20,000$ psi, and $E = 1$. **Figure 3-3** illustrates the relationship between the MAWP and the RSF.

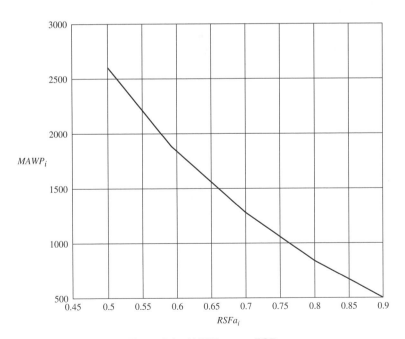

Figure 3-3. *MAWP versus RSF$_a$.*

For the cases when $\lambda \le 0.3475$, $R_t = 0.2$, or $\lambda \ge 10.0$, $R_t = 0.885$, letting $a = 0.3475$ or 0.885, it can be seen that the equation for MAWP becomes

$$MAWP = \frac{(t_{mm} - FCA)(S_a)(E)}{a(R_C)\left(1 + \dfrac{0.6(FCA)}{a} \right)}$$
Eq. 3-18(a)

Note that Eq. 3-18 can be used for both general metal loss and local metal loss. It has been proved effective and works in field practice.

Assessment Techniques and Acceptance Criteria

The failure assessment diagram (FAD) is a graphical representation of two parameter failure criteria: unstable fracture and plastic collapse. Its application is primarily for crack-like flaws. It is used for HIC and SOHIC mechanisms. We will see FADs in crack assessments.

FFS assessments in API 579 are deterministic; the following types are given:

- Sensitivity assessments—multiple assessments are performed to find if small changes in input parameters significantly change the assessment results (e.g., evaluating the effect of toughness on a crack). Spreadsheet software is ideal for this.
- Probabilistic analysis—The probability of a failure is determined.
- Partial safety factors—They are used to account for uncertain parameters (e.g., crack-like flaws).

Remaining Life Assessment

FFS assessments must include an evaluation of the remaining life of a component. A damaged component may be acceptable in the present, but there must be an established remaining life. This assessment is necessary to establish inspection intervals and a basis for reliability-based inspection (RBI). RBI is the assessment technique used to determine what equipment is more likely to fail first and is used to establish a priority system of equipment inspection during a turnaround. We are not going to deal much with RBI in this chapter, but it is an invaluable technique to make turnarounds more efficient and productive. Gone are the days when every single piece of equipment is opened for inspection during a turnaround. The RBI technique assesses existing inspection records and FFS assessments and the evaluations of operations personnel who rank the various pieces of equipment. Low-ranked equipment is not opened during a turnaround, thus saving many resources. Remaining life is a critical parameter in RBI, which is based on the applicable code.

Remaining life is not always easily determined. One such circumstance occurs when there is no corrosion rate data. Another circumstance occurs when there is no reliable crack growth rate. As a consequence, monitoring or remediation is used to account for any uncertainty. When there is little or no remaining life, repair is the obvious option. API 579 sets criteria for acceptance, but little is said about defining remaining life. Hence Eq. 3-18 is developed to make the process simpler.

To see how to compute the remaining life, see "Performing the Remaining Life Assessment" later in this chapter.

Remediation

Remediation is necessary when a flaw is not acceptable in its current condition, the remaining life is minimal or uncertain, or the state-of-the-art assessment is not sufficient to predict the behavior of the flaw. API 579 covers remediation methods in each part (section). Remediation methods can consist of grinding out cracks on the weld overlay or welding on sleeves or pads. Some other examples follow:

- Corrosion probes
- Hydrogen probes
- Coupons and physical probes
- UT measurements and scanning

In-Service Monitoring

Because of uncertainties mentioned earlier, it sometimes is necessary to monitor in-service operating equipment. This can consist of NDE techniques to determine remaining wall thicknesses.

Documentation

As a general rule, we should be able to repeat all FFS analyses without consulting the original analyst. This requires good record keeping and accurate description of all assumptions. Such assumptions would be the boundary conditions used to model a component using finite element

(FE), as the next analyst may use a different FE software package. It is highly preferable to make all documentation electronic. Level 1 and Level 2 assessments are best recorded on spreadsheets. Cracks are difficult using spreadsheets, so defining the critical parameters is essential. *It is the responsibility of the owner to document all FFS assessments. This is critical if RBI assessments are going to be performed.*

Damage Mechanisms

Before any FFS assessment is made, it is absolutely necessary to identify the damage mechanism. Not taking this critical step can lead to failure. Damage defects or flaws that are not identified can lead to false conclusions from a stress perspective. This involves the FFS integrity triangle mentioned in Chapter 1—inspection, materials, and mechanical form the integrity triangle that makes the FFS assessment complete. If we were to remove one of those disciplines from the effort, the entire FFS process would fall apart.

Identifying the damage mechanism requires the appropriate NDE method, an estimate of the future damage rate to find the remaining life, and the proper monitoring and mitigation methods.

A proactive approach is the best defense against failure by damage mechanisms. Questions addressing the potential degradation mechanism, the level of degradation that is tolerable, which remaining safety factor is acceptable, the consequence of a failure, and the consequence of a leak versus a rupture must be addressed. Note that API 579 does not describe how to find corrosion, but what to do when you find it.

General and specific material information including heat treatment, chemistry, and strength level are some of the data required in damage assessment. Material toughness is a function of grain size, which is a function of heat treatment. Most damage mechanisms are chemistry dependent. Normally a good rule of thumb is 5 mils per year for the upper shelf of tolerable corrosion. If this rate is exceeded, then a higher alloy is required. Corrosion occurs most dramatically during operational excursions – how fast it occurs during an upset. Corrosion is not an "on-off" type of phenomenon. Service exposure—general and specific, normal and upset—which can lead to trace amounts of corrosives; operation cycles; leaking valves; and the vicissitudes of human behavior are all factors affecting corrosion. The topic of concentration becomes a critical element in dealing with caustic (sodium hydroxide) service. To learn more about trace amounts of corrosives, refer to Table G.1 in API 579 in Appendix G. Also, because not all chemicals are totally cleaned out, corrosives can be carried over to another operational unit in a facility with various processes. One of the most temperamental substances in this regard is caustic. It is hard to decontaminate and is easily carried in steam and other substances. Caustic is fairly innocuous at ambient conditions, but when carried over and exposed to high temperatures, it is extremely corrosive, especially in another unit not designed for it. To defend against such events, corrosion probes (coupons), hydrogen probes, or pH probes can be installed to warn of such possibilities. Previous inspections and their effectiveness at discovering particular mechanisms are vital in an FFS program. Finally, the morphology of the damage is vital in damage diagnosis.

Damage mechanisms can be categorized into two prime categories—preservice flaws and in-service flaws. Preservice flaws are caused by the following:

- *Material flaws caused by production*—laminations, voids, shrinks, and cracks. Some of these were discussed in the beginning of this chapter.
- *Welding-induced flaws*—weld undercutting caused by a lack of penetration and fusion, weld porosity, induced hydrogen cracking.
- *Fabrication fit-up*—out-of-roundness, lamellar tearing.
- *Heat treatment flaws resulting in embrittlement*—flaws that can be induced by reheat cracking, sigma phase embrittlement, 885°F embrittlement, and sensitization. Sensitization occurs in austenitic stainless steels when carbides precipitate at grain boundaries during heat treatment.

In-service flaws are caused by the following:

- *General corrosion* proceeds without appreciable localization of attack, which leads to relatively uniform thinning on the entire exposed surface.
- *Localized corrosion* is a corrosive attack limited to a specific, relatively small surface area.
- *Galvanic corrosion* occurs when a metal is joined or coupled to a more noble metal or conducting nonmetal in the same electrolyte.
- *Environmental cracking* is a brittle fracture of a normally ductile material in which the corrosive effect of the environment is causing the embrittlement. Examples of this would be stress corrosion cracking, causing branch cracks or planar cracks. Caustic stress corrosion cracking, chloride stress corrosion cracking, sulfide stress corrosion cracking, and hydrogen blistering are other examples.
- *Erosion-corrosion, cavitation, and fretting* occur when a corrosion reaction is caused by the relative movement of the corrosive fluid and the metal surface.
- *Intergranular corrosion* is preferential attack at or adjacent to the grain boundaries of a metal or alloy. For example, sensitization is caused by precipitation of constituents at grain boundaries of a metal caused by adverse thermal treatment, whether accidental, intentional or incidental

(e.g., welding), resulting in susceptibility to intergranular corrosion. Another example is polythionic acid attack, where aqueous sulfur-based oxidized products of iron sulfide corrosion scale result in the rapid intergranular attack of sensitized austenitic stainless steels and other austenitic alloys. Attack occurs below the aqueous dew point.

- *Dealloying* is selective corrosion whereby one constituent of an alloy is preferentially removed leaving an altered residue structure.
- *High-temperature corrosion or scaling* is the formation of thick corrosion product layers on a metal surface.
- *Internal attack* is the alteration of metal properties caused by the entry of an environmental constituent at high temperature.
- *Carburization* is carbon entry.
- *Hydrogen attack* is hydrogen fusing into the metal.

Mitigation strategies can be devised to avoid some damage mechanisms. A polythionic acid attack resulting in an intergranular attack can be avoided by making efforts to avoid air; water (maintained above the aqueous dew point) can be neutralized with alkaline solution; and avoid the use of sensitized materials Its necessary ingredients are sensitized austenitic steel; iron sulfide or sulfur oxides; water; and oxygen. Mitigation, can also involve physically modifying the process—altering the temperature, adjusting the velocity, and removing fractions. This is much easier to say than accomplish. Chemically modifying the process can be mitigation, as can isolating the environment from the flaw or damage. This can be accomplished with organic coatings, metallic linings, or weld overlay. We must use caution with weld overlays, as normally four times the nominal wall is the limit on a weld overlay. However, adding damage allowance (i.e., making the component thicker with a weld overlay) is a form of mitigation. Finally, putting the component into a different stress state by stress relieving or physical surface treatment (e.g., shot peening) is another example of mitigation.

Damage mechanisms will be treated in API 571, which is not available at the time of this writing. This document will handle in detail the types and causes of damage mechanisms. Other API documents (e.g., API 939, API 941, and API 579 Appendix G) cover this vast subject. Also the Materials Properties Council has published many documents on damage mechanisms. Finally, NACE (National Association of Corrosion Engineers) has countless documents on the subject. For a concise well-documented source on damage mechanisms, the interested reader is referred to Hansen and Puyear [Reference 3].

A word is in order for dents, grooves, and gouges, which typically occur when mechanical equipment strikes the piping, forming these defects. This is normally more common in pipelines, where machinery is used to excavate the land or poor construction techniques have been practiced, than in-plant piping. When objects strike a pipe, metallurgical damage can occur, particularly when the equipment slides along the pipe. Within the scrape or gouge, plastic flow, metal transfer, or even pipe melting can result at the point of contact. As a dent pushes out, very high tensile strains can result at the root of the defect and cause cracks to form in the cold-worked, reduced ductility of the pipe metal. Cracks that form may often not be seen by the naked eye. Depending on the geometry of the defect, cracks may grow. Even for shallow dents, the defect never fully rebounds; it fluctuates in and out of the pipe as the pressure changes. Because it may take many cycles for the dent to shake down to elastic response, the plastic strain can continue to grow in the damaged area. When these defects are discovered, the area needs to be inspected for surrounding cracks. There will be further discussion about these defects later. See the discussion at the end of the chapter about grooves, dents, and gouges with dents.

Blisters and Laminations

Hydrogen blisters are the result of HIC, and, per API 579 paragraph 2.5.3.2, the following is stated:

> The Remaining Life Cannot be Established With Reasonable Certainty – Examples may be a stress corrosion cracking mechanism where there is no reliable crack growth rate available or hydrogen blistering where a future damage rate cannot be estimated. In this case remediation methods should be employed, such as application of a lining or coating to isolate the environment, drilling of blisters, or monitoring. Inspection would then be limited to assuring remediation method acceptability, such as lining or coating integrity.

Note that coatings are not an acceptable mitigation method for avoidance of stress corrosion cracking, but it may be an acceptable method to minimize HIC damage.

Thus hydrogen blisters, or HIC damage, cannot be analyzed. Blisters are formed when the hydrogen proton, H^+, diffuses into the steel. The temperature (between 32°F to 130°F) allows the hydrogen proton to diffuse through the steel at a higher rate. The diffusion of the proton stops when two hydrogen protons, H^+, recombine into the hydrogen molecule. The H_2 molecule is too large to diffuse through the lattice and becomes trapped in the steel. As the temperature increases, the metal lattice opens up more, and the H^+ proton diffuses through the metal wall into the atmosphere. Hydrogen panels on the outside surface of the pipe are used to detect hydrogen diffusion.

HIC damage occurs between 32°F and 130°F. It can occur at 160°F, but certainly below 200°F. The size of the blister is irrelevant if there is cracking in or around the blister. Also when a blister is found, we must find what is under the blister (e.g., a lamination or other defects).

Some treat blisters on the inside wall as local metal loss; however, if the blister is not vented, one has a double-walled pipe (or vessel). Just because the pipe is blistered does not mean that there is a loss of strength.

Blisters can occur in high temperature where the hydrogen can react with the carbon in the steel to form methane. This form of blistering is another form of hydrogen damage and can be assessed if it is vented to the inside or outside surface of the pipe.

Laminations are acceptable if they are not operating in sour service (hydrogen charging service). Laminations are defined as a type of discontinuity with separation or weakness in a pipe or vessel wall, generally aligned parallel to the worked surface. A lamination may be the result of seams, nonmetallic inclusions (MnS), or alloy segregation that is made directional by working the material. One specific cause is high sulfur content (<0.10) in a carbon steel. The higher sulfur content allows for the formation of manganese sulfide inclusions, which when worked from elongated stringers can coalesce into laminations. This has been a problem with steels made in some Third World countries where SA-105 flange material has laminations caused by higher sulfur content in the steel.

With respect to SOHIC, 90% of SOHIC initiates adjacent to a weld seams because of the higher stress levels (e.g., residual stresses in the HAZ region). (SOHIC that initiates in the HAZ is termed soft zone cracking.) Thus the weld seams should be inspected for this mechanism because they cannot be visually seen, whereas HIC, as surface blisters, are visible.

Note that HIC damage can occur at the mid-wall of a pressure vessel or plate steel and will not be visible as blisters. Stepwise cracking (a through-wall cracking phenomena) associated with HIC damage may or may not produce blistering. The length of the through-wall crack from hydrogen damage needs to be assessed.

Assessment of Local Thin Areas

The assessment of LTAs is divided into two parts (sections) in API 579—General Metal Loss (Part 4; formerly Section 4) and Localized Metal Loss (Part 5; formerly Section 5). As of this writing, the second edition of the API 579 is being developed. In the first edition, the term "Section" is being replaced by "Part" to avoid confusion with the ASME BPV code names Section II, III, and so on. The exact distinction between uniform metal loss and local metal loss cannot be made without assessing the details of the metal loss profile. The Part 4 [Section 4] rules are structured to be consistent with the Part 5 [Section 5] rules. The inspection requirements for Part 4 are also consistent with Part 5, with the exception of groove-like flaws. Generally it is recommended to perform a general metal loss assessment before proceeding to a localized metal loss assessment. Each section gives acceptance criteria, but it is difficult to find the remaining life using these criteria without countless iterations. That is why we derived Eq. 3-18 to determine the MAWP. This equation was developed from both parts (sections) on general and localized metal loss. Each part (section) is divided into three assessment levels, as mentioned earlier. We will get into the details for each part, but the vast majority of cases can be handled at Level 1 or Level 2. Level 3 nonlinear elastic-plastic analysis is reserved for critical applications where a major impact in operations is anticipated. However, oftentimes a finite element linear elastic analysis can be applied on a routine basis, as will be discussed later.

General Metal Loss Assessment

The distinction between general metal loss (Part 4 of API 579) and local metal loss (Part 5 of API 579) is not always clear. API 579, Paragraph 4.3.3.2(b), states, "A minimum of 15 thickness readings is recommended unless the level of NDE utilized can be used to confirm that the metal loss is general." The word "general" implies uniform metal loss with minimum variation of the remaining wall. This fact is supported in Paragraph 4.2.1 where it states that "The assessment procedures in this section [Part 4—Assessment of General Metal Loss] which exceeds or is predicted to exceed the corrosion allowance before the next scheduled inspection." Paragraph 4.1.2 states that if local areas of metal loss are found on the component, the thickness averaging approach may produce conservative results. For these cases, the procedures in Part 5—Assessment of Local Metal Loss, which require detailed thickness profiles, will be utilized to reduce the conservatism in the analysis. The exact distinction between uniform and local metal loss cannot be made without knowing the characteristics of the metal loss profile. Thus general metal loss implies "uniform" metal loss, and the guidelines based on the characteristics of the thickness profile are incorporated into the rules to direct the user to Part 5 when appropriate. API 579 concludes in Paragraph 4.1.2 that for most evaluations, it is recommended to first perform an assessment using Part 4—Assessment of General Metal Loss. Thus, the formulation in Eq. 3-18 is very appropriate and helpful in the assessment process. It is generally accepted that extensive pitting is one example of general metal loss.

To have a corroded region with minimum variation of remaining wall resulting in general (uniform) metal loss is defined in Part 4 under the coefficient of variation (COV). The COV is defined in Part 4 and is illustrated in Example 3-3. If the COV is greater than 10%, then the corroded region is local, not general. However, the COV test requires a minimum of 15 data points. In many applications, this is not possible because the inspection frequently is done long before it reaches the hands of the engineer. Generally, general, or uniform, metal loss accounts for 20% of the cases encountered. In corrosion services (e.g., seawater), it is a rare event. General corrosion is normal in corrosion under insulation with vessels or large pipelines. General corrosion internal to piping or pipelines it is not as common as local metal loss. Often the assessment methods of both general and local metal loss are used in an assessment to ensure a comprehensive assessment. Figure 4.2 of API 579 gives an assessment procedure to FFS to evaluate a component with metal loss using Part 4 and Part 5, as shown in **Figure 3-4**. From this figure, it is not entirely clear whether an LTA falls under General Metal Loss (Part 4) or Local Metal Loss (Part 5). It does, however, spell out from the inspection data collection whether an LTA is general or localized.

Generally, general metal loss areas consist of approximately 20% of the LTAs found. Normally, an analysis is concurrently performed for both general and local metal loss. One cannot obtain a higher MAWP from a general metal loss assessment than from a local metal loss assessment. Also, a Level 1 assessment under a general metal loss procedure does not consider supplemental loads, just internal pressure. Caution is recommended in systems that may have significant temperatures that result in large thermal stresses. In these situations, a Level 2 analysis must be performed. Parts 4 and 5 are attempts to present FFS in closed-form solutions using classic code equations. Closed-form solutions are not as flexible as numeric solutions (e.g., finite element) because they require well-defined parameters. If one is to seriously venture into FFS, learning to apply the finite element method becomes a necessity. One will discover that jumping to a Level 3 assessment often may be a quicker and more efficient approach to FFS. One can jump to a higher level and bypass the procedure to find a more precise answer. We will review this approach later.

The reader is referred to **Figure 3-5**, which shows the LTA and a grid in the circumferential and meridian (axial) directions. The figure may appear abstract to some; however, it will be clearer in the examples. It is a mathematical generalization of the inspection matrix and how t_{mm} is found. As seen in the figure, the inspection grid is constructed with remaining wall thickness values in both the axial and circumferential directions. The parameters c and s are used to calculate t_{mm}. Once this parameter is found, we can compute the algorithm given in the following discussion.

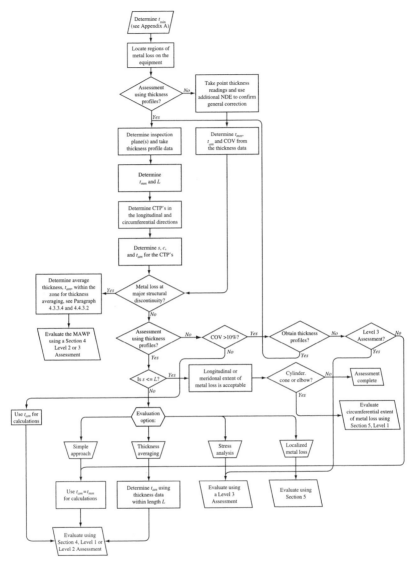

Figure 3-4. Assessment procedure to evaluate a component with metal loss using Parts 4 and 5. Courtesy of the American Petroleum Institute.

The critical thickness profile (CTP) represents the variation of thickness readings in the longitudinal and circumferential directions developed from the matrix in **Figure 3-5**. The CTP is discussed in more detail later. First, there is information gathered from the inspectors. This entails data collection. There are two options: (1) individual point readings and

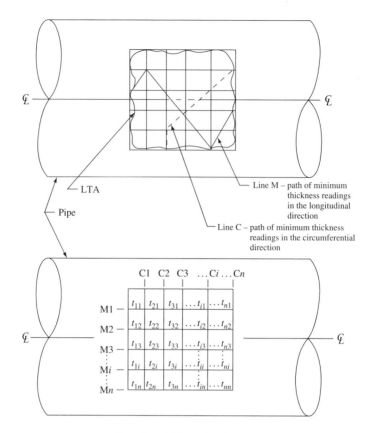

Circumferential inspection planes

Meridional inspection planes	C1	C2	C3	... Ci	... Cn	Circumferential CTP
M1	t_{11}	t_{21}	t_{31}	... t_{i1}	... t_{n1}	t_{c1min}
M2	t_{12}	t_{22}	t_{32}	... t_{i2}	... t_{n2}	t_{c2min}
M3	t_{13}	t_{23}	t_{33}	... t_{i3}	... t_{n3}	t_{c3min}
Mi	t_{1i}	t_{2i}	t_{3i}	... t_{ii}	... t_{ni}	... t_{cimin}
Mn	t_{1n}	t_{2n}	t_{3n}	... t_{in}	... t_{nn}	... t_{cnmin}
Meridional CTP	t_{1mmin}	t_{2mmin}	t_{3mmin}	... t_{immin}	... t_{nmmin}	

Where t_{1mmin} = the minimum thickness reading of t_{11}, t_{12}, t_{13}, t_{1i}, and t_{1n}, and the same for t_{2mmin}, t_{3mmin}, t_{immin}, and t_{nmmin}.

Similarly, t_{c1min} = minimum value of t_{11}, t_{21}, t_{31}, ... t_{i1}, ... t_{n1}, and likewise for the other circumferential CTPs.

Now CTPc = circumferential CTP = MIN(t_{c1min}, t_{c2min}, t_{c3min}, ... t_{cimin}, ... t_{cnmin})

And CTPm = meridional CTP = MIN(t_{1mmin}, t_{2mmin}, t_{3mmin}, ... t_{immin}, t_{nmmin})

Where CTPc = CTPm, then this value is t_{mm}

Figure 3-5. Inspection grid and data.

(2) thickness profiles. The grid spacing shown in **Figure 3-5** is based on an industrial formula, as follows:

$$L_s = MIN \left[0.36\sqrt{(OD)t_{min}} \; ; 2t_{nom} \right]$$

where L_s = grid spacing, in.
$\quad OD$ = outside diameter of pipe, in.
$\quad t_{min}$ = minimum required code wall thickness, in.
$\quad t_{nom}$ = pipe nominal wall thickness, in.

Note that even though this equation is helpful in setting grid readings, it is becoming obsolete with the emergence of inspection technology. There are inspection tools with highly sophisticated electronics that can give wall loss data down to fractions of an inch (or millimeter). For example, intelligent pigs can scan for wall loss in pipelines (and now plant piping) to give a 360° scan of wall loss down the length of the pipeline. This inspection data can be fed into a computer where an FFS assessment is performed. Similar tools exist for scanning the outside surface of a pipe. However, it is not within the scope of this book to discuss inspection technology and these tools.

Individual Point Readings

Individual point thickness readings can be used to characterize the metal loss on a component as general if there are no significant differences among the values obtained during inspection. If there is a significant variation in the thickness readings, the metal loss may be localized, and thickness profiles (thickness readings on a prescribed grid) should be used to characterize the remaining thickness and size of LTA. If this approach is used, a minimum of 15 thickness readings is recommended unless a level of NDE can be used to confirm that the metal loss is general. In some cases, additional readings may be required. A sample data sheet to record thickness readings is given in the API 579 in Table 4.2.

If point readings are taken, a COV of the thickness readings minus the future corrosion allowance is greater than 10%, and then the use of thickness profiles should be considered. The COV is defined as the standard deviation divided by the average. A template for calculating the COV is in API 579 Table 4.3 and in Example 3-3 below.

Thickness Profiles

This is the preferred approach because if point readings produce COV >10%, this approach has to be used anyway. This approach used in the part (section) for the assessment of general metal loss involves average measured thickness and the minimum measured thickness. If the thickness readings indicate that the metal loss is general, the procedures in Part 4 (Section 4) will provide an adequate assessment. However, *if the metal loss is localized and thickness profiles are obtained*, the assessment procedures of Part 4 (Section 4) may produce results that are too conservative, and the option for performing the evaluation using the assessment procedures of Part 5 (Section 5) is required. Throughout this discussion, we will assume that thickness profiles are used.

For cylindrical shells (straight runs of pipe), conical shells, or elbows, the critical inspection plane(s) are meridional (longitudinal) if the circumferential stress due to pressure governs; if the longitudinal stress due to pressure and supplemental loads governs (see Figure 4.4 of the API 579), the inspection plane(s) are circumferential.

Structural Discontinuities

If the region of metal loss is close to or at a major structural discontinuity, the remaining thickness can be established using the procedures described earlier. However, additional thickness readings that include sufficient data points should be taken in the region close to the major structural discontinuity. Sufficient thickness readings should be taken within the zones defined as follows for the components listed below, remembering that the emphasis here is piping:

- Branch connection **Figure 3-6a** (Figure 4.9 for the thickness zone, L_v, L_{no}, and L_{ni})
- Conical reducers **Figure 3-6b** (Figure 4.10 for the thickness zone, L_v)
- Flange connections **Figure 3-6c** (Figure 4.12 for the thickness zone, L_{vh} and L_{vt})

Level 1 Assessment

1. Calculate the minimum required thickness, t_{min}

$$t_{min}^C = \frac{PR_C}{SE - 0.6P}$$ Eq. 3-11

$$t_{min}^L = \frac{PR_C}{2SE + 0.4P} + t_{sl}$$ Eq. 3-12

where t_{sl} is thickness required for supplemental load based on longitudinal stress (see API 579 par. A.7)

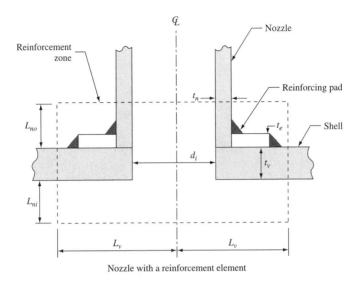

Figure 3-6a. Branch connection thickness zone. Courtesy of the American Petroleum Institute.

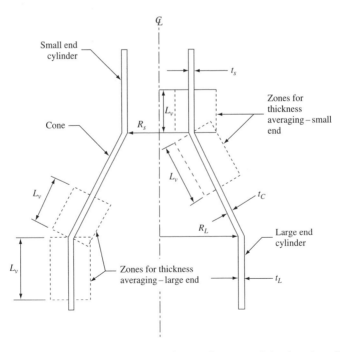

Figure 3-6b. Thickness zone for conical reducers. Courtesy of the American Petroleum Institute.

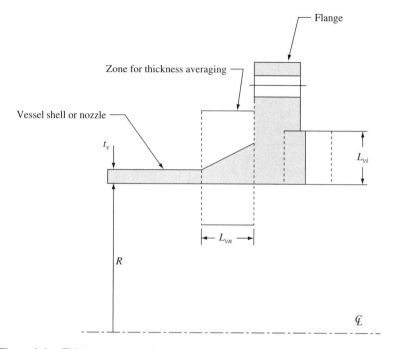

Figure 3-6c. Thickness zone for flanges. Courtesy of the American Petroleum Institute.

2. Determine the length for thickness averaging, L.
3. Compute the remaining thickness ratio, R_t

$$R_t = \left(\frac{t_{mm} - FCA}{t_{min}} \right)$$

where FCA = future corrosion allowance—in API 579 the term "corrosion allowance" is replaced by "metal loss," and "future corrosion allowance" is based on actual data and is not a guess. The FCA is used to predict remaining life, in. (mm).

t_{min} = Minimum required thickness, in. (mm)

t_{mm} = Minimum measured thickness, in. (mm)

$$L = Q\sqrt{Dt_{min}} \qquad \text{Eq. 3-19}$$

where D = inside diameter of pipe, in. (mm)

Q = factor from Table 4.4 of API 579, based on the Remaining Strength Ratio (RSF) and the remaining thickness ratio (R_t)

$$Q = 1.123 \sqrt{\left[\frac{1 - R_t}{1 - \frac{R_t}{RSF_a}}\right]^2 - 1} \qquad \text{Eq. 3-20}$$

$$Q = 50.0 \text{ for } R_t \geq RSF_a \qquad \text{Eq. 3-21}$$

Establish the critical thickness profiles (CTPs) from the thickness profile data, and determine s and c, the dimensions which define the region of metal loss in the longitudinal and circumferential directions, respectively. The dimensions s and c are found from their respective *CTP* and t_{min}.

1. For $s \leq L$, the meridional or longitudinal extent of metal loss is acceptable if the limiting flaw size criteria in Part 5 (Section 5) Paragraph 5.4.2.2.d are satisfied, that is:

   $$R_t \geq 0.20 \qquad \text{Eq. 3-22}$$

   $$t_{mm} - FCA \geq 0.10 \text{ in. (2.5 mm)} \qquad \text{Eq. 3-23}$$

   $$L_{msd} \geq 1.8\sqrt{Dt_{min}} \qquad \text{Eq. 3-24}$$

   where L_{msd} = distance from the edge of the region of local metal loss under investigation to the nearest major structural discontinuity (in., mm)
       Equation 3-24 is far too conservative. Normally a value of $6t_{nom}$ or $7t_{nom}$ is more realistic. The parameter, t_{nom}, is the nominal wall thickness. This may be reflected in the upcoming editions.
       For cylindrical shells (piping), conical shells, and elbows, the circumferential extent of the metal loss must be checked by using Part 5 (Section 5) Paragraph 5.4.2.2.g to complete the assessment, that is, to evaluate the circumferential extent of the flaw using Figure 5.7 with the calculated values of c/D and R_t. If the point defined by the intersection of these values is on or above this figure, then the circumferential extent of the flaw is acceptable; otherwise, the circumferential extent of the flaw is unacceptable.
2. For $s > L$, one of the following assessment methods may be used:
   a. A simple approach is to set the average thickness equal to the measured minimum thickness, or $t_{am} = t_{mm}$, and to proceed to the acceptance criteria below (Level 1 or Level 2, as applicable). This approach facilitates the FFS assessment; however, the

results may be too conservative if the remaining thickness is too small.

    b. Determine the average and minimum measured thickness for the meridional and circumferential CTPs as described later, and then proceed to the acceptance criteria, Level 1 or Level 2 as applicable, to complete the assessment.

3. Determine the minimum measured thickness, t_{mm}, considering all points on the longitudinal and circumferential CTPs.

4. Compute the average measured thickness, t_{am}, from the CTP in the meridional (longitudinal for cylindrical and conical shells) and circumferential directions and designate these values as t_{am}^s and t_{am}^c, respectively. The average thickness is computed by numerically averaging the thickness readings over length L. The center of midpoint of averaging, L, should be located at t_{mm}.

5. For cylindrical and conical shells and pipe bends, $t_{am} = t_{am}^s$ in a Level 1 assessment. In a Level 2 assessment, t_{am}^s and t_{am}^c are used directly in the assessment to account for supplemental loads.

The region of metal loss can be evaluated using a Level 3 Assessment, or a Part 5 (Section 5) Assessment for local metal loss.

Level 1 Part 4 Acceptance Criteria

The average measured wall thickness should satisfy the following thickness criteria. Alternatively, the MAWP calculated based on the thickness $(t_{am} - FCA)$, should be equal to or greater than the current MAWP.

$$t_{am} - FCA \geq t_{min} \qquad\qquad \text{Eq. 3-15}$$
$$\text{(Eq. 4.4 of API 579)}$$

The minimum measured wall thickness (t_{mm}) should satisfy the following thickness criterion for piping (and vessels):

$$t_{mm} - FCA \geq MAX[0.5\, t_{min}, 2.5 \text{ mm } (0.10 \text{ in.})] \qquad \text{Eq. 3-23}$$
$$\text{(Eq. 4.5 of API 579)}$$

If both of these conditions are not met, then the LTA does not pass the Part 4 Level 1 criteria.

If the component does not meet the Level 1 Assessment requirements, then the following, or combinations thereof, can be considered:

1. Re-rate, repair, replace, or retire the component.
2. Adjust the FCA by applying remediation techniques.

3. Adjust the weld joint efficiency or quality factor (*E*) by conducting additional examination and repeat the assessment. (Note: To raise the value of *E* from 0.7 to 0.85, or from 0.85 to 1.0, would require that the weld seams be spot welded or 100% radiographed, respectively, and the examinations may reveal additional flaws that will have to be evaluated.)

Level 2 Assessments

The Level 2 Assessment procedure can be used to evaluate components described in Paragraphs 4.2.3.1.f and 4.2.3.1.g subject to the loads defined in Paragraph 4.2.3.1.h in API 579. If the flaw is found to be unacceptable, the procedure can be used to establish a new MAWP.

The following assessment procedure can be used to evaluate components described in Paragraph 4.2.3.1.f subject to the loads defined in Paragraph 4.2.3.1.h (supplemental loads).

Step 1—Calculate the thickness required for supplemental loads (t_{sl}) and the minimum required thickness (t_{min}).

Step 2—Locate regions of metal loss on the component and determine the type of thickness data that will be recorded. Determine the minimum measured thickness (t_{mm}). If the thickness profile data are used, then proceed to Step 3. If point readings are used, then complete the assessment following the methodology in Paragraph 4.4.2.1.b, just as in Level 1.

Step 3—Determine the length of thickness averaging (*L*), just as in Level 1.

Step 4—Establish the critical thickness profiles and determine *s* and *c*, just as in Level 1.

Step 5—Perform the FFS assessment of the region of metal loss using one of the methods in the Level 1 Assessment.

Step 6—The acceptability for continued operation can be established using the following criteria:

For piping systems and pressure vessels, the average wall thickness for the CTPs should satisfy the following thickness criteria. Alternatively, the MAWP calculated based on the thicknesses,

$$\frac{(t_{am} - FCA)}{RSF_a} \text{ and } \frac{(t_{am} - FCA - t_{sl})}{RSF_a}$$

should be equal to or exceed the design MAWP. The allowable remaining strength factor (RSF_a) can be determined as shown earlier.

Level 2 Part 4 Acceptance Criteria

$$t_{am}^S = FCA \geq RSF_a(t_{\min}^C)$$
Eq. 3-25

$$t_{am}^C = FCA \geq RSF_a(t_{\min}^L)$$
Eq. 3-26

where t_{am} = average wall thickness of component determined at the time of inspection

If either condition in Eq. 3-18 or Eq. 3-19 is not satisfied, then the assessment does not pass Level 2. If this happens, it is acceptable to perform a local LTA assessment per Part 5. However, if the LTA is a large region governed by general membrane stresses, then an assessment per Part 5 would be difficult to justify. A Level 3 Assessment would be more appropriate. The reader is again referred to **Figure 3-4** for guidance; however, experience is the best tool for any engineering application.

The following assessment procedure can be used to evaluate components described in Paragraph 4.2.3.1.g (pressure vessel nozzles, tank nozzles and piping branch connections, the reinforcement zone of conical transitions, and piping systems) subject to supplemental loads (Paragraph 4.2.3.1.h).

1. Design rules for components at a major structural discontinuity typically involve the satisfaction of a local reinforcement requirement (e.g., nozzle reinforcement area) or require the computation of a stress level based upon a given load condition and geometry and thickness configuration (e.g., flange design). These rules typically result in one component with a thickness that is dependent upon that of another component (examples were listed earlier). Design rules of this nature have thickness interdependency, and the definition of a minimum thickness for a component is ambiguous.

2. To evaluate components with a thickness interdependency, compute the MAWP based on the average measured thickness minus the future corrosion allowance ($t_{am} - FCA$) and the thickness required for supplemental loads (Appendix A Paragraph A.2.6) for each component using the equations in the original construction. The calculated MAWP should be equal to or exceed the design MAWP.

3. The average thickness of the region (t_{am}) can be obtained as follows for components with thickness interdependency:

   Nozzles and Branch Connections—Determine the average thickness within the nozzle reinforcement zone, as shown in **Figure 3-6** (API 579 Figure 4.9). The assessment procedures in Appendix A, Paragraphs A.3.11 and A.5.7 can be utilized to evaluate metal loss at a nozzle or piping branch connection, respectively. The weld load path analysis in this paragraph should be also checked, particularly if the metal loss has occurred in the weldments of the connection.

Piping Systems—Piping systems have thickness interdependency because of the relationship between the component thickness, piping flexibility, and resulting stress. For straight sections of piping, determine L using the procedure discussed earlier and compute the average thickness to represent the section of pipe with metal loss in the piping analysis. For elbows or bends, the thickness readings should be averaged within the bend, and a single thickness should be used in the piping analysis (i.e., a pipe stress analysis where the flexibility factor, system stiffness, and stress intensification factor are computed). For branch connections, the thickness should be averaged within the reinforcement zones for the branch and header, and these thicknesses should be used in the piping model (to compute the stress intensification factor). An alternative assumption is to use the minimum measured thickness to represent the component thickness in the piping model. This approach may be warranted if the metal loss is localized; however, this may result in an overly conservative evaluation. In these cases, a Level 3 Assessment may be required to reduce the conservatism in the assessment—see "Level 3 Assessments" in this chapter (API 579 Paragraph 4.4.4.4).

4. The minimum measured metal wall thickness (t_{mm}) meets Eq. 3-23 (Eq. 4.5 of API 579).

If the component does not meet the Level 2 Assessment requirements, then the following, or combinations thereof, can be considered:

1. Re-rate, repair, replace, or retire the component.
2. Adjust the FCA by applying remediation techniques (see API 579 Paragraph 4.6).
3. Adjust the weld joint efficiency factor (E) by conducting additional examination and repeat the assessment—see preceding discussion in this chapter under Level 1 (see Level 1 Part 4 Acceptance Criteria, Step 3).
4. Conduct a Level 3 Assessment.

Local Metal Loss Assessment

The API 579 provides two screening tools for a Part 5, or local metal loss, assessment. **Figure 3-7** (Figure 5.6 in API 579) is the Level 1 screening criteria for local metal loss in a shell (e.g., piping). **Figure 3-8** (Figure 5.7 in API 579) is the Level 1 screening criteria for the maximum allowable circumferential extent of local metal loss in a cylinder. Note that **Figure 3-8** is valid *only* for pressure loads and *not* for supplemental loads (e.g., external bending moments). Supplemental loads are quite common in piping.

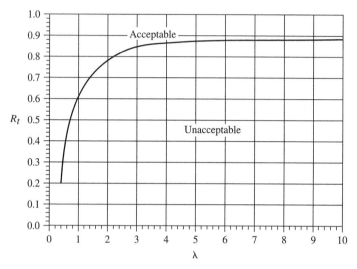

Figure 3-7. Screening criteria for local metal loss in a shell. Courtesy of the American Petroleum Institute.

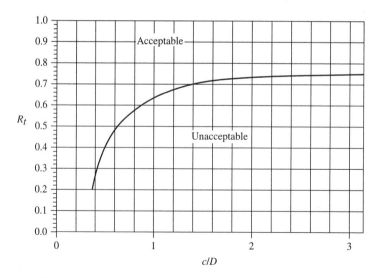

Figure 3-8. Screening criteria for the maximum allowable circumferential extent of local metal loss in a cylinder. Courtesy of the American Petroleum Institute.

The acceptance criteria for Level 1 and Level 2 are summarized next. The Level 1 and Level 2 Assessment procedures in Part 5 of the API 579 apply only if the following conditions are met:

1. The original design criteria were in accordance with a recognized code or standard.
2. The component is not operating in the creep range.
3. The component is made of material with sufficient toughness. This will be covered later in the discussion about brittle fracture. Appendix G of API 579 covers temperature and/or process conditions that result in material embrittlement.
4. The component is not subject to cyclic service.
5. The component does not contain crack-like flaws. We will discuss these in another section on crack-like flaws below.
6. The component is not subjected to external pressure. This subject will be covered in future editions of API 579.
7. The original design criteria were based on a recognized code or standard that has a design equation that specifically relates pressure and/or other loads to a required wall thickness.
8. When the Level 1 and/or Level 2 Assessment procedures do not apply, or when they produce conservative results (e.g., would not permit operation at the desired design or operating conditions), a Level 3 Assessment procedure may be performed. We will discuss Level 3 Assessments later.
9. The Level 2 Assessment procedure for components that do not have a design equation that specifically relates pressure and/or other loads, as applicable, to a required wall thickness is limited to the components listed, as discussed in "General Metal Loss Assessment" in this chapter.
10. The following limitations on applied loads must be satisfied when using the assessment procedures of the local metal loss part (section):
    - *Level 1 Assessment*—components have a design equation that specifically relates pressure and/or loads, as applicable, to a required wall thickness (e.g., piping, conical shell sections, elbows or pipe bends that do not have structural attachments)—Paragraph 4.2.3.1.f of API 579, subject to *internal pressure*.
    - *Level 2 Assessment*—components listed in Paragraph 4.2.3.1.f subject to internal pressure; cylinders subject to internal pressure and/or supplemental loads; components listed above (Paragraph 4.2.3.1.g) subject to internal pressure and/or external pressure and/or supplemental loads.

Determining the LTA Boundary

The methodology for defining what is an LTA is shown in **Figure 3-9**. Two LTAs are shown in **Figure 3-9a** (Step 1). In **Figure 3-9b** (Step 2), an LTA has a rectangle drawn around it. **Figure 3-9c** (Step 3) shows the rectangle drawn in **Figure 3-9b** doubled in size. The rectangle of double size overlaps another LTA. This other LTA must be added in to the original LTA and a new rectangle is drawn, as shown in **Figure 3-9d**. This is the boundary of the

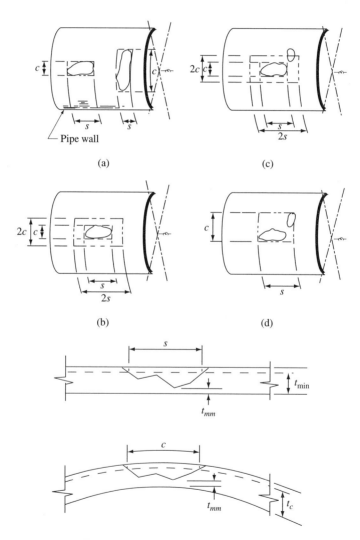

(a)

(c)

(b)

(d)

Figure 3-9. Determining the LTA boundary.

LTA used in the analysis for determining the parameters s, c, and t_{mm} in the assessment. A structural discontinuity cannot fall into the bounds of the LTA.

If a structural discontinuity falls within an LTA, it depends on the type of discontinuity. A weld is actually not a discontinuity, but if a weld extends up to or across an LTA, this must be considered (see Example 3-1). To accomplish this, assess the LTA with the methodology presented and then assess the area next to the weld separately because the heat-affected zone next to the weld has different properties than the parent metal. The material properties of the HAZ must be accounted for, and using the section on material properties along with the algorithms for cracks presented later does this. If the discontinuity is a nozzle, conventional ASME (B31.3) area replacement algorithms can be used, but a Level 3 Assessment is more appropriate. A branch connection is like a nozzle because it must meet the area replacement rules. If the discontinuity is a stiffening ring or structural support welded to the pipe, then we are talking about a weld, which follows the rules mentioned earlier.

Level 1 Part 5 Acceptance Criteria

Level 1 Assessment procedures can be used to evaluate with local metal loss subject to *internal pressure*. The procedures can be used to determine the acceptability and/or to re-rate a component with a flaw. If there are significant thickness variations over the length of the flaw or if a network of flaws is closely spaced, this procedure may produce conservative results, and a Level 2 Assessment is recommended.

The following assessment procedure can be used to evaluate components that meet these conditions.

Step 1—Determine the critical thickness profile(s), using parameters defined previously.

Step 2—Determine the minimum required thickness (t_{min}).

Step 3—Determine the minimum remaining thickness (from inspection) (t_{mm}), the remaining thickness ratio (R_t) using Eq. 3-3, and λ, defined in Eq. 3-4.

Step 4—Check the limiting flaw size criteria; if the following requirements are satisfied, proceed to Step 5; otherwise, the flaw is not acceptable per the Level 1 Assessment procedure.

$$R_t \geq 0.20 \qquad\qquad \text{Eq. 3-27}$$

$$t_{mm} - FCA \geq 0.10 \text{ in. (2.5 mm)} \qquad\qquad \text{Eq. 3-28}$$

$$L_{msd} \geq 1.8\sqrt{Dt_{min}} \qquad\qquad \text{Eq. 3-29}$$

This equation for L_{msd} is far too conservative. Normally a value of $6t_{nom}$ or $7t_{nom}$ is more realistic. The parameter t_{nom} is the nominal wall thickness. This may be reflected in the upcoming editions of API 579.

Step 5—If the region of metal loss is categorized as an LTA (a groove is not present in the LTA) then proceed to Step 6; otherwise, check the following criteria for a groove-like flaw.

Step 5.1—Per API 579 Paragraph 5.4.2.2(e), if there is a gouge or groove defect, then compute the critical groove radius (g_r^c) using the following equation:

$$g_r^c = MAX[0.25t_{min}, 6.4 \text{ mm} (0.25 \text{ in.})] \qquad \text{Eq. 3-30}$$

Step 5.2—If both of the following are satisfied, then proceed to the next step; otherwise, the groove-like flaw may be reevaluated as an equivalent crack-like flaw using Part 9 (Section 9) Level 1 Assessment criteria, or another acceptable method for cracks (e.g., the BS 7910). In this evaluation, the maximum depth and length of the groove-like flaw should be used to determine the equivalent crack-like flaw.

Referring to **Figures 3-10a, 3-10b**, and **3-10c**, the following equations must be satisfied for the flaw to be categorized as a groove:

$$g_r \geq g_r^c \qquad \text{Eq. 3-31}$$

$$\frac{g_r}{(1 - R_t)t_{min}} \geq 1.0 \qquad \text{Eq. 3-32}$$

Step 5.3—If the flaw is categorized as a groove, then proceed to a Level 1 or Level 2 Assessment procedure, as applicable, to complete the assessment. Otherwise, characterize the flaw as a gouge and determine the critical exposure temperature (CET) based on operating and design conditions (see the part (section) on brittle fracture for cold temperature applications and Chapter 4).

Step 5.4—Determine the minimum allowable temperature (MAT) using the section in the brittle fracture assessment, which is presented in Chapter 4. In this section, the curve in ASME B31.3, Figure 323.2.2, and design metal temperature versus nominal thickness is used. Instead of using the design metal temperature, the MAT is used instead. This curve will be discussed later under "Brittle Fracture Concepts" in Chapter 4. The thickness of the plate containing the gouge and the material specification must be known to utilize this figure (if the material specification is not known, Curve A of this figure should be used). For example, a 25.4 mm (1 in.) thick plate of API 51 × 40, nonnormalized, would have an $MAT = -1°C$ (30°F).

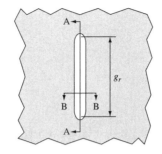

(a) Groove-like flaw – plan view

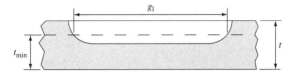

(b) Groove-like flaw length – section A-A

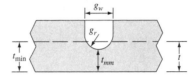

(c) Groove-like flaw width – section B-B

Figures 3-10. Groove-like flaw dimensions—flaw profile. Courtesy of the American Petroleum Institute.

Step 5.5—If $CET > MAT + 14°C$ ($MAT + 25°F$), then proceed to Step 5.6; otherwise, proceed to Step 5.7.

Step 5.6—Proceed to Step 6 (Level 1 or Level 2 Assessment procedure, as applicable) and complete the assessment.

Step 5.7—The groove-like flaw is not acceptable per the Level 1 Assessment procedure. Alternatively, the groove-like flaw may be reevaluated as an equivalent crack-like flaw (see "Crack-like Flaws" in this chapter. In this evaluation, the maximum depth and length of the groove-like flaw should be used to determine the equivalent crack-like flaw.

Step 6—For a Level 1 Assessment, the procedure is the same as for an LTA. Enter **Figure 3-7** (API 579 Figure 5.6) with the calculated values for λ and R_t. If the point defined by the intersection of these values is on or above and to the left of the curve, then the longitudinal extent of the flaw is acceptable per Level 1. If the point falls into the unacceptable zone, then the RSF can be reevaluated, using Eq. 3-33, or in line with the following recommendations.

$$RSF = \frac{R_t}{1 - \frac{1}{M_t}(1 - R_t)}$$ Eq. 3-33

where $M_t = (1 + 0.48\lambda^2)^{0.5}$

Step 7—For cylindrical and conical shells, evaluate the circumferential extent of the flaw using **Figure 3-8** (Figure 5.7 in API 579). To evaluate the circumferential extent of the flaw, enter **Figure 3-8** with the calculated values of c/D and R_t. If the point defined by the intersection of these values is on or above the curve in this figure, then the circumferential extent of the flaw is acceptable; otherwise, the circumferential extent of the flaw is unacceptable.

If a component does not meet the Level 1 Assessment requirements, the other recommendations are as follows:

1. Re-rate, repair, replace, or retire the component.
2. Adjust the future corrosion allowance by applying remediation techniques (weld overlay, strip linings, metal spray linings, increasing or decreasing the process temperature and/or pressure, or changing the process stream velocity, or installing scrubbers, treaters, coalescers, and filters to remove certain fractions and/or contaminants in a stream).
3. Adjust the weld joint efficiency factor (E) by conducting additional radiography and repeating the assessment.
4. Conduct a Level 2 or Level 3 Assessment.

Figure 3-5 (Figure 5-6 in API 579) has been quantified in the following criteria:

$R_t \geq 0.2$ for $\lambda \leq 0.3475$ Eq. 3-34

$$R_t \geq \frac{\left(RSF_a - \frac{RSF_a}{M_t}\right)}{\left(1.0 - \frac{RSF_a}{M_t}\right)} \text{ for } 0.3475 < \lambda < 10.0$$ Eq. 3-35

$R_t \geq 0.885$ for $\lambda \geq 10.0$ Eq. 3-36

The following criteria are for the circumferential extent of an LTA:

$R_t \geq 0.2$ for $c/D \leq 0.348$ Eq. 3-37

$$R_t = \frac{-0.73589 + 10.511\left(\dfrac{c}{D}\right)^2}{1.0 + 13.838\left(\dfrac{c}{D}\right)^2} \text{ for } \frac{c}{D} \geq 0.348 \qquad \text{Eq. 3-38}$$

$$L_{msd} \geq 1.8\sqrt{Dt_{min}} \qquad \text{Eq. 3-29}$$

where L_{msd} = distance from the region of local metal loss to the nearest structural discontinuity.

If either of the above sets of equations, Eqs. 3-34 through 3-38, is not satisfied, then the assessment fails the Part 5 (Section 5) Level 1 criteria.

Equation 3-27 is controversial, and many, including members of the API 579 team, believe that it is too conservative. A more realistic value would be as follows:

$$L_{msd} > 6t_{nom} \text{ or } 7t_{nom} \qquad \text{Eq. 3-39}$$

where t_{nom} = nominal thickness of the component, in. (mm)

The API 579 is vague in regard to structural discontinuities. In Paragraph 4.4.3.3.a, it states the following:

> Design rules for components at a major structural discontinuity typically involve the satisfaction of a local reinforcement requirement (e.g. nozzle reinforcement area), or necessitates the computation of a stress level based upon a given load condition and geometry and thickness configuration (e.g. flange design). These rules typically result in one component with a thickness which is dependent upon that of another component (for examples, see paragraph 4.2.3.1.g). Design rules of this type have a thickness interdependency, and the definition of a minimum thickness for a component is ambiguous.

If an LTA on a pipe is located next to a weld connecting a flange or valve, for example, the minimum thickness is not ambiguous but must be assessed on the LTA that is in proximity to a structural discontinuity. Stresses at structural discontinuities tend to be magnified by stress intensification factors, and a true assessment should require a Level 3 Assessment, unless the SIFs are known. A Level 3 Assessment is recommended because closed-form solutions are based on assumptions that are too simplistic.

As of this writing, the API 579 does not give criteria for external pressure, but it is planned in upcoming editions.

Level 2 Part 5 Acceptance Criteria

The assessment procedures in Level 2 provide a better estimate of the remaining strength factor than computed in Level 1 for local metal loss in a component subject to internal pressure loading if there are significant variations in the thickness profile. These procedures account for the local reinforcement effects of the varying wall thickness in the region of the local metal loss and ensure that the weakest ligament is identified and properly evaluated. The procedures can also be directly used to evaluate closely spaced regions of local metal loss as well as cylindrical and conical shells with supplemental loads.

The following assessment procedure can be used to evaluate components subjected to loads described earlier. If the flaw is found to be unacceptable, the procedure can be used to establish a new MAWP.

Step 1—Determine the critical thickness profile(s) and the parameters defined previously.

Step 2—Calculate the minimum required thickness (t_{min}), including the thickness required for supplemental loads (t_{sl}).

Step 3—Determine the minimum measured thickness (t_{mm}) and the remaining thickness ratio (R_t) using Eq. 3-3, and the flaw dimensions (s and c) and the shell parameter (λ) using Eq. 3-4.

Step 4—Check the limiting flaw size criteria defined below:

$$R_t \geq 0.20 \qquad\qquad\qquad \text{Eq. 3-40}$$

$$(t_{mm} - FCA) \geq 0.10 \qquad\qquad\qquad \text{Eq. 3-41}$$

$$L_{msd} \geq 1.8\sqrt{Dt_{min}} \qquad\qquad\qquad \text{Eq. 3-42}$$

In addition, the length of the flaw must satisfy the relationship $\lambda \leq 5.0$. If all these requirements are satisfied, then proceed to Step 5; otherwise, the flaw is not acceptable per the Level 2 Assessment procedure.

Step 5—If the region of metal loss is categorized as an LTA (a groove is not present in the LTA), then proceed to Step 6; otherwise, check the groove-like flaw criteria in the section on grooves and continue the assessment.

Step 6—Determine the remaining strength factor for the longitudinal CTP. If there are significant variations in the thickness profile, then the following procedure can be used to compute a less conservative value for the RSF when compared to the procedures of Level 1.

Step 6.1—Rank the thickness readings in ascending order based on metal loss.

Step 6.2—Set the initial evaluation starting point as the location of maximum metal loss; this is the location in the thickness profile where t_{mm} is recorded. Subsequent starting points should be in accordance with the ranking in Step 6.1.

Step 6.3—At the current evaluation starting point, subdivide the thickness profile into a series of subsections, shown in **Figure 3-11** (API 579 Fig. 5.8). The number and extent of the subsections should be chosen based on the desired accuracy and should encompass the variations in metal loss.

Step 6.4—For each subsection, compute the remaining strength factor for the longitudinal CTP using the following equations. *Paragraph 5.4.3.2f states that if there are significant variations in the thickness profile, then the following equations can be used to compute a less conservative RSF when compared to the procedures of Level 1. Note that it does not define what significant variations in the thickness profile are.*

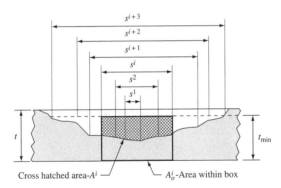

(a) Subdivision process for determining the *RSF*

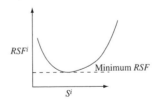

(b) Determining the minimum *RSF* value

Notes:
A^i = Area of metal loss associated with length s^i (cross-hatched area). This area can be evaluated using a numerical integration technique (e.g. Simpson's or Trapezoidal Rule).
A_o^i = Total original area associated with length s^i and thickness t_{min}, or $A_o^i = s^i t_{min}$.

Figure 3-11. Definition of areas to compute the RSF for a region of local metal loss for a Level 2 Assessment. Courtesy of the American Petroleum Institute.

Step 6.5—Alternatively, the RSF can be computed using the equations of API 579 Appendix D, Paragraph D.2.3.3 (for definition of terms, the reader is referred to the discussion in Appendix D, Paragraph D.2.3.3).

Where

$$RSF^i = \frac{1.0}{(M_s^{NS})^i} \qquad \text{Eq. 3-43}$$

$$RSF^i = \frac{1 - \left(\dfrac{A^i}{A_o^i}\right)}{1 - \dfrac{1}{M_t^i}\left(\dfrac{A^i}{A_o^i}\right)} \qquad \text{Eq. 3-44}$$

with

$$A_o^i = s^i t_{min}$$

$$M_t^i = \left(\frac{1.02 + 0.4411(\lambda^i)^2 + 0.006124(\lambda^i)^4}{1.0 + 0.02642(\lambda^i)^2 + 1.533(10^{-6})(\lambda^i)^4}\right) \qquad \text{Eq. 3-45}$$

Equation 3-45 is valid only for cylindrical shells.

And A^i = area of metal loss based on s^i, including the effect of FCA (see **Figure 3-11**), in.^{2} (mm^2)
A_o^i = original metal area based on s^i, in.^{2} (mm^2)
s^i = length increment of metal loss (see **Figure 3-11** [Figure 5.8 in API 579]), in. (mm)
λ^i = shell parameter computed using Eq. 3-7 with $s = s^i$

Step 6.6—Determine the minimum value of the remaining strength factors (RSF_i) found in Step 6.4 for all subsections (see **Figure 3-11**). This is the minimum value of the RSF for the current evaluation point.
Step 6.7—Repeat Steps 6.3 through 6.5 of this calculation for the next evaluation point, which corresponds to the next thickness reading in the ranked thickness profile list.
Step 6.8—The remaining strength factor to be used in the assessment is the minimum value determined for all evaluation points.
Step 7—For cylindrical and conical shells, evaluate the circumferential extent of the flaw using the following criteria. If supplemental loads are not present or are not significant, then the circumferential

dimension (c) of the flaw determined from the circumferential CTP should satisfy the criterion in using **Figure 3-8** (Figure 5.7 in API 579). If the supplemental loads are significant, then the circumferential extent of the region of local metal loss shall be evaluated using the procedures in the following section.

The assessment procedure of this section can be used to find the acceptability of the circumferential extent of a flaw in a cylindrical or conical shell subject to pressure and/or supplemental loads. Note that the acceptability of the longitudinal extent of the flaw is evaluated using procedures described above.

Assessing Supplemental Loads

Supplemental loads are those that produce a net section axial force, external bending moment, torsion, and shear applied to the cross section of the piping component containing the flaw. Supplemental loads are those that act in addition to the longitudinal and circumferential (hoop) membrane stress caused by internal pressure.

To consider supplemental loads, the reader is referred to Chapter 2. One must always remember that pipe stress software is based on linear elastic analysis. Hence, the assessment should include both load-controlled and strain-controlled conditions. Thus, the net section axial force, bending moment, torsion, and shear should be computed on two load cases: (1) weight and pressure (includes the weight of the component and occasional loads from wind or earthquake) and (2) weight plus thermal (includes the results of the weight case plus the results of the thermal case, which encompasses the effects of temperature, support displacements and other loads that are considered strained-controlled). As will be discussed in Level 3, this procedure does not include supplemental loadings in the plastic range.

Piping systems contain supplemental loads that affect the relationship between the component thickness, piping flexibility, and/or stiffness and the resulting stress state of the piping system. In performing this assessment, allowance must be made in the pipe stress assessment that considers the LTA. Pipe stress software is such that LTAs cannot be modeled into them without oversimplifications. The average wall thickness of the component should be lowered to compensate for the metal loss. Thus, the pipe wall thickness is modified by lowering the pipe wall thickness and rerunning the analysis to simulate the loads on the anchor points.

If the metal loss in the circumferential plane can be approximated by a single area, the API 579 gives closed-form algorithms to assess the acceptability of the component. If not, a numerical technique (e.g., finite element) must be used.

Table 3-1 outlines the evaluation of the circumferential inspection plane being assessed, using **Figures 3-12a** and **3-12b**, with

$$\bar{y}_{LX} = \frac{2R \sin\theta}{3\theta} \left(1 - \frac{d}{R} + \frac{1}{2 - \frac{d}{R}} \right)$$

Eq. 3-46

Table 3-1
Section Properties of a Cylinder with a Region of Local Metal Loss

Parameter	LTA on Inside Surface	LTA on Outside Surface
	$D_f = D_o - 2(t_{mm} - FCA)$	$D_f = D_o + 2(t_{mm} - FCA)$
Circumferential angular extent of LTA (θ = degrees) of cylinder without metal loss	$A_{tf} = \dfrac{c(D_o + D_f)}{8}$	$\theta = \dfrac{c}{D_f}\left(\dfrac{180}{\pi}\right)$
Cylinder aperture cross section (in.^{2}, mm^2) of cylinder without metal loss	$A_f = \dfrac{\pi}{4} D_i^2$	$A_f = \dfrac{\pi}{4} D_i^2$
Cylinder metal cross section (in.^{2}, mm^2) of cylinder without metal loss	$A_m = \dfrac{\pi}{4}\left(D_o^4 - D_i^4\right)$	$A_m = \dfrac{\pi}{4}\left(D_o^4 - D_i^4\right)$
Cylinder moment of inertia (in.^{2}, mm^2) of cylinder without metal loss	$I_x = I_y = \dfrac{\pi}{64}\left(D_o^4 - D_i^4\right)$	$I_x = I_y = \dfrac{\pi}{64}\left(D_o^4 - D_i^4\right)$
For LTA	$A_f = \dfrac{\theta}{4}\left(D_f^2 - D_i^2\right)$	$A_f = \dfrac{\theta}{4}\left(D_f^2 - D_i^2\right)$
Effective area on which pressure acts (in.^{2}, mm^2)	$A_w = A_a + A_f$	$A_w = A_a + A_f$
Location of the neutral axis (see **Figure 3-12**) (in., mm)	$\bar{y} = \dfrac{1}{12}\dfrac{\sin\theta\left(D_f^3 - D_i^3\right)}{A_m - A_f}$	$\bar{y} = \dfrac{1}{12}\dfrac{\sin\theta\left(D_0^3 - D_F^3\right)}{A_m - A_f}$

Table 3-1
Section Properties of a Cylinder with a Region of Local Metal Loss—cont'd

Parameter	LTA on Inside Surface	LTA on Outside Surface
Distance along the x-axis to Point A on the cross section shown in **Figure 3-12** (in., mm)	$x_A = 0.0$	$x_A = 0.0$
Distance from the $\bar{x} - \bar{x}$ axis measured along the y-axis to point A on the cross section shown in **Figure 3-12** (in., mm) $\bar{x} - \bar{x}$	$y_A = \bar{y} + \dfrac{D_o}{2}$	$y_A = \bar{y} + \dfrac{D_f}{2}$
Distance along the x-axis to point B on the cross section shown in **Figure 3-12** (in., mm)	$x_B = \dfrac{D_o}{2}\sin\theta$	$x_B = \dfrac{D_f}{2}\sin\theta$
Distance from the $\bar{x} - \bar{x}$ axis measured along the y-axis to point B on the cross section shown in **Figure 3-12** (in., mm)	$y_B = \bar{y} + \dfrac{D_o}{2}\cos\theta$	$y_B = \bar{y} + \dfrac{D_f}{2}\cos\theta$
Location of the centroid of area A_w, measured from the x-x axis (in., mm)	$b = \dfrac{1}{12}\dfrac{\sin\theta\left(D_f^3 - D_i^3\right)}{A_a + A_f}$	$b = 0$
Outside radius of area, A_f (in., mm)	$R = \dfrac{D_f}{2}$	$R = \dfrac{D_o}{2}$
Maximum depth of the region of local metal loss (in., mm)	$d = \dfrac{\left(D_f - D_i\right)}{2}$	$d = \dfrac{\left(D_o - D_f\right)}{2}$
Mean area to compute torsion stress for the region of the cross section with metal loss (in.^{2}, mm^2)	$A_{tf} = \dfrac{c\left(D_o + D_f\right)}{8}$	$A_{tf} = \dfrac{c\left(D_i + D_f\right)}{8}$

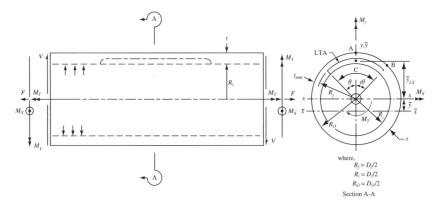

Figure 3-12a. Parameters for permissible bending moment, axial force, and pressure for a cylinder with an LTA. Courtesy of the American Petroleum Institute.

$$I_{LX} = R^3 d(C_1 + C_2)$$ Eq. 3-47

where

$$C_1 = \left(1 - \frac{3d}{2R} + \frac{d^2}{R^2} - \frac{d^3}{4R^3}\right)\left(\theta + \sin\theta\cos\theta - \frac{2\sin^2\theta}{\theta}\right)$$

$$C_2 = \frac{d^2\sin^2\theta}{3R^2\theta\left(1 - \frac{d}{R} + \frac{d^2}{6R^2}\right)}$$

$$I_{\bar{X}} = I_X + A_m\bar{y}^2 - I_{LX} - A_f(\bar{y}_{LX} + \bar{y})^2$$ Eq. 3-48

$$A_f = \frac{[0.5\pi(D_i + D_o) - c](D_i + D_o)}{8}$$ Eq. 3-49

$$I_{LY} = R^3 d\left[\left(1 - \frac{3d}{2R} + \frac{d^2}{R^2} - \frac{d^3}{4R^3}\right)(\theta - \sin\theta\cos\theta)\right]$$ Eq. 3-50

$$I_{\bar{Y}} = I_Y - I_{LY}$$ Eq. 3-34

$$A_f = \frac{[0.5\pi(D_i + D_o) - c](D_i + D_o)}{8}$$ Eq. 3-51

where A_t = cross-sectional area of the region of local metal loss, in.^{2} (mm^2)

$$\sigma_{lm} = MAX[\sigma_{lm}^A, \sigma_{lm}^B]$$

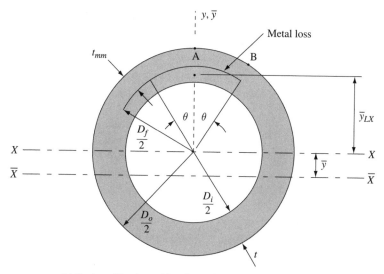

(a) Region of local metal loss located on the inside surface

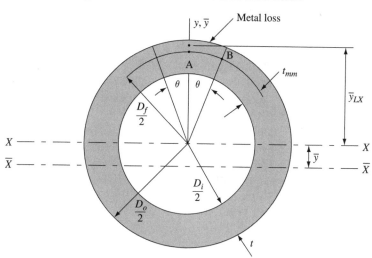

(b) Region of local metal loss located on the outside surface

Figure 3-12b. Parameters for determining section properties of a cylinder with an LTA. Courtesy of the American Petroleum Institute.

$I_{\bar{x}}$ = moment of inertia of the cross section with the region of local metal loss about the $\bar{x}$-axis, in.^{4} (mm^4)

$I_{\bar{y}}$ = moment of inertia of the cross section with the region of local metal loss about the $\bar{y}$-axis, in.^{4} (mm^4)

I_{LX} = moment of inertia of area (A_f) about a local x-axis, in.^{4} (mm^4)

I_{LY} = moment of inertia of area (A_f) about a local y-axis, in.^{4} (mm^4)

Now we compute the maximum section longitudinal membrane stress for both the weight and weight plus thermal cases considering points A and B in **Figure 3-12a** and **3-12b**.

$$\sigma_{lm}^A = \frac{A_w}{A_m - A_f}(MAWP_r) + \frac{F}{A_m - A_f}$$

$$+ \frac{y_A}{I_{\bar{x}}}\left[(\bar{y} + b)(MAWP_r)A_w + M_x\right] + \frac{x_A}{I_{\bar{y}}}M_y \qquad \text{Eq. 3-52}$$

$$\sigma_{lm}^B = \frac{A_w}{A_m - A_f}(MAWP_r) + \frac{F}{A_m - A_f}$$

$$+ \frac{y_B}{I_{\bar{x}}}\times\left[(\bar{y} + b)(MAWP_r)A_w + M_x\right] + \frac{x_B}{I_{\bar{y}}}M_y \qquad \text{Eq. 3-53}$$

$$\sigma_{lm} = \text{MAX}\left[\sigma_{lm}^A, \sigma_{lm}^B\right] \qquad \text{Eq. 3-38}$$

where F = applied section axial force determined from the pipe stress analysis for the weight or weight plus thermal case (refer to Paragraph 319.2.3(c) of ASME B31.3), as applicable, lb_f (N)

M_x = applied section bending moment determined above for the weight or weight plus thermal load case about the x-axis (see **Figure 3-12a**), as applicable, lb_f (N)

M_y = applied section bending moment determined above for the weight or weight plus thermal load case about the y-axis (see **Figure 3-12a**), as applicable, lb_f (N)

σ_{lm} = maximum longitudinal membrane stress, computed for both the weight and weight plus thermal load cases in the pipe stress analysis, psi (MPa)

The API 579 gives the following criterion that should be satisfied for either a tensile or compressive stress for both the weight and weight plus thermal load cases:

$$\sqrt{\sigma_{cm}^2 - \sigma_{cm}\sigma_{lm} + \sigma_{lm}^2 + 3\tau^2} \leq H\sigma_{ys} \qquad \text{Eq. 3-54}$$

where

$$\sigma_{cm} = \frac{MAWP_r}{E_L RSF}\left(\frac{D_i}{D_o - D_i} + 0.6\right) \qquad \text{Eq. 3-55}$$

$$\tau = \frac{M_T}{2(A_t + A_{tf})d} + \frac{V}{A_m - A_f} \qquad \text{Eq. 3-56}$$

E_L = longitudinal weld joint efficiency

H = allowable stress factor, $H = 0.75$ for weight only case, and $H = 1.5$ for weight plus thermal case, in.-lb$_f$ (N-mm)

$MAWP_r$ = maximum allowable working pressure computed from Eq. 3-18, psi (MPa)

M_T = applied net-section torsion for the weight and weight plus thermal case, as applicable, ft-lb$_f$ (N-mm)

RSF = remaining strength factor

V = Applied net-section shear force for the weight or weight plus thermal case, as applicable, lb$_f$ (N)

σ_{cm} = maximum circumferential stress, typically the hoop stress from pressure loading for the weight and weight plus thermal case, as applicable, psi (MPa)

σ_{lm} = maximum longitudinal membrane stress computed for the weight and weight plus thermal case, psi (MPa)

σ_{ys} = yield stress, psi (MPa)

τ = maximum shear stress in the region of local metal loss for the weight and weight plus thermal load case, as applicable, (psi, MPa)

Level 3 Assessments

The Level 1 and Level 2 Assessments are for closed-form solutions. Closed-form solutions are valid only with simplified assumptions, often at the expense of assuming a simplistic geometry or analysis. Handbook solutions can be used if the problem parameters match the component geometry and loading condition. Level 3 is either for those assessments that do not pass Level 1 and Level 2 or for those that desire a more accurate solution from the start. To accomplish a Level 3 Assessment, knowledge of a numerical method is necessary. Such numerical methods are boundary elements, finite difference, or finite element, the most common approach. Recall that we denote finite element as FE.

Per Paragraph 4.4.4.1 of the API 579, a Level 3 Assessment can be based on a linear stress analysis, with acceptability determined using stress categorization, or a nonlinear stress analysis, with acceptability determined using a plastic collapse load. The latter are recommended to provide the best estimate of the acceptable load-carrying capacity of the component; however, the linear analysis is often quicker. In the field, speed is essential, especially if one has a large number of problems to solve in a time-limiting work effort like a plant turnaround. We will briefly describe both approaches.

Elastic-Plastic Analysis of LTAs

As has already been mentioned, this approach can be a significantly more accurate estimate of the remaining strength factor than a linear elastic analysis for pressure-only loading. This methodology is not described in the API 579, but a common approach is presented. This approach is based on, as of this writing, unpublished research reports by the Pipeline Research Council International (PRCI) and international bodies that developed a failure criterion that was developed and validated against the results of 93 ring-tension tests and pipe-burst tests. The accuracy of the failure prediction depends on several factors. One is the judgment that is required at every step. Other factors are knowledge of structural analysis and material behavior and experience with nonlinear finite element analysis. This method consists of the following:

1. Create an FE mesh appropriate for the analysis. Model the undam-aged shell for a distance in each direction of at least $10\sqrt{r_m t_r}$ (where r_m = mean radius, t_r = required thickness) from the edge of the thin area(s). Either plate (depending on the FE code used—shell or continuum elements) may be used, depending on the geometry and the need to account for eccentricity bending effects. Typically, solid second-order elements, such as 20-node hexahedral elements, are used. Normally it is not necessary to model small local discontinu-ities such as exact surface profiles of thickness transitions (e.g., the boundary of the LTA). Such discontinuities can be modeled with a sufficient number of elements (or integration points) through the thickness to model the nonlinear through thickness stress distribu-tion. The analyst should use judgment and experience on the defect shape simplification, the mesh density through the metal-loss liga-ment, and the extent of the mesh refinement area. The defect simu-lated in the finite element model can be placed in a convenient position around the circumference of the pipe if the actual defect is not likely to be affected by any specific supporting position or con-straint, as mentioned previously. Actual defect shapes are mostly difficult to model, so the shape can be modeled conservatively. Its maximum dimensions (length, width and depth) need to be used in the model. One should account for possible uncertainty in the measurement of the defect shape and dimensions. The minimum value of the wall thickness measured in the local region around the

defect should be used. Mesh precision should be highest in the region of the defect, which uses coarse mesh in areas distant from the defect. Precision of the mesh enhancement is available in several FE codes.

2. Apply the appropriate boundary conditions to the model, keeping in mind the general deformation of the damaged pipe under the given operating conditions. For buried or anchored pipeline sections, a boundary condition to restrict axial displacement of the pipe should be applied.

3. A nonlinear, large deformation FE stress analysis requires the use of true stress and true strain material behavior up to but not limited to the point corresponding to the ultimate tensile strength. *A representative true stress-strain curve is required. It can be constructed by fitting the results of true stress and true strain measured from a tensile test of round bar specimen or converted from the engineering stress-strain curves measured from a standard tensile test at the temperature of operation, or desired condition. Typically, a coupon is cut from the subject pipe and a tensile test is performed in a reputable test laboratory to develop a true stress-strain curve representative of the heat of steel of the pipe in question. The true stress value corresponding to the ultimate strength of the material will be used when determining failure pressure. Typically a coupon is obtained by performing a hot tap. See note 1 in "Common Mistakes Made in Level 3 Assessments" later in this chapter.*

4. The FE assessment should be done using validated FE software, with a nonlinear analysis procedure that includes both incremental plasticity and large displacement theory. It is recommended that the von Mises yield criterion and associated flaw rule be used in the analysis. If the pipe is subjected to a quasi-static loading, select isotropic hardening if this option is available. Static stress procedures (ignoring dynamic or inertia effects) should be used. In the static stress analysis, the increment of the pressure load applied to the pipe wall has to be sufficiently small. As an alternate, automatic load incrementation algorithms may be used.

5. The FE results should give the results of the von Mises equivalent stress variations and/or equivalent plastic strain variations against pressure values. The stress and/or strain values from one or more positions within the metal-loss area, at which high von Mises equivalent stresses exist, should be examined.

6. Generally the stress variations with increased pressure load will show three distinct stages. The first stage is a linear-elastic

response progressing to a point when the elastic limit is obtained. At this point, a second stage is evident (i.e., a stage where plasticity spreads through the section while the von Mises equivalent stress remains approximately constant or slowly increases). This is caused by a constraint from the surrounding pipe wall. The third stage is manifested or dominated by the material's hardening and begins when the von Mises equivalent stress in the entire section exceeds the material's yield strength. Upon reaching this stage, the whole section deforms plastically, but failure does not occur immediately due to strain hardening. The overall stress and deformation in the model should be carefully checked. Errors may occur due to the application of incorrect constraints or the use of inappropriate elements.

7. The failure pressure can be determined from the von Mises equivalent stress through the minimum section of the metal-loss area. The failure pressure is considered to be the pressure that causes the averaged stress in the section to be equal to the material's tensile strength from a uniaxial test. The accuracy of the preceding failure prediction depends on the accuracy of the stress analysis. The analyst should not only justify the adequacy of the analysis methods and the computer software but also ensure that the pipe modeling and stress analysis results are correct.

8. The MAWP for the corroded (or eroded) pipe section can be determined by multiplying the predicted failure pressure by an appropriate safety margin factor (e.g., that used by the code governing the pipe) as follows:

$$MAWP = F_s P_f$$

where $MAWP$ = maximum allowable working pressure, psig (KPag)
F_s = design factor of safety
P_f = predicted failure pressure, psig (KPag)

The safety margin factor can be equal to the original design factor applied to the piping section as follows:

$$MAWP = F_c P_f$$

where F_c = factor of safety used by the applied code

As has already been stated several places in the text, we use the term "MAWP" instead of "design pressure" to be consistent with the API 579, even though we are discussing piping.

Another approach to a Level 3 Assessment is to use Steps 1, 2, and 3 along with the following steps:

1. Select a minimum increment size that will allow load increments of no less than approximately 2% of the limit load for the undamaged component to limit analysis time.
2. At the completion of the analysis, examine the results to locate the element(s) with the largest values of equivalent plastic strain. Interpolate between load increments as necessary to determine the pressure to first cause 5% strain at any element. This pressure is considered to be the theoretical collapse pressure. If the analysis stopped before any element reached 5% strain value, restart the analysis with a smaller minimum increment size. The analysis may be run to a lower strain level to save computing time. However, the results of this lower level must be used, as they may not be extrapolated to the 5% strain limit (see Paragraph 3.4.1 of API 579).
3. Multiply the collapse pressure found in Step 2 by the ratio of the allowable stress at design temperature from the original design code to the yield strength used in the analysis. If this pressure is equal to or greater than 90% of the MAWP of the pipe, then thin area or LTA is acceptable. If not, the LTA must be repaired, or the equipment must be re-rated.

Note that multiplying the collapse pressure by the ratio of the allowable stress to the yield strength in Step 3 is intended to provide a factor of safety equivalent to that used in the original design of the pipe. The strength is permitted to be reduced by up to 10% from this value as stated in Step 3. This strength reduction is consistent with provisions in the new design/construction codes, which permit unreinforced openings up to a certain size or allow higher stresses in LTAs (e.g., ASME Section VIII, Division 1, Paragraph UG-36 and Section VII, Division 2, Paragraphs 4-112 and 4-132).

The first method presented is preferable to the second method because it has been validated with many tests in an international environment. The second method uses the minimum tensile strength of the applicable code, which is many times less than that of the heat of metal of the component being assessed and will produce more conservative results.

As previously mentioned, the elastic-plastic analysis is best for pressure-only loads. The reason for this is that when external loads from a pipe stress analysis are applied to a nonlinear FE model, these external loads will shift significantly when the component enters the plastic range and distortion sets in. How these loads will shift with strain-hardening effects and the exposure of the component to a problematic state of distortion starts becoming speculation. One must not forget that pipe stress programs are *linear-elastic*. The forces and moments acting on the piping component are taken from a linear-elastic analysis, so extrapolating them into the plastic region with any great deal of certainty is difficult. The piping codes want to keep the piping in the elastic range, thus imposing the piping to a hypothetical state. To perform an elastic-plastic analysis for piping, the pressure loading must be considered separately, and then a linear-elastic analysis can be made for the external loading case. In a great number of piping applications, the thermal case can be very significant with respect to the overall stress state, making such a study questionable. The main idea of the elastic-plastic analysis is for pressure vessels or piping without supplemental loadings where a burst test is quantified in a numerical simulation. Kiefner et al. [Reference 2] is the only known source to burst test a pipe with supplemental loads, so this case could possibly be used to predict a failure test using a numerical method.

Common Mistakes Made in Level 3 Assessments

One of the most common mistakes made in Level 3 Assessments is not using a true stress-strain curve indicative of the heat of metal of the component being assessed. Selecting a stress-strain curve from a reference book (particularly a textbook) can lead to inaccurate and erroneous conclusions.

Another pitfall is assuming that an appropriate stress-strain curve is not required because one will perform an assessment of the damaged component and then of the undamaged component and the ratio of the two in computing the RSF will "cancel out." It has been shown that in the first method, the appropriate stress-strain curve will yield the most accurate results.

Note 1: Material properties vary as a result of the mill process. Carbon steels strengths are heat-treated to meet certain SMYS. Depending on the heat treatment, whether they are annealed, normalized, quench and tempered, and so on will determine their strengths and hence their stress-strain curves. All carbon steel specifications contain heat treatment specifications. Austenitic stainless steels and nickel-based alloy strengths are obtained by cold working in the mill. Depending on the amount of cold work, the stress-strain curve will vary. When a pipe is in operation, there are operations that can change the material properties so that it is different from that delivered from the mill. For this reason, a coupon is essential to verify the properties of the steel in a test lab. For solid components (e.g., valves), obtaining coupons is mostly impractical, so a Level 3 Assessment cannot be accomplished. Reference is made to the ANSI/API 6A, "Specification for Wellhead and Christmas Tree Equipment," Paragraph 5.7, Qualification Test Coupon (QTC), Paragraph 7.7.1, where it states, "The properties exhibited by the QTC shall represent the properties of the thermal response of the material comprising the production parts it qualifies. Depending upon the hardenability of a given material, the QTC results may not always correspond with the properties of the actual components at all locations throughout their cross-section." Even though the ANSI/API 6A may not be a direct application of in-plant or pipeline applications, the material basics are the same when dealing with material properties. Some construct a "stress-strain" diagram by drawing a straight line from the SMYS to the UTS in performing a Level 3 Assessment. For a material to have a straight line from the SMYS to the UTS, it would have to be brittle, like glass. Because the ASME code materials are ductile, such an assumption is invalid. *A Level 3 Assessment should always involve an experienced materials specialist knowledgeable with the metals being assessed to verify the appropriate stress-strain curve. Note that there is no such thing as "theoretical metallurgy."*

Paragraph F.2.1.1.b of the API 579 states the following:

Hardness tests can be used to estimate the tensile strength (see Table F.1). The conversions found in this table may be used for carbon and alloy steels in the annealed, normalized, and quench-and-tempered conditions. The conversions are not applicable for cold worked materials, austenitic steels, or nonferrous materials.

Hardness has nothing to do with alloy steels, only carbon steels. One has to develop curves for hardness in alloy steels. Of course the term "alloy" is not defined. In the NACE Standard MR0175-2002, "Standard Material Requirements," the term "low alloy" is defined as follows: "steel with a total alloying element content of less than 5%, but more than specified for carbon steel."

Table F.2 of the API 579 contains yield and tensile strength properties for Type 304, 310, 316, 316L, 321, and 347 austenitic steels. These materials are cold worked, so there is no method of finding the yield and ultimate strengths without tensile tests. It also lists Alloy 800, 800H, 800HT, and HK-40, which also must be tested.

In Paragraph F.2.3.1, the Ramberg-Osgood relationship to attempt to find the stress-strain curve for a material is highly suspect by the materials community, especially for austenitic stainless steels. One cannot calculate material properties.

With respect to using Paragraph F.2.1.1 in determining the stress-strain curves for a material, a tensile test is mandatory by new PRCI reports in assessing the remaining strength of corroded pipelines when performing Level 3 Assessments—there is no other option.

The integrity triangle shown in Figure 1-1 indicates that three disciplines—Inspection, Mechanical, and Materials—are necessary for an FFS assessment. Leaving out any one of these disciplines invalidates, or makes questionable, an FFS assessment.

Performing the Remaining Life Assessment

The MAWP Approach

There are two approaches to calculating the remaining life—the MAWP approach and the thickness approach. The MAWP approach is best suited in finite element use, particularly where a linear-elastic finite element (LEFE) model is constructed of the pipe and corroded region and Eq. 3-18 is used. The method is quite simple. It is as follows:

Let FCA_1 = the future corrosion allowance for the initial condition, such that

$$MAWP > MAWP_r$$

where $MAWP_r$ = the required MAWP for the pipe component

Let FCA_2 = the future corrosion allowance such that $MAWP = MAWP_r$. Now

$$R_{life} = \frac{(FCA_1 - FCA_2)(\text{in.})}{C_{rate}\left(\dfrac{\text{in.}}{\text{yr}}\right)} = \text{years of remaining life} \qquad \text{Eq. 3-57}$$

where C_{rate} = corrosion rate, in./yr (mm/yr)
FCA = inches or millimeters

In using Eq. 3-18 on spreadsheet software, the MAWP can be calculated quickly using the values of FCA_1 and FCA_2, respectively. In using the LEFE, we calculate the stress level at the initial condition using FCA_1 and then consider the FCA_2 that will give the allowable stress level. Then the remaining life can be calculated using Eq. 3-57. The LEFE normally gives less conservative results, about 2–3% higher in most cases. If the difference is significant (e.g., 10% or greater), the finite element mesh may need to be refined. It is encouraged that Eq. 3-18 be solved using spreadsheet software to eliminate errors. If one does not have access to finite element software, then use Eq. 3-18, which has proven to be very accurate for hundreds of cases. If one is to engage in fitness-for-service assessment work, learning the finite element method is strongly encouraged. The LEFE is not mentioned that much in the API 579, as the emphasis is on the Level 3 non linear finite element method, but in field practice the LEFE is considerably faster and, in the vast majority of cases, adequately accurate.

It is mentioned in the API 579, Paragraph 4.5.2.2.c, that the MAWP can be plotted versus time and where *MAWP* intersects $MAWP_r$ will give the remaining life. Such a curve is fine, but it is not necessary. As mentioned above, the remaining life can be quickly calculated by finding FCA_1 and FCA_2.

The Thickness Approach

For a Part 4, General Metal Loss Assessment, the remaining life can be found based upon the computation of a minimum required thickness for the intended service conditions, thickness measurements from an inspection, and an estimate of the anticipated corrosion rate. The method is adequate for finding the remaining life if the component does not have thickness interdependency. Such cases of thickness interdependency are shown in **Figures 3-6a**, **b**, and **c**. Such thickness interdependency exists if there is a relationship between the component thickness, piping flexibility, and the resulting stress. Straight portions of pipe can be assessed using the thickness averaging method, using *L*, as described above. For elbows and bends, the thickness readings should be averaged within the bend, and a single thickness reading should be used in the assessment. For branch connections, the thickness should be averaged within the reinforcement zone (see **Figure 3-6a**), for the branch and header, and these thicknesses should be used in the piping flexibility model. Note that the LEFE with nozzle-header templates is a much faster approach using the MAWP method for branch connections, as described previously. As mentioned in the API 579, Paragraph 4.5.1.2, the MAWP method is best for situations of thickness interdependency.

For the General Metal Loss Assessment, the remaining life by the thickness approach is found by using the following equation:

$$R_{life} = \frac{t_{am} = Kt_{min}}{C_{rate}}$$
Eq. 3-58

where C_{rate} = anticipated (future) corrosion rate, in./yr (mm/yr)

K = factor depending on the assessment level; for a level 1 assessment, $K = 1.0$; for a Level 2 assessment, $K = RSF_a$ for piping components

R_{life} = remaining life, years
t_{am} = average wall thickness of the component found at the time of the inspection, in. (mm)
t_{min} = minimum required wall thickness required by code

For a Local Metal Loss Assessment, the remaining life is computed by using the method for a Level 1 Assessment procedure based upon computation of a minimum required thickness for the intended service conditions, the actual thickness and region size measurements from an inspection, and an estimate of the anticipated corrosion/erosion rate and the rate of the size of the flaw. An assessment considering the defect as an LTA can be used if the parameters and previous equations above are known, by substituting the following:

$$RSF \rightarrow RSF_a$$

$$R_t \rightarrow \frac{t_{mm} - (C_{rate})(\text{time})}{t_{min}}$$

For an LTA or a groove-like flaw evaluated as an LTA,

$$s \rightarrow s + C_{rate}^s (\text{time})$$

$$c \rightarrow c + C_{rate}^c (\text{time})$$

where C_{rate} = anticipated (future) corrosion rate, in./yr (mm/yr)
C_{rate}^s = estimated rate of change of the longitudinal length of the region of local metal loss, in./yr (mm/yr)
C_{rate}^c = estimated rate of change of the circumferential length of the region of local metal loss, in./yr (mm/yr)
c = circumferential length of the region of local metal loss at the time of the inspection, in. (mm)
RSF = computed remaining strength factor
RSF_a = allowable remaining strength factor
R_t = remaining thickness ratio
s = longitudinal length of the region of local metal loss at the time of the inspection, in./yr (mm/yr)
t_{min} = minimum required code thickness for the component that governs the MAWP, in. (mm)
t_{mm} = minimum remaining wall thickness determined at the time of inspection, in. (mm)
time = time in the future, years

In the API 579, Paragraph 5.5.1.3, the remaining life can be found on the thickness based method only if the local metal loss is characterized by a single thickness; otherwise, the MAWP approach must be used.

Material Property Data

The FFS triangle of integrity consists of Inspection, Mechanical, and Materials, the three disciplines required for an FFS assessment. These three groups form an FFS assessment. The materials properties of significance to an FFS assessment are toughness and strength. Fracture toughness data are required to assess crack-like flaws, and the strength data may be needed for advanced assessment of LTAs or other flaws, like blisters. Material property data vary, especially for fracture toughness data. This variability can affect an FFS assessment, so one must be aware of the available data and how the variability can affect the assessment.

It is necessary to identify the material property data needed for an FFS assessment and to explain how that data are used. In addition, the following must be considered:

1. Lower bound toughness data for cases where actual test data are not available on the material being assessed
2. A procedure for using Charpy impact data to estimate fracture toughness properties
3. The variability of material property data and its effect on an FFS evaluation
4. Methods for approaching assessments of crack-like flaws in services prone to environmental cracking

Material Property Data Required for Assessment

The significant material properties for assessment are as follows:

1. *Yield and Tensile Strength*—Yield and tensile strength data are available on mill certification documents for each heat of steel. These data should be kept with the permanent equipment records. In case these data are not available, then one has to use minimum specified properties for the material specification. Also required is the material allowable stress values based on the design/construction code used at the time the equipment was designed. It is permissible to use the alternative allowable stress basis in

FFS assessments constructed to the ASME Section VIII, Division 1, 1968 or later edition based on the ASME Code Case 2278 of the ASME Section VIII, Division 1 with certain restrictions. Refer to the code case for specific details.

2. *Properties for Fracture Mechanics Assessments*—These properties are required when assessing crack-like flaws. These two properties are the critical stress intensity factor (K_{IC}) and the crack tip opening displacement (CTOD). The K_{IC} is used when the total stress (including the residual welding stress) in the vicinity of a crack is less than the yield strength. The use of the K_{IC} yields conservative results when there is significant plastic deformation. It can be used for material of any thickness. The CTOD is used when the total stress (including the residual welding stress) in the vicinity of a crack tip exceeds the yield strength. The CTOD value depends on the component thickness. It must be obtained from material of comparable thickness to the component being assessed because it is very difficult to obtain by testing.

Thus, an approach to obtaining toughness data consists of actual tests, which must represent the flaw location (i.e., weld, HAZ, base metal). Testing is usually either not an option or is not available. Another approach to obtaining toughness data is to convert it from other toughness tests, if available, or to use lower bound data from the literature. We will discuss these three approaches.

Determining Fracture Toughness Levels

1. *Determining K_{IC}*—The standard testing procedures for finding the K_{IC} value for a material are specified in the ASTM E399. Unfortunately, for almost all steels, a valid K_{IC} value cannot be measured because the area of plastic deformation is significant compared to the material thickness, requiring extremely thick test specimens for valid results. Alternatives to direct testing for the K_{IC} must be considered. These alternatives include the correlation of J_{IC} data, the correlation of Charpy V-notch data, and, finally, the use of lower bound K_{IC} data from the literature.

2. *Determining the CTOD*—The procedure for performing CTOD tests is specified in the ASTM E1290. CTOD tests should be performed on material of the same nominal thickness as the component being assessed because CTOD varies with thickness.

3. *Testing for Both the K_{IC} and CTOD*—Testing must be performed at a temperature appropriate for the equipment being assessed. In most cases, this will be the critical exposure temperature, which we will discuss in Chapter 4 on brittle fracture. The CET will be

used unless other considerations define a different minimum pressurization temperature. The tests must be performed on material that is representative of the material that contains the flaw. The test specimen should be from the base metal, weld metal, or HAZ as appropriate. It is desirable that the orientation of the specimen be representative of the defect that is being assessed. For example, a specimen in the transverse direction may yield different results from the one in the longitudinal direction of the pipe or cylinder.

4. *Determining K_{IC} from J_{IC} Data*—J_{IC}, or the J integral, is a measure of the strain energy required to cause crack initiation in the area affected by the crack and operating stresses. The procedure for measuring the J_{IC} is given in the ASTM E-183. The advantage of using this test procedure is that a valid J_{IC} test can be performed on thinner sections close to the section thickness used in typical plant equipment. Once the J_{IC} toughness has been measured, the K_{IC} value can be calculated using the following:

$$K_{IC} = \sqrt{J_{IC}E}$$
Eq. 3-59

where E = modulus of elasticity, ksi (MPa)
J_{IC} = J integral, in.-ksi (m-MPa)
K_{IC} = fracture toughness, ksi$\sqrt{\text{in.}}$ (MPa$\sqrt{\text{mm}}$)

A word about SI Units in the API 579

Throughout the API 579 document (e.g. Eqs. 9.22 and 9.39), the SI unit of toughness is given the unit $MPa\sqrt{m}$, where m = meters. The critical value of the mode stress intensity, K_I, at which fracture occurs, is a function of the maximum uniform membrane stress. In the SI system, stress is usually denoted as MPa (N/mm^2). Since the stress unit (MPa) is 1.0 N/mm^2, the unit for toughness becomes

$$\frac{\text{lb}_f}{\text{in.}^2}\sqrt{\text{in.}} = \frac{\text{lb}_f}{\text{in.}^2}\left(\frac{4.448\,N}{1\,\text{lb}_f}\right)\left(\frac{\text{in.}}{25.4\,\text{mm}}\right)^2\left[\text{in.}\left(\frac{25.4\,\text{mm}}{1\,\text{in.}}\right)\right]^{0.5}$$

Thus,

$$1.0\,\frac{\text{lb}_f}{\text{in.}^2}\sqrt{\text{in.}} = 0.0347\,\frac{N\sqrt{\text{mm}}}{\text{mm}^2}$$

Since 1 MPa $= 1\dfrac{N}{mm^2}$, then

1.0 MPa$\sqrt{mm}$ = 28.78 psi $\sqrt{in.}$

The unit MPa$\sqrt{mm}$, or N(mm)$^{-1.5}$, and not MPa$\sqrt{m}$, is used in SI. API 579 is a U.S. document, and SI units are not used on a regular basis by some of its authors. This is to avoid confusion among SI readers and those who use the English or Imperial system of units, the American Engineering System (AES)—See Chapter 8 for units.

5. *Using Charpy V-Notch Data to Assess Toughness*—Most likely a K_{IC}, CTOD, or J_{IC} test is not available on the equipment we are assessing. In most circumstances, the only toughness data is the Charpy V-notch (C_V) data collected during original fabrication. On older equipment, Charpy U data may be the only toughness information available. We will discuss later how to apply Charpy U data and when it is necessary to use this parameter as the basis of toughness correlations. Even when the Charpy V-notch testing is performed, the actual values may not be known; only the information that the component was impact tested and met a specific standard may be available. When C_V is known, it is possible to correlate it to an appropriate fracture toughness parameter for an FFS evaluation.

The following correlation provides a safe assessment in all situations tested:

$$K_{IC} = \frac{\sqrt{5C_V E}}{1000} \text{ ksi}\sqrt{in.}$$ Eq. 3-60

$$K_{IC} = \frac{\sqrt{650C_V E}}{1000} \text{ MPa}\sqrt{mm}$$ Eq. 3-61

where C_V = Charpy V-notch impact energy, ft-lb (J)
      E = modulus of elasticity, psi (MPa)
      K_{IC} = critical stress intensity factor, ksi$\sqrt{in.}$ (MPa$\sqrt{mm}$)

The scatter of the C_V data is wide, so the preceding correlation is conservative in many situations. The following correlation between K_{IC} and CTOD is conservative. This is given for information only, since the BS 7910 (a standard accepted under the API 579) permits the use of K_{IC} for all situations (this includes stresses above yield).

$$CTOD = \frac{K_{IC}^2(1e^6)}{2\sigma_y E}, \text{ in.}$$

Eq. 3-62

$$CTOD = \frac{500K_{IC}^2(1e^6)}{E\sigma_y}, \text{ mm}$$

Eq. 3-63

where $CTOD$ = crack tip opening displacement, in. (mm)
E = modulus of elasticity, psi (MPa)
K_{IC} = critical stress intensity factor, ksi$\sqrt{\text{in.}}$ (MPa$\sqrt{\text{mm}}$)
σ_y = yield strength, psi (MPa)

The CTOD value obtained by this correlation is a conservative estimate of fracture toughness.

In addition to applying the listed correlations between the Charpy V-notch and fracture toughness, it is also possible to index the transition curve using Charpy V-notch data. As is described later, the K_{IC} or K_{IR} (defined later) versus the temperature curve can be established after the reference temperature (T_{REF}) is known. T_{REF} can be defined in terms of the temperature at which a specified Charpy V energy is achieved. The appropriate Charpy value adopted for indexing may vary according to the intended application and the way in which safety factors are to be applied in the assessment. A minimum value of 15 ft-lb (20 J) is appropriate for most MAT evaluations on carbon steel materials, as will be discussed in "Brittle Fracture Concepts" in Chapter 4. The MAT is defined in Paragraph 3.1.4 of the API 579 as the minimum allowable temperature. This is the permissible lower temperature limit for a given material at a thickness based on its resistance to brittle fracture. It may be a single temperature, or an envelope of allowable operating temperatures as a function of pressure. The MAT is derived from mechanical design information, materials specifications, and/or materials data. However, it may be appropriate to use 30 ft-lb (40 J) for consistency with BS 7910 [Reference 4] in the case of flaw assessments as referenced in the section concerning crack-like flaws or the section about brittle fracture.

6. *Use of Extrapolated Charpy V-Notch Data*—There are occasions when it may be useful to extrapolate Charpy V-notch data from one temperature to a lower temperature, especially when the measured impact values exceed the minimum toughness requirements at the specified test temperature. In this situation, one can take advantage of this extra toughness by qualifying the equipment for a lower temperature operation if required. If, for any reason, the measured

toughness is lower than that required, then the equipment may be qualified at a higher temperature. After the extrapolated Charpy V-notch value is identified, a correlated fracture toughness value can be determined as described below. Thus,

The toughness may be extrapolated to a higher or lower temperature, as long as the following conditions are met:

1. For extrapolation to lower temperatures, the starting average/ minimum toughness must be greater than 15/12 ft-lb (20/16 J).
2. For extrapolation to higher temperatures, both the starting and ending toughness values must be between 15/12 to 35/28 ft-lb (20/16 to 47/38 J).

Measured Charpy V-notch toughness values may be extrapolated using the slope of 0.6 ft-lb/°F (1.5 J/°C) to arrive at a toughness for the lower or higher temperature. The extrapolation begins at both the average and minimum Charpy V-notch toughness values measured, and ends at the average and minimum values allowed. This is represented in equation form in the following:

$$\overline{T}_L = T_1 \left(\frac{\overline{C}_{VI} - \overline{C}_{VR}}{C} \right) \qquad \text{Eq. 3-64}$$

$$T_{LMIN} = T_1 \left(\frac{C_{VIMIN} - C_{VRMIN}}{C} \right) \qquad \text{Eq. 3-65}$$

$$MAT = \text{higher of } \overline{T}_L \text{ or } T_{LMIN}$$

where $\overline{T}_L$ = acceptable lower temperature based on average of *average* C_V values, °F (°C)

T_{LMIN} = acceptable lower temperature based on minimum of *measured* C_V values, °F (°C)

$\overline{Ti}$ = temperature at which the impact testing was done, °F (°C)

$\overline{C}_{VI}$ = average of three C_V values measured at T_1, ft-lb (J)

$\overline{C}_{VR}$ = average of three C_V values at minimum temperatures as recommended below in Charpy impact testing, ft-lb (J)

C_{VIMIN} = lowest of three C_V values measured at T_1, ft-lb (J)

C_{VRMIN} = lowest of three C_V values at minimum temperatures as recommended below in Charpy impact testing, ft-lb (J)

C = 0.6 for temperature in °F and C_V in ft-lb

C = 1.5 for temperature in °C and C_V in J

Table 3-2
Charpy V Impact Toughness for Piping and Vessels That Will Be Hydrotested and for Other Equipment

(See Table 3-5)	Required Impact Values for Full-Size Specimens at the CET[1] (ft-lb)			
	Class of Steel Reference Thickness, *t* (in.)			
	$t \leq 1/2$	$1/2 < t \leq 1$	$1 < t \leq 2$	$t > 2$
1a	15/12[2,3]	15/12	15/12	20/16
1b	15/12	15/12	20/16	25/20
2a	13/10	15/12	15/12	25/20
2b	15/12	20/16	25/20	35/28
2c	20/16	25/20	35/28	45/36
3a	25/20	30/24	40/32	55/44
3b	35/28	40/32	50/40	60/48
4	20/16	20/16	20/16	20/16
5	25/20	25/20	25/20	25/20

Notes:
1. When subsize specimens are necessary, the requirements given in UG 84 of the ASME Code Section VIII, Div. 1, regarding subsize specimens shall be followed.
2. In the notation such as 15/12, the first number is the minimum average energy of three specimens in the impact determination.
3. Acceptable metric equivalents for the values in Tables 3-2 or 3-3 are given in Table 3-4.

Values for Charpy V impact values for various applications and steel are given in **Tables 3-2** through **3-9**.

7. *Use of Charpy U-Notch Data*—When assessing a component for crack-like flaws, no attempt should be made to correlate Charpy U energy directly with fracture toughness. If actual fracture toughness cannot be found using test samples, then lower bound values provided in this chapter should be used. The Charpy U can be used as the basis for MAT in a brittle fracture assessment. In this assessment, 30°F (17°C) shall be added to the impact test temperature to obtain an equivalent Charpy V-notch test temperature. The equivalent temperature obtained in this manner should be compared to the impact test temperature obtained by UCS 66 in Section VII Division I. The lower of the two temperatures may be adopted as the initial point for development of the MAT curve. For advanced assessments, such as a Level 3, excluding the

Table 3-3
Charpy V Impact Requirements for Piping and Vessels That Will Be Pneumatically Tested

Required Impact Values for Full-Size Specimens at the CET[1] (ft-lb)

(See Table 3-5)	$t \leq 1/2$	$1/2 < t \leq 1$	$1 < t \leq 2$	$t > 2$
		Class of Steel Reference Thickness, t (in.)		
2a	20/16[2,3]	20/16	25/20	35/29
2b	20/16	25/20	35/28	45/30
2c	25/20	35/28	45/30	50/40

Notes:
1. When subsize specimens are necessary, the requirements given in UG 84 of the ASME Code Section VIII, Div. 1, regarding subsize specimens shall be followed.
2. In the notation such as 15/12, the first number is the minimum average energy of three specimen in the impact determination.
3. Acceptable metric equivalents for the values in Tables 3-2 or 3-3 are given in Table 3-4.

Table 3-4
Acceptable Metric Equivalents for the Values of Tables 3-2 or 3-3

Thickness in.	mm	Energy ft-lb	Values J
1/2	13	15/12	20/16
1	25	20/16	27/22
2	50	25/20	34/27
		30/24	40/32
		35/28	47/38
		40/32	54/43
		45/36	61/48
		50/40	67/54
		55/44	74/59
		60/48	81/65

assessment of known flaws, any adjustment from Charpy U to Charpy V using a temperature shift basis should include full details of assumptions. For low carbon steels only, the 30°F (17°C) shift should be used only if the Charpy U values are known to

Table 3-5
Classes of Commonly Used Steels (for use with Tables 3-2 and 3-3)

Type of Steel	Min. YS, ksi (MPa)	Max. TS, ksi (MPa)	Class
Rimmed and semi-killed steels	≤35 (241)	≤65 (448)	Ia
	≤45 (310)	≤80 (551)	Ib
Fully killed and medium strength carbon and low alloy steels	≤35 (241)	≤75 (517)	IIa
	≤55 (379)	≤90 (620)	IIb
	≤65 (448)	≤105 (724)	IIc
High-strength steels[1]	>70 (483)	≤125 (861)	IIIa
	≤100 (689)	≤135 (930)	IIIb
2½ and 3½ nickel steels			IV
9 nickel steel			V

Class	Material Specification[2]	Class	Material Specification[2]
1a	SA 53 Grs A & B	2b	SA 335 Gr P 11
1a	SA 283 Grs A, B, & C	2b	SA 333 Gr 8
1a	SA 333 Gr 1	2b	SA 336 Cls F1, F12 & F22A
1a	SA 285 Gr A, B, & C	2b	SA 350 Gr LF2
1b	SA 283 Gr D	2b	SA 372 Type II
2a	SA 106 Grs A & B/ API 5L Gr B	2b	SA 387 Cl1 Grs 2, 12, 11,
2a	SA 181 Class 60		21, 22, &5
2a	SA 216 Gr WCA	2b	SA 387 Cl 2 Grs 2 & 12
2a	SA 217 Gr WC1	2b	SA 515 Grs 65 & 70
2a	SA 333 Gr 6	2b	SA 516 Grs 65 & 70
2a	SA 335 Grs P1 & P2	2b	SA 537 Cl 1
2a	SA 350 Gr LF1	2b	SA 541 Cls 1 & 2
2a	SA 352 Grs LCB & LC1	2b	SA 662 Grs A, B, & C
2a	SA 372 Type 1	2c	SA 182 Gr F22
2a	SA 442 Grs 55 & 60	2c	SA 202 Gr B
2a	SA 515 Grs 55 & 60	2c	SA 372 Type III
2a	SA 516 Grs 55 & 60	2c	SA 203 Gr B
2b	SA 217 Grs WC6 & WC9	2c	SA 266 Class 2 & 4
2b	SA 266 Class 1	2c	SA 302 Grs A, B, C, &D
2b	SA-105	2c	SA 336 Cls F6 & F22
2b	SA-106 Gr C	2c	SA 387 Cl 2 Grs 11, 21, 22, & 5
2b	SA 181 Class 70	2c	SA 533 Cl 1
2b	SA 182 Gr F1, F2, F11 & F12	2c	SA 537 Cl 2
2b	SA 202 Gr A	2c	SA 541 Cl 3
2b	SA 203 Gr A	2c	Sa 612
2b	SA 204 Grs A, B, & C	3a	SA 266 Class 3
2b	SA 216 Grs WCB & WCC	3a	SA 372 Types IV & V

(Table continued on next page)

Table 3-5
Classes of Commonly Used Steels
(for use with Tables 3-2 and 3-3)—cont'd

Class	Material Specification[2]	Class	Material Specification[2]
3a	SA 533 Cls 2 & 3	4	SA 333 Grs 3 & 7
3a	SA 542 Cl 2	4	SA 350 Gr LF3
3a	SA 517	4	SA352 Grs LC2 & LC3
3b	SA 537 Cl 2	5	SA 553 Type 1
3b	SA 542 Cl 2	5	SA 333 Gr 8
4	SA 203 Grs D & E	5	SA 353

Notes:
1. Additional toughness testing is required for these materials.
2. Abbreviations: Grade of material (Gr), Class (Cl).

Table 3-6
Material Classifications for Impact Testing Exemption

	Heat Treatment	Condition
ASME specification	Normalized, normalized & tempered, or quenched & tempered	As rolled (or annealed)
SA 105	B	A
SA 106 Gr A, B	Note 4	B
SA 106 Gr C	Note 4	A
SA 266	B	A
SA 333 Gr 1, 6	D	Note 5
SA 350 Gr LF2	D	Note 5
SA 352 Gr LCB, LC1	D	Note 5

Notes:
1. Materials with P-numbers of 1–5 of the ASME Code that are not listed shall be exempt as Class A.
2. Materials that have been Charpy V tested as part of a national materials standard are accepted without further testing provided the toughness meets or exceeds the requirement of Table 3-2 or 3-3 (as applicable) at a temperature equal to or below the CET.
3. This specification does not accommodate normalized, normalized and tempered, or quenched and tempered heat treatments.
4. Specification requires a normalizing, normalizing and tempering, or quenching and tempering heat treatment.
5. Specification requires a normalizing, normalizing and tempering, or quench and tempering heat treatment.

Table 3-7
Impact Requirements for Machinery Components

CET, °F (°C)	Maximum Casing Working Pressure, psi (KPa)	Impact Requirement[1]
≤−20°F (−29°C)	All	15/12 ft-lb (20/16 J)
≥−20°F < 60°F (≥−29°C > 16°C)	>1000 psi (6900 KPa)	15/12 ft-lb (20/16 J)
≥−20°F < 60°F (≥−29°C < 16°C)	≤1000 psi (6900 KPa)	None
≥60°F (16°C)	All	None

Note:
1. In the notation such as 15/12, the first number is the minimum average energy of three specimens in the impact determination.

Table 3-8
Lower Bound Fracture Data for Materials Without Charpy V-Notch Data ASTM A516/A105

Temperature, °C	Temperature, °F	Charpy, ft-lb	K1C ksi√in.	CTOD
−29	−20	6	31	0.0005
−20	−4	9	36	0.0006
0	32	16	48	0.0011
20	68	24	60	0.0018
40	104	32	70	0.0023
60	140	41	78	0.0030
80	176	49	85	0.0035
100	212	57	92	0.0040
120	248	65	99	0.0047
140	284	70	103	0.0052
160	320	74	105	0.0060

equal or exceed 15 ft-lb. After applying the shift, the Charpy V energy should be assumed as 15 ft-lb, even if Charpy U values are greater. This can then be used as a basis for establishing a T_{ref} and toughness as described later.

8. *Fracture Toughness When No Data Are Available*—Often the material of the component being assessed does not have specific toughness data. If strain age embrittlement, temper embrittlement, or hydrogen-assisted crack growth is possible, then the following

Table 3-9
Lower Bound Fracture Data for Materials Without Charpy V-Notch
Data ASTM A515/A106/A285

Temperature, °C	Temperature, °F	Charpy, ft-lb	K1C, ksi√in.	CTOD
−29	−20	3	21	0.0002
−20	−4	4	25	0.0003
0	32	5	27	0.0004
20	68	8	34	0.0006
40	104	16	48	0.0011
60	140	24	60	0.0018
80	176	32	70	0.0023
100	212	41	78	0.0030
120	248	49	85	0.0035
140	284	57	92	0.0040
160	320	65	99	0.0047

sections should be considered. If these are not of concern and the material is a carbon or low alloy steel, sample K_{IC} values can be obtained from the following equation to be used in an advanced assessment:

$$K_{IC} = 33.2 + 2.806e^{[0.02(T - T_{ref} + 100)]}, \text{ ksi}\sqrt{\text{in.}} \qquad \text{Eq. 3-66}$$

where T is the temperature at which the toughness is required and T_{REF} is the MAT as read from UCS 66, using the equipment thickness and material specification. If the K_{IC} value calculated using this formula is greater than 100 ksi√in., then use $K_{IC} = 100$ ksi√in.

If the FFS assessment requires the use of an arrest fracture toughness measurement (K_{IR}), this can also be estimated using a similar equation:

$$K_{IR} = 26.8 + 1.223e^{[0.0144(T - T_{ref} + 160)]}, \text{ ksi}\sqrt{\text{in.}} \qquad \text{Eq. 3-67}$$

Again, T is the temperature at which the toughness is required and T_{REF} is the MAT as read from UCS 66, using the equipment thickness and materials specification. Alternatively, the 15 ft-lb (20 J) transition temperature can be used for T_{REF} if it is known. If the K_{IR} value calculated using this equation is greater than 100 ksi√in., then use $K_{IC} = 100$ ksi√in.

Crack-like Flaws

Crack-like flaw assessments are highly dependent upon material properties because they are based on fracture mechanics principles. It is in this area that the fitness-for-service integrity triangle becomes obvious—Inspection, Materials, and Mechanical (Stress) all play an important role. Without any one of the three the entire process falls apart.

There have been many approaches to crack assessment. The API 579 gives its own methodology but recognizes other developments as acceptable, namely the British Standard, BS 7910, as mentioned earlier. This standard was developed mainly for the offshore structures in the North Sea, not for pressurized equipment. However, the BS 7910 methodology has been used for many years in pressurized equipment.

Our focus here is an overview of crack assessment. A discussion of the entire subject would comprise many volumes. Also we are not attempting to regurgitate the API 579 Recommended Practice but rather augment it with discussions pertinent to crack assessment. The reader is encouraged to have a copy of API 579 to follow this discussion.

There are several software packages for crack assessment. The most common approach is the use of the failure assessment diagram. On this diagram, the ordinate axis is a progression toward fracture failure and the abscissa is a progression toward plastic collapse. Another solution form for cracks is solving for the maximum crack length versus various crack depths. The FAD approach is used by the API 579 and is shown in **Figure 3-13a**.

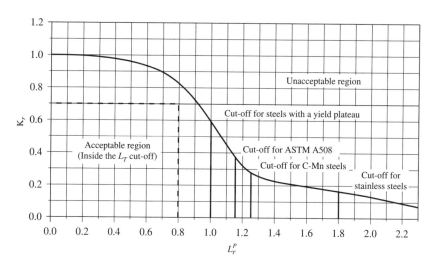

Figure 3-13a. The failure assessment diagram. Courtesy of the American Petroleum Institute.

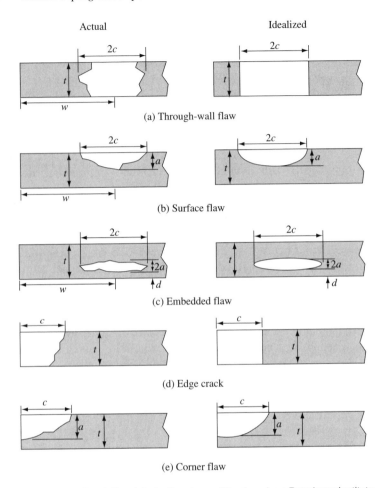

Actual   Idealized

(a) Through-wall flaw

(b) Surface flaw

(c) Embedded flaw

(d) Edge crack

(e) Corner flaw

Figure 3-13b. Crack-like defects. Courtesy of the American Petroleum Institute.

An overview of crack-like flaws begins with the nomenclature, and idealized shapes used to assess them are shown in **Figure 3-13b**.

The procedure of crack assessment will be presented in steps. Characterization of flaw length is characterized by two approaches. the conservative option uses the entire length of the crack (c_o), which is oriented to be normal to the maximum principal tensile stress. The equivalent flaw length option projects the flaw onto a principal plane. See **Figure 3-13c** (Figure 9.2 in the API 579).

Step 1: Project the flaw onto a principal plane. In the case of uniaxial loading, there is only one possible principal plane; however, when the loading is biaxial (e.g., a pressurized component which is

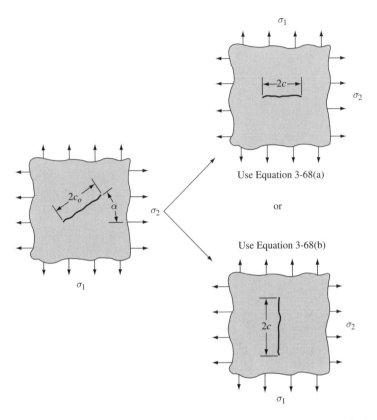

Figure 3-13c. Procedure for determining the effective flaw length on a principal stress plane (Courtesy of the American Petroleum Institute).

subject to a hoop stress and an axial stress), there is a choice of principal planes on which to project the flaw. In most cases, the flaw should be projected to the plane normal to the maximum principal tensile stress (the σ_1 plane), but there are instances where the σ_2 plane would be more appropriate (e.g., when the angle between the flaw and the principal plane (α) is greater than 45°).

Step 2: Computing the equivalent flaw length

    (a) For the plane of the flaw projected onto the plane normal to the maximum principal tensile stress (σ_1)

$$\frac{c}{c_o} = \cos^2\alpha + \frac{(1 - B)\sin\alpha\cos\alpha}{2} + B^2\sin^2\alpha$$

Eq. 3-68(a)
(Equation 9.1 in the API 579)

(b) For the plane of the flaw projected onto the plane normal to σ_2:

$$\frac{c}{c_o} = \frac{\cos^2\alpha}{B^2} + \frac{(1-B)\sin\alpha\,\cos\alpha}{2B^2} + \sin^2\alpha$$

Eq. 3-68(b)
(Equation 9.2 in the API 579)

(c) In Eqs. 3-68(a) and 3-68(b), the dimension c corresponds to the half flaw length (or total length for corner or edge cracks) to be used in calculations, c_o is the measured half length for the flaw oriented at an angle α from the σ_1 plane, and B is the biaxiality ratio, defined as follows:

$$B = \frac{\sigma_2}{\sigma_1} \text{ where } \sigma_1 > \sigma_2 \text{ and } 0.0 \leq B \leq 1.0$$

Eq. 3-69
(Equation 9.3 in the API 579)

(d) Equation 3-69 is valid only when both σ_1 and σ_2 are positive—tensile. If σ_2 is compressive, B should be set to zero, and Eq. 3-68 should be applied to compute the equivalent flaw length. If stress gradients occur in more than one direction, the sum of membrane and bending stress components should be summed for the purpose of computing σ_1 and σ_2. For uniaxial loading, $B = 0$ and Eq. 3-68a reduces to

$$\frac{c}{c_o} = \cos^2\alpha + \frac{\sin\alpha\,\cos\alpha}{2} \qquad \text{Eq. 3-70}$$

The reader is referred to the API 579 Appendix F for the relationships between the c/c_o and the biaxial stress ratio in Figure 9-3 of the standard.

API 579 gives comprehensive methods for crack assessment. We can't describe all the methods and their applications here; however, we will concentrate on the Level 2 approach, as the Level 3 Assessment is built largely on this approach. The reader is referred to the characterization of flaw depth, which is significantly more difficult to estimate than the length. Throughout the years, inspectors in the field often have difficulty in finding the crack depth. In these situations the engineer has to be conservative in the flaw depth in crack assessments.

In the Level 2 approach, we must evaluate the operating conditions, temperature, and supplemental loading combinations to be evaluated. Next, we must determine the stress distributions at the location of the crack based on the applied loads. The next three steps are critical; the rest are basic and must be classified into the following stress categories:

- Primary stress
- Maximum primary stress (proof load or hydro test load)
- Secondary stress
- Residual stress

We need to determine the material properties—yield stress, tensile strength and fracture toughness (K_{mat})—for the above conditions that exist in the pipe. The yield and tensile strength should be established using nominal values (i.e., minimum specified per the material specification if actual values are unknown), and the toughness should be based on the mean values. The toughness is perhaps one of the most important parameters in crack assessment. Its value will have a direct effect on whether a crack is acceptable. **Tables 3-2** through **3-9** provide lower bound values that we can use if this parameter is not known. Likewise if Charpy-V data are known, use of the previously discussed tables and equations will help us determine the material toughness.

The next step is to determine the crack-like flaw from inspection data. The flaw can be either a surface flaw, an embedded flaw, or a through-wall crack. The API 579 has a complete categorization of the various types of cracks.

Modify the primary stress, material fracture toughness, and the flaw size using the partial safety factors (PSFs). The PSFs are used for each dependent variable (e.g., toughness). The term "partial safety factor" is used in lieu of "safety factor" because standard and code members do not like the term "safety" used in a single context. These PSFs are applied as follows:

1. *Primary Membrane Stress and Bending Stress*—Modify the primary membrane and bending stress components determined earlier (P_m and P_b, respectively) using the PSF for stress.

$$P_m = P_m * PSF \qquad\qquad \text{Eq. 3-71}$$

$$P_b = P_b * PSF \qquad\qquad \text{Eq. 3-72}$$

2. *Material Toughness*—Modify the mean value of the material fracture toughness determined earlier using the PSF for fracture toughness. Use the mean value if PSFs are used. If you bypass the partial safety factors, use the lower bound toughness value.

$$K_{mat} = \frac{K_{mat}}{PSF_k}$$ Eq. 3-73

3. *Flaw Size*—Modify the flaw depth determined earlier using the PSF for flaw size. If the factored flaw depth exceeds the wall thickness of the component, then the flaw should be recategorized as a through-wall crack.

$a = a*PSF_a$ (for a surface flaw)   Eq. 3-74

$2a = 2a*PSF_a$ (for an embedded flaw)   Eq. 3-75

$2c = 2c*PSF_a$ (for a through-wall flaw)   Eq. 3-76

Note that if a given input value is known to be a conservative estimate (e.g., lower-bound toughness or upper-bound flaw size), a PSF of 1.0 may be applied.

Next one must compute the reference stress for primary stresses (σ_{ref}^P), based on the factored primary stress distribution and factored flaw size from above and the reference stress solutions in API 579 Appendix D.

Compute the load ratio or the *abscissa* of the FAD (failure assessment diagram) using the reference stress for primary loads and yield stress as discussed earlier.

$$L_r^p = \frac{\sigma_{ref}^P}{\sigma_{ys}}$$ Eq. 3-77

Compute the stress intensity attributed to the primary loads (K_1^P), using the factored primary stress distribution and factored flaw size discussed earlier, as well as the stress intensity factor solutions in the API 579 Appendix C. If $K_1^P < 0.0$, then set $K_1^P = 0.0$.

Next compute the reference stress for secondary and residual stresses (σ_{ref}^{SR}), based on the secondary and residual stress distributions and factored flaw size discussed earlier, as well as the reference stress solutions from the API 579 Appendix D.

Compute the secondary and residual stress reduction factor (S_{srf}) using the following:

$$S_{srf} = \min\left[\left\{1.4 - \frac{\sigma_{ref}^P}{\sigma_f}\right\}, 1.0\right] \qquad \sigma_{ref}^{SR} > \sigma_{ys} \qquad \text{Eq. 3-78}$$

$$S_{srf} = 1.0 \text{ when } \sigma_{ref}^{SR} \leq \sigma_{ys} \qquad\qquad\qquad\qquad \text{Eq. 3-79}$$

where σ_{ref}^P = reference stress associated with the primary stress or maximum primary stress, as applicable (see the following note), psi (MPa)

σ_{ref}^{SR} = reference stress associated with the secondary and residual stress from above, psi (MPa)

σ_f = flow stress, normally membrane stress plus 10,000 psi (68.97 MPa), psi (MPa)

Note that when computing S_{srf}, the following two conditions must be considered:

- If the crack is present before the application of the load associated with the maximum primary stress, then σ_{ref}^P to be used in Eq. 3-61 can be based on the maximum primary stress and the reference stress solutions in Appendix D of API 579.
- If the crack occurs after the application of the load associated with the maximum primary stress, then σ_{ref}^P to be used in Eq. 3-61 can be based on the factored primary stress distribution and flaw size discussed earlier and the reference stress solutions in Appendix D, or the maximum primary stress discussed earlier and the reference stress solutions in Appendix D of API 579 using zero flaw dimensions, whichever results in the largest secondary and residual stress reduction factor.

Normally $0.9 \leq S_{srf} \leq 1.0$. S_{rf} is the knockdown factor for residual stress.

Next compute the reference stress intensity attributed to the secondary stress and residual stress $\left(K_1^{SR}\right)$, using the secondary and residual stress distributions, the factored flaw size, the secondary and residual stress reduction factor from earlier, and the stress intensity factor solutions in Appendix C of API 579. If $K_1^{SR} < 0.0$, then set $K_1^{SR} = 0.0$. The value of K_1^{SR} should be determined at the same location along the crack front as that used to determine K_1^P.

Next compute the plasticity interaction factor (Φ). Note that the plasticity interaction factor (Φ) is a fudge factor to account for the fact that one cannot add nonlinear parameters. The following procedure is to compute Φ:

If $K_1^{SR} = 0.0$, then set $\Phi = 1.0$ and solve for K_r described later. Otherwise, compute L_r^{SR}, using the following equation with σ_{ref}^{SR}, S_{srf}, and σ_{ys} from before.

$$L_r^{SR} = \frac{\sigma_{ref}^{SR}}{\sigma_{ys}} S_{srf} \qquad\qquad \text{Eq. 3-80}$$

Determine ψ and ϕ using Tables 9.3 to 9.6 in API 579 and compute Φ/Φ_o using the following equation, or from Figure 9.19 in API 579:

$$\frac{\Phi}{\Phi_o} = 1 + \frac{\psi}{\phi} \qquad\qquad \text{Eq. 3-81}$$

Next, compute the plasticity interaction factor Φ. If $0 < L_r^{SR} \leq 4.0$, then set $\Phi_0 = 1.0$ and

$$\Phi = 1 + \frac{\psi}{\varphi} \qquad\qquad \text{Eq. 3-82}$$

Otherwise, compute the stress intensity factor for secondary and residual stresses corrected for the plasticity effects $\left(K_{1P}^{SR}\right)$ and compute Φ_o and Φ using the following:

$$\Phi_o = \frac{K_{1P}^{SR}}{K_1^{SR}} \qquad\qquad \text{Eq. 3-83}$$

$$\Phi = \Phi_o\left(1 + \frac{\psi}{\phi}\right) \qquad\qquad \text{Eq. 3-84}$$

The following simplified method may be used to compute Φ_o; however, this method may produce overly conservative results.

$$\Phi_o = \left(\frac{a_{eff}}{a}\right)^{0.5} \qquad\qquad \text{Eq. 3-85}$$

with

$$a_{eff} = a + \left(\frac{1}{2\pi\tau}\right)\left(\frac{K_1^{SR}}{\sigma_{ys}}\right)^2 \qquad\qquad \text{Eq. 3-86}$$

where a = depth of the crack from above, in. (mm)

$\quad a_{eff}$ = effective depth of the crack, in. (mm)

$\quad K_1^{SR} = K_1^{SR}$ based on crack depth from above, ksi$\sqrt{\text{in.}}$ (MPa$\sqrt{\text{mm}}$)

$\quad \tau$ = factor equal to 1.0 for plane stress and 3.0 for plane strain

$\quad \sigma_{ys}$ = yield stress at the assessment temperature (see Appendix F of API 579), psi (MPa)

The final step is to determine the toughness ratio or *ordinate* of the FAD assessment point where K_1^P is the applied stress intensity due to the primary stress distribution from earlier, K_1^{SR} is the applied stress intensity due to the secondary and residual stress distributions from earlier, K_{mat} is the factored material toughness, and Φ is the plasticity correction factor from earlier.

$$K_r = \frac{K_1^P + \Phi K_1^{SR}}{K_{mat}}$$

Once we get K_r, then we know the ordinate for the FAD.

Now we must evaluate the results—the FAD assessment point for the current crack size and operating conditions (stress levels) is defined as $\left(K_r, L_r^P\right)$.

1. Determine the cut-off for the L_r^P-axis of the FAD (see **Figure 3-13a**—Figure 9.2 of the API 579).
2. Plot the point on the FAD shown in **Figure 3-13a** (Figure 9.2 in API 579). If the point is on or inside the FAD (on or below and to the left), then the component is acceptable per the Level 2 Assessment procedure. If the point is outside of the FAD (above and to the right), then the component is unacceptable per the Level 2 Assessment procedure. Note that the values of K_1^P and K_1^{SR} will vary along the crack front; therefore, the assessment may have to be repeated at a number of points along the crack front to ensure that the critical location is found.

Remediation of Crack Defects

Cracks can be repaired by various techniques. A hole may be drilled on each side of the crack, but often the crack may continue to propagate past the holes. The hole sizes must be calculated. Grinding is also another repair option. Here again, the repair may result in a worst option. When areas are ground out, an LTA is formed. Thus, a crack is replaced with an LTA, and this should be assessed by the methods presented earlier.

An empirical formula has been proposed in [Reference 5] to determine the safe amount of grinding. This equation is as follows:

$$L \leq 1.12 \, (Dt) \left[\left[\left(\frac{\frac{a}{t}}{1.1\left(\frac{a}{t}\right) - 0.11} \right)^2 - 1 \right] \right]^{0.5}$$

Eq. 3-87

where L = length of grinding
a = depth of grinding
D = pipe diameter
t = pipe wall thickness

Equation 3-87 may be used with any consistent system of units.

Equation 3-87 is depicted in **Figure 3-14** for a 24 in. ϕ pipe with a nominal wall of 0.5 in.

The depths of shallow cracks are very difficult to determine, especially anything under 1/8 in. (3 mm). These types of cracks are almost always ground out by inspection and maintenance, often without the engineer knowing about them. Also drilling of holes to arrest crack growth has mixed results—many do not work. The holes must be calculated and

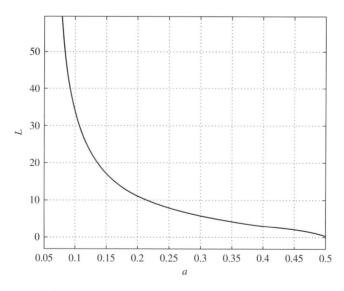

Figure 3-14. Eq. 3-87 is depicted for a 24 in. ϕ pipe with a nominal wall of 0.5 in.

sized—most holes are not large enough, and cracks will pass right through them. Typically this is also done in turnarounds with engineers knowing of the flaw, which is not good.

Grooves, Plain Dents, and Dents with Gouges

In Part 5 (Section 5) of the API 579, Paragraph 5.2.1.1, a groove is defined as a local elongated thin spot *caused by directional erosion or corrosion*; the length of the metal loss is significantly greater than the width. An example of such a flaw would be corrosion in a horizontal pipeline at the bottom, or six o'clock position. A groove is assessed by the methods in API 579 Part 5—Local Metal Loss.

A gouge is defined as a local mechanical removal/relocation of material from the surface of a component, causing a reduction in wall thickness at the defect. The length of a gouge is much greater than the width, and the material may have been cold worked in the development of the flaw. Dents are typically caused by mechanical damage (e.g., a crane lifting a pipe and the pipe swinging in the air and striking an object). This is quite common. Gouges have been caused by explosions where airborne objects (metal fragments, wood, and piping) strike the pipe, vessel, or tank. Gouges typically contain dents because of the nature of the mechanical damage. The API 579 treats the case of the gouge containing a dent in Part 8. We will also discuss plain dents.

Plain Dents

The API 579 treats plain dents in Part (Section) 8.4.3.3—Out-of-Roundness—Cylindrical Shells and Pipe Elbows. We will offer a simpler approach, which is presented in a detailed discussion in the WRC Bulletin 465, "Technologies for the Evaluation of Non-Crack-Like Flaws" [Reference 6]. The American Gas Association Pipeline Research Committee (now the PRCI), along with Battelle Memorial Institute, performed burst tests on 44 pipes containing plain dents. In the tests involving dents, the pipes failed at a pressure level equivalent to the ultimate tensile strength of the material and at a location remote from the dent [Reference 7]. This fact indicated that these plain dents did not influence the failure of the pipe samples. Without a sharp increase in the presence of stress intensification, the yielding occurred over a large area such that the pipe had sufficient ductility to yield and accept the plastic flow without failure. Dents are dangerous if they occurred on longitudinal weld seams because then cracks can develop. Several sources report that dented seam welds can have very low burst pressures. The low burst test pressure is caused by the weld cracking during indentation, spring back, or rerounding. The burst strength of a dented weld is very dependent on whether the weld cracks occurred during the denting process. There are no known methods for predicting the burst pressures of a smooth dent on a weld. For this reason, dented welds are typically repaired if found on an operational pipeline. It may be possible that if a dented weld is tough and free from defects, it is satisfactory.

The reader is referred to **Figure 3-15**. A later study was made by the European Pipeline Research Group (EPRG) (summarized in [Reference 6]), which discovered that for plain smooth dents located away from pipe weld seams, dent depths up to 10% of the pipe outside diameter will not fail at membrane stress levels less than 72% of the SMYS, or

$$\frac{d_d}{D_o} \leq 10\% \qquad\qquad \text{Eq. 3-88}$$

where d_d = depth of the dent in the nonpressurized condition, in. (mm)
D_o = pipe outside diameter, in. (mm)

Internal pressure in the pipe tends to push out the dent, thus reducing the dent depth (spring-back phenomenon). The measured depth on the operational pipeline must be corrected before this criterion can be applied. EPRG found the correlation between the dent depth on a non-pressurized pipe and a pressurized pipe to be as follows:

$$d_d = 1.43 d_d^p \qquad\qquad \text{Eq. 3-89}$$

Therefore, the EPRG limit for plain dents in a pressurized pipe is

$$\frac{d_d^p}{D_o} \leq 7\% \qquad\qquad \text{Eq. 3-90}$$

In Eq. 3-90, d_d^p is the depth of the dent in the pressurized condition and D_o is the pipe outside diameter. For a detailed description, the reader is referred to [Reference 5, pp. 105–107].

Depending on the geometry of the dent and pipe diameter and wall thickness, the spring-back phenomenon is *not* guaranteed to happen in all cases. See the comments at the end of this section.

Dent-gouge combinations present a different set of criteria.

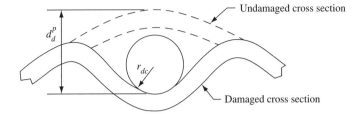

(a) Cylinder circumferential cross section

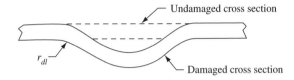

(b) Cylinder longitudinal cross section

Damage parameters for plain dents using the EPRG method

Figure 3-15. EPRG plain dent parameters. Courtesy of the Welding Research Council.

Dents and Gouge Combination Type Flaws

The API 579 gives an assessment procedure for the dent and gouge combination type flaw in Part (Section) 8.4.3.7. Refer to **Figure 3-16** to see that the depth of the dent to be used in the assessment is the depth that occurs at the instant of damage.

After the damaged area is cleared of the impact device that caused the damage (or damage tool), the dent in a pressurized pipe *may* rebound. The API 579 says that the dent will rebound, but this has not been observed in the field in all cases. If the dent is found while the pipe or

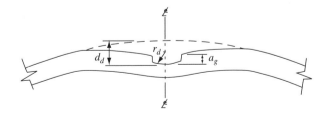

Figure 3-16. Dent and gouge EPRG parameters.

pipeline is in service, then the depth of the dent to be used in the assessment is computed using the following equation:

$$d_d = d_{dp}\left(-0.22 \ln\left[\frac{\sigma_p}{\sigma_f}\right]\right)^{-1}$$ Eq. 3-91

The limiting circumferential stress is computed from the following:

$$\sigma_{cl} = \frac{[Q_d - 300]^{0.6}}{300} \sigma_f$$ Eq. 3-92

where

$$Q_d = \max\left[\left\{\frac{\left(\dfrac{C_{vt}}{C_{uc}}\right)}{\left(\dfrac{d_d}{D}\right)\left(\dfrac{a_g + FCA}{t - FCA}\right)\left(\dfrac{s}{C_{us}}\right)}\right\}, 300.0\right]$$ Eq. 3-93

where a_g = depth of the gouge, in. (mm)

C_{uc} = constant for units conversion; C_{us} = 1.0 if C_{vt} is ft-lb and C_{uc} = 1.355818 if C_{vt} is expressed in Joules

C_{us} = constant for units conversion; C_{us} = 1.0 if s is expressed in inches and C_{us} = 25.4 if s is expressed in millimeters

C_{vt} = two thirds of Charpy energy, required only if the dent has a groove-like flaw, ft-lb

d_d = maximum depth of the dent at the instance of the damage, in. (mm)

d_{dp} = depth of the dent after removal of damaging tool, in. (mm)

D = inside diameter, in. (mm)

FCA = future corrosion allowance, in. (mm)

L_{msd} = distance from the edge of the dent under investigation to the nearest major structural discontinuity or the adjacent flaw, in. (mm)

r_d = local radius of the dent or groove-like flaw located at the base of the dent at the point of impact, in. (mm)

s = length of the groove-like flaw, required only if the dent has a groove-like flaw, in. (mm). The reader is referred to **Figure 3-11a** and **3-11b** on computing s.

Also the reader is referred to Eqs. 3-30, 3-31, and 3-32 for grooves.

t = current thickness, typically the nominal thickness minus the metal loss, in. (mm)

σ_c = circumferential or hoop stress, required only if the dent has a groove-like flaw, psi (MPa)

σ_f = flow stress equal to SYS + 10,000 psi (SYS + 69 MPa), psi (MPa)

σ_p = circumferential or hoop stress when the measurement of the dent is taken, psi (MPa)

σ_{ys} = yield stress at the assessment temperature, psi (MPa)

The results are evaluated by satisfying the following criteria:

1. $\sigma_c \leq \dfrac{\sigma_{cl}}{1.5}$ Eq. 3-94

2. $r_d \geq 0.25(t - FCA)$ Eq. 3-95

3. $\dfrac{d_d}{D + (t - FCA)} \leq 0.05$ Eq. 3-96

4. $L_{msd} \geq 1.8\sqrt{Dt}$ Eq. 3-97

5. All parts of the deformed shell at the location of the dent do not contain a weld seam.
6. The loading of the component is internal pressure, and the stress due to the supplemental loads is insignificant.
7. If the dent contains a groove-like flaw, then the pressure fluctuations are not permitted; otherwise, pressure fluctuations are limited to start-up and shutdown cycles, which will not exceed 500 for the duration the component is in service.
8. The deformed surface, including the groove-like flaw if one is present, does not contain any crack-like flaws.

The API 579 (Paragraph 8.4.3.8a) recommends that if $RSF \geq RSF_a$, the component is acceptable per Level 2. If this criterion is not satisfied, then the component can be re-rated. The API 579 (Paragraph 8.4.3.8b) recommends that a Level 3 analysis be performed to re-rate a component with a dent. A Level 3 would be justified only if the re-rate increased the internal pressure. In most piping and pipeline operations, time is of the essence, and an answer is necessary in two to three hours. If a pipe or pipeline has a dent and it is under construction and a single spool is involved, welding weld caps on the end and hydro testing will suffice. If a new line is put in, then a hydro test should resolve whether a pipe or pipeline with a dent is satisfactory.

Example 3-1: API 579 Example 5.11.1 Revisited

We will consider Example 5.11.1 in the API 579 with one exception; the LTA will have a weld seam passing through it with a joint efficiency of 0.85. The example in API 579 is a pressure vessel with an inside diameter of 96 in. with the following data:

$FCA = 0.125$ in.
$UML = 0.10$
$MAWP = 300$ psig @ 650°F
$t = 1.25$ in. = thickness of the shell

From the inspection data, shown in **Figures 3-17a** and **3-17b**, $t_{mm} = 0.45$ in. What is the new RSF based on the new condition of the weld seam? Before there was no weld seam in the shell, and the calculated RSF was 0.93. Also with the new situation, what is the maximum allowable pressure for the component?

$$t_{min}^C = \frac{(300)(48.225)}{(17500)(0.85) - 0.6(300)} = 0.9845$$

$$t_{min}^L = \frac{(300)(48.225)}{2(17500)(0.85) - 0.6(300)} = 0.4843$$

$$\Rightarrow t_{min} = 0.9845$$

$$R_t = \frac{0.45 - 0.125}{0.9845} = 0.330$$

Calculating the longitudinal length of the LTA:
With $t_{min} = 0.9845$ in., and referring to **Figure 3-17c**, we have

$$s = 6(0.5) + \left(\frac{0.9845 - 0.685}{1.025 - 0.685}\right)(0.5)$$

$$+ \left(\frac{0.9845 - 0.775}{1.025 - 0.775}\right)(0.5) = 3.859 \text{ in.}$$

$$\lambda = \frac{1.285(3.859)}{\sqrt{(96)(0.9845)}} = 0.510$$

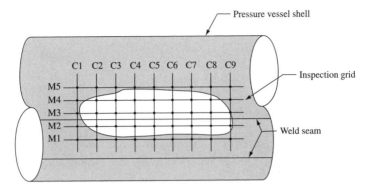

Figure 3-17a. View of weld seam located in LTA. Courtesy of the American Petroleum Institute.

Inspection data (inches)

Longitudinal inspection planes	Circumferential inspection planes									
	C1	C2	C3	C4	C5	C6	C7	C8	C9	Circumferential CTP
M1	1.15	1.15	1.15	1.15	1.15	1.15	1.15	1.15	1.15	1.15
M2	1.15	0.87	0.75	0.70	0.76	0.80	0.85	0.94	1.15	0.70
M3	1.15	0.81	0.82	0.84	0.62	0.45	0.65	0.90	1.15	0.45
M4	1.15	0.85	0.88	0.81	0.84	0.83	0.90	0.91	1.15	0.81
M5	1.15	1.15	1.15	1.15	1.15	1.15	1.15	1.15	1.15	1.15
Longitudinal CTP	1.15	0.81	0.75	0.70	0.62	0.45	0.65	0.90	1.15	

Notes:
1. Spacing of thickness readings in longitudinal direction is 1/2 inch.
2. Spacing of thickness readings in circumferential direction is 1.0 inch.
3. The localized corrosion is located away from all weld seams.

Figure 3-17b. Corrosion data for LTA. Courtesy of the American Petroleum Institute.

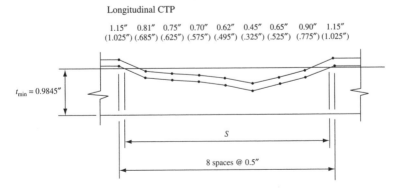

Figure 3-17c. Computation of the parameter *s*.

$R_t = 0.330 > 0.20$

$t_{mm} - FCA = 0.45 - 0.125 = 0.325 \geq 0.10$

$L_{msd} = 60 \text{ in.} \geq 1.8\sqrt{(96)(0.9845)} = 17.50$

From Figure 5.6, $\lambda = 0.441$, $R_t = 0.330 \Rightarrow$ *Flaw is acceptable for this criterion.*

Now, in this assessment, the remaining strength factor will be based on a Level 1 assessment (for precedence, see the API 579, 1st edition, Example 5.11.3, page 5-42).

$$M_t = \sqrt{1 + 0.48(0.510)^2} = 1.0606$$

$$RSF = \frac{0.330}{1 - \left(\dfrac{1}{1.0606}\right)(1 - 0.330)}$$

$$= 0.896 < 0.90 \Rightarrow \text{Flaw is not acceptable.}$$

At what pressure is the component acceptable? Using Eq. 3-18,

$$MAWP = \frac{\left[(t_{mm} - FCA)\left(1.0 - \dfrac{RSFa}{M_t}\right)\right] S_a E}{R_C\left(RSF - \dfrac{RSFa}{M_t}\right) + 0.6\left[(t_{mm} - FCA)\left(1.0 - \dfrac{RSFa}{M_t}\right)\right]}$$

Substituting values,

$$MAWP = \frac{(0.325)(0.1514)(17500)(0.85)}{(48.225)(0.0514) + 0.6(0.325)(0.1514)} = 291.8 \text{ psig}$$

Example 3-2: Dents and Gouge Combination Example

A 36 in. ϕ pipe with 0.5 in. wall is made of API 5L Gr B has a dent and gouge defect. Referring to **Figure 3-18**, the parameters are as follows:

$D_o = 36.0$ in.
$R_o = 18.0$ in.

$UML = 0.0625$ in.
$FCA = 0.0625$ in.
$P = 300$ psig
$T = 200°F$
$t = 0.5$
$E = 1.0$
$d_{dp} = 0.5$ in.
$\sigma_f = 45{,}000$ psi
$a_g = 0.125$
$r_d = 0.1875$

where UML = uniform metal loss

$R_c = R_o + UML + FCA = 18.0 + 0.0625 + 0.0625 = 18.125$ in.
$t_c = t - UML - FCA = 0.5 - 0.0625 - 0.0625 = 0.375$ in.

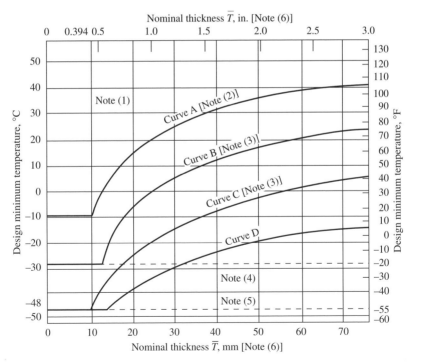

Figure 3-18. Minimum temperatures without impact testing for carbon steel materials. Courtesy of the American Society of Mechanical Engineers.

$$\sigma_c = \left(\frac{P}{E}\right)\left[\left(\frac{R_c}{t_c}\right) + 0.6\right] = \left(\frac{300}{1.0}\right)\left[\left(\frac{18.125}{0.375}\right) + 0.6\right]$$

$$= 14{,}680 \text{ psi}$$

$$d_d = d_{dp}\left(-0.22 \ln\left(\frac{\sigma_c}{\sigma_f}\right)\right) = 0.50\left(-0.22 \ln\left(\frac{14{,}680}{45{,}000}\right)\right)$$

$$= 0.123 \text{ in.}$$

The Charpy impact value for the pipe was found to be 40 ft-lb. Thus,

$$C_{vt} = 40\left(\frac{2}{3}\right) = 26.667 \text{ ft-lb}$$

For English (Imperial) units:

$$C_{uc} = 1.0 \text{ and } C_{us} = 1.0$$

The length of the groove defect beneath the minimum required wall thickness is

$$s = 8.0 \text{ in.}$$

Using Eq. 3-93,

$$Q_{d1} = \frac{\left(\dfrac{26.667}{1.0}\right)}{\left(\dfrac{0.123}{36.0}\right)\left[\dfrac{0.125 + 0.0625}{0.5 - 0.0625}\right]\left(\dfrac{8.0}{1.0}\right)} = 2{,}272$$

Now $Q_d = \text{MAX}(Q_{d1}, 300) = 2{,}272$

The criteria for the dent-gouge combination are as follows:

$$\sigma_{cl} = \frac{(Q_d - 300)^{0.6}}{90} \sigma_f = \frac{(2{,}272 - 300)^{0.6}}{90}(45{,}000) = 47{,}410 \text{ psi}$$

$$\sigma_c = 14.680 \text{ psi} \leq \frac{\sigma_{cl}}{1.5} = 31{,}610 \text{ psi } \textbf{TRUE}$$

$$r_d = 0.1875 \text{ in.} \geq 0.25(t - FCA) = 0.25(0.5 - 0.0625)$$
$$= 0.109 \text{ in. } \textbf{TRUE}$$

$$\frac{d_d}{D_o + (t + FCA)} = \frac{0.123}{36 + (0.5 - 0.0625)}$$

$$= 0.003382 \leq 0.05 \textbf{ TRUE}$$

$$L_{msd} = 12.0 \text{ in.} \geq 1.8\sqrt{D_o t} = 1.8\sqrt{(36)(0.5)} = 7.637 \text{ in. } \textbf{TRUE}$$

The dent gouge combination flaw does not contain a weld seam. The only loading is internal pressure. There are no pressure fluctuations in the pipe, and the flaw does not contain any crack-like flaws.

Thus, the flaw is acceptable under Level 1 Assessment.

Example 3-3: Testing for General Metal Loss

A pipeline is in crude oil service, and UT measurements were taken. It is desirable to determine if the corroded region falls under the category of general metal loss. The minimum thickness reading is 6.8 mm, and *FCA* is zero. A table based on actual field measurements follows:

Thickness Reading Number	Thickness Reading, mm	$(t - FCA)$	$(t - FCA)^2$
C10	7.0	7.0	49
C11	9.4	9.4	88.36
C12	12.2	12.2	148.84
C13	11.9	11.9	141.61
C14	9.8	9.8	96.04
C15	10.4	10.4	108.16
C16	11.4	11.4	129.96
C17	9.0	9.0	81.0
C18	6.8	6.8	46.24
C19	7.6	7.6	57.76
C20	11.0	11.0	121.00
C21	7.3	7.3	53.29
C22	10.6	10.6	112.36
C23	12.3	12.3	151.29
C24	10.6	10.6	112.36
		S1 = 147.30	S2 = 1497.27

$$t_{mm} - FCA = 6.8 \text{ mm}$$

$$N = \text{number of readings}$$

$$t_{avg} - FCA = \frac{S_1}{N} = \frac{147.30}{15} = 9.82 \text{ mm}$$

$$t_{SD} = \left[\left\{\frac{S_2}{N} - (t_{avg} - FCA)^2\right\}\left\{\frac{N}{N-1}\right\}\right]^{0.5} \qquad \text{Eq. 4.13 in API 579}$$

$$t_{SD} = \left[\left\{\frac{1497.27}{15} - (9.82)^2\right\}\left\{\frac{15}{15-1}\right\}\right]^{0.5}$$

$$t_{SD} = 1.904$$

From Eq. 4.12 in the API 579 the COV is defined as follows:

$$COV = \frac{t_{SD}}{t_{avg} - FCA} = \frac{1.904 \text{ mm}}{6.8 \text{ mm}} = 0.28$$

or, $COV = 28\% > 10\%$. Thus, the average thickness above cannot be used; the thickness profiles must be applied in the calculations.

Example 3-4: Surface Crack-like Flaw in a Pipe

A 24 in. ϕ process pipe designed to ASME B31.3 is made of API5l Gr B and has STD wall thickness (0.375 in.). The pipe operates at an internal pressure of 200 psig at 750°F. The pipe becomes fully pressurized at 100°F. The actual yield strength of the pipe is unknown, so we will have to use the SMYS of 35,000 psi. The pipe was not post-weld heat-treated (PWHT) during fabrication. Inspection found an outside surface crack in the HAZ weld region of the single butt welded pipe oriented parallel to the pipe axis that was 0.2 in. deep and 3.2 in. long. The weld joint efficiency is 0.85. There is no uniform metal loss or future corrosion allowance, so $UML = 0$ and $FCA = 0$. Assess the crack using the API 579 Level 2 methodology.

The reference temperature is found using the Figure 3.3 in the API 579, which is adopted from the ASME Section VIII Division 1 UCS-66.1. The materials are oriented for vessel plates, so we use the ASME B31.3 Figure 323.2.2A for the same purpose for piping, shown in **Figure 3-18**.

The pipe material, API5L Gr B, is a Curve B material, defined in Table A-1 of the ASME B31.3. With a thickness of 0.375 in., the design minimum temperature without impact testing is $-29°F$. Thus the reference temperature (T_{ref}) is $-29°F$.

The inside radius (R_c) is as follows:

$$R_c = OD/2 - 2(t_{nom} - UML - FCA), \text{ where } t_{nom} = 0.375 \text{ in.}$$

Thus, $R_c = 24/2 - 2(0.375 - 0 - 0) = 11.25$ in.

The flaw is located away from all major structural discontinuities. Therefore, the primary stress at the weld perpendicular to the crack face is a membrane hoop stress. From Appendix A:

With $t_c = t_{nom} - UML - FCA = 0.375$ in.

$$P_m = \left(\frac{P}{E}\right)\left(\frac{R_c}{t_c} + 0.6\right) = 7{,}200 \text{ psi}$$

$P_b = 0 \text{ psi}$

Now solving for the maximum primary stress, it has been verified that the crack was in the pipe during a field hydrostatic test previously performed. Therefore, the maximum primary stress is

$S_h = 13{,}000$ psi @ 750°F

$S_c = 20{,}000$ psi @ ambient

$$P_m^{max} = 1.5(7200)\left(\frac{20.0}{13.0}\right) = 16{,}615.38 \text{ psi}$$

Secondary Stress: Thermal gradients do not exist in the pipe at the location of the flaw, and the flaw is located from all structural discontinuities. Therefore, there are no secondary stresses and $P_b = 0$.

Residual Stress: The flaw is located at a weldment in a pipe that was not subject to PWHT at the time of fabrication. From Appendix E Paragraphs E.3 and E.4 of the API 579, we have

$\sigma_{residual} = 35 \text{ ksi} + 10 \text{ ksi} = 45 \text{ ksi}$

This is a conservative assessment of residual stress. The reader is referred to [Reference 8] for a more detailed discussion of residual stresses. In many approaches to predicting residual stresses, weld heat is a factor. Most typical heat input for welds is 25–30 KJ/in. Also some companies use 15% of the yield stress as a minimum value of mechanically reduced residual stress used in fracture assessment. Justification of this is given by the API 579 in the discussion of brittle fracture in Paragraph 3.6.2.3, which states that the beneficial effect of a hydrostatic test is that crack-like flaws located in the component are blunted, which results in an increase in brittle fracture resistance. We will follow the more conservative approach in the API 579 Appendix E in this example.

Next we determine the following material properties: yield stress, tensile strength, and fracture toughness. Material properties for the pipe containing the flaw are not available; therefore, the specified minimum yield strength and tensile strength are used. Thus,

$$S_{uts} = 60 \text{ ksi}$$
$$S_{ys} = 35 \text{ ksi}$$
$$T_{ref} = 100°F$$
$$K_{1c1} = 33.2 + 2.806e^{(0.02-(T_{ref}+100))}$$
$$K_{1c1} = 306.83 \text{ ksi}\sqrt{\text{in.}}$$

Since $K_{1c1} > 100 \text{ ksi}\sqrt{\text{in.}}$ then $K_{1c} = 100 \text{ ksi}\sqrt{\text{in.}}$
From the inspection data,

$$a = 0.20 \text{ in.}$$
$$2c = 3.2 \text{ in.}$$

Next we modify the primary stress, material fracture toughness, and flaw size using partial safety factors. Based on a risk assessment, it was decided that the most appropriate probability of failure to apply in the FFS assessment would be $p_f = 10^{-3}$. The mean fracture toughness to specified minimum yield stress ratio (R_{ky}) is required to determine the PSF. The information in Notes 5 and 6 of Table 9.2 (note $\sigma = 1$) are used in the following calculations. First, the notes state the following:

Note 5: R_{ky} is used in conjunction with R_{c1} (we use R_{c1} to prevent its being confused with the corroded inside radius, R_c) to determine the Partial Safety Factors to be applied in an assessment. R_{c1} is a cut-off value used to define the regions of brittle fracture/plastic collapse and estimates the corresponding category of Partial Safety Factors to be used in an assessment. The definition of R_{ky} is given by the following equation:

$$R_{ky} = \frac{K_{mat}^{mean}}{\sigma_{ys}} C_u$$

where C_u = conversion factor; if the units of K_{mat}^{mean} are $\text{ksi}\sqrt{\text{in.}}$ and σ_{ys} are psi, then $C_u = 1.0$. If the units of K_{mat}^{mean} are $\text{MPa}\sqrt{\text{mm}}$ and σ_{ys} are MPa then $C_u = 6.268$.

K_{mat}^{mean} = average value of the material fracture toughness ($\text{MPa}\sqrt{\text{mm}}$; $\text{ksi}\sqrt{\text{in.}}$), and

σ_{ys} = nominal yield stress taken as the specified minimum value (MPa; psi).

Note 6: If the only source of fracture toughness data is the lower bound estimate in Appendix F, Paragraph F.4.4, then the mean value of toughness described in Paragraph F.4.4.1.e should be used in the assessment. The mean value of fracture toughness is used because the Partial Safety Factors are calibrated against the mean fracture toughness.

Now performing the calculations we have the following:

$$\Delta T = T - T_{ref} = 100°F - (-29°F) = 129°F$$

$$K_{mean} = \frac{K_{mat}^{mean}}{K_{1c}} = \frac{1.0}{100} = 0.01$$

Note that the preceding equation for K_{mean} is valid only for $-200°F \le \Delta T \le 400°F$. Since $\Delta T = 129°F$, the equation is valid for use.

$$R_{ky} = \frac{K_{mat}}{\left(\dfrac{SMYS}{1000}\right)} = 4.66$$

From Table 9.2 of the API 579 (**Figure 3-19**), with $(R_{ky} = 4.66) > (R_{c1} = 1.9)$, the partial safety factors are as follows:

$$\begin{cases} (a = 0.20") \ge 0.20" \\ COV_s = 0.10 \\ R_{c1} = 1.9 \end{cases} \Rightarrow \begin{cases} PSF_s = 1.5 \\ PSF_k = 1.0 \\ PSF_a = 1.0 \end{cases}$$

In note 5 the COV_s of 0.10 occurs when the primary loads and corresponding primary stresses in the region of the flaw are computed or measured, and are well known. A value of the COV_s of 10% is a typical variation on safety pressure valve pop tests.

The primary stress, fracture toughness, and flaw size are factored by the PSF as follows:

$$P_{m1} = PSF_s(P_m) = (1.5)(7200 \text{ psi}) = 10,800 \text{ psi}$$

$$P_{b1} = PSF_s(P_b) = 0$$

$$K_{mat1} = PSF_a(K_{mat}) = (1.0)(163) = 163 \text{ ksi} \sqrt{\text{in.}}$$

$$a = (0.20")(1.0) = 0.20"$$

We now compute the reference stress for the primary stress. From Appendix C, Table C.1, of the API 579, the flaw geometry, component

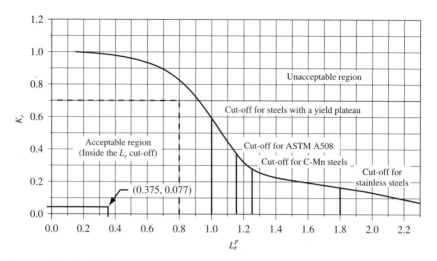

Figure 3-19. The FAD diagram shows that the Level 2 Assessment is satisfied at the base and edge of the flaw. Courtesy of the American Petroleum Institute.

geometry, and loading condition correspond to the stress solution RCSCLE1. The reference stress solution for RCSCLE1 is provided in the API 579 Appendix D, Paragraph D.5.10.

$$a = 0.20''$$

$$c = \frac{3.2}{2} = 1.6''$$

$$t = t_c = 0.375''$$

$$\lambda = \frac{1.818c}{\sqrt{(R_c)(t_c)}} = 1.416$$

$$M_t = \sqrt{\frac{1.02 + 0.4411\lambda^2 + 0.006124\lambda^4}{1.0 + 0.02642\lambda^2 + [1.533(10^{-6})]\lambda^4}}$$

$$M_t = 1.35$$

$$\alpha = \frac{\dfrac{a}{t_c}}{1 + \dfrac{t_c}{c}}$$

$$\alpha = 0.432$$

$$g = 1 - 20\left[\frac{a}{2c}\right]^{0.75}(\alpha^3) = 0.798$$

$$M_s = \frac{1 - 0.85\left(\dfrac{a}{t_c}\right)\left(\dfrac{1}{M_t}\right)}{1 - 0.85\left(\dfrac{a}{t_c}\right)} = 1.217$$

$$\sigma_{ref} = \frac{gP_b\sqrt{(g)(P_{b1}) + 9[(M_s)(P_{m1})]^2}}{3}$$

$$\sigma_{ref} = 13{,}139.55 \text{ psi}$$

Now computing the load ratio (L_r) or abscissa of the FAD, we have

$$\sigma_{ys} = 35 \text{ ksi}$$

$$L_r = \frac{\sigma_{ref}}{SMYS} = \frac{13{,}139.5}{35{,}000} = 0.375$$

We now compute stress intensity at the base and edge of the flaw. Based on Appendix C, Table C.1, the flaw geometry, component geometry, and loading condition correspond to the stress intensity solution KCSCLE1 in Paragraph C.5.10.

The flaw ratios and parameters to determine the G_o influence coefficient come from Table C.11 in the API 579. The angle of $0°$ is at the edge of the flaw and $90°$ is at the base of the flaw. We obtain coefficients for both locations. Thus we have the following:

$$G_{o0} = A_{0,0} + A_{1,0}\beta + A_{2,0}\beta^2 + A_{3,0}\beta^3 + A_{4,0}\beta^4 + A_{5,0}\beta^5 + A_{6,0}\beta^6$$

$$G_{o90} = A_{0,0} + A_{1,0}\beta_1 + A_{2,0}\beta_1^2 + A_{3,0}\beta_1^3 + A_{4,0}\beta_1^4 + A_{5,0}\beta_1^5 + A_{6,0}\beta_1^6$$

$$G_{10} = A_{0,1} + A_{1,1}\beta + A_{2,1}\beta^2 + A_{3,1}\beta^3 + A_{4,1}\beta^4 + A_{5,1}\beta^5 + A_{6,1}\beta^6$$

$$G_{190} = A_{0,1} + A_{1,1}\beta_1 + A_{2,1}\beta_1^2 + A_{3,1}\beta_1^3 + A_{4,1}\beta_1^4 + A_{5,1}\beta_1^5 + A_{6,1}\beta_1^6$$

The coefficients are as follows:

$$\begin{Bmatrix} \dfrac{R}{t} = \dfrac{12.0}{0.375} = 32.0 \\[2mm] \dfrac{c}{a} = \dfrac{1.6}{0.2} = 8 \\[2mm] \dfrac{a}{t} = \dfrac{0.2}{0.375} = 0.533 \end{Bmatrix} \Rightarrow \begin{Bmatrix} A_{0,0} = 0.57 \\ A_{1,0} = 1.70 \\ A_{2,0} = 0.52 \\ A_{3,0} = -1.50 \\ A_{4,0} = -1.20 \\ A_{5,0} = 1.50 \\ A_{6,0} = -0.70 \end{Bmatrix}, \begin{Bmatrix} A_{0,1} = 0.086 \\ A_{1,1} = 0.418 \\ A_{2,1} = 1.77 \\ A_{3,1} = -0.60 \\ A_{4,1} = -1.70 \\ A_{5,1} = 1.15 \\ A_{6,1} = -0.27 \end{Bmatrix}$$

At the edge of the flaw, $\varphi = 0°$ in which,

$$\varphi = 0° \Rightarrow \beta = 0°$$

At the base of the flaw, $\varphi = 90°$ in which,

$$\varphi = 90°\left(\frac{\pi}{2}\right) \Rightarrow \beta = \frac{2}{\pi}\left(\frac{\pi}{2}\right) = 1$$

From which

$$G_{o0} = 0.057; \ G_{o90} = 1.29; \ G_{10} = 0.086; \ G_{190} = 1.404$$

Now,

$$Q = 1.0 + 1.464\left(\frac{a}{c}\right)^{1.65} \quad \text{for } a/c \le 1.0$$

And

$$Q = 1.0 + 1.464\left(\frac{c}{a}\right)^{1.65} \quad \text{for } a/c > 1.0$$

Since $a/c = 0.125$, thus $Q = 1.087$.

At the edge of the flaw,

$$K_{1,0}^P = \frac{\left(\frac{(P)(R_c)^2}{R_o^2 - R_c^2}\right)(2G_{o0} + 2G_{10})\sqrt{\dfrac{\pi a}{Q}}}{1000}$$

At the base of the flaw,

$$K_{1,90}^P = \frac{\left(\frac{(P)(R_c)^2}{R_o^2 - R_c^2}\right)(2G_{o90} + 2G_{190})\sqrt{\dfrac{\pi a}{Q}}}{1000}$$

From which

$$K_{10}^P = 1.45 \text{ ksi }\sqrt{\text{in.}} \text{ and } K_{190}^P = 5.95 \text{ ksi }\sqrt{\text{in.}}$$

Now we compute the reference stress for secondary stresses. Note that the following calculation is based on the residual stress computed above. Hence,

$$\sigma_{ref}^{SR} = \frac{gP_b + \sqrt{(gP_{b1})^2 + 9(M_s\sigma_{residual}(1000))^2}}{3}$$

$$\sigma_{ref}^{SR} = \frac{(0.798)(0) + \sqrt{((0.798)(0))^2 + 9((1.217)(45)(1000))^2}}{3}$$

From which

$$\sigma_{ref}^{SR} = 54,748.12 \text{ psi}$$

Note that the numbers were computed with a spreadsheet program and that they were not rounded-off, as shown above. This results in a discrepancy if calculated manually using the preceding equations.

Now,

$$\sigma_{ref}^{P} = \frac{gP_b + \sqrt{(gP_{b1})^2 + 9(M_sP_{mmax})^2}}{3}$$

$$\sigma_{ref}^{P} = \frac{(0.798)(0) + \sqrt{((0.798)(0))^2 + 9((1.217)(16,615.385))^2}}{3}$$

$$\sigma_{ref}^{P} = 20,214.689 \text{ psi}$$

$$\sigma_{ys} = 35,000 \text{ psi and } \sigma_{uts} = 60,000 \text{ psi}$$

Thus,

$$\sigma_f = \frac{\sigma_{ys} + \sigma_{uts}}{2} = 47,500 \text{ psi}$$

Since $\left(\sigma_{ref}^{SR} = 47,500 \text{ psi}\right) > \left(\sigma_{ys} = 35,000 \text{ psi}\right)$, then

$$S_{srf} = \min\left[\left\{1.4 - \frac{20,214.689}{47,500}\right\}, 1.0\right] = 0.974$$

We now compute K_1^{SR}. The details regarding the calculation of the stress intensity factor were provided earlier. Note that $S_{srf} = 0.974$ as computed previously is applied to the secondary stress.

Thus, at the base of the flaw, $\varphi = 0°$,

$$K^P_{1,0c} = S_{srf}K^P_{1,0} = (0.974)(1.45) = 1.410 \text{ ksi}\sqrt{\text{in.}}$$

And at the edge of the flaw, $\varphi = 90°$,

$$K^P_{1,90c} = S_{srf}K^P_{1,90} = (0.974)(5.945) = 5.793 \text{ ksi}\sqrt{\text{in.}}$$
$$K^P_{10} = 1.45 \text{ ksi}\sqrt{\text{in.}} \text{ and } K^P_{190} = 5.95 \text{ ksi}\sqrt{\text{in.}}$$

We now compute the plasticity factor (Φ) as follows. The plasticity factor is a correction ("fudge") factor to account for the nonlinear parameters because one cannot directly add nonlinear parameters as linear ones. So the plasticity factor is used to account for the nonlinearity of the problem.

$$L^{SR}_r = \frac{54,748.115}{35,000}(0.974) = 1.524$$

After finding ψ in Table 9.3 and ϕ in Table 9.5 from API 579, we have the following:

$$\begin{Bmatrix} L_r = 0.375 \\ L^{SR}_r = 1.524 \end{Bmatrix} \Rightarrow \begin{Bmatrix} \psi = 0.094 \\ \phi = 1.135 \end{Bmatrix}$$

$$RATIO\Phi = \frac{\Phi}{\Phi_o} = 1.0 + \frac{0.094}{0.696} = 1.135$$

Since $0 < (L^{SR}_r = 1.524) \le 4.0$, then $\Phi_o = 1.0$ and $\Phi = 1.135$

To find the toughness ratio or ordinate of the FAD assessment point, At the base of the flaw, $\varphi = 0°$,

$$K_{r0} = \frac{K^P_{10} + \Phi(K^P_{10c})}{K_{mat}} = \frac{1.488 + (1.135)1.411}{163} = 0.0187$$

At the edge of the flaw, $\varphi = 90°$,

$$K_{r90} = \frac{K^P_{190} + \Phi(K^P_{190c})}{K_{mat}} = \frac{5.945 + (1.135)5.793}{163} = 0.077$$

To evaluate the results, we determine the cutoff of the L_r-axis of the FAD. Because the hardening characteristics of the material are not known, the following value can be used in **Figure 3-19** (Figure 9.20 of the API 579).

$$L_{rmax} = 1.0$$

We plot the assessment point on the FAD shown in **Figure 3-19** (Figure 9.20). Because the partial safety factors are used in the preceding assessment, the full FAD may be used.

At the base of the flaw $\varphi = 90°$:

$(L_r, K_r) = (0.375, 0.0187)$; the point is inside the FAD

At the edge of the flaw $\varphi = 0°$:

$(L_r, K_r) = (0.375, 0.077)$; the point is inside the FAD

The Level 2 Assessment criterion are satisfied.

Verification of Results: The crack was checked with the methodology in the British Standard BS 7910 and was found acceptable. This methodology is recognized in the API 579 Paragraph 9.4.4.1.e. It gives criteria for computing residual stresses in Paragraph 7.3.4.2, Eq. (14a), which is similar to that of the API 579. The API 579, although tedious, gives a reasonable methodology to work with unknown residual stresses. However, working with data (e.g., the influence coefficients in Appendix C) requires many interpolations. The gap between a 20 in. cylinder and 60 in. cylinder is quite large. There is much work being done on residual stresses and predicting their magnitude. Typically for areas where the membrane stress dominates and the flaw is remote from discontinuities, the relaxation in residual stress may be approximated by calculating the stress above the upper bound yield strength the pipe will experience under maximum loading (e.g., a hydrostatic test). The residual stress may then be reduced by the difference between the maximum loading stress and the upper bound yield strength. Note that the minimum value of mechanically reduced residual stress that may be applied in a fracture assessment is 15% of the yield strength. This is a criterion used in many companies. However, no credit may be taken for compressive residual stress. As for the effects of PWHT, values for the residual stress are given in Appendix E of the API 579 for various types of welds. The BS 7910 gives a criterion of the effects of proof testing in Annex O.2 Eq. (O.1). The equation reads as follows:

$$\sigma_{residual} = \text{either } \sigma'_Y \text{ or } \left(1.4 - \frac{\sigma_{ref}}{\sigma'_f} \right) \sigma'_Y$$

where σ_{ref} = the maximum reference stress under the proof load conditions

σ'_Y = the appropriate material yield strength at the proof test temperature

σ'_f = the flow strength, which is the average of the yield and the tensile strengths at the proof test temperature. Here σ'_f is not limited to 1.2 times the yield strength.

This equation is used to account for the mechanical stress relief under proof test or prior overload, on the assumption that the same equation is equally applicable under any loading conditions and that the reduction in residual stress caused by a proof load or prior overload remains after the load is removed. Where a crack in a proof-loaded pipe is believed to have been initiated in service, after the proof loading, the residual stress level in fracture analysis should be taken as a uniform stress equal to the lower of the following factors:

$$\sigma_{residual} = \sigma'_Y \text{ or } \left(1.1\sigma'_Y - 0.8\sigma_a \right)$$

where σ_a = the applied stress due to proof loads at the location of interest

References

1. API Recommended Practice 579, *Fitness-for-Service*, 2001, American Petroleum Institute.
2. John F. Kiefner, Patrick H. Vieth, and Itta Roytman, "Continued Validation of RSTRENG," Prepared for the Line Pipe Research Supervisory Committee Pipeline Research Committee, December 20, 1996.
3. David A. Hansen and Robert B. Puyear, *Materials Selection for Hydrocarbon and Chemical Plants*, 1996 Marcel Dekker, New York.
4. BS 7910, *Guide to Methods for Assessing the Acceptability of Flaws in Metallic Structures*, 2005 British Standards Institute.
5. Michael J. Rosenfeld, "Here Are Factors That Govern Evaluation of Mechanical Damage to Pipelines," *Oil and Gas Journal*, September 9, 2002.
6. WRC Bulletin 465, "Technologies for the Evaluation of Non-Crack-Like Flaws in Pressurized Components," September 2001, Welding Research Council.
7. Cosham, Andrew, and Kirkwood, "Best Practice in Pipeline Defect Assessment," Proceedings of IPC 2000, International Pipeline Conference, October 2000, Calgary, Alberta, Canada, pp. 4–5.
8. Koppenhoefer, Kyle, Feng, Zhili, Cheng, and Wentao, "Incorporation of Residual Stresses Caused by Welding Into Fracture Assessment Procedures," Project No. 40568-CSP, Material Properties Council, June 2, 1999, New York.

Chapter Four

Fitness-for-Service for Brittle Fracture Concerns

Cold safety is for low-temperature design, assessment, and operation of operating plants. Cold temperatures can drastically affect the reliability and safety of a unit by causing brittle fracture. One cannot overemphasize cold safety because it is responsible for many accidents around the world that result in the loss of life and property.

The material in this chapter has been presented, taught, and practiced successfully at other places. It represents many years of practical use of code rules and fracture mechanics principles. A basis for this discussion is API 579 [Reference 1], discussed in the last chapter.

Introduction

Low-temperature service requires special consideration in the design of process equipment due to the possibility of a failure mode known as brittle fracture. The principal characteristic of brittle fracture is the very rapid catastrophic failure of equipment (generally without warning) that can lead to large openings or total collapse at loads below the anticipated design levels. Definitions and guidelines for low-temperature service are in ASME B31.3 Chapter III [Reference 2] for piping.

Traditional design codes (ASME, BS, etc.) that evolved in principle from the boiler industry have traditionally been more concerned with sufficient integrity at higher temperature where the main concern is adequate strength. Under these conditions, materials are expected to stretch elastically under overload, and small local cracks or bursts are expected to occur initially rather than as a sudden catastrophic failure, unless the structure is significantly overloaded beyond design.

The special issues of brittle fracture were not really put into perspective until significant industrial failures occurred in the 1940s and 1950s. These failures coupled with the new technology of fabrication by welding stimulated the development of fracture mechanics through the 1950s, 1960s, and 1970s. One of the early conclusions, however, was that something supplementary was required because strength alone provided inadequate criteria for safe design.

Many design codes were slow to address these issues, so some companies introduced their own design requirements for preventing brittle fracture. In Chapter 3 we discussed the importance of the material property toughness. In this chapter we will introduce additional terms. One example involves the two principle issues of defining a *lower temperature* limit or design condition known as the critical exposure temperature (CET) and supplementing materials requirements by defining a minimum brittle fracture resisting property known as *toughness*. By specifying a minimum level of toughness (the Charpy Impact Energy, ft-lb/Joules) at the coldest significant process condition (CET), we achieve the basic ingredients for preventing brittle fracture. This aspect of optimum materials selection and testing at an appropriate CET will be discussed later in this chapter.

Design codes generally now reflect the latest brittle fracture technology and industry experience. These codes are specific to process equipment types. We will, however, recommend additional and more restrictive requirements on piping, resulting in increased integrity.

Recent advances in the fracture aspects of the ASME code have, among other things, formalized relationships between pressure and temperature such that the concept of a single safe minimum temperature (CET) can, if desired, be replaced by a locus of pressure-dependent temperatures specific to a piece of equipment. Such a locus forms a *safe operating envelope* within which it is safe to operate without risk of brittle fracture. This range provides a particularly useful tool for establishing operating limits for existing equipment that may have a brittle fracture concern or that may be required to operate or withstand upset conditions not originally envisaged. This aspect of safe operation will be discussed later in this chapter.

This chapter is intended to be an awareness document. Its aim is to encourage cooperation between process designers, safety engineers, and mechanical/materials/inspection personnel to achieve maximum integrity of equipment. Without adequate cooperation it will not be possible to consider fully all aspects of safe operation.

Brittle Fracture Concepts

For brittle or "fast" fracture to occur, three conditions must act simultaneously and in sufficient magnitude.

1. A defect as a stress raiser, such as a crack or notch due to mechanical damage or careless welding
2. Sufficient stress to provide a driving force on the defect to cause it to grow
3. A low level of material toughness or resistance to crack propagation

These conditions, shown schematically in **Figure 4-1**, make up the *integrity triangle*, and all three components must exist to have brittle fracture.

Defect

All structures contain defects of some form, but from a practical viewpoint most significant defects are associated with welds. These usually take the form of cracks, lack of fusion, slag, and poorly designed, specified, or controlled welding techniques. Defects of this type are generally inbuilt during fabrication and, if they are small enough, may not be a significant risk to brittle fracture, provided they are not subject to cyclic loading.

To ensure that unacceptably large defects are not allowed in new piping, the codes require that nondestructive examination (NDE) be carried out with sufficient guaranteed sensitivity such that a predetermined defect size should be detected and excluded. Under these "steady state" conditions there are valid reasons to assume that, provided the service of a pipe does not become more severe, brittle fracture will not subsequently occur. This is the basis of "grandfathering."

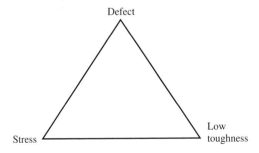

Figure 4-1. The three components required for brittle fracture. Without all three, brittle fracture is not possible.

In many process environments, however, deterioration can occur with time. Caustic MEA and DEA, for example, can cause stress corrosion cracking, and fatigue loads can cause small harmless defects to grow to critical proportions. Under these circumstances grandfathering is not possible, and more sensitive NDE is generally applied at regular intervals to ensure that any defects are below critical proportions.

From the foregoing obviously only limited control exists over the defect side of the fitness-for-service integrity triangle (see **Figure 4-1**), and so the codes assume that some readily detectable crack could exist (typically 70% of thickness) to enable minimum toughness and maximum stress levels to be specified as more direct control against fracture.

Stress

Stresses arise from various loads applied to equipment from sources such as pressure, thermal expansion, and piping reactions. The most critical load is generally internal pressure, as the others often give rise only to localized effects that can be treated in a less conservative manner with respect to maximum allowable levels and consequences of failure. Also internal pressure, like weight loads, is a primary stress. Secondary loadings (e.g., thermal stresses) must be accounted for, and this was discussed in Chapter 3.

Design codes require stresses from anticipated loads to be computed and kept below allowable levels, keeping the stresses in the elastic range. These codes include the SIFs (stress intensification factors) that account for stress risers at elbows, tees, and branch connections. Because the ASME piping codes, namely B31.3 and B31.1, do not categorize stress types—other than sustained, displacement, and occasional loads—any supplemental loads (e.g., external forces such as wind or seismic conditions) are considered.

As the integrity triangle shows, reducing stress (pressure is one primary cause of stress) reduces the chances of brittle fracture. Because all materials exhibit some lower limit of residual toughness, it follows that brittle fracture cannot occur if the stress is low enough. The exact level of minimum stress is debatable (6 ksi or 41 MPa is often quoted); however, values of 6–8 ksi (35–55 MPa) actually represent levels sufficiently low not only to prevent fracture initiation but also to arrest crack propagation in regions of higher stress (e.g., localized stressed areas). This concept is important and supports the fracture prevention philosophy presented, namely that the goal is to prevent catastrophic brittle failure as opposed to local fracture.

We will apply this principle by stating that brittle fracture will not be a concern if the lowest metal temperature which a component will be subject to is either 30% of the MAWP or a combined total longitudinal stress equal to 8 ksi (55.2 MPa) due to pressure, weight effects, and displacement strains. This lowest metal temperature is known as the CET, critical exposure temperature (Reference 1, Paragraph. 3.1.3.2). Furthermore, this level of stress defines a lower cutoff value for coincident pressure and low operating temperature. This is one of many important considerations in setting the CET. Clearly then, 30% of 20.0 ksi = 6.0 ksi is in the 6–8 ksi threshold, which meets the acceptance criteria.

In practice the assumptions about local and other loads may not be quite as simple as previously implied, particularly where they can remain high even when pressure is reduced. It is thus important to review the assumptions whenever safe temperature calculations are made. On the other hand if general stresses from pressure are limited to 30% of the MAWP via pressure, then any local cracking should be arrested before it leads to a catastrophic failure. This is the criterion adopted in developing safe operating envelopes based only on the membrane pressure stresses in the wall of a pipe.

Toughness

The single important material property in resisting brittle fracture is toughness. It was discussed in Chapter 3. Toughness can best be defined as the ability of a material to absorb energy and is commonly expressed as an energy value (Joules) as measured by a Charpy impact test.

The Charpy test consists of raising a pendulum to a predetermined height and letting it swing onto a test specimen (10 mm square by 50 mm long with a 2 mm notch machined into one face). The notch that is now used is a V-notch, but a U notch was used in the past. The specimen breaks on impact, and the resulting loss of height on the follow-through swing gives a measure of the energy absorbed in breaking the specimen.

Even though the Charpy test is not a direct measure of fracture resistance, it provides an inexpensive and easy test that correlates reasonably well to actual fracture properties as measured, for instance, by crack tip opening displacement (CTOD) tests.

Toughness is inversely proportional to strength and is improved by fine-grained structure, chemistry control, and heat treatment among other factors. It is characterized by an "impact transition curve" such as that shown in **Figure 4-2**. This illustrates that toughness (energy absorption capacity) decreases with decreasing temperature.

At higher temperatures, high toughness produces ductile behavior and is characteristic of the elastic design assumptions discussed in the introduction. At lower temperatures, brittle behavior occurs, and failures resemble the breaking of glass. An important feature of the brittle region (known as the lower shelf) is that it does not go to zero but retains some

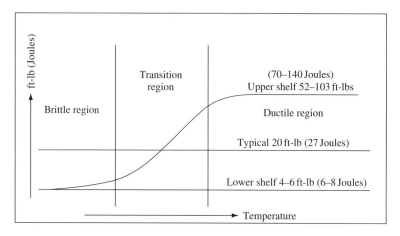

Figure 4-2. Typical impact transition curve.

nominal toughness. This is the fact that is exploited in terms of a "harmless stress" (30%) as previously discussed. The transition region represents the change between the extremes, and it is in this region that most low temperature material selection is done.

The transition is characterized not only by changing energy absorption but also by the fracture appearance of the test specimen. Because of this, among many other factors, the common term "transition temperature" (often taken to represent the temperature of lowest acceptable toughness) must be qualified. This qualification is done by specifying an energy value or percentage of lateral contraction of the test specimen to define a unique transition temperature.

The process of material selection, heat treatment specification, and so on has the effect of moving the transition region to the left for improved toughness. On the other hand, some forms of in-service deterioration such as temper embrittlement can significantly move the curve to the right, indicating that a higher temperature is required for the same toughness. This forms the basis of controlled start-up procedures as sometimes applied to heavy-walled piping for instance. Here pressure is kept below 30% of MAWP until the safe temperature is reached based on a modified transition temperature. This is an example of a special CET situation and illustrates the "safe operating envelope" in its simplest form.

Toughness levels are set out in the design codes based principally on experience but backed by fracture mechanics considerations that balance the integrity triangle. Chapter 3 specifies energy values based on specific material type and thickness and generally requires higher levels of toughness than required in the code. The normal 20–27 J levels found in

Chapter 3 would probably lie somewhere between half and two thirds down the transition curve toward the lower shelf.

The process designer's careful selection of a realistic CET is essential because it allows optimum material selection, balancing cost and integrity. The minimum requirement is to select a material whose transition temperature is on or to the left of the CET and then to verify this by actual impact tests.

These aspects are addressed in more detail in selecting materials and defining impact requirements for new equipment and in determining the CET. In principle, the lower the CET is, the quality of steel that is required is higher and the steel itself costs more. However, readily available carbon steels can be used with minimum increasing cost to around $-46°C$ ($-51°F$). It should be noted that the common practice of calling for a "killed carbon steel" for low-temperature service does not guarantee good toughness because the requirement for a fine-grained steel is more important. This fact should be made clear on the specification, and it should be well understood that a killed steel could equally be a coarse-grained high-temperature steel totally unsuited for low temperatures.

At temperatures below $-46°C$ ($-50°F$), specially enhanced carbon steels can get to $-60°C$ ($-76°F$), below which special and significantly more expensive low nickel alloys will be required for temperatures around $-100°C$ ($-148°F$). These have transition properties similar to the carbon steel grades. For services below $-100°C$ ($-148°F$), austenitic stainless steel, aluminum, and some high nickel alloys are common—the latter have been used less in recent years. These materials, except the high nickel alloys, do not exhibit an appreciable transition region and are suitable for most of the lowest temperature services in the chemical plant commonly to around $-196°C$ ($-321°F$). The welding of these materials does, however, require some special considerations and qualifications.

In applying fracture mechanics principles using the codes, it will be seen that thickness is an important parameter fundamental to impact test exemption and required toughness. Thickness, however, does not feature directly in the integrity triangle but can be interchanged with the defect parameter a, one half the length of a crack. When this is done, the specific case of a through-wall thickness crack is assumed. If this can be shown to be acceptable, then we have the special condition of leak-before-break. Clearly, therefore, if thickness, as a, is increased, with a constant crack size to thickness ratio, the tendency for brittle fracture increases. Conversely, thin sections have good resistance to brittle fracture.

The explanation for this is in the degree of constraint of the crack. In thicker sections, this is greater, so, all things considered, the crack experiences greater stress intensity. For this reason and the fact that larger defects could remain undetected in larger sections and that chemical and physical properties may be become nonuniform, the code requires higher toughness for thicker sections.

Definitions

CET: As defined earlier, the critical exposure temperature (CET) is defined as the lowest metal temperature derived from either the operating or atmospheric conditions. The CET may be a single temperature at an operating pressure or a locus of temperatures and pressures. It is the temperature that is the lowest that the piping designed to ASME B31.3 Piping Code will see coincident with a pressure (stress) greater than 30% of the maximum allowable working pressure (MAWP), or a combined longitudinal stress that equals 55.2 MPa (8 ksi) due to pressure, weight effects, and displacement strains. It is derived from anticipated or contingency process or atmospheric conditions. No knowledge of the geometric form of the equipment or the materials of construction is required to determine CET for the criterion of 30% of the MAWP or a combined longitudinal stress that equals 55.2 MPa (8 ksi). However, the second criterion, 30% of the MAWP or 8 ksi/in. maximum combined longitudinal stress, requires knowledge of the geometric form. The CET is derived from the process engineer.

MAT: The minimum allowable temperature (minimum safe operating temperature) is the permissible lower metal temperature limit for a given material at a thickness based on its resistance to brittle fracture. It may be a single temperature or a locus of allowable operating temperatures as a function of pressure. The MAT term is chosen to represent the minimum temperature a piece of equipment or component part thereof may be safely exposed to at any process pressure up to the equipment strength capabilities. The MAT is derived *purely* from mechanical design information, materials specifications, and/or materials data and can be made without any knowledge of process conditions, contingencies, or contained fluid. It can be based on Charpy impact test data. The MAT is the permissible lower metal temperature limit for a given material at a specified thickness based on its resistance to brittle fracture.

MDMT: Minimum design metal temperature is a single point on the MAT curve that defines the minimum temperature that can be sustained at full design pressure (stress). This is deemed to include normal safety valve accumulation. The MDMT is derived as an integral part of the MAT curve and is essentially the value that would be chosen for CET on a new vessel if design conditions governed. The MDMT can be the MAT.

While the CET is determined by the process engineer, the stationary (fixed) mechanical/materials engineer is best suited to determine the MAT. The two should work in partnership to achieve optimum design and integrity.

Safe Operation at Low Temperatures Existing Equipment

Safe Operating Envelopes

This section is intended to provide an understanding of the concept of safe envelopes as an MAT and how they can be defined, developed, and influenced by various common variables. The methodology assumes that the equipment is in sound condition, which can/should be verified by personnel and the appropriate nondestructive examination (NDE).

The brittle fracture assessment is performed by constructing a MAT curve. This curve is a plot of the MAT points for various temperatures over the design, or fully expected operating conditions, including excursions and upsets.

The conclusions of an assessment often result in a nonleak before break condition requiring the application of sensitive nondestructive testing to verify potential defects will remain below critical proportions during the life of the equipment or its inspection intervals.

As seen, cold safety principles cannot be directly applied to piping without a good understanding of the stress pattern in the system. ANSI B31.3 that is the design basis for *new* process piping is one of the criteria that we use to address brittle fracture. API 579 refers one to ASME BPVC Section 8 Division 1 Figures UCS-66 and UCS-66.1. The B31.3 uses similar curves in Figures 323.2.2A and 323.2.2B, respectively. The main differences are that the B31.3 curves provide more detailed treatment of piping materials, and curve B, which covers many carbon steel piping materials, was altered so that the minimum temperature was $-20°F$ $(-29°C)$ through a $\frac{1}{2}$-in. (13-mm) wall thickness. It is the author's opinion that these curves can safely be used for piping systems, even though the API 579 uses the Figures UCS 66 and 66.1. Use of the B31.3 curves for piping has been in successful practice for many years for existing piping systems. This fact is augmented by API 579 Paragraph 3.4.2.2, "Piping Systems," by stating:

> Piping systems should meet the toughness requirements contained in ASME B31.3 at the time the piping system was *designed* (or an equivalent piping design code if that contains material toughness

requirements). Piping systems should be evaluated on a component basis; the MAT for a piping system is the highest MAT obtained for all the components in the system.

Also, API 579 uses different terminology in regard to the stress ratio curve of both UCS 66.1 and Figure 323.2.2B. The stress ratio as defined in API 579 Table 3.2 is as follows:

$$R_{ts} = \frac{t_r E^*}{t_g - LOSS - FCA} = \frac{S^* E^*}{SE} \qquad \text{Eq. 4-1}$$

where, t_g = term governing nominal uncorroded thickness, in. (mm)
t_r = required thickness, in. (mm)
E^* = joint efficiency used in the calculation of t_r (E^* shall not be less than 0.8)
E = joint efficiency
S = allowable stress value in tension, psi (MPa)

The term S^* is defined as the applied primary stress. API 579 refers one to Appendix A, Table A.1, "Loads, Load Cases, and Allowable Design Stresses." In this table, piping loads including pressure thrust (such as bellows expansion joints), pressure and fluid loading during normal operation, and thermal loads are included. In piping, this would encompass three cases: the pressure, weight, and thermal case; the pressure and weight case; and the thermal case. Thermal stresses are not primary, but secondary. They are secondary because they are self-relieving. As the pipe deforms it self-relieves. During this process, brittle fracture could easily occur if there is a flaw and low toughness, making the critical three components present to trigger brittle fracture at the low enough temperature. Thus, in piping, the S^* stress would be the maximum that would occur during one of the three cases described.

The B31.3 code precludes brittle fracture as a concern down to −20°F (−29°C) (using common pipe materials) provided its design requirements are met in all other respects. It also contains a 30% rule for pressure in Table 323.2.2 Note (2) in Chapter 3 on Materials. This rule states the following:

Impact testing is not required if the design temperature is below −20°F (−29°C) but at or above −155°F (−104°C), and the Stress Ratio defined in Fig. 323.2.2B does not exceed 0.3 times S.

Unfortunately the pressure stress in a pipe is not the one that generally governs the 6 ksi limit but rather thermal expansion, weight, and the like that are not always easy to estimate without a detailed pipe stress analysis.

In practice, piping tends to be relatively tolerant of brittle fracture via thin sections, and if good design/layout has been applied, this should minimize piping stresses. Local fracture at supports, restraints, and intersections, however, is a real possibility. On balance, this means that it is unlikely that piping materials would have to be upgraded (e.g., to handle the common case of auto refrigeration in propylene systems), but a detailed review is recommended for confirmation.

For existing piping, API 579, the "Fitness-for-Service," governs. Brittle fracture concerns were enhanced in the 1993 addenda of the B31.3 piping code. The code committee added curves similar to the ASME Section 8 Division 1 Figure UCS 66. Shown in **Figure 4-3** (also **Figure 3-18**) are curves A, B, C, C, and D of Figure 323.2.2A.

Finding the MDMT for the material is done just like it is with Figure UCS 66. Table A-1 lists the material curves. Listed under the "Minimum Temperature" column is either a number or a letter (A, B, C, or D). If

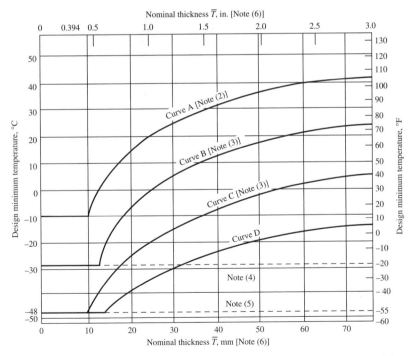

Figure 4-3. Minimum temperatures without impact testing for carbon steel materials.

a number is listed, then there is no need to use the curves given in
Figure 4-3; that number is the MDMT.

The piping code gives no temperature reduction curve, such as Figure
UCS 66.1 in ASME. The intent for curves in the B31.3 code is for *new*
piping. For existing piping, the API 579 must be followed.

In piping systems, unlike pressure vessels where internal pressure
may be the dominant stress, Charpy testing to a lower temperature
is only necessary if one anticipates the CET to drop. As seen above,
as the temperature decreases the thermal stresses normally increase
in piping systems. Thus if the CET is expected to drop a piping
stress analysis is required to assess the colder temperature.

Note that, as mentioned previously, the CET is defined as the lowest
metal temperature at which a component will be subject to either 30% of the
MAWP or a combined total longitudinal stress equal to 55.2 MPa (8 ksi)
due to pressure, weight effects, and displacement strains. Because the internal pressure stress is well below the 8 ksi limit and the total combined longitudinal stress due to pressure, weight effects, and displacement strains
increases because of the thermal effects, the CET must be equal to or greater
than the limit of −0°F. *The CET should be higher than the MAT. If the CET
is lower than the MAT, then one is vulnerable to brittle fracture. The ASME
criterion of the 8 ksi (55.2 MPa) is very practical for existing piping systems,
where the longitudinal stresses can be determined. For new piping systems,
it is usually more expensive to limit the flexibility stresses (i.e., making the
system more flexible) than to provide adequate toughness in the material at
the lower temperature. It is recommended practice in all piping systems that
the pipe material meet the toughness requirements in ASME B31.3 (or an
equivalent piping design code if that code contains material toughness
requirements) or possess a CET equal to or warmer than −20°F (−29°C).*
Low-alloy steel (e.g., 2-1/4 Cr-Mo) and ferritic steels (e.g., Type 430
stainless steel) may lose ambient temperature ductility if exposed to
temperatures above 750°F (400°C) for long periods of time. These piping
systems may require special consideration if a hydrostatic test is required
because the temperature of the water could be critical. Also if piping
made of these materials is exposed to low-temperature pressurization,
special precautions may be required.

It should be noted that throughout MAT work it is assumed that equipment is designed to ASME B31.3 or B31.1 and hence the "design-
by-rule" concepts of the code with its inherent safety factors apply. With
this assumption, the brittle fracture prevention concepts of ASME can be
used directly (with some conservative modifications from Chapter 3) in
such a way that the normal effects of safety valve accumulation, piping,

and other attachment loads will be considered. Any deviations from these assumptions need special evaluation. Technically, as of this writing, the pipeline ASME codes B31.4 and 31.8 do not yet fall under API 579, although this is expected to change.

That the material would be exempt from impact test above this threshold of $-20°F$ ($-29°C$) implies adequate inherent toughness (code minimum level); conversely, the temperature may be regarded as a lower limit without impact testing. It is worth noting that curve A in **Figure 4-3** is deemed conservative and is suitable also to assess equipment that may have suffered environmental attack during its life. (An exception to this is temper embrittlement.) This fact is useful because it allows a lower bound assessment in cases where properties or conditions are uncertain.

Note that only curves A and D in **Figure 4-3** have a calculated basis. Curve B is the one under scrutiny. These may be revised with later editions of the code and API 579.

Example 4-1: Determining the Basic MAT and Constructing the MAT Curve

The concept of MAT can readily be demonstrated by an example. Consider a pipe that has a possible minimum temperature of $-50°F$, based on process data, with the following data:

Design code ASME B31.3, 20 in. ϕ pipe, STD wall (0.375 in.), API 5L Gr B, $P = 200$ psig, $T = 30°F$
seamless pipe (mill tolerance of 12.5%)
$t = 0.375(1.0 - 0.125) = 0.328$ in.
OD $= 20.0$ in.; $t = 0.375$ in.; $CA = 0.0$ in.; $E = 1.0$
ID $= 19.25$ in.; $S_{all} = 20,000$ psi

where CA = corrosion allowance, in.
$\quad\quad E$ = joint efficiency (seamless pipe) $= 1.0$
$\quad\quad$ ID = inside diameter $= 19.25$ in.
$\quad\quad P$ = pressure, psig
$\quad\quad$ OD = outside diameter, in.
$\quad\quad S_{all}$ = allowable stress of shell material (API 5L Gr B), at ambient conditions (cold temperature) and hot temperature, psi
$\quad\quad t$ = vessel shell wall thickness (adjusted with 12.5% mill tolerance), in.
$\quad\quad t = 0.375(1.0 - 0.125) = 0.328$ in.

From the API 579 Paragraph 3.1.3.2, the maximum pressure for the CET is 0.3(MAWP) or a pressure that results in a membrane stress of 8 ksi.

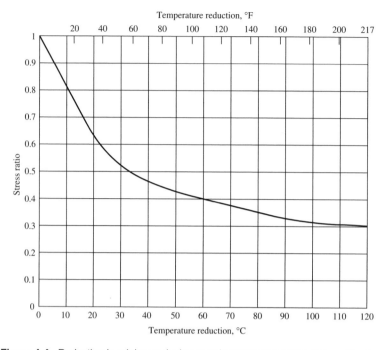

Figure 4-4. Reduction in minimum design metal temperature without impact testing.

This rule is augmented by the ASME B31.3 Figures 323.2.2A (**Figure 4-3**) and 323.2.2B (**Figure 4-4**), the latter of which has the maximum temperature reduction credit extending down to 30% of the allowable stress.

We assume here that the 200 psig is equal to or less than the MAWP of the pipe. Thus,

$$P = \frac{S_{all}Et}{R_i + 0.6t}$$

$$P = \frac{(20000)(1)(0.328)}{\left(\dfrac{19.25}{2}\right) + 0.6(0.328)}$$

$$P = 667.90 \text{ psi}$$

$$P_{RF} = \text{risk-free pressure} = (0.3)(667.90) = 200.4 \text{ psig}$$

The corresponding pressure stress is as follows:

$$S = \frac{P(R_i + 0.6t)}{Et}$$

$$S = \frac{(200)\left[\left(\dfrac{19.25}{2}\right) + 0.6(0.328)\right]}{(1)(0.328)}$$

$$S = 5988.9 \text{ psi}$$

The ratio of the operating stress (assuming pressure is the only acting load) to the allowable stress ratio (S_R) (see **Figure 4-3**) is

$$S_R = \frac{5988.9}{20000} = 0.2994 < 0.3$$

From Table A-1 in ASME B31.3, Figure 323.2.2A, the minimum temperature is seen to be on Curve B in **Figure 4-3** as $-20°F$ without impact testing.

Now from the ASME B31.3, Table 323.2.2, Note 3, impact testing is not required if the design minimum temperature is below $-29°C$ $(-20°F)$ but at or above $-104°C$ $(-155°F)$ and the stress ratio defined in Figure 323.2.2B does not exceed 0.3 times S. Use of stress identification factors is not required when making these calculations. In this case, the combined longitudinal stress due to pressure, dead weight, and displacement strain governs.

The process conditions are as follows:

T, °F	P, psig
30	200
10	150
0	125
-10	110
-20	100
-30	50
-40	
-50	20

where S_1 = dead weight stress + pressure stress + thermal stress, psi

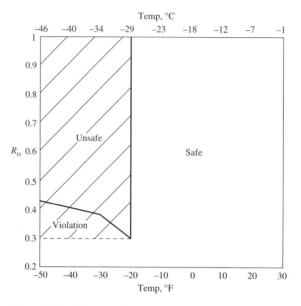

Figure 4-5. MAT curve for the process pipe being assessed.

The pipe stress run reveals the following results:

T, °F	S_1, psi
30	5243
10	5952
0	6080
−10	6233
−20	7421
−30	7628
−40	8075
−50	8608

The stress ratio (S_R) in **Figure 4-4** (Figure 323.2.2B in ASME B31.3) is $0.3S$ before the temperature reaches 0°F. Since the material is Curve B in Figure 323.2.2A, and the minimum temperature is −20°F (−30°C) without Charpy impact testing, Charpy impact testing is required below this threshold when S_R increases. This is because as the temperature decreases, the pipe wall gets colder, and the thermal stresses increase. The opposite is usually true with pressure vessels when internal pressure is the dominant stress and the S_R continues to drop with decreasing temperature. The MAT curve is shown in **Figure 4-5**. If the stress ratio (S_R) stayed at 0.3, then Charpy testing could be avoided to −155°F (−104°C). We are just concerned with going down to −50°F.

Determining the MAT Using Fracture Mechanics

NOTE: Even though temperature differentials are very basic, They can be a source of confusion when working with the MAT curve. A temperature differential of 50°C, for example, must be converted in the following way:

$$C = (F - 32)\frac{5}{9} \qquad\qquad\text{Eq. 4-2}$$

$$F = \frac{9}{5}C + 32 \qquad\qquad\text{Eq. 4-3}$$

C = degrees Centigrade
F = degrees Fahrenheit

Note that

$$\frac{5}{9} = 0.5556 \text{ and } \frac{9}{5} = 1.800$$

From Eq. 4-2,

$$C = \left(\frac{5}{9}\right)F - 32\left(\frac{5}{9}\right) = \frac{5F}{9} - 17.78$$

$$\Delta C = (F_1 - 32)\left(\frac{5}{9}\right) - (F_2 - 32)\left(\frac{5}{9}\right) = \left(\frac{5}{9}\right)(F_1 - F_2)$$

$$\Delta F = \left(\frac{9}{5}\right)C_1 + 32 - \left[\left(\frac{9}{5}\right)C_2 + 32\right] = \left(\frac{9}{5}\right)(C_1 - C_2)$$

From the above value of 50°C, we can calculate a temperature differential in Fahrenheit,

$$\Delta F = \left(\frac{9}{5}\right)50°C = 90°F$$

As already mentioned, even though this calculation is very basic, there have been many questions in presentations about it.

Variations to MAT

Considering the Maximum Allowable Working Pressure

We use the term "maximum allowable working pressure" here because the API 579 uses it for all types of equipment. In ASME B31.3, Paragraph 301.2.1 states that the design pressure of each component in a piping system shall not be less than the pressure at the most severe condition of coincident internal or external pressure and temperature (minimum or maximum) expected during service except allowances for pressure and temperature variations, Paragraph 302.2.4. This last clause, which covers pressure and temperature variations, must be used with extreme caution with brittle fracture because there are no exceptions if there is a violation of the CET; in other words, if the exposed temperature is less than the CET at any time, then the system is in danger of brittle fracture.

The MAWP for the piping system is basically the *maximum allowable pressure for the weakest component* of the system—be it the piping, heads, flanges, nozzle reinforcements, or something else. In Example 4-1, we just considered the pipe shell. To find the MAWP of the piping system, we have to make calculations for all of the piping components and use the minimum value. This minimum value becomes the vessel MAWP and is the value used in the analysis. In practical terms, therefore, MAWP can be regarded as *maximum* design pressure.

Material That Is Already Impact Tested

Provided the impact tests were done to the recommendations in Chapter 3, the toughness will equal or exceed that required by the ASME code. It cannot be assumed, however, that old equipment (pre-1971) will meet today's generally tougher standards found in the Charpy V test, as original testing would most probably have been done to Charpy U or keyhole standards. However, given that the impact test requirements are satisfied, the MDMT is the same as the test temperature.

As safeguarded in Chapter 3, a Charpy test should be done for full-size specimens. In the past at the mill, full-size specimens sometimes failed the Charpy test, while the subsize (thinner) specimens passed the Charpy test but still were within the code requirements. A Charpy test in the longitudinal direction results in potential cracks along the circumferential direction. A Charpy test along the transverse axis results in potential cracks along the longitudinal axis. The cracks develop (should the Charpy test fail) perpendicular to the direction of the impact test plane. Thus, a Charpy test along the transverse axis results in the safest test. These factors are considered in Chapter 3.

Pressure Reduction

The preceding discussion illustrated the way in which reduced pressure will not in itself guarantee reduced stress in piping systems. It is easy to see that a whole locus of points can be developed for any pressure between the maximum design pressure and the 30% risk-free pressure. Unlike pressure vessels, as the temperature drops, the thermal stresses can increase. This is not necessarily true in all piping systems, but for many in-plant piping this is true. Thus, temperature credits for reduced pressure are not possible for this example, as described. Because of this, temperature credits for post-weld heat treatment and impact testing to lower temperatures than required are normally not allowed in piping system, contrary to pressure vessels and tanks.

Hydrostatic testing of piping, in general, provides benefits by blunting cracks due to local yielding of highly stressed areas. Hydrostatic testing provides a proven safe datum point for an MAT to be evaluated or grandfathering to be applied.

Charpy Exemption Pitfalls—Words of Caution

In common available carbon steels, both killed and non-killed, there are occurrences of lamination and lamellar tearing failures. These failures are the result of stresses parallel to the laminations in the metal causing the metal to laminate and lose structural integrity. Residual stresses induced in butt welds are a leading cause. This phenomenon is a result of the sulfur content in the metal. The ASTM standards for some commonly used carbon steels allow maximum sulfur content as follows:

A 516	0.04% sulfur
A 105	0.05% sulfur
A 106B	0.025% sulfur (was 0.058% until the late 1980s)

Many mills, particularly in some Third World countries, are using the maximum permitted sulfur levels. When a material is exempted from Charpy impact testing using the methodology described earlier, if the sulfur level is close or at the maximum sulfur content allowed by the ASTM specification, lamination can result in a breakdown of a material and an eventual failure. With these commonly used carbon steels, the sulfur content can vary widely. A through tensile test can detect this phenomenon. Radiography will *not* detect the laminations in a metal specimen; however, shear wave UT test will detect a lamination (normal UT compression wave test will not detect a lamination). Another test that will guarantee this problem will not happen is the Charpy impact test through the wall specimens. For more details on Charpy testing and how it is to be administered, it is strongly recommended that the Chapter 3 guidelines be applied in all specifications. The following section has additional details on problems associated with welding.

Welding

When welding carbon steel in the $-20°F$ ($-29°C$) to $40°F$ ($4°C$) temperature range, the carbon equivalent (CE) calculated using the following IIW (International Institute of Welding) equation should be less than 0.43.

$$CE = C + \left(\frac{Mn}{6}\right) + \frac{(Cr + Mo + V)}{5} + \frac{(Ni + Cu)}{15} \qquad \text{Eq. 4-4}$$

where C = carbon
   Cr = chromium
   Cu = copper
   Mn = manganese
   Mo = molybdenum
   Ni = nickel
   V = vanadium

Note that the values in Eq. 4-4 are weight percents of the constituent components.

   Low hydrogen electrodes (such as #7018) and controls should be used. Special weld-bead sequencing should also be considered.

Considerations for Design Codes Other Than ASME

All the considerations in the preceding section only strictly apply to equipment designed and built to ASME B31.3 because of the balanced assumptions relating likely stresses, minimum toughness, potential defect sizes, and normal materials of construction.

In principle, the methodology can also be applied to equipment designed to other national standards; however, this requires care and understanding of its implications. Most other codes will result in higher pressure stresses such that the 0.3S level could exceed the nominal 6 ksi, and some require less demanding toughness levels particularly for thicker sections. The degree of conservatism applied to such assessments will depend on the quality of the data available (e.g., whether impact results are available; whether new conditions will be sustained, and, if not, the frequency at which they will they occur; what the consequences of a local fracture will be; and what inspection data we have).

As a first pass at an MAT for a non-ASME vessel, a curve of the type shown above can be developed by allowing only a temperature credit after pressure stresses have fallen below the ASME maximum of 20.0 ksi. The result will be short vertical drop from the MDMT before the inclined slope occurs. In practice, the MAT will be based on judgment.

Selecting Materials and Defining Impact Requirements—New Piping and Components

The principles discussed in the preceding section are generally applied in the assessment of existing piping from which we are trying to obtain extra capacity or accommodate some previously undefined contingency.

This section addresses the philosophy that should be adopted for new equipment on the basis that future reevaluations as per above are not necessary for initial design.

The fact that material selection may not be solely dependant on low-temperature considerations should not be overlooked. In specifying new equipment, we should initially be more conservative in terms of ignoring the potential credits discussed earlier and specify materials with inherently adequate toughness. This philosophy should not, in general, prove prohibitively expensive and should result in greater freedom to cover CET uncertainties by impact testing at lower temperatures. There will be situations, particularly when a CET is less than $-50°F$ ($-46°C$), and based on contingency events, in which the application of these principles could result in significant material cost savings.

This is where the choice of a realistic CET is important not only to ensure that impact testing is done at an appropriate temperature but that the optimum material selection is made and the costs minimized.

The schematic diagram in **Figure 4-6** illustrates the typical temperature range and cost of steels. It was recently obtained from British Steel and is useful in understanding some important concepts.

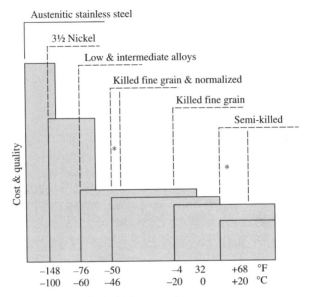

* Lower limit obtained with impact testing

Figure 4-6. Material selection for low temperature (based on a 1-in. thick plate subject to 20 ft-lb).

Use Good Quality Steel in the Base Case

Because most of our equipment has a CET at least as low as the lowest one-day mean, the use of semi-killed steels is not recommended. The cost penalty for using a higher quality grade of approximately 8% should be regarded as the base case. Clearly there will be exceptions to this, and many pieces of existing equipment are made of the lower quality steel, exempted from concern mainly by virtue of their small thickness.

Impact Test Temperature

The loosely defined boundaries from **Figure 4-6** provide guidance on how the estimated CET should or could have an "uncertainty factor" or margin applied to it. For example:

1. If the CET will be limited by the lowest one-day mean and consequently the degree of uncertainty in the CET is low, then a killed steel such as A516 without any special low temperature enhancements should be adequate possibly as low as −4°F (−20°C). This needs to

be verified by impact testing if required by the code. Additionally, the cost difference of testing to say $-20°C$ should be negligible, and so we gain an advantage from this fact.

2. If the CET is controlled by process conditions below the minimum one-day mean and the degree of uncertainty is higher (a good chance of going below $-20°C$), then steel with enhanced toughness (normalized) should be used. The cost premium might be approximately 5–6% on the basic grade; however, acceptable impact properties should be achievable down to around $-40°C$, with little additional cost.

3. If the CET is lower than $-40°F$ $(-40°C)$, a high-quality steel is required, such as the low and intermediate alloy steel (A-333 series), where $-100°F$ $(-73°C)$ can be regarded as the lower limit, at a cost premium of 6–7%. However, the availability may be more restrictive.

Therefore, the closer the CET is to the boundaries of 32°F (0°C), $-4°F$ $(-20°C)$, $-50°F$ $(-46°C)$, and $-76°F$ $(-60°C)$, the more incentive there is to have accurate CET values to avoid a material change. Similarly, the use of steels close to the boundaries could result in requirements for additional testing and the potential for retests and rejections. Therefore, the process engineer should talk to the mechanical/materials engineer(s) using data such as CET + UNCERTAINTY. Conversely, midrange CETs are not as critical, and UNCERTAINTY may be covered in the toughness flexibility by testing to the lowest adjacent boundary.

CETs below $-76°F$ $(-60°C)$ clearly require a step change in materials, and design in this region should fully involve the appropriate considerations.

Recall that both the parent plate and the welds need to be tough. This implies that welders, inspectors, and procedures need to be qualified and strictly controlled, so that there will be increased quality control requirements. The implications also extend to field maintenance where similar standards must be maintained.

Determining the CET

As previously discussed, the key to our system of brittle fracture management is the setting of the CET so that we can select a sufficiently tough material. In Chapter 3, there are requirements for proper toughness.

The Lowest Temperature Expected in Normal Operation
The normal operating temperature is the lowest temperature you can reasonably anticipate during normal operation including nondesign feeds and typical instrument malfunctions.

*Start-up, Shutdown, and Upset Conditions and Pressure Tightness
Testing*
This condition is the most challenging to the process engineer. The following is a partial list of questions you must answer when setting
the CET. Before all the questions is the phrase "How cold would it
get if . . . ."

The warming stream failed? This can easily happen if a pump stops, a
control valve shuts or the flow is diverted in another direction. One
must not forget that both the hot stream and the cold stream outlet
temperatures will equalize and reach the cold inlet temperature.

The warming stream was colder than anticipated? This condition
could exist because the pressure of condensation was lower (and
therefore the temperature was lower) or the exchanger upstream was
more effective than anticipated due perhaps to a lower flow rate.

The reboiler on the side of the process column stops functioning? This
could be due to loss of reboiling medium (refrigerant compressor
stops), failure of a control valve, or just loss of thermosyphon
reboiler on the side of the process column. The demethanizer in
steam crackers is typical of this case. You need to consider cases
where the feed stops and where the feeds continue, but the tower
dumps its contents. There is a history of assessing this scenario
incorrectly in several companies. For an illustration showing a thermosyphon reboiler on the side of a process column, the interested
reader is referred to the author's book [Reference 3].

*The system while still containing liquid is depressurized to 30% of the
design temperature?* This needs to be considered when it can happen so quickly that the operator has no time to stop it (typically
quicker than 15 minutes) or where the operator is not necessarily
aware of it).
   The draining and depressurizing of a vessel for maintenance work
is clearly a case where the system can sometimes get cold if it is
depressurized before the liquid is removed. Should this special case
be allowed to set the CET? The answer is No! The procedures will
say to remove the liquid first. The physics will encourage liquid
removal first (it will go faster if there is pressure), and the operators
are trained to drain first. The system will be isolated before draining
and the procedure will be critical from a safety concern if the system is large enough for a release to be a major concern.
   The case of an operator deliberately depressurizing a system to
below its CET in a maintenance operation is an event that need not
be considered by a site that complies with the safety procedures

practiced by most companies because it is a remote possibility and the release will be limited.

A pressure safety valve locks in the open position? Pressure safety valves do stick open with reasonable frequency. This is clearly one possible route to depressurizing the system as described previously. Once the valve has opened (which is good) and stuck open (which is bad), what happens then? Will the operator know that the pressure is falling (priority one alarm); will the operator be able to do anything about it (turn up the vapor generation source and isolate the safety valve)?

Note that if a pressure safety valve (PSV) malfunctions and remains open, immediate action is required. Normally a share PSV is used and the malfunctioning valve is brought into a maintenance shop and repaired.

Cold liquid is introduced to the system during start-up? One way to do this is to put cold liquid into a nitrogen-filled system where it will evaporate at very low temperatures. If you cannot find a way to pressurize with the vapor of the cold liquid, then there may be a problem unless this is done extremely slowly. The scenario in **Figure 4-7** is an example of this case where brittle fracture occurred.

Figure 4-7a. View of a nitrogen gas line accidentally loaded with liquid nitrogen.

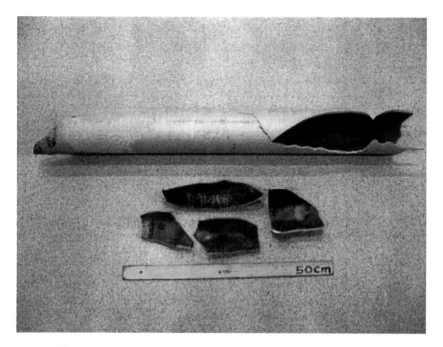

Figure 4-7b. A closer view of the remaining pieces of the nitrogen line.

Repressure some equipment that has become cold by depressurization?
The CET is only critical when there is a stress above 30% of design
pressure, except under the special considerations associated with piping. As shown in Example 4-1, thermal stresses usually increase with
a decrease in temperature. This pressure does not need to come from
vapor pressure; it can well come from imposed pressure such as
nitrogen or a process gas. You must ensure that operators cannot
pressurize a stream that is at less than 30% of design pressure while
it is below the CET. If they can, and you are not confident it can be
managed, then this will set the CET. One obvious concern is that a
quick restart after an emergency trip that would require repressuriz-
ing will give a high likelihood of this occurrence.

*Cold liquid falls the wrong way into the system rather than mixing
with the warm gas?* Designers should not allow this occurrence to
set the CET. They should be aware of it and ensure that the follow-
up prevents this happening by use of seal legs.

A flowing stream chills down by depressurizing? The depressurization
could occur at a pressure safety valve or at a control valve or at a

manual block valve. It will normally be shown on the heat and mass balance for process streams, but it also occurs in nonprocess streams (e.g., blow-downs, jump-overs, safety valve releases). It is particularly severe if the usual vapor let-down (with minor Joules Thompson cooling) is replaced by liquid let-down (with significant cooling). This often happens when vapor is carried over in a separator that is pushed above design or when a separator vessel overfills. It can also occur when liquid condenses in the "vapor" stream (at cold ambient conditions) immediately upstream of a pressure let-down device.

Shock Chilling

Shock chilling is clearly defined. It is the lowest temperature of liquid that can be introduced into the system under the closely defined conditions that follow:

1. The liquid causing the shock must be at a temperature less than $-20°F$ ($-29°C$).
2. The liquid must be $100°F$ ($56°C$) below the temperature of the system immediately before the shock.

With shock chilling, the thermal stress sets up transients effects that are potentially significant such that the liquid temperature is the CET irrespective of pressure. See Example 4-2, which concerns thermal transient stresses with shock chilling.

If the temperature is $-18.4°F$ ($-28°C$) or the shock only $97.2°F$ ($54°C$) below the pipe wall temperature, the shock chilling rule does not apply based on accepted practice. If, however, the situation is close to the limit, caution should be taken. *Note that failures induced by high thermal transient stresses are more common in piping systems. Pressure vessels, exchangers, or tanks are less likely to fail by high thermal transient stresses than piping systems.* See the following discussion about transient heat transfer.

Hydrostatic Test Temperature Minus 10.8°F (6°C) for a 2 in. or Thinner Pipe

If the hydrostatic test conditions set the CET, which is most unlikely, then this too can be alleviated by using warm water. Brittle failures during hydrostatic tests have resulted in fatalities, so it is not just an expensive embarrassment if this goes wrong. Recall that Chapter 3 requires extra toughness if you propose to pneumatically test, reflecting the additional risk. *Note that most brittle fractures occur during hydrostatic test.*

Should all propylene piping be fabricated from "killed carbon steel" to cover auto refrigeration contingencies? First, there is a general confusion about the term "killed carbon steel." Refer to the earlier discussion where it was pointed out that this is not the correct definition of steel for cold service. Standard grade "pipe" such as A106 Gr B is a killed steel but only good for $-20°F$ ($-29°C$). We discussed the considerations in going below $-20°F$ ($-29°C$) and the fact that the 30% rule must be applied with caution. The question clearly alludes to the need for a material upgrade to low-temperature steel (usually A333 Gr 6) that is killed, fine-grained, and normalized, the latter properties making the difference.

The answer to the question, however, is "probably not," as discussed previously. A106 type material should be acceptable provided the line is sufficiently flexible.

Managing Potential CET Violations

For new piping, control of potential CET excursions by instrumentation should be actively discouraged and is not an acceptable substitute for proper selection of DP (design pressure), DT (design temperature) and CET.

In situations where it is necessary to supplement good design assumptions, those systems should be deemed safety critical and should be fully approved by the piping engineer.

For existing piping that cannot meet current MAT/CET criteria, it may be possible to justify a CET management system. At a minimum, this would include documented data covering the following:

1. A full detailed inspection of current conditions; a complete fitness-for-service review of piping design
2. A complete process and safety review of the system
3. A safety critical instrument system meeting the minimum safety and fire protection requirements

Cases of Brittle Fracture

When brittle fracture occurs, there are cleavage marks in the area of fracture. The fracture is dramatic, with sections of metal being removed during the fracture as shown in **Figure 4-7a** and **4-7b**. In this case, a 4 ϕ nitrogen gas line was mistakenly fed liquid nitrogen, which was considerably below the pipe's CET. Operation mistakes like this one make brittle fracture a reality.

Transient Thermal Stresses

Failures resulting from thermal transients are more common in in-plant piping than in pressure vessels, heat exchangers, and other process equipment. The reason for this is the relatively smaller metal area to dissipate the thermal energy and the larger number of structural discontinuities with valves and piping intersections (e.g., tees and elbows). Transient heat transfer solutions are often difficult to obtain with closed-form solutions. The reason for this is that the time steps required for a solution (solution time step) may be very small, and many iterations are required to converge on a solution. Several algorithms have been proposed for solving the stress in a body that is subjected to thermal transients. These algorithms involve simplistic assumptions that may be valid for only a one-dimensional case. Additional discussion of the closed-form approximation will follow. The interested reader is referred to the author's book [Reference 4] regarding a finite element assessment of transient heat transfer in a manufacturing process of roof shingles that gives a detailed description of transient heat transfer in an actual application.

In transient heat transfer, the key parameter in determining the solution time step is the Biot number, denoted by β. The Biot number is defined as follows:

$$\beta = \frac{hL}{K} \qquad\qquad \text{Eq. 4-5}$$

where h = convection coefficient, Btu/hr-ft^2-°F (W/m^2-°C)
    L = characteristic length, ft (m)
    K = thermal conductivity of the material, Btu/hr-ft-°F (W/m-°C)

Finite element software that handles transient heat transfer has time load modules that are used for both heat transfer and dynamic applications, where the solution is a function of time. The key to such a study is determining the number of time steps required to converge on a solution. If the number of time steps is incorrect, the solution will appear as a series of interconnected straight lines. A correct solution will appear as a continuous curve.

A formulation used in many successful applications is

$$\Delta\theta = \frac{1}{(100)\beta}$$

Eq. 4-6

This relationship is valid except when very low values of solution time steps are involved (e.g., 1.0×10^{-5} sec). This would involve a thermal shock application, where the transient effects are about instantaneous. In these problems, the user will have to rely on trial and error to arrive at acceptable results.

There have been several formulations proposed for estimating time steps. One is

$$\Delta\theta = 0.001\left(\frac{L^2}{\alpha_d}\right)$$

Eq. 4-7

where α_d = thermal diffusivity, ft^2/hr (m^2/hr)
L = characteristic length, ft (m)

The term for thermal diffusivity is denoted as α_d to prevent it from being confused with the coefficient of thermal expansion, denoted as α. The equation for the thermal diffusivity is as follows:

$$\alpha_d = \frac{K}{\rho C_p}$$

Eq. 4-8

where K = thermal conductivity of the material, Btu/hr-ft-°F (W/m-°C)
ρ = material density, lb$_m$/ft^3 (kg/m^3)
C_p = specific heat of material, Btu/lb$_f$-°F (J/kg-°C)

Thus, the higher the thermal diffusivity is, the more uniform the heat and temperature distributions are. If we compare low carbon steels (<0.3% C) with austenitic stainless steel at ambient temperature, the low carbon steel has a higher thermal conductivity than austenitic stainless steel, but both have about the same specific heat. Thus, heat travels slower through the austenitic stainless steel, with larger amounts of heat being absorbed throughout the wall of the component. This means larger surface-to-interior temperature differentials and slower heating of the interior.

In some software packages, the solution time step is found by using the following expression:

$$\Delta\theta = \frac{\theta_{ss}}{10} = \frac{\left(\dfrac{\alpha t}{L}\right)}{10} = \frac{\alpha t}{10L^2} \qquad \text{Eq. 4-9}$$

where θ_{ss} = time step for steady state (ss = steady state)

t = time, sec

In [Reference 3], Eq. 4-6 $\Delta\theta = \dfrac{1}{(100)\beta}$ was used successfully for the solution time step, where the example involved the time-varying heat transfer in an infinite sheet. If we were to use this varying time-dependent distribution to solve for stresses in a body with superimposed boundary conditions, we would do so with each time step. The stress obtained for the various time steps would then be compared to find the maximum stress value. The algorithms used to approximate the stress magnitude take the form

$$\sigma = \frac{E\alpha\Delta T}{f(\beta)(1 - \chi\mu)} \qquad \text{Eq. 4-10}$$

where E = the modulus of elasticity of the material, psi (MPa)

α = coefficient of thermal expansion of the material, in./in.-°F (mm/mm-°C)

μ = Poisson's ratio

χ = 1 for a one-dimensional case; 2 for a two-dimensional case; 3 for a three-dimensional case, such as thick-walled shells

ΔT = temperature difference between initial temperature of component and temperature of fluid injected into the component

The function $f(\beta)$ was derived by Boley and Weiner [Reference 5] for simplistic cases. For the simple case of a flat plate of thickness $2L$ that is suddenly exposed to an ambient temperature of T_a through a boundary layer of film conductance h on each surface, as described in the example in [Reference 3], where water was being sprayed onto a hot sheet of roof composite sheet to cool it, we have

For $\beta \leq 10$,

$$f(\beta) = 1.5 + \frac{3.25}{\beta} - e^{\left(\frac{-16}{\beta}\right)} \qquad \text{Eq. 4-11}$$

For $10 \leq \beta < 1000$,

$$f(\beta) = 2.0 - e^{\left(\frac{-16}{\beta}\right)}$$

Eq. 4-12

As mentioned, Eq. 4-10 was derived by Boley and Weiner. It was adopted as an industrial standard and used in the API 579 as follows:

$$f(\beta) = 1.5 + \frac{3.25}{\beta} - 0.5e^{\left(\frac{-16}{\beta}\right)}$$

Eq. 4-13

The 0.5 term in Eq. 4-13 was a mistaken carryover from the original in the Boley and Weiner text. It doesn't make a significant difference until the Biot number exceeds 10, which is quite large for most applications. Also, the API 579 uses $c = 1.0$ for the one-dimensional case to compute the transient thermal stress in Paragraph 3.11.6 (example problem 6).

The middle term of Eq. 4-11 is called the thermal shock parameter and is most pronounced for lower values of the Biot number. The stress solved by substituting Eq. 4-11 into Eq. 4-10 is the failure stress for a cooled material. For high values of the Biot number, the thermal shock parameter becomes insignificant, and the large Biot number corresponds to a large film coefficient at the surface. The parameters—film coefficient, plate thickness, and the material conductivity—do not alter the value of D_t in Eq. 4-11 but do influence the magnitude of the Biot number.

Thermal transients that occur at very small time steps (e.g., thermal shock) generate maximum stresses at the surface of the body. These stresses are tensile on the surface if the process involves cooling and compressive on the surface if the process involves heating. For brittle materials that are weak in tension, surface cracks can result in rapid cooling. For rapid heating of such materials, tensile stresses develop in the interior to counter the compressive stresses on the surface. This process can result in internal cracks.

For thermal transients, it is hard to generalize the heat flux varying with time with a closed-form algorithm. Normally equations like Eq. 4-11 are valid only for one-dimensional slabs or two-dimensional bodies. For complex geometries, such as a circular nozzle on a cylindrical shell, a finite element solution gives a more realistic solution and is the best option.

Example 4-2: Thermal Transients in a Pressure Relief Piping System

A safety valve on a 6 in. ϕ Schedule 40 API 5L Grade B liquid ethylene header relieves into a blow-down drum that operates at 5 psig

(34.5 KPa$_g$), which equals 19.7 psia, or 1.34 Bars absolute, shown in **Figure 4-8**. The flashing ethylene will be at $-140.8°F$ $(-96°C)$. Find the CET and the stresses resulting from thermal transient.

A pressure safety valve relieves at a very high velocity, sometimes approaching sonic speed. Consequently, the film coefficient inside the pipe is very high. The process group simulated the fluid flow and computed an approximate film coefficient of 7826 Btu/hr-ft²-°F. The thermal conductivity of the pipe is 30.0 Btu/hr-ft-°F. Calculating the Biot number, we have the following:

$$\beta = \frac{hL}{K} = \frac{(7826)\left(\dfrac{\text{Btu}}{\text{hr-ft}^2\text{-}°F}\right)(0.023)\ \text{ft}}{30\left(\dfrac{\text{Btu}}{\text{hr-ft-}°F}\right)} = 6.0$$

where L = 0.280 in. pipe wall = 0.023 ft.

Since $\beta < 10$, from Eq. 4-13 we have

$$f(\beta) = 1.5 + \frac{3.25}{6} - e^{\left(\frac{-16}{6}\right)} = 1.972$$

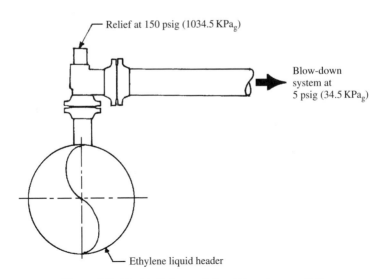

Relief at 150 psig (1034.5 KPa$_g$)

Blow-down system at 5 psig (34.5 KPa$_g$)

Ethylene liquid header

Figure 4-8. Liquid ethylene relief and blow-down system.

From Eq. 4-10 using the following material properties (a), the thermal transient stress is as follows:

mean coefficient of thermal expansion as $a = 6.25 \times 10^{-6}$ in./in.-°F
modulus of elasticity $= 29 \times 10^6$ psi

With 100°F as ambient temperature, $\Delta T = (100 - (-142)) = 240.8$°F

$$\sigma = \frac{(30 \times 10^6)\left(\dfrac{\text{lb}_f}{\text{in.}^2}\right)(6.50 \times 10^{-6})\left(\dfrac{\text{in.}}{\text{in.-}°F}\right)(240.8)°F}{(1.972)\left(1 - \dfrac{1}{3}\right)}$$

$$= 34{,}521.4 \text{ psi}$$

Now considering the stress concentration factor of 1.3 in the assessment, the thermal transient stress becomes

$$\sigma = 44{,}877.8 \text{ psi}$$

The CET will be set at the flashing ethylene, shown in **Figure 4-9** in the vapor pressure curve, using the 1.34 bars absolute at 142°F.

This case illustrates shock chilling of the discharge line because the liquid is colder than -20°F, and the differential is more than 100°F, with the ambient temperature at 100°F. Thus, the effective CET in this case is -142°F. In practice, the CET specified would be -150°F because a material that is good for -142°F would also be good for -150°F. This would alleviate concerns that the blow-down pressure would be lower than 5 psig. However, with an allowable differential of 100°F, the CET would be 0°F, giving a thermal stress of 18,637 psi using Eq. 4-10. The maximum stress is a through thickness bending stress with tension on the inside surface. The resultant transient stress is considered to be a primary stress, and for further conservatism, it is categorized into equal membrane and bending components. Thus, with the stress from internal pressure negligible at 5 psig, the stresses for $\Delta T = 100$°F (37.8°C) are as follows:

$$\text{primary membrane stress} = P_m = \left(\frac{18{,}637}{2}\right)(1.3) = 12{,}114 \text{ psi}$$

Similarly,

$$\text{primary bending stress} = P_b = 12{,}114 \text{ psi}$$

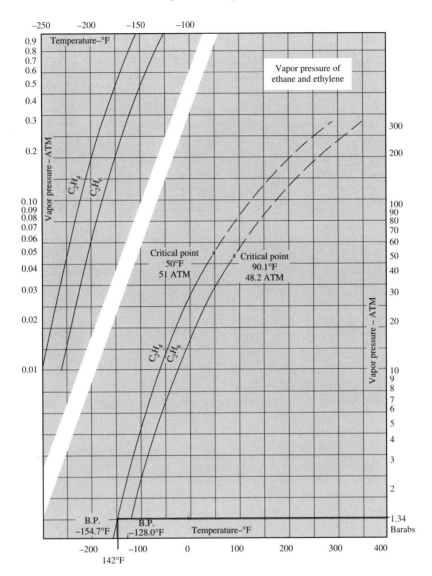

Figure 4-9. Vapor pressure curve for ethylene. Courtesy of Robert E. Krieger Publishing Company, Inc.

The allowable stress for the material is 20,000 psi at ambient relieving condition. The solution is to avoid shock chilling and to change the relieving conditions to comply with the $\Delta T = 100°F$ (37.8°C) criterion.

References

1. API 579, *Fitness-For-Service*, 2000, American Petroleum Institute.
2. ASME B31.3, 2004, American Society of Mechanical Engineers.
3. A. Keith Escoe, *Mechanical Design of Process Systems*, Vol. 2, 2nd edition, 1995, Gulf Publishing Company, Houston, Texas, p. 97.
4. A. Keith Escoe, *Mechanical Design of Process Systems*, Vol. 1, Gulf Publishing Company, Houston, Texas, pp. 148–152, Appendix F.
5. Bruno A. Boley and Jerome H. Weiner, *Theory of Thermal Stresses*, 1960, John Wiley and Sons, New York.

Chapter Five
Piping Support Systems for Process Plants

The support scheme of a piping system is critical to its function and the equipment it is connected to. When equipment and piping increase in temperature, they expand; likewise, for cold temperatures, they contract. We have outlined the various failure mechanisms in piping in the last two chapters. Now we will address what the plant engineer and inspector routinely work with—pipe supports.

Spring Supports

Spring supports sustain a pipe that has undergone displacement. Simple supports are no longer useful if the pipe rises off and loads are transferred to other supports or fragile equipment nozzles. To ensure support for the pipe as it moves, a support to compensate movement is desired. One very practical device for this is the spring.

Springs come in two basic types—variable springs and constant springs. The *variable spring*, which is by far the most common, provides loading to a pipe at a constant spring rate, lb/in. (N/mm), but the amount of force required to compress the spring varies with the amount of compression– hence the name variable spring. The *constant spring* is a spring that will provide the same spring rate for any force great enough to cause initial displacement. Constant springs are used in critical installations where loadings or displacements induced on or by the piping system are critical. We will go into more detail later in the chapter.

237

Variable Springs

Variable springs are used where a variation in piping loads can be tolerated. They are easier to adjust and more forgiving than constant springs, which require more precision. They are also much cheaper than constant springs and should be preferred in any piping design. As an example, consider the configuration shown in **Figure 5-1**.

The spring in **Figure 5-1** is attached to it with a rod and clevis. The arrangement is known as a spring hanger. As shown the spring supports the weight of the pipe and insulation. As the pipe heats up and expands it moves upward (in this example). The amount of deflection (Δ) relates to the amount of differential force transferred to the spring as

$$F_e = \Delta K, \text{ lb}_f \text{ (N)} \qquad\qquad \text{Eq. 5-1}$$

where K = spring constant of spring, lb/in. (N/mm)
Δ = deflection or displacement, in. (mm)

It is common practice to calibrate the spring in such a manner that when the piping is at its operating (hot or cold) condition, the supporting force of the spring is equal to the weight of the pipe. This means that when the maximum variation in the supporting force occurs, the pipe is at its lowest position compressing the spring. Depending on the support layout,

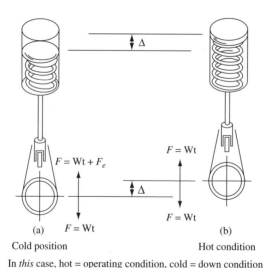

(a) Cold position

(b) Hot condition

In *this* case, hot = operating condition, cold = down condition

Figure 5-1. The "cold" and "hot" operating positions of a variable spring support.

this can occur in the design condition or when the spring is in the non-operating position ("cold" position in the case of hot operating piping). This force is

$$F = F_e + WP$$ Eq. 5-2

where WP = pipe, pipe contents, and insulation weight

    To reduce the amount of variability, it is desirable to use the smallest type of spring available that is required in the function used. The piping movements, or *deflections*, must never exceed those of the range of the spring. Typical spring sizes and ranges are shown in **Table 5-1**.

    Depending on the amount of pipe movement and loading, the spring type is selected from the table. Variable springs are typically used for pipe movements up to 2 in., but some can accommodate movements up to 5 in., as shown in **Table 5-1**. The more the pipe moves, the taller and larger the spring required will be. Variable springs are made by using a single spring, or by using two springs stacked on top each other (decreasing the spring constant by half), or by stacking three or four springs on top of each other. The spring constant decreases by the number of springs added (or basically, using a longer single spring). Usually when a pipe moves over 2 in., constant springs should be considered, depending on the application.

    The variable spring is selected using a spring manufacturer's table like that in **Table 5-1**. The engineer selects the "cold" load to be in the middle of the table. Then the operating, or "hot" load, must be in the range of the spring size selected. Variable springs typically have 14% displacement reserved for spring loading, that is, the spring must compress about 14% for the load indicator to register "zero." This indicator is shown in **Figure 5-2** where the load indicator is a plate attached to a hole made in the spring can. Similarly, there is approximately 14% of reserve displacement for excessive spring movements and the higher load end. When the spring manufacturer loads the spring into the can, a calibrated compression column compresses the spring into the can. When an operating company is not familiar with a new spring vendor, the inspectors working for the operating company need to verify that the machine has been correctly calibrated before the new springs are installed.

    As an example of sizing a variable spring, we consider the following. A computer analysis shows that a pipe at a given location in the piping system, a "node" in the computer analysis, moves 0.247 in. upward from its cold position to the hot position. The amount of load required in the cold position is 1627 lb. This is the theoretical installed load. Note that the pipe weight does not change throughout its cold to hot cycle, whereas the support load varies. The theoretical installed load is equal to the weight

Table 5-1
Typical Spring Sizes and Ranges

hanger size	0	1	2	3	4	5	6	7	8	9	10	11	12	13	14	15	16	17	18	19	20	21	22
	43	63	81	105	141	189	252	336	450	600	780	1020	1350	1800	2400	3240	4500	6000	7990	10610	14100	18750	25005
	44	66	84	109	147	197	263	350	469	625	813	1063	1406	1875	2500	3375	4688	6250	8322	11053	14688	19531	26047
	46	68	88	114	153	206	273	364	488	650	845	1105	1463	1950	2600	3510	4875	6500	8655	11495	15275	20313	27089
	48	71	91	118	159	213	284	378	506	675	878	1148	1519	2025	2700	3645	5063	6750	8987	11938	15863	21094	28131
	50	74	95	123	165	221	294	392	525	700	910	1190	1575	2100	2800	3780	5250	7000	9320	12380	16450	21875	29173
	52	76	98	127	170	228	305	406	544	725	943	1233	1631	2175	2900	3915	5438	7250	9652	12823	17038	22656	30215
	54	79	101	131	176	236	315	420	563	750	975	1275	1688	2250	3000	4050	5625	7500	9985	13265	17625	23438	31256
	56	81	105	136	182	244	326	434	581	775	1008	1318	1744	2325	3100	4185	5813	7750	10317	13708	18213	24219	32298
	58	84	108	140	188	252	336	448	600	800	1040	1360	1800	2400	3200	4320	6000	8000	10650	14150	18800	25000	33340
	59	87	111	144	194	260	347	462	619	825	1073	1403	1856	2475	3300	4455	6188	8250	10982	14592	19388	25781	34382
	61	89	115	149	200	268	357	476	638	850	1105	1445	1913	2550	3400	4590	6375	8500	11315	15035	19975	26563	35424
	63	92	118	153	206	276	368	490	656	875	1138	1488	1969	2625	3500	4725	6563	8750	11647	15477	20563	27344	36466
	65	95	122	158	212	284	378	504	675	900	1170	1530	2025	2700	3600	4860	6750	9000	11980	15920	21150	28125	37508
	67	97	125	162	217	291	389	518	694	925	1203	1573	2081	2775	3700	4995	6938	9250	12312	16362	21738	28906	38549
	69	100	128	166	223	299	399	532	713	950	1235	1615	2138	2850	3800	5130	7125	9500	12645	16805	22325	29688	39591
	71	102	132	171	229	307	410	546	731	975	1268	1658	2194	2925	3900	5265	7313	9750	12977	17247	22913	30469	40633
	73	105	135	175	235	315	420	560	750	1000	1300	1700	2250	3000	4000	5400	7500	10000	13310	17690	23500	31250	41675
	74	108	138	179	241	323	431	574	769	1025	1333	1743	2306	3075	4100	5535	7688	10250	13642	18132	24088	32031	42717
	76	110	142	184	247	331	441	588	788	1050	1365	1785	2363	3150	4200	5670	7875	10500	13975	18575	24675	32813	43759
	78	113	145	188	253	339	452	602	806	1075	1398	1828	2419	3225	4300	5805	8063	10750	14307	19017	25263	33594	44801
	80	116	149	193	258	347	462	616	825	1100	1430	1870	2475	3300	4400	5940	8250	11000	14640	19460	25850	34375	45843
	82	118	152	197	264	354	473	630	844	1125	1463	1913	2531	3375	4500	6075	8438	11250	14972	19902	26438	35156	46885
	84	121	155	201	270	362	483	644	863	1150	1495	1955	2588	3450	4600	6210	8625	11500	15305	20345	27025	35938	47926
	86	123	159	206	276	370	494	658	881	1175	1528	1998	2644	3525	4700	6345	8813	11750	15637	20787	27613	36719	48968
	88	126	162	210	282	378	504	672	900	1200	1560	2040	2700	3600	4800	6480	9000	12000	15970	21230	28200	37500	50010
	90	129	165	214	288	386	515	686	919	1225	1593	2083	2756	3675	4900	6615	9188	12250	16302	21672	28788	38281	51052
	91	131	169	219	294	394	525	700	938	1250	1625	2125	2813	3750	5000	6750	9375	12500	16635	22115	29375	39063	52094
	93	134	172	223	300	402	536	714	956	1275	1658	2168	2869	3825	5100	6885	9563	12750	16967	22557	29963	39844	53136
	95	137	176	228	306	410	546	728	975	1300	1690	2210	2925	3900	5200	7020	9750	13000	17300	23000	30550	40625	54178
spring scale — lb. per in.	30	42	54	70	94	126	168	224	300	400	520	680	900	1200	1600	2160	3000	4000	5320	7080	9400	12500	16670
	15	21	27	35	47	63	84	112	150	200	260	340	450	600	800	1080	1500	2000	2660	3540	4700	6250	8335
	7	10	13	17	23	31	42	56	75	100	130	170	225	300	400	540	750	1000	1330	1770	2350	3125	4167

Working range, in.

fig. 98	fig. B-268	fig. 82
0	0	0
1	1/2	1/4
2	1	1/2
3	1 1/2	3/4
4	2	1
5	2 1/2	1 1/4

Spring deflection, in.

fig. 82	fig. B-268	fig. 98
0	0	0
1/4	1/2	1
1/2	1	2
3/4	1 1/2	3
1	2	4
1 1/4	2 1/2	5
1 1/2	3	6
1 3/4	3 1/2	7

Typical travel scale

Figure 5-2. Travel scale that is mounted on spring supports.

of the pipe. Thus, the hot load on the spring hanger from the computer analysis is 1459 lb. We refer to **Table 5-1** and see that a Size 11 spring is required, as 1627 lb is about midway down the spring load scale. Because the movement is 0.247 in., we can use a Fig. 82 spring—a single spring. As shown in **Table 5-1**, there is a movement of 0.247 in. from 1627 lb to 1459 lb (also, (1627–1459)/680 = 0.247 in.). With a spring constant of 680 lb/in., the spring has a variation of 0.247 in. × (680) lb/in. = 167.96 lb. The spring variation, expressed as a percentage, is 167.96/1459 × 100 = 11.51% < 25%. The accepted rule is to limit the variability to 25% for critical systems in process plants. In some applications, the use of an upper limit of variability is 10% for very critical piping systems. This is rather uncommon in most situations in the process industries. Variability is defined as the absolute value of the hot load minus the cold load divided by the hot load times 100 to give the percent of variability. One wants to limit the variability. The reason for this is that a considerable amount of supporting force change would occur if a variable spring was used when the variability exceeded 25%. When this 25% variability rule can't be met, then a constant spring needs to be considered. Some computer programs that use piping flexibility (pipe stress) software allow the user to specify the 25% limit for variable springs.

The next step one would take would be to refer to the spring manufacturer's catalog to obtain the physical dimensions of the spring. *Note that it is good practice to select a hanger—both variable and constant— having a total travel rating somewhat larger than that which is expected. In those cases where travel cannot be accurately calculated, a more generous over travel allowable should be included.* This strategy compensates for process excursions and upsets.

Springs should be sized using computer analysis. These days, computers are in ubiquitous use all over the world. Springs not appropriately sized can result in rigid hangers when the spring bottoms out and compresses its maximum amount, or when there isn't enough force to compress the spring. Normally, this can happen when a new piping system is being installed and either the springs are not sized correctly, or the wrong spring is placed in the wrong location. Both events have happened and do happen, and the plant engineer must act immediately to remedy the situation. Such events can be quite traumatic, so springs must be treated very seriously by all parties.

We have mentioned spring hangers. There are times when springs are mounted below the pipe, usually called Type F springs. These springs have a load flange on top. If the pipe moves more than $\frac{1}{4}$ in. (6.35 mm), then rollers need to be placed on top of the load flange. With this kind of movement, guided load columns are placed inside the spring can to prevent the spring from jamming. This is shown in **Figure 5-3**.

Occasionally springs are used as moment-resisting devices, as shown in **Figure 5-4**. In such an application, the spring preloads the pipe in a specific direction. As the pipe expands or contracts, the spring counters the force created by the movement and, thus, reduces the movement at an end connection, such as a nozzle. Such a system in normal practice

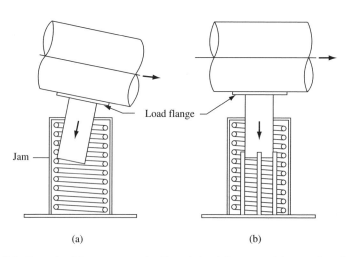

(a) (b)

Figure 5-3. Enough piping movement will cock load flange and jam spring. Note that arrows indicate direction of movement. The guide load column shown here in (B) will prevent the situation in (A). There are various designs for guide load columns, but for pipe movement greater than $\frac{1}{4}$ in. (6.4 mm), one should consider a column with rollers or Teflon on top of the load flange.

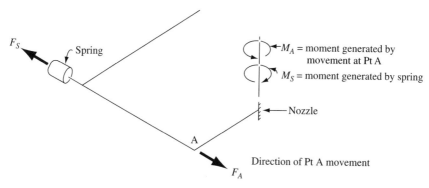

Figure 5-4. Utilizing a spring to counter a moment generated by piping is appropriate only when the spring movement does not result in the nozzle being overstressed in the down condition and no movement at point A. This condition is required after the operating condition is met.

usually works in the operating mode, but when the system shuts down, the spring can overload the nozzle. Thus, if such a scheme is used, care must be used to ensure that the protected items are safe in both the operating and shutdown conditions.

Constant Springs

Constant springs provide constant supporting force for the pipe throughout its full range of movement. As shown in **Figure 5-5**, the constant support mechanism consists of a helical coil spring working in conjunction with a bell crank lever in such a manner that the spring force times its distance to the lever pivot is always equal to the pipe load times its distance to the lower pivot. Thus, the constant spring is used where it is not desirable for piping loads to be transferred to connecting equipment or other supports, although this is often not achieved even under the most ideal circumstances.

Constant springs are mostly used where large displacements are encountered, such as very high temperature piping. Displacements exceeding 2 in. justify their consideration, but they are not used as much in the process industries as in the power industry. They are also less forgiving than variable springs to mistakes made in design or installation.

Most constant spring designs operate on the principle of the offset-slider crank mechanism; even though the pivot doesn't "crank" or rotate 360 degrees, the basics are the same. The basic principle involves counter-moment arms. As shown in **Figure 5-5**, the lever arm moves from the maximum to a lower position (crank). The spring decreases in

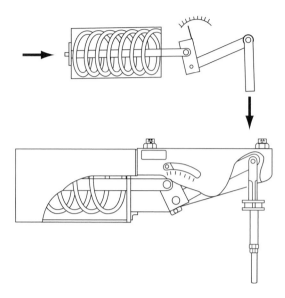

Figure 5-5. Schematic of a constant spring. A constant spring support provides constant support loading in critical situations.

length by increasing the counter-force acting opposite to the lever arm. This balancing action allows for the lever arm to move with a constant load.

The constant spring is more accurately perceived as a piece of machinery than a structural support. The spring itself must be fabricated to a closer tolerance (<6%) than a spring used for a variable hanger (±15%). Also, the pivoting mechanism operates more efficiently with bearings machined onto the pivot rod, eliminating washers to contain the bearings. Elimination of these washers reduces the distance between lug plates shown in **Figure 5-5** and results in misalignment that can cause additional friction and wear—just like an automobile engine. Thus, the closer tolerances and machining required with the offset slider-crank mechanism qualify the constant spring support as a piece of machinery.

In constant spring hangers, all spring coils are designed to have the same amount of travel for all sizes of spring hangers—normally about 2.0 in. (51 mm). The total vertical movement is also held constant for all sizes—normally about 5/8 in. (16 mm). Also, the coils normally have the same outside diameter in hangers that utilize single springs. A travel scale on the rear of the spring is available for 10% travel adjustment. All springs are designed such that the 10% adjustment is obtained by moving the spring by 5/16 in. (8 mm) in either direction.

Usually constant vertical hangers require spring coils that are a different size than the coils required for horizontal hangers because of differing amounts of travel. In either case, the springs can be calibrated by the spring coil manufacturer (the one supplying the spring coils to the spring support manufacturer). Because of the extra machining and closer tolerances, constant spring hangers are necessarily more expensive than variable springs.

The mechanics of constant springs built to the offset-slider crank mechanism is shown in **Figure 5-6**. This figure shows a rod of fixed length (L). The rod is connected at point A to D. The fixed pivot piece, AOB, rotates about point O. Point A rotates to point A$'$, and point B rotates to point B$'$. The vertical distance from point B to B$'$ is the amount of travel (T)—that is, the amount of movement the spring is expected to move. Behind the spring is a round plate that compresses the spring as the rod, or crank length (L), moving from DA to D$'$A$'$. This distance is ΔX. The distance from point O to point A and from point O to point A$'$ is P, the pivot length. Thus, the spring moves in the direction along the x-axis ΔX to produce a pivot about point O of a length P to produce a vertical travel T. The term θ is the angle height, which is the distance from the can centerline to the pivot point O. The term a_1 is the initial moment arm perpendicular to the rod length (L) extending from the

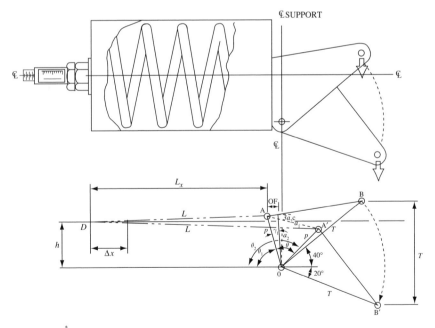

Figure 5-6. The mechanics of the offset-slider crank mechanism common in many constant spring designs.

points D to A, and the term a_2 is the final moment arm perpendicular to the rod length (L) extending from points D$'$ to A$'$. The load on the spring is W. The term ω_1 is the initial crank angle—the angle the length of the rod (L) initially makes with the x-axis; and ω_2 is the final crank angle—the angle the length of the rod (L) makes with the x-axis extending from point D$'$ to A$'$. The chord length (C) is the line that extends from point A to point A$'$. The dimension OF is the offset eccentricity that is usually a constant value for each design. The equations for the mechanism can be written as follows:

$$\Delta X = P(\cos \theta_1 - \cos \theta_2) + \sqrt{L^2 - (-P \sin \theta_1 + h)^2}$$
$$- \sqrt{L^2 - (-P \sin \theta_2 + h)^2} \qquad \text{Eq. 5-3}$$

$$\theta = \theta_2 - \theta_1 = \text{angle of travel} \qquad \text{Eq. 5-4}$$

$$\gamma_1 = \arcsin\left(\frac{OF}{P}\right) = \text{offset angle} \qquad \text{Eq. 5-4(a)}$$

$$\omega_1 = \arcsin\left[\frac{(P \sin(\theta_1) - h)}{L}\right] = \text{initial crank angle} \qquad \text{Eq. 5-5}$$

$$a_1 = P \cos(\gamma_1 - \omega_1) = \text{initial moment arm} \qquad \text{Eq. 5-6}$$

$$c = 2P \sin\left(\frac{\theta}{2}\right) = \text{chord length, in. (mm)} \qquad \text{Eq. 5-7}$$

$$x' = C \sin\left[\frac{(180 - \theta)}{2} + \gamma_1\right] = \text{horizontal chord length, in. (mm)}$$
$$\text{Eq. 5-8}$$

$$\beta_1 = \arccos\left(\frac{x'}{C}\right) = \text{chord angle} \qquad \text{Eq. 5-9}$$

$$y' = P \sin(\beta_1) = \text{vertical chord length, in. (mm)} \qquad \text{Eq. 5-10}$$

$$\omega_2 = \arcsin\left[\frac{(y' - y_1)}{L}\right] = \text{final crank angle} \qquad \text{Eq. 5-11}$$

$$a_2 = P \sin(90 - \theta + \gamma_1 + \omega_2) = \text{final crank angle} \qquad \text{Eq. 5-12}$$

T = spring vertical travel, in.

$b_1 = T\cos(40) = 0.766T$ = initial lug moment arm, in. (mm)

Eq. 5-13

$b_2 = T\cos(20) = 0.94T$ = final lug moment arm, in. (mm)

Eq. 5-14

W = spring support load (pipe weight), lb_m (kg)

$$F_1 = \frac{Wb_1}{a_1} = \text{initial spring force, } lb_f \text{ (N)}$$

Eq. 5-15

$$F_2 = \frac{Wb_2}{a_2} = \text{final spring force, } lb_f \text{ (N)}$$

Eq. 5-16

K = spring constant, lb_f /in. (N/mm)

$$K = \frac{(F_2 - F_1)}{\Delta X}$$

Eq. 5-17

If the spring is a horizontal type,

$$TVRM = P(1 - \cos(\gamma_1)) + C\sin(\beta_1)$$
$$= \text{total vertical rod movement, in. (mm)}$$

Eq. 5-18

If the spring is a vertical type,

$$TVRM = 2P\sin\left(\frac{\theta}{2}\right)\sin(180 - (\theta + \theta_2))$$

Eq. 5-19

A constant spring that was designed and installed in a field application, an unusual circumstance, is an example. The spring is to carry 1700 lb_f of load on the pipe and compensate for 4.50 in. of vertical movement. A local spring coil fabricator can provide an actual spring $OD = 9'' - 1.25'' = 7.75$ in. The spring constant given is $K = 437\ lb_f$/in. There are seven coils over an initial spring length of 12.5 in. Using this spring, we use the preceding equations to design the constant hanger. Using a spreadsheet program, we have the following results:

The equations used are the same as above. We use the following values:

$$P = 3.0 \text{ in.}; \ \theta = 60°; \ OF = \frac{13}{16} \text{ in.}; \ L = 14\frac{13}{16} \text{ in.}; \ T = 4\frac{1}{2} \text{ in.}$$

$\theta_1 = 74°$; $\theta_2 = 134°$; $h = 2.50$ in.

$$\gamma 1 = ASIND\left[\frac{OF}{P}\right]$$

$$\omega 1 = ASIND\left[\frac{P \cdot SIND(\theta 1) - h}{L}\right]$$

$$a1 = P \cdot COSD(\gamma 1 - \omega 1)$$

$$C = 2 \cdot P \cdot SIND\left[\frac{\theta}{2}\right]$$

$$XPRIME = C \cdot SIND\left[\frac{180 - \theta}{2} + \gamma 1\right]$$

$$\beta 1 = ACOSD\left[\frac{XPRIME}{C}\right]$$

$$YPRIME = P \cdot SIND(\beta 1)$$

$$Y1 = L \cdot SIND(\omega 1)$$

$$\omega 2 = ASIND\left[\frac{YPRIME - Y1}{L}\right]$$

$$a2 = P \cdot SIND(90 - \theta + \gamma 1 + \omega 2)$$

$$b1 = T \cdot COSD(40)$$

$$b2 = T \cdot COSD(20)$$

$$F1 = \frac{W \cdot b1}{a1}$$

$$F2 = \frac{W \cdot b2}{a2}$$

$$\theta = \theta 2 - \theta 1$$

$$\Delta X = P \cdot (COSD(\theta 1) - COSD(\theta 2)) + \sqrt{L^2 - (-P \cdot SIND(\theta 1) + h)^2} - \sqrt{L^2 - (-P \cdot SIND(\theta 2) + h)^2}$$

$$K = \frac{F2 - F1}{\Delta X}$$

The results are shown in the following table:

Input	Name	Output	Unit	Comment
	$\gamma 1$	15.713861	Deg.	OFF SET ANGLE
.8125	OF		in.	OFF SET
3	P		in.	PIVOT RADIUS
	$\omega 1$	1.48467354	deg.	INITIAL CRANK ANGLE
74	$\theta 1$		deg.	PRESET ANGLE
2.5	h		in.	OFF SET HEIGHT
14.8125	L		in.	CRANK ROD LENGTH
	a1	2.90796078	in.	INITIAL MOMENT ARM
	C	3	in.	CHORD LENGTH
	θ	60	deg.	ANGLE OF TRAVEL
	XPRIME	2.90722637	in.	HORIZONTAL CHORD LENGTH
	$\beta 1$	14.286139	deg.	CHORD ANGLE
	YPRIME	.740293741	in.	VERTICAL CHORD LENGTH
	Y1	.383785088	in.	INITIAL CRANK LENGTH
	$\omega 2$	1.37913342	deg.	FINAL CRANK ANGLE
	a2	2.19737899	in.	FINAL MOMENT ARM
	b1	3.44719999	in.	INITIAL LUG MOMENT ARM
4.5	T		in.	SPRING VERTICAL TRAVEL
	b2	4.22861679	in.	FINAL SPRING MOMENT ARM
	F1	2015.24038	lbf	INITIAL SPRING FORCE
1700	W		lbf	SUPPORT LOAD
	F2	3271.46505	lbf	FINAL SPRING FORCE
	ΔX	2.90986273	in.	HORIZONTAL TRAVEL OF SPRING IN CAN
134	$\theta 2$		deg.	POST SET ANGLE
	K	431.712691	lbf/in.	SPRING CONSTANT
	TVRM	.852414978	in.	TOTAL VERTICAL ROD MOVEMENT

Note: The above is output from a computer program. The software allows no subscripts.

These values are very typical of those used in constant hanger design. This was and is a real constant hanger and its photograph is shown in **Figures 5-7a** and **5-7b**. The spring has been used successfully for many years.

Figure 5-7a. Fabricated constant spring tested in shop.

Figure 5-7b. Fabricated constant spring tested in field shop before installation.

You ask why an operating facility built a constant spring hanger. The answer is simple—someone or several people made mistakes. One pipe was at 1650°F and was 16 ft 6 in. high, coming out of a heat exchanger going straight up and elbowing into another connecting tower above.

One could see the cherry red of the hot pipe through one cut in the insulation. The stress engineer on the other side of the world entered 6 ft instead of 16 ft 6 in. Another problem in the same facility was even more severe. A 36 in. header at 890°C (1634°F) was designed for a 0.312 in. wall and 6 in. of insulation. Someone in the contractor's purchasing group got a "good deal" and bought 300 ft of 36 in. 304 SS pipe with 0.375 in. wall. To compound matters further, the insulation was the correct 6 in. thick, but was 14.5 lb/ft^3 in density versus the 11.0 lb/ft^3 used in the pipe stress runs. So the overweight pipe and insulation caused seven constant spring hangers to bottom out. New springs were eventually put in, but that is another story. Adjusting constant spring hangers can be a challenge. One pipe shown in **Figure 5-8** has the correct constant hangers, after incorrect size hangers were initially installed. One can see how it is very flexible and "floats," as piping at high temperatures should respond.

When constant springs are installed in series, they act like a waterbed—any slight change in one throws all the others off. So one must have a crew at each spring and make the adjustments together. For very hot services, the pipe is best to let "float" and be very flexible.

Figure 5-8a. A typical piping supported by constant spring hangers. The 36″ ϕ header is an ethyl benzene line that operates at 893°C (1639.4°F). The line has large thermal growth and "floats" with the constant springs. Adjusting the springs was like the "waterbed" phenomenon, which is described later. These springs were known as "the seven sisters." The pipe material is Incoloy.

Figure 5-8b. Insert of one of the seven constant spring hangers of the 36″ ϕ header shown in **Figure 5-8a**.

One has to be careful of wind loads, as normally big headers don't always sit on the ground. This one was 120 ft (36.58 meters) above grade, so there was plenty of wind at the time. With the prospect of waiting several months for the delivery of new springs and facing an angry management, one can build constant springs with a good machine shop, a spring coil manufacturer (or automobile shop), and common sense. However, it is preferred to purchase the springs from a reputable manufacturer. Unless all the checks and balances are in place, you may be building a constant spring hanger.

In some applications, such as below a superheated steam line below a furnace on grade, it is not uncommon for the constant spring supports below the pipe to bottom out. This is not preferred, but if the system has been in operation and has been through a start-up and shutdown and has started back up again, the springs are usually acceptable. This is not always true for all piping systems. Note that when a piping system has operated satisfactory and is supported by springs, the springs should not be reset once they have been in operation, unless one or more of the springs is dysfunctional. The pipe usually deforms over several start-up and shutdown cycles and is not the same as it was when it was new. Thus, when the system is shut down, it is best not to adjust the springs, unless there is an obvious spring misalignment or one was installed incorrectly. This rule has worked all over the world.

Another approach to spring problems is to maintain a good junk yard. Don't throw old springs away, because they can be reused again for emergencies. An example at the end of the chapter will show how.

Piping Nozzle Loads on Rotating Equipment

The problem of designing appropriate piping for rotating equipment can be challenging. Almost always the piping attached to rotating equipment is well under the allowable code stress, but the difficulty is in the nozzle loads. Rotating equipment cannot withstand the nozzle loads that a pressure vessel or heat exchanger can withstand. Standards have been developed that provide guidelines for this purpose. The American Petroleum Institute (API) has standards for rotating equipment:

API 610—Centrifugal Pumps for Petroleum, Heavy Duty Chemical, and Gas Industries
API 611—General Purpose Steam Turbines for Refinery Service
API 612—Special Purpose Steam Turbines for Refinery Service
API 617—Centrifugal Compressors for Refinery Service
API 618—Reciprocating Compressors for General Refinery Service

For turbines, the NEMA (National Electrical Manufacturers Association) SM-23 provides guidelines for nozzle loadings on turbines. We will discuss the NEMA standards later.

The reader is encouraged to review these referenced standards. The API 610 nozzle loads values have changed over the years, so the reader must refer to the one that is applicable if an older pump was designed to an earlier edition of API 610, unless it is agreed to use the current version. Often rotating equipment manufacturers give a set of allowable forces and moments for the suction and discharge nozzles, but one would be safe in using that required by the API 610. This rule holds for all the previously mentioned standards.

Pump Nozzle Loads

Oftentimes the pump nozzles may not always be subject to the maximum allowable resultant force and moment simultaneously. Thus, an increase in either the resultant applied force or the resultant applied moment may be permitted if the following limitation can be satisfied at the individual nozzle:

$$\frac{F_a}{F_r} + \frac{M_a}{M_r} \leq 2 \qquad\qquad \text{Eq. 5-20}$$

$$\frac{F_a}{F_r} \leq 2 \ \& \ \frac{M_a}{M_r} \leq C \qquad\qquad \text{Eq. 5-21}$$

where $C = 2$ for nozzles < 6 in. (152.4 mm)
$$C = \frac{(D + 6)}{D} \text{ for nozzles} \leq 8 \text{ in. (203.2 mm)}$$
D = nominal diameter of nozzle flange, in. (mm)
M_a = resultant applied moment, ft-lb (Joule = N-m)
M_r = resultant applied force, ft-lb (Joule = N-m)
F_a = resultant applied force, lb (N)
F_r = resultant allowable force, lb (N)

Basically expecting zero forces and moments on nozzles is unrealistic. Loadings can be minimized with proper piping design, but obtaining zero loads is unpractical. Typical allowable forces and moments are shown in **Table 5-2**. Another example is shown in **Figure 5-9**.

The NEMA SM-23 mentioned earlier is for nozzle loads on steam turbines. Section 8.4.6.2 states: "The combined resultants of the forces and moments on the inlet, extraction, and exhaust connections, resolved at the centerlines of the exhaust connection should not exceed the values per Limit 2." The word "centerlines" is used, not "centerline." Moreover, the example in SM-23 is unclear about the resolution of the forces and moments. When the summation is performed, the moment arms are all zero, meaning that all connections have zero distance from the point of resolution. The NEMA committee responded and said: "The forces and moment on the steam turbine connections are to be resolved about the intersection of the centerline of the turbine exhaust and the centerline of the turbine shaft." This line had been missing from the standard. Thus, one must input distances relative to the intersection point of the centerline of the discharge nozzle and the turbine shaft. When using the distances described, the forces and moments on the steam turbine connections are resolved at the intersection of the centerline of the discharge nozzle and the centerline of the turbine shaft simply using the distances from the connections to the force/moment resolution point.

The NEMA SM-23 committee has been reviewing these issues to provide clarification of the point of resolution and a better example. There are two

Table 5-2
Typical Manufacturer Allowables for Nozzle Loadings
for Inline Pumps

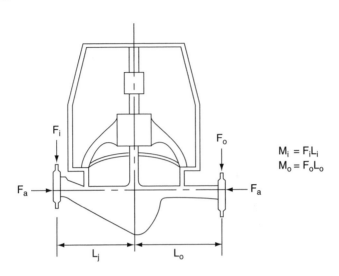

$$M_i = F_i L_i$$
$$M_o = F_o L_o$$

PUMP SIZE (in)	F_a lb	$M_{i_{max}}$ ft-lb	$M_{o_{max}}$ ft-lb
2 × 3 × 6	4000	5000	4000
3 × 4 × 6	6000	6000	5000
2 × 3 × 8	4000	5000	4000
3 × 4 × 8	5000	6000	5000
4 × 6 × 8	6000	7000	6000
4 × 6 × 10	6000	7000	6000
6 × 8 × 10	8000	9000	8000
6 × 6 × 20	5000	6000	5000
10 × 10 × 20	8000	9000	6000
12 × 12 × 20	12000	13000	10000

$$\frac{F}{F_a} + \frac{M_{i_{act}}}{M_{i_{max}}} + \frac{M_{o_{act}}}{M_{o_{max}}} < 2.0$$

Where,

F = resultant of actual force applied, lb
$M_{i_{act}}$ = actual bending moment on suction nozzle, ft-lb
$M_{o_{act}}$ = actual bending moment on discharge nozzle, ft-lb

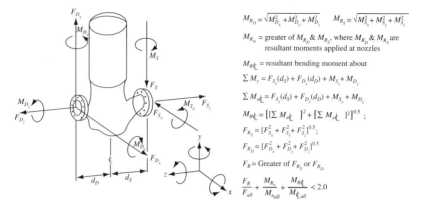

$$M_{R_D} = \sqrt{M_{D_x}^2 + M_{D_y}^2 + M_{D_z}^2} \qquad M_{R_S} = \sqrt{M_{S_x}^2 + M_{S_y}^2 + M_{S_z}^2}$$

M_{R_N} = greater of M_{R_D} & M_{R_S}, where M_{R_D} & M_{R_S} are
resultant moments applied at nozzles

$M_{R\Phi}$ = resultant bending moment about

$$\Sigma M_y = F_{S_x}(d_S) + F_{D_x}(d_D) + M_{S_y} + M_{D_y}$$

$$\Sigma M_{y\Phi} = F_{S_y}(d_S) + F_{D_y}(d_D) + M_{S_y} + M_{D_y}$$

$$M_{R\Phi} = [[\Sigma M_{x\Phi}]^2 + [\Sigma M_{y\Phi}]^2]^{0.5} ;$$

$$F_{R_S} = [F_{S_x}^2 + F_{S_y}^2 + F_{S_z}^2]^{0.5} ;$$

$$F_{R_D} = [F_{D_x}^2 + F_{D_y}^2 + F_{D_z}^2]^{0.5}$$

F_R = Greater of F_{R_S} or F_{R_D}

$$\frac{F_R}{F_{all}} + \frac{M_{R_n}}{M_{n_{all}}} + \frac{M_{R\Phi}}{M_{\Phi all}} < 2.0$$

Figure 5-9. Generalization of forces, moments, and allowable nozzle loadings.

interpretations on where the combined forces and moments should be resolved. They are as follows:

1. The face of the flange at the discharge nozzle connection. To resolve the forces and moments at the discharge nozzle connection, the distance from the discharge nozzle to each connection should be used.
2. The intersection point of the discharge nozzle centerline and the equipment shaft centerline. To resolve the forces and moments at the intersection point of the discharge nozzle and the shaft centerlines, the distance from the intersection point to each connection should be used.

Piping Layout Schemes for Rotating Equipment

The function of piping designers is an invaluable one in that they ensure the appropriate layout of any piping systems. In many operating facilities, these professionals are normally not employed, so the plant engineer must be cognizant of various classical acceptable piping arrangements, particularly in approving contractor work. **Table 5-3** categorizes preferred piping schemes for (A) pumps supported at the nozzle centerline, (B) pumps supported above nozzle centerlines, (C) pumps supported below the nozzles, and (D) pumps supported below the nozzle centerline. The piping configurations for the suction are noted as "S-1, S-2, etc." For the discharge piping, the piping configuration is noted as "D-1, D-2, etc." Even though these configurations do not guarantee satisfactory nozzle loads for all situations, they are a good start as to what a preferred piping scheme should look like at the suction and discharge nozzles. Often a spring hanger is used to support the weight of the pipe, particularly at the suction nozzle. One must use caution with rotating equipment, as piping loads can and do affect the performance of pumps, as well as all rotating equipment. Note that a piece of rotating equipment should never be used as an anchor. For example, a vessel may be considered an anchor in a pipe stress analysis, but a vessel can take higher loads than machinery, which has very small clearances between moving parts and can be easily overloaded. Although the pump nozzle is coded as an "anchor" in the pipe stress, adhering to this guideline is the conservative and safe approach to good piping design.

Arrangements with the equipment manufacturer at times are made for an agreement on the nozzle loads (e.g., twice the API 610 allowables). When this is done, extreme care should be used because there is no guarantee that even if it worked in one case it will work again in another. Following the API 610 allowables is encouraged, unless the pump vendor has designed for higher loads. Verbal guarantee should never be accepted. The vendor should

---

Table 5-3
Acceptable Piping Configurations for Pumps

A. Pumps supported at nozzle centerline

Configuration 1

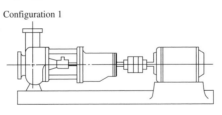

Note: P = Preferred/A = Alternative

Suction	S-3
Discharge	D-1 (P), D-2 (A), D-3 (A)

Configuration 2

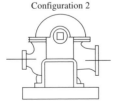 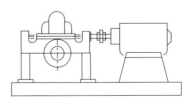

Suction	S-3
Discharge	D-4

Table 5-3
Acceptable Piping Configurations for Pumps—cont'd

B. Pumps supported above nozzle centerlines

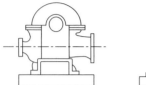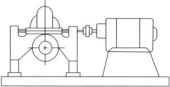

Suction	S-1
Discharge	D-5

C. Pumps supported below nozzles

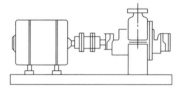

Suction	S-5 (P), S-6 (A)
Discharge	D-1 (P), D-2 (A), D-3 (A)

D. Pumps supported below nozzle centerline

Configuration 1

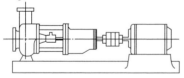

Suction	S-1, S-3
Discharge	D-1 (P), D-2 (A), D-3 (A)

Configuration 2

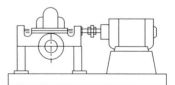

Suction	S-4 (P), S-1 (A)
Discharge	D-6 (P), D-5 (A)

Table 5-3
Acceptable Piping Configurations for Pumps—cont'd

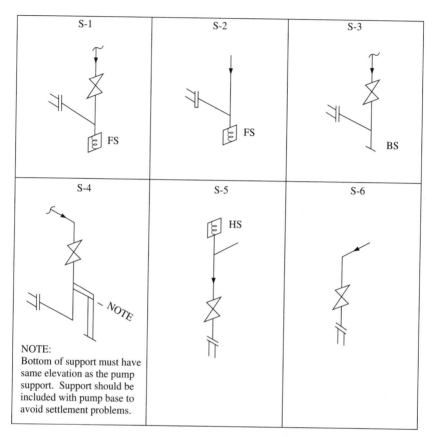

S-1	S-2	S-3
FS	FS	BS

S-4	S-5	S-6
NOTE	HS	

NOTE:
Bottom of support must have same elevation as the pump support. Support should be included with pump base to avoid settlement problems.

FS = TYPE "F" BOTTOM SUPPORT SPRING
HS = HANGER TYPE SPRING
BS = BASE SUPPORT

Table 5-3
Acceptable Piping Configurations for Pumps—cont'd

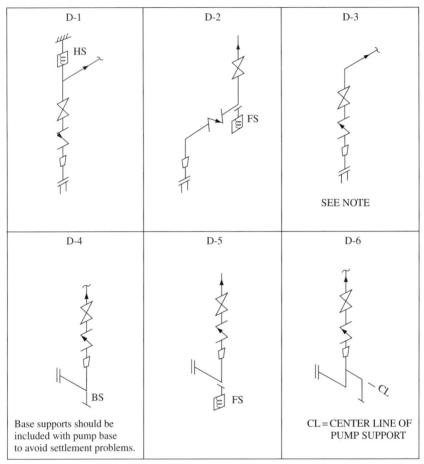

NOTE: Configuration D-3 is acceptable alternate if nozzle loads are within allowables. This
configuration will require temporary support during piping alignment or pump
removal/maintenance.

make design changes to accommodate higher nozzle loads, or trouble could be ahead. It is unusual for the vendor to make a special provision for high nozzle loads.

Compressor Nozzle Loads

Compressors are generally less forgiving than pumps about nozzle loads. The API 617 gives guidelines for centrifugal compressors for refinery service, and API 618 does the same for reciprocating compressors for general refinery service. Centrifugal compressors are much more common and are in general use for supplying gases throughout the plant. Reciprocating compressors are basically low flow with high head and are not used as much as centrifugals; nevertheless, they have unique features, such as pulsation bottles for damping out pulsations to eliminate piping resonance. They can induce a pulsation response in the attached piping systems.

The piping engineer working for a contractor should take special care in not subjecting compressors to high nozzle loads. They can be quite difficult if subjected to high nozzle loads. Shown in **Table 5-4** are the typical compressors in use. Some acceptable piping configurations, similar to

---

Table 5-4
Acceptable Piping Configuration for Compressors

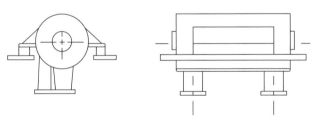

Compressor with nozzles below the centerline

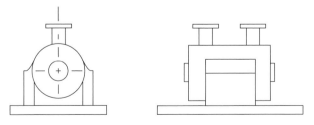

Compressor with nozzles above the centerline

---

those shown for pumps, are shown in **Figure 5-10**. Shown in **Figure 5-10a** are piping schemes for compressors with nozzles above centerlines, and in **Figure 5-10b** are piping schemes for compressors with nozzles below centerlines.

Like pumps, when layout requirements restrict the locations of spring supports and offsets from the nozzles are necessary, the spring supports should be designed to carry the entire piping dead weight. The piping

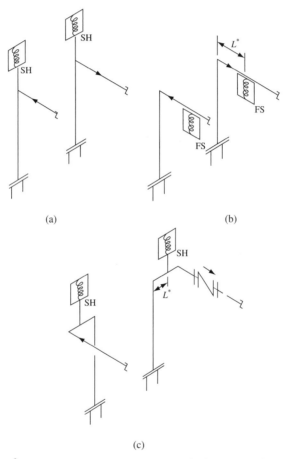

L^* – NOTE: When the layout limits the location of spring supports and an offset from the nozzle is required, like in (b) and (c), the pipe supports should be designed to carry the total dead weight of the pipe. The figure in (a) is preferred with SH (spring hangers). Type FS (Type F springs that fit below the pipe) should be avoided for systems above 400°F (204°C) or pipe displacements greater than ¼" (6.35 mm).

Figure 5-10a. Typical pipe support arrangements for compressors with nozzles above the centerline.

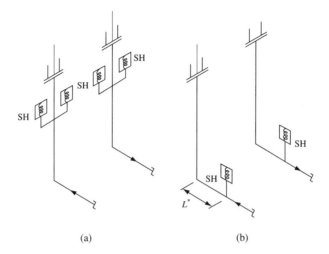

(a)   (b)

L^* – NOTE: When the layout limits the location of spring supports and an offset from the nozzle is required, like in (b), the pipe supports should be designed to carry the total dead weight of the pipe. The figure in (a) is preferred with SH (spring hangers).

Figure 5-10b. Piping support schemes for compressors with nozzles below centerlines.

should be supported such that there will be negligible dead weight deflection and rotations of the pipe flange mating to the equipment prior to bolt-up. When deflection or rotations are critical, then one may consider having the pipe flange tack welded in the shop with final fit-up and welding in the field. Ideally, spring supports should be located directly above, below, or adjacent to the equipment nozzles with additional supports, if necessary, to try to achieve "zero" dead weight loadings on the nozzles. The three acceptable methods of modeling dead weight analysis of compressor and turbine piping systems are as follows:

1. A flange may be modeled as fully restrained if the spring support is directly above or below the equipment nozzle. The vertical load on the nozzle, including the pipe flange weight, must then be added to the spring design load. All other loadings on the equipment should be negligible.
2. A flange may be modeled as restrained in all directions except the vertical with all dead weight vertical loads carried by the supports. Any loads or deflections at this flange should be negligible.
3. A flange may be modeled as a free end with all dead weight loads carried by the supports. Any deflections or rotations at this free end should be negligible.

Typically, when verifying the nozzle loads, the pipe flange at the equipment nozzle is considered to be an anchor when assessing thermal loads or during the thermal, pressure, and weight analysis to see how much loading the nozzle has. In existing systems that have been in operation for some time, when the pipe is decoupled from the equipment nozzle, it may move from the ideal location as described previously for new systems. This indicates that the piping has deformed and assumed a different configuration during operation, which is quite normal. As mentioned earlier, it is not wise to adjust the spring supports in this situation because the piping has changed. Note that it is best for the piping to be aligned in the hot, or operating, case because one wants the loadings to be less during the hot, or operation, loads. Here we are using the word "hot" as meaning operational; however, in many cases, the operating case is actually a cold service rather than a hot service. This is just to prevent any confusion for cold service piping.

In existing systems that have been in service for some time, the piping should not be expected to meet the criteria in steps 1, 2, and 3 if thermal distortion has occurred. Typically a common problem with steam turbines is making certain all the equipment supports are installed. Frequently steam turbine vendors leave out the key way under the turbine, which causes excessive nozzle loads or vibration. In this case, regardless of the piping configuration, the steam turbine will not operate properly. Thus, it is desirable for the operating company to have a representative review the steam turbine installation checklist and verify the installation steps are executed correctly. When the key way is left out, the piping is always faulted until the piping is detached and the turbine is checked. Even though this is a major inconvenience and is quite embarrassing for the equipment vendor, it has happened many times. The proper installation of rotating equipment is very serious and deserves top priority to prevent such mishaps.

Other items to check on the pipe stress analysis are as follows:

1. Piping with mission-duo check valves should be well supported because of rotational creeping of the flange face caused by long stud bolts.
2. Take care that all valves entered into the pipe stress analysis, particularly old out-of-spec valves, have the correct weight.
3. Adequate pipe supports and guides must accommodate steam blowdown or chemical cleaning of the piping.
4. When using struts for large diameter lines, the radial expansion of the pipe should be considered at the points of attachment.
5. Structural members, especially those close to the equipment nozzle, should be checked as to their stiffness. The flexural spring rate of the supporting steel must be considered in the analysis.
6. Struts should have sufficient pin-to-pin dimensions for the calculated pipe deflection in order to lessen the reaction on the equipment nozzle.

7. Take into consideration frictional loads on equipment from sliding base supports near equipment. Friction loads on directional anchors also need to be considered.
8. The effects of PSV thrusts upon the valve going off must be considered in the piping evaluation on nozzle loading.
9. The design of spring supports must include the weight of pipe flanges at equipment nozzles.
10. Variable spring support variability should be minimized as much as practical to reduce the cold loading on the rotating equipment nozzles.
11. Horizontal deflections at load flanges on F type spring supports should be minimized to prevent spring from binding and avoid friction loads. This can be avoided with use of guided load columns that fit inside the cans. If movements exceed 0.25 in. (6.35 mm), then spring hangers should be used. Normally, F type springs that fit under the pipe are not desired for piping that moves more than $\frac{1}{4}$ in. or is subjected to temperatures higher than 400°F (204°C).
12. Turbine piping analysis should include a by-pass line hot with trip and throttle valve closed (i.e., the equipment being cold), as well as all operating and upset conditions.
13. Do not use the austenitic stainless steel type of supports if horizontal deflections exceed $\frac{1}{4}$ in. because the austenitic stainless steel has low yield strength and may deflect.
14. Often the turbine does not have a trip and throttle valve that is supplied by the vendor. The valve should be supported and considered in the analysis. If the vendor does supply this support, all design data on the support (e.g., design load, spring rate, spring setting, spring capacity) should be supplied by the manufacturer.
15. If hot piping is being designed or added during an expansion, the heat transfer through the support member should be considered if Teflon slide plates are being used, as Teflon melts at 400°F (204°C).

Shown in **Table 5-5** are typical acceptable nozzle load values for turbo expanders that have been proven in practice. This table may be helpful in determining reasonable loads for this type of equipment.

Nozzle Stiffness and Elastic End Conditions

To treat a restraint with elastic end conditions, only rotations are considered significant. Deformations induced by radial force and other translations are ignored because their influence is insignificant. The use of such factors is helpful in more accurately assessing nozzle loads on pressure vessels and heat exchangers. *Note that they are not recommended for use with rotating*

Table 5-5
Reasonable Turbo Expander Nozzle Loadings

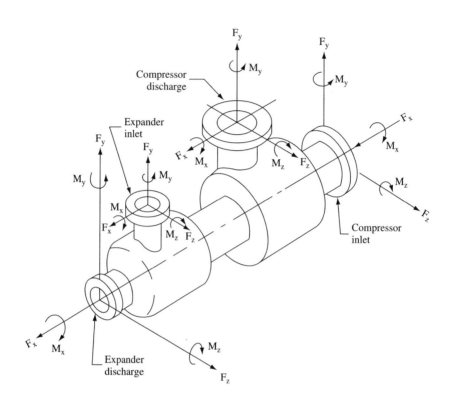

Expander Inlet

Nozzle Size (in.)	F_x	F_y	F_z	F_R	M_x	M_y	M_z	M_R
4	649	1,299	1,299	1,948	1,624	2,436	3,383	4,474
6	974	1,948	1,948	2,922	2,436	3,654	5,074	6,710
8	1,299	2,597	2,597	3,896	3,247	4,870	6,764	8,947
10	1,623	3,246	3,246	4,869	4,059	6,088	8,455	11,184
12	1,948	3,895	3,895	5,843	4,871	7,306	10,146	13,421
14	2,272	4,545	4,545	6,817	5,683	8,524	11,838	15,658
16	2,595	5,189	5,189	7,784	6,486	9,730	13,513	17,870

Table 5-5
Reasonable Turbo Expander Nozzle Loadings—cont'd

Expander Discharge

Nozzle Size (in.)	F_x	F_y	F_z	F_R	M_x	M_y	M_z	M_R
4	649	1,299	1,299	1,948	1,624	3,383	2,436	4,474
6	974	1,948	1,948	2,922	2,436	5,074	3,654	6,710
8	1,299	2,597	2,597	3,896	3,247	6,764	4,870	8,947
10	1,623	3,246	3,246	4,869	4,059	8,455	6,088	11,184
12	1,948	3,895	3,895	5,843	4,871	10,146	7,306	13,421
14	2,272	4,545	4,545	6,817	5,683	11,838	8,524	15,658
16	2,595	5,189	5,189	7,784	6,486	13,513	9,730	17,870

Compressor Inlet

Nozzle Size (in.)	F_x	F_y	F_z	F_R	M_x	M_y	M_z	M_R
4	648	1,080	1,080	1,659	1,620	2,699	2,699	4,147
6	972	1,620	1,620	2,488	2,429	4,049	4,049	6,220
8	1,296	2,160	2,160	3,318	3,239	5,399	5,399	8,294
10	1,620	2,699	2,699	4,147	4,049	6,748	6,748	10,367
12	1,944	3,239	3,239	4,976	4,859	8,098	8,098	12,441
14	2,268	3,779	3,779	5,806	5,669	9,448	9,448	14,514
16	2,592	4,319	4,319	6,635	6,479	10,798	10,798	16,588
18	2,915	4,859	4,859	7,464	7,289	12,147	12,147	18,661
20	3,240	5,399	5,399	8,294	8,099	13,497	13,497	20,735
24	3,892	6,486	6,486	9,964	9,730	16,216	16,216	24,912

Note: Forces indicated are in lb_f and moments are ft-lb.

equipment! The basic relationship for rotational deformation of nozzle ends is

$$K = \frac{P}{U} = \frac{M}{\theta} = \frac{\pi}{180} \left[\frac{EI}{D_N k_f} \right]$$

Eq. 5-22

where $K = KRX$ or KRY, ft-lb/deg
where KRX = rotational stiffness about the x-axis and
KRY = rotational stiffness about the y-axis
M = moment, ft-lb (N-m)
θ = angle of rotation, deg

E = modulus of elasticity of vessel metal at ambient temperature, psi (MPa)

I = moment of inertia of vessel nozzle, in.^{4} (mm^4)

D_N = diameter of vessel nozzle, in. (mm)

K_f = flexibility factor, referred to in the piping codes as "k"

The flexibility factor (K_f) is a parameter that has had several formulations over the years. One widely used variant was that proposed by the "Oak Ridge ORNL Phase 3 Report-115-3-1966." Since this document was published in 1966, several revisions have been made. The current ASME Section III Division 1 code gives detailed discussions on the flexibility factor. Individuals who are designing piping for nuclear systems should only consult that code. Outside the nuclear industry, the piping engineer rarely knows all the parameters that are necessary to compute the flexibility factor of Section III. Also, the piping engineer in nonnuclear work rarely knows which vendor will supply the piping components, thereby making the Section III parameters unknown.

The WRC Bulletin 329 (December 1987) gives several formulations for flexibility and SIF factors for unreinforced nozzles and various types of reinforced nozzles. For a simple unreinforced pipe on a header, Equations 2-52 through 2-53 can be rewritten in more convenient forms as follows:

Flexibility Factor

Longitudinal = $K_L = C_L F$ Eq. 5-23

Circumferential = $K_C = C_C F$ Eq. 5-24

where $C_L = 0.1$ and $C_C = 0.2$ and F = flexibility constant, with

$$F = \left(\frac{D}{t}\right)^{1.5}\left[\left(\frac{t}{t_B}\right)\left(\frac{D_B}{D}\right)\right]^{0.5}\left(\frac{t_B}{t}\right)$$

The flexibility constant was conceived when an error was discovered in the WRC Bulletin 329. This error was also corrected in Eqs. 2-52 and 2-53. It should be emphasized that the flexibility factor is *not* the flexibility factor in the ASME B31.3 piping code.

Rotational Spring Rate

Similarly Equations 2-54 and 2-55 can be rewritten as follows:

Longitudinal = $R_L = \dfrac{R_s}{K_L}\left(\dfrac{\text{ft-lb}_f}{\text{deg}}\right)$ Eq. 5-25

$$\text{Circumferential} = R_C = \frac{R_s}{K_C} \left(\frac{\text{ft-lb}_f}{\text{deg}} \right)$$ Eq. 5-26

$$\text{where } R_s = \frac{EI\pi}{2160D_B} \left(\frac{\text{ft-lb}_f}{\text{deg}} \right)$$

Angle of Twist

$$\text{Longitudinal} = \theta_L = \frac{MD_BK_L}{EI} \text{ (radians)}$$ Eq. 5-27

$$\text{Circumferential} = \theta_C = \frac{MD_BK_C}{EI} \text{ (radians)}$$ Eq. 5-28

where C_L = 0.09 for in-plane bending
C_C = 0.27 for out-of-plane bending
D = diameter of vessel or pipe header, in.
D_B = diameter of branch, in.
E = modulus of elasticity, psi
I = moment of inertia of branch, in.[4]
K_L = longitudinal flexibility factor
K_C = circumferential flexibility factor
M = applied moment, in.-lb
θ_L = longitudinal angle of twist, radians
θ_C = circumferential angle of twist, radians
t = wall thickness, in.
t_B = wall thickness of branch, in.
t_n = t_B + reinforcement; for example, t_n would be the branch (nozzle) hub thickness at the base of a reinforced nozzle; for a nozzle with no nozzle wall reinforcement, $t_n = t_B$

The reader is referred to **Figure 5-11** (ASME B31.3 Fig. 319.4.4A) and **Figure 5-12** (ASME B31.3 Fig. 319.4.4B) for the orientation of in-plane and out-of-plane moments.

Piping Systems Without Springs

Anyone who has worked in a process plant or refinery with high or low temperatures knows that springs can't be totally avoided. Tolerance toward their use falls off the closer one gets to the oil patch. Likewise, shim plates are tolerated in operational facilities, particularly pipelines, whereas they

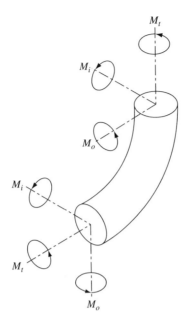

Figure 5-11. ASME B31.3 loadings on a 90° elbow showing in-plane (M_i) and out-of-plane (M_o) bending moments. Courtesy of the American Society of Mechanical Engineers.

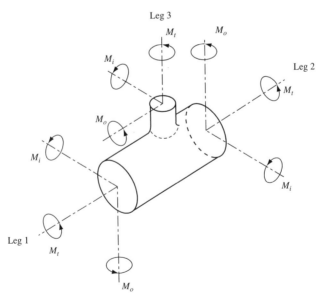

Figure 5-12. ASME B31.3 showing in-plane (M_i) and out-of-plane (M_o) bending moments. Courtesy of the American Society of Mechanical Engineers.

are usually not tolerated in engineering and construction firms. Their use depends on the application. The reason for the engineering and construction firms not liking their use is fear that they may not be installed, or at least not in the right location, or both. These fears are not without merit, but by using shim plates designers can avoid the use of an expensive spring if the installation and use of the shim plates is monitored. Ensuring that shim plates are installed in the correct location is easier said than done, so its application mainly lies in the operating plant side and pipelines, and engineering and construction companies frown on their use. The point is that if they are used correctly, they can be used in lieu of a spring in some special instances, especially when the movement is very small, say 1/16″ (1.59 mm) to 1/8″ (3.17 mm). Shim plates work well with simple supports, that is, supports resting on steel or concrete. Shim plates normally do not work well with restraints where the pipe moves. In these cases the shim plate will either become dislodged or obstruct the pipe movement. The type of restraints being referred to are multiple restraint supports shown in Figure 2-9 in Chapter 2. Shim plates should be avoided in these kind of restraints, as verified by field practice.

Supporting a pipe while it is moving can be accomplished with another simple device, called the *flexible beam support*, shown in **Figure 5-13**. This support is used to provide flexibility in situations where generally small piping is attached to a piece of equipment that has thermal expansion and the use of springs would be expensive or impractical. For example, in an MTBE plant, many ¾″ and 1½″ nitrogen lines were to be connected to a piece of equipment that was thermally expanding downward about one inch. The connected nitrogen lines were moving with the equipment. Using the flexible beam support was a simple solution that worked. They were bolted to the floor grid of the unit and provided ample support, allowing

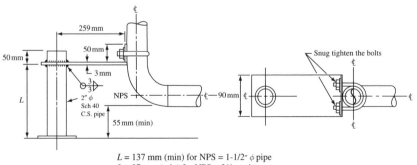

L = 137 mm (min) for NPS = 1-1/2″ ϕ pipe
L = 97 mm (min) for NPS = 3/4″ ϕ pipe

Figure 5-13. Flexible beam support.

the nitrogen piping to move. The nitrogen piping connected off the equipment like a bowl of spaghetti. These supports eliminated the use of springs, which would have been more expensive. **Figure 5-14** shows the support in the installed condition, and **Figure 5-15** shows the support in the operating condition with the support in the displaced condition.

The flexible plate thickness is critical in determining the stiffness. It performs the same function as a spring. In this application, the flexible plate

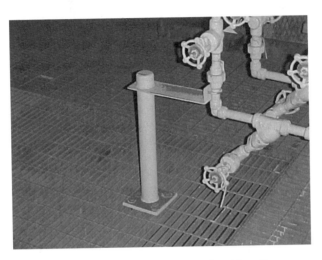

Figure 5-14. Flexible beam support in cold position.

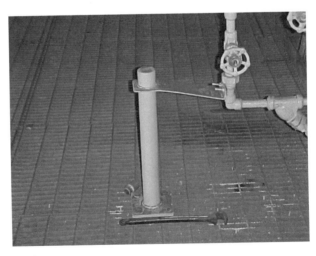

Figure 5-15. Flexible beam support in operating hot condition.

had to accommodate downward expansion of approximately 20 to 25 mm (0.788 to 1.0 in.) downward without causing the nitrogen lines to become overstressed. Linear elastic finite element results with the von Mises stress and displacements are shown in **Figures 5-16** and **5-17**, respectively.

The stress is high on the inside of the pipe support for the 3 mm plate. This localized stress will be much lower at the inside outside edge of the plate facing the supported pipe because the fillet weld (not modeled) will reduce the stress concentration at the corner considerably. The preceding models do not consider the elastic response of the vertical pipe connecting

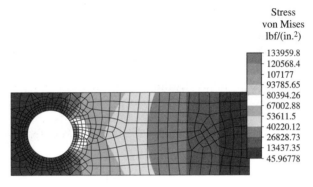

Stress
von Mises
lbf/(in.^{2})

133959.8
120568.4
107177
93785.65
80394.26
67002.88
53611.5
40220.12
26828.73
13437.35
45.96778

Load case: 1 of 1
Maximum value: 133960 lbf/(in.^{2}) (Peak stress-shakes down to lower value in elastic range as described in the text)
Minimum value: 45.9678 lbf/(in.^{2})

Figure 5-16. Stress profile in flexible beam support.

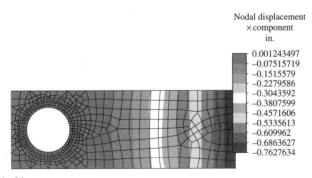

Nodal displacement
×component
in.

0.001243497
−0.07515719
−0.1515579
−0.2279586
−0.3043592
−0.3807599
−0.4571606
−0.5335613
−0.609962
−0.6863627
−0.7627634

Load case: 1 of 1
Maximum value: 0.0012435 in.
Minimum value: −0.762763 in.

Figure 5-17. Displacement in flexible beam support.

the flexible beam to the floor. With this pipe bending, the stresses shake down to the elastic range. These supports have been in use for many years and work where space prohibits the use of springs. They have been used in services that are not cyclic. If cyclic loads exist, then the beam or plate element needs to be tapered so as to reduce the stress concentration at the vertical pipe support-plate juncture.

Sizing the flexible beam support is not done accurately with beam equations because it is a plate. One can get a crude answer with beam formulas accurate enough to compute the stiffness of the support. It would be better to use a quick finite element linear elastic study to more accurately compute the support stiffness.

Fluid Forces Acting on Piping Systems

There have been many formulations predicting the reaction of PSV (pressure safety valve) and rupture discs over the years. One of the difficulties the piping engineer usually faces is obtaining the physical properties of the fluid being considered. The reader is referred to **Figure 5-18**.

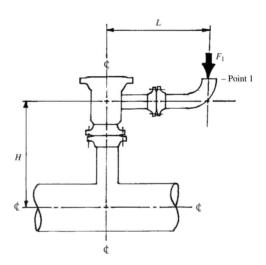

Figure 5-18. The *L* dimension is to be kept to a minimum. The PSV vendor normally has support requirements. If this dimension is less than 24″ (610 mm) and the stack height is 6′0″ (1.8 m) or less, no support is generally required. The reaction force is opposite to the open pipe end. The *H* dimension should be kept to a minimum. Reactions may be ignored in closed flare relief systems.

Moments from piping reactions must be considered in the pipe stress analysis. One of the simplest formulas, used by the ASME B31.1, Power Piping Code, in Paragraph 2.3.1.1, is the reaction force at the exit flow area at the end of a 90° elbow attached to the discharge of the PSV, as follows:

$$F_1 = \frac{W}{g_c}V_1 + (P_1 - P_a)A_1 \qquad \text{Eq. 5-29}$$

where F_1 = reaction force at point 1, the exit of the elbow, lb_f
W = mass flow rate, (relieving capacity stamped on the valve $\times$ 1.11), lb_m/sec
g_c = gravitational constant = 32.2 lb_m-ft/lb_f-sec^2
V_1 = exit velocity at the exit gas conditions, ft/sec
P_1 = static pressure at exit gas conditions, psia
A_1 = pipe exit flow area, in.^{2}
P_a = atmospheric pressure, psia = 14.7 psia

To consider the dynamic effects of the shock of the PSV opening, F_1 is multiplied by the dynamic load factor (DLF), often referred to as the impact factor. The B31.1 gives a curve in Figure 3-2 of the standard that gives the DLF as a function of the ratio of the safety valve opening time to installation period. For all practical purposes, the *DLF* = 2.0 for maximum dynamic effects.

Equation 5-29 is mainly for steam in power systems. For process fluids in general the following equation can be used, which includes an impact factor (*DLF*) of 2.0:

$$F = \left[\frac{CC_d}{183}\sqrt{\frac{\gamma}{\gamma + 1}}\right]AP \qquad \text{Eq. 5-30}$$

where F = reaction force at orifice
γ = ratio of specific heats C_p/C_v
C_d = discharge coefficient (usually 1.15)
A = orifice area, in.^{2}
P = set pressure of PSV, psia

The constant C in Eq. 5-30 is given by the following relationship:

$$C = 520\left[\gamma\left(\frac{2}{\gamma + 1}\right)^{\frac{\gamma+1}{\gamma-1}}\right]^{0.5}$$

The mass flow rate for the PSV can be found by using the following:

$$W = CC_d AP \sqrt{\frac{M}{T}}$$ Eq. 5-31

where M = the molecular weight of the gas or vapor; see **Table 5-6** for selected fluids

T = temperature of the relieving fluid, °R (°F + 460)

Table 5-6
Gas Properties and Physical Constants

Gas	Formula	Molecular Weight, M	Critical Constants, T_c/P_c	Specific Heats C_p/C_v	Ratio of Specific Heats, γ
Air	$N_2 + O_2$	28.97	238.39/547	0.241/0.171	1.41
Acetylene	C_2H_2	26.04	557.09/905	0.397/0.320	1.24
Argon	A	39.94	272.40/705	0.124/0.074	1.67
Ammonia	NH_3	17.03	731.09/1657	0.523/0.399	1.31
Benzene	C_6H_6	78.11	1012.70/714	0.240/0.215	1.12
Benzene (ethyl)	C_8H_{10}	106.16	1114.49/560	0.280/0.261	1.07
Butane (normal)	C_4H_{10}	58.12	765.31/551	0.398/0.363	1.09
Carbon monoxide	CO	28.01	241.69/510	0.248/0.177	1.40
Carbon dioxide	CO_2	44.01	1073.0/0.0348	0.199/0.153	1.30
Chlorine	Cl_2	70.91	1120.0/0.0280	0.115/0.084	1.37
Ethane	C_2H_6	30.07	549.78/708.0	0.410/0.343	1.20
Ethylene	C_2H_4	28.05	509.51/742.0	0.362/0.291	1.24
Helium	He	4.00	27.69/33.0	1.25/0.754	1.66
Heptane (normal)	C_7H_{16}	100.20	972.31/370.0	0.399/0.379	1.05
Hexane (normal)	C_6H_{14}	86.17	914.19/440.0	0.398/0.375	1.06
Hydrogen	H_2	2.02	59.89/188.0	3.408/2.420	1.41
Hydrogen chloride	HCl	36.50	584.19/1200.0	0.191/0.137	1.40
Hydrogen sulfide	H_2S	34.08	672.39/1306.0	0.254/0.192	1.32
Methane	CH_4	16.04	343.19/673.0	0.527/0.402	1.31
Methyl chloride	CH_3Cl	50.50	749.69/930.0	0.240/0.201	1.20
Nitrogen	N_2	28.01	226.89/492.0	0.248/0.177	1.40
Nitrous oxide	N_2O	44.00	–/–	0.221/0.176	1.26
Octane (normal)	C_8H_{18}	114.22	1024.89/362	0.400/0.382	1.05
Oxygen	O_2	32.0	277.89/730	0.219/0.156	1.40
Pentane (normal)	C_5H_{12}	72.15	845.61/490	0.397/0.370	1.07
Pentane (iso)	C_5H_{12}	72.15	829.69/483	0.388/0.361	1.08
Propane	C_3H_8	44.09	665.95/617	0.389/0.342	1.14
Sulfur dioxide	SO_2	64.06	774.69/1142	0.147/0.118	1.25
Water	H_2O	18.01	1165.09/3206	0.445/0.332	1.34

With no impact factor, the reaction force is

$$T = \frac{W\sqrt{\dfrac{\gamma T}{(\gamma + 1)M}}}{366} \qquad\qquad \text{Eq. 5-32}$$

Note that $\gamma \geq 1.41$ (air) should be used in any system where it is possible for air to be in the relieving gas or vapor.

Simplified formulas for predicting reaction forces may be used, but they normally are not as accurate in predicting the force for gases in PSVs. They are as follows:

For gases:

$$F = 0.6(\gamma + 1)AP \qquad\qquad \text{Eq. 5-33}$$

For liquids:

$$F = 0.8\left(\frac{A^2}{A_P}\right)P \qquad\qquad \text{Eq. 5-34}$$

For rupture discs:

$$F = 0.375(\gamma + 1)\, AP \qquad\qquad \text{Eq. 5-35}$$

Example: A PSV in a steam line has a line pressure of 2800 psia. The ID of the orifice valve is 2.141 in. and the ratio of specific heats is $\gamma = 1.3$. The pipe stress software used predicts the valve exit gas conditions are as follows:

$$P = 46.1 \text{ psia}; \ V = 1848.8 \text{ fps}; \ T = 809.6°F$$

The gas conditions at the orifice are as follows:

$$P = 1528 \text{ psia}; \ V = 2135.4 \text{ fps}; \ T = 809.6°F$$

The software computes the mass flow rate as being 388,229.30 lb_m/hr and a reaction force of 17,658.85 lb_f. Check these results with other methods.

Using Eq. 5-29 we have the following:

$$F_1 = \frac{388229.30 \left(\dfrac{lb_m}{hr}\right)\left(\dfrac{1hr}{3600\ sec}\right)}{32.2 \left(\dfrac{lb_m\text{-}ft}{lb_f\text{-}sec^2}\right)} (1848.8) \left(\dfrac{ft}{sec}\right)$$

$$+ (46.1 - 14.7) \left(\dfrac{lb_f}{in.^2}\right) (28.89)\ in.^2$$

$$F_1 = 6811.02\ lb_f + 907.15\ lb_f = 7718.17\ lbf$$

Multiplying by $DLF = 2.0$, the reaction force becomes

$$F_1 = 15{,}436.34\ lb_f$$

Using the methodology of Eq. 5-31, we have the following:

Computing the constant (C)

$$C = 520 \left[(1.3) \left(\frac{2}{1.3 + 1}\right)^{\frac{1.3+1}{1.3-1}} \right]^{0.5} = 346.9$$

The molecular weight of the steam $= 18.01$. The steam temperature $= 1000°F = 1460°R$.

$$W = (346.9)(1.15)(3.60)\ in.^2\ (3080)\ \frac{lb_f}{in.^2} \sqrt{\frac{18.01}{1460}}$$

$$W = 491{,}287.58\ lb_m/hr$$

Now the reaction force is

$$F = \left[\frac{(346.9)(1.15)}{183} \sqrt{\frac{1.3}{2.3}} \right](3.60)\ in.^2\ (2800)(1.1)\ \frac{lb_f}{in.^2}$$

$$= 18{,}172.4\ lb_f$$

Using the simplified formula of Eq. 5-33, we have

$$F = 0.6(1.3 + 1)(3.60)\ in.^2\ (2800)(1.1)\ \frac{lb_f}{in.^2} = 15{,}301.4\ lb_f$$

So the reaction force could be taken as approximately 18,000 lb$_f$. There is another method presented in [Reference 1] that uses a method for PSV with gases that is somewhat conservative, but it predicts a reaction force of 21,405.37 lb$_f$ in this case.

Nozzle Movements and Thermal Displacement

When nozzles are attached to process columns and equipment, the thermal displacement of the nozzle is necessary for the pipe stress (flexibility) analysis. This is somewhat straightforward with the equipment to predict the thermal movements. When it becomes less clear is when the component is a vessel skirt or pipe support. To compensate for the unknown, simplistic figures and algorithms have been created to make an estimate, but, as we shall see, they are often inaccurate. One such figure, which is shown in **Figure 5-19**, is an empirical approximation curve fit, that "predicts" the average temperature in a vessel skirt. Its developer gave the author the method back in 1984 after the author published the method presented below, which was later republished in [Reference 2]. There is no compensation for the skirt or pipe support material, type of insulation, or thickness of insulation in **Figure 5-19**. Because so many different materials

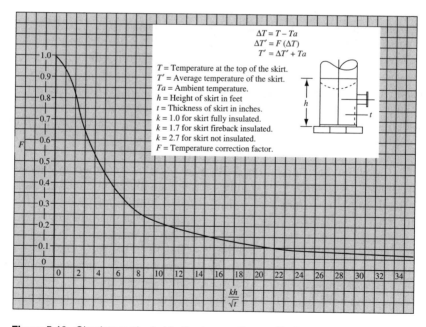

Figure 5-19. Obsolete method of finding temperature profile down length of vessel skirt.

are used in construction, **Figure 5-19** may not necessarily always be conservative. Such simplifications can lead to wrong skirt temperatures and nozzle movements. For brittle fracture concerns, these temperatures can be critical. This figure has found its way around several companies.

When vessels operate at temperatures outside of the carbon steel envelope (i.e., the temperature exceeds the 750 to 800°F [399 to 427°C], or below -20 to 50°F [-29 to -46°C]), alloy materials are often used for the skirt or pipe support. For skirts over 4–6 ft in height, it is common practice to fabricate the skirt from alloy steel attached to the vessel and then weld a carbon steel section on the bottom of the alloy portion to make the overall required skirt length. For example, a skirt 20 ft tall would be cost prohibitive to make entirely of austenitic stainless steel (18Cr-8Ni—like 304 SS or 316 SS). Also knowing the temperature down the length of the skirt will allow one to determine where to weld structural steel attachments. As mentioned previously, austenitic stainless supports are undesirable for thermal movements exceeding $\frac{1}{4}$ in. (6.35 mm). Carbon steels are much more desirable as structural members than austenitic stainless steels because of the relative low yield strengths of the latter.

Discussed in this section are the procedures for analyzing heat transfer in such residual components as vessel skirts and pipe supports. These methods have been tested with empirical data and have been used for many years (since they were developed by the author in 1982 and published in 1983). They have been extended and published with ASME and later in 1986 and 1994 [Reference 2]. They can be used for any materials of construction.

Vessel skirts are sometimes insulated on the inside and outside as depicted in **Figure 5-20**. In cryogenic applications, there are many reasons why a heat transfer analysis of the skirt is required. The primary reason is to protect carbon steel components from brittle fracture (see Chapter 4). Another reason, as mentioned earlier, is economics—a tall skirt made of alloy steel is much more expensive than a similar skirt made mostly of carbon steel. Also, we will see how the skirt can actually thermally displace as a result of this heat exchange.

Consider the skirt in **Figure 5-20**. The vessel is at either an elevated temperature or a cold temperature denoted at the shell juncture as t_s. Thermal conduction and convection are the controlling modes of heat transfer. The convection can either be considered as natural convection or free convection or, in the case of wind, forced convection. It has been found that using the free convection coefficient is the most desirable in many cases because vessels are normally surrounded by other equipment and structures, making the free convection more applicable.

Assume that the temperature inside the skirt is the same as the ambient temperature and that wind chill factors are not present. Air seepage under the skirt and open apertures on the shell allow for equilibrium to be established with the outside temperature.

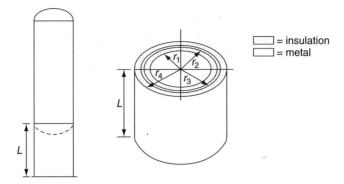

Figure 5-20. Skirt geometry for vessel skirt.

The first step is to determine the free convection film coefficient for the outside surface of the pressure vessel skirt insulation. In normal conditions, the air temperature inside the vessel skirt (t_i) is assumed 5°F (2.78°C) lower than the outside ambient temperature (t_5). The free convection film coefficient is found by iteration using the following equations:

$$U_4 = \left[\frac{r_1 \ln\left(\frac{r_2}{r_1}\right)}{k_{2-1}} + \frac{r_4 \ln\left(\frac{r_3}{r_2}\right)}{k_{3-2}} + \frac{r_4 \ln\left(\frac{r_4}{r_3}\right)}{k_{3-4}} + \frac{1}{h_{4-5}} \right] \quad \text{Eq. 5-36}$$

$$t_4 = \left(\frac{U_4}{h_{4-5}}\right)(t_i - t_5) + t_5 \quad \text{Eq. 5-37}$$

$$N_{Gr} = \frac{[d^3 \gamma^2 g \beta |\Delta t| (3600)^2]}{\mu^2} \quad \text{Eq. 5-38}$$

where N_{Gr} = Grashof number, dimensionless
   g = 32.2 ft/sec^2
   α = coefficient of thermal expansion, in./ft
   d = outside diameter (OD) of insulation, ft
   β = 1/(460 + $t_{ambient}$) °R^{-1}
   β = volume coefficient of expansion
   μ = absolute viscosity, lbm/ft-hr.

where $\beta \triangleq \frac{1}{T}$, where T = the absolute temperature of the gas, °R (°K)
(see [Reference 3], pp. 335 and 338, or [Reference 4], p. 106, or any reference on heat transfer for more detail)

γ = specific weight of insulation, lb_m/ft^3
k_{1-2} = thermal conductivity of insulation inside vessel skirt, Btu/hr-ft²-°F
k_{2-3} = thermal conductivity of vessel skirt metal, Btu/hr-ft²-°F
k_{3-4} = thermal conductivity of insulation outside vessel skirt, Btu/hr-ft²-°F
h_{4-5} = film coefficient of air outside of skirt, Btu/hr-ft²-°F
$\Delta t = t_4 - t_5$

$$N_{Nu} = C(N_{Gr}N_{Pr})^m \qquad \text{Eq. 5-39}$$

$$h'_{4-5} = \frac{(k_{air}N_{Nu})}{d} \qquad \text{Eq. 5-40}$$

N_{Pr} = Prandtl number, dimensionless
$N_{Pr} = \mu C_p/k = 0.712$ for atmospheric air with these applications
C_p = Specific heat, Btu/lb²-°F
N_{Nu} = Nusselt number, dimensionless

For free convection of cross flow around cylinders, the following constants hold [Reference 5]:

$10^4 < N_{Gr}N_{Pr} < 10^9$, $C = 0.525$, $m = \frac{1}{4}$
$10^9 < N_{Gr}N_{Pr} < 10^{12}$, $C = 0.129$, $m = \frac{1}{3}$

These relationships are valid for applications for the refining, petrochemical, and gas processing industries.

Now, for a cylinder with insulation on both sides, we use the final value of h_{4-5} after performing iterations from Eqs. 5-36 through 5-40 in the following equations:

$$Q = \left(\frac{2\pi k_{2-1}}{k_m A_m}\right)\left[\frac{1}{\left|\ln\left(\frac{r_1}{r_2}\right)\right|} + \frac{1}{\ln\left(\frac{r_4}{r_3}\right)}\right] \qquad \text{Eq. 5-41}$$

$$Z = \left|\left(\frac{2\pi}{k_m A_m}\right)\left[r_4 h_{4-5}(t_4 - t_5) - k_{2-1}\left[\frac{t_4}{\ln\left(\frac{r_4}{r_3}\right)} - \frac{t_i}{\ln\left(\frac{r_1}{r_2}\right)}\right]\right]\right| \qquad \text{Eq. 5-42}$$

$$\bar{Z} = \frac{Z}{Q} \qquad \text{Eq. 5-43}$$

A_m = cross-sectional metal area of skirt, ft²

Substituting these parameters into the following equation, we obtain the temperature distribution down the skirt length:

$$t_x = \frac{2(t_s - \overline{Z})e^{xQ^{0.5}}}{1 + e^{xQ^{0.5}}} + \overline{Z} \qquad \qquad \text{Eq. 5-44}$$

The difference between the process temperature inside the vessel and the outside ambient temperature is the main driving force of heat transfer. It is analogous to the electromagnetic force (EMF) or the potential energy of height differential from which a fluid is dropped and turned into kinetic energy.

The degree in significance of convection is inversely proportional to the insulation thickness. The air around the outside insulation surface is in a state of local turbulence. For this reason, the variance of the Grashof number down the outside insulation wall is insignificant. Experimental measurements confirm this fact. The reader will see the example that follows on how to apply this method.

Piping that is supported by piping sections is treated in a similar manner to vessel skirts. Such piping supports are shown in **Figure 5-21** in which pipe supports and branch lines are subject to thermal gradients from a hot or cold process header. **Figure 5-21A** shows a stub piece used as a piping header support. The temperature gradient through the stub piece must be analyzed to determine if the slide plates coated with Teflon will be protected from the elevated temperature inside the process header. If the process header is in cryogenic service, the stub piece must be analyzed to assure the engineer that the carbon steel structural members are adequately protected from temperatures below the transition temperature.

Figure 5-21B illustrates a common situation in which a process line connected to a turbo expander is supported by a section of pipe welded to a base plate that is anchored to the foundation. The pipe stubs displacement (shown by δ_1) caused by low temperature in the process pipe could induce a sufficient bending moment on the turbine to cause damage. Even though this is very unlikely, being able to assess can be valuable for certain specific applications (e.g., module skids where space is very limited).

Figure 5-21C is a branch line running from a hot or cold pipe header to a fragile piece of equipment. Even though the valve on the branch line is closed, the residual temperature distribution through the branch line may of interest.

Referring to **Figure 5-22**, the procedure for determining the temperature distribution through the empty branch pipe or pipe support is similar to the case of a vessel skirt. First, solve for the free convection

Heat transfer in piping and equipment

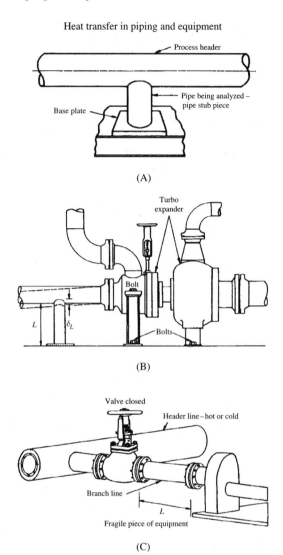

(A)

(B)

(C)

Figure 5-21. (A) The stub piece is used as a header support. (B) The process line is connected to a turbo expander; the line is supported by a short section of pipe welded to a base plate. (C) The branch line from a header (hot or cold) is connected through a shut-off valve to a fragile piece of process equipment.

film coefficient on the exterior surface of the pipe insulation (h_o, Btu/hr-ft^2-°F). To do this, use the equation for the overall heat transfer coefficient:

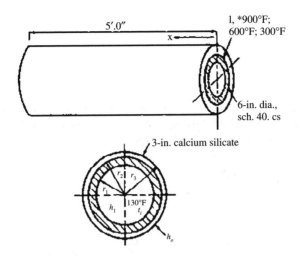

Figure 5-22. Empty branch pipe with one end uniformly subjected to three temperatures at separate times.

$$U_3 = \left(\frac{r_3 \ln\left(\dfrac{r_2}{r_1}\right)}{k_m} + \frac{r_3 \ln\left(\dfrac{r_3}{r_2}\right)}{k_1} + \frac{1}{h_o} \right)$$ Eq. 5-45

$$t_3 = \left(\frac{U_3}{h_o} \right)(t_i - t_o) + t_o$$ Eq. 5-46

$$\Delta t = t_3 - t_o$$

$$N_{Gr} = \frac{[d^3 \gamma^2 g \beta(|\Delta t|)(3600)^2]}{\mu^2}$$ Eq. 5-38

$$N_{Nu} = C(N_{Gr}N_{Pr})^m$$ Eq. 5-39

where C and m are determined previously for skirts:

$$h'_o = \frac{k_{air}N_{Nu}}{d}$$ Eq. 5-47

where k_{air} = thermal conductivity of air, Btu/hr-ft^2-°F

$$t'_3 = \left(\frac{U_3}{h'_o}\right)(t_i - t_o) + t_o \qquad \text{Eq. 5-48}$$

$$\Delta t'_3 = t_3 - t'_3 \le 2°F \qquad \text{Eq. 5-49}$$

Once the criterion in Eq. 5-49 is met, we can proceed with the final iterative value for the film coefficient (h_o). With this final value, we solve for the parameters Q, Z, and $\bar{Z}$ as follows:

$$Q = \frac{h_i \pi d_i}{k_m A_m} + \frac{2\pi k_i}{k_m A_m \ln\left(\dfrac{r_3}{r_2}\right)} \qquad \text{Eq. 5-50}$$

$$Z = \left(\frac{\pi}{k_m A_m}\right)\left[\left[2r_3 h_o(t_3 - t_o) - \frac{2k_i t_3}{\ln\left(\dfrac{r_3}{r_2}\right)} - d_i h_i t_i\right]\right] \qquad \text{Eq. 5-51}$$

where d_i = inside diameter of support pipe, ft
   h_i = natural convection coefficient at inside of the pipe wall, Btu/hr-ft²-°F

$$\bar{Z} = \frac{Z}{Q} \qquad \text{Eq. 5-43}$$

Once Q and $\bar{Z}$ are known, we solve for the temperature distribution with

$$t_x = \frac{2(t_s - \bar{Z})e^{xQ^{0.5}}}{1 + e^{2xQ^{0.5}}} + \bar{Z} \qquad \text{Eq. 5-52}$$

Notice that the form of the final solution in Eq. 5-52 is the same for the skirt problem with insulation on the inside and outside shell surfaces as for the pipe problem with insulation on only the outside surface. The difference in the solutions is the boundary conditions (i.e., a cylinder with insulation on both inside and outside surfaces versus a cylinder with just insulation on the outside surface alone). The solutions to the basic differential equations are affected by these differences in boundary conditions.

For cases of tapered shirts (cone shaped), the cylinder section can be approximated by using an average diameter. This approximation is very close to actual results because the skirts should not have a taper of more than 15°.

As a consequence of heat transfer along vessel skirts and pipe connections, thermal deflections will occur. The deflection equations are the same regardless of what case is considered, whether it is a shell with insulation on the inside and outside surfaces or a shell with only external insulation. The values of Q and $\overline{Z}$ are determined from the appropriate equations of each respective case.

The thermal deflection equations are dependent on the type of material considered since the coefficient of thermal conductivity is the governing property of the particular material being considered. Taking a differential element of a shell, we solve for the amount of thermal deflection by

$$dl = \alpha(t)t(x)\,dx \qquad\qquad \text{Eq. 5-53}$$

Since the temperature varies over the shell length, we integrate Eq. 5-53 to obtain the total deflection (δ) as

$$\delta = \int dL = \int_0^L \alpha(x)\,t(x)\,dx \qquad\qquad \text{Eq. 5-54}$$

The function ($\alpha(t)$) is the coefficient of thermal expansion for the particular material being considered. Values of the thermal expansion were curve-fitted over a large range of temperature, and a relation in terms of temperature was obtained for various materials. The function for $t(x)$ is obtained from Eq. 5-44 and is substituted with $\alpha(t)$ in Eq. 5-54. Then, the product of $\alpha(t)t(x)$ is integrated over a length L, and we obtain the thermal deflection function for each particular material. For carbon steel, the expanded thermal deflection equation is as follows:

$$\delta_{cs} = C_1 + C_2 + C_3 + C_4 - C_5\,[C_6 + C_7 + C_8 + C_9], \text{ in.} \qquad \text{Eq. 5-55}$$

where

$$C_1 = \frac{2[5.89 + (2.496 \times 10^{-3})2\overline{Z}]\,(t_s - \overline{Z})\arctan(e^{LQ^{0.5}})}{10^6 Q^{0.5}}$$

$$C_2 = \frac{(2.496 \times 10^{-3})(t_s - \overline{Z})^2(e^{2LQ^{0.5}} - 1)}{(10^6)Q^{0.5}(1 + e^{2LQ^{0.5}})}$$

$$C_3 = \frac{4\overline{Z}(2.496 \times 10^{-3})(t_s - \overline{Z})}{(10^6)Q^{0.5}}\arctan(e^{LQ^{0.5}})$$

$$C_4 = \frac{L\overline{Z}^2(2.496 \times 10^{-3})}{10^6}$$

$$C_5 = \frac{(6.536 \times 10^{-7})}{10^6}$$

$$C_6 = 8(t_s - \overline{Z})\left[\frac{sech(LQ^{0.5})\tanh(LQ^{0.5}) + \arctan[\sinh(LQ^{0.5})]}{2Q^{0.5}}\right]$$

$$C_7 = \frac{8\overline{Z}(t_s - \overline{Z})^2}{2Q^{0.5}}\left(\frac{e^{2LQ^{0.5}}}{1 + e^{2LQ^{0.5}}}\right)$$

$$C_8 = 4\overline{Z}(t_s - \overline{Z})\left[\frac{2}{Q^{0.5}}\arctan(e^{LQ^{0.5}})\right]$$

$$C_9 = \overline{Z}^3 L$$

Even though these equations don't look innocuous, each term must be solved, which can easily be done on a spreadsheet program.

Similarly, for austenitic stainless steel, the thermal deflection equation is as follows:

$$\delta_{ss} = C_{s1} + C_{s2} + C_{s3} + C_{s4}$$
$$- C_{s5}[C_{s6} + C_{s7} + C_{s8} + C_{s9}], \text{ in.} \qquad \text{Eq. 5-56}$$

where

$$C_{s1} = \frac{2(t_s - \overline{Z})[8.96 + (4.11 \times 10^{-3})\overline{Z}]\arctan(e^{LQ^{0.5}})}{10^6 Q^{0.5}}$$

$$C_{s2} = \frac{4(2.055 \times 10^{-3})\overline{Z}(t_s - \overline{Z})\arctan(e^{LQ^{0.5}})}{(10^6)Q^{0.5}}$$

$$C_{s3} = \frac{4(2.496 \times 10^{-3})L\overline{Z}^2}{10^6}$$

$$C_{s4} = \frac{(2.055 \times 10^{-3})(t_s - \overline{Z})^2(e^{2LQ^{0.5}} - 1)}{(10^6)Q^{0.5}(1 + e^{2LQ^{0.5}})}$$

$$C_{s5} = \frac{(1.06 \times 10^{-6})}{10^6}$$

$$C_{s6} = 8(t_s - \overline{Z})^3 \left[\frac{\text{sech}(LQ^{0.5})\tanh(LQ^{0.5}) + \arctan[\sinh(LQ^{0.5})]}{2Q^{0.5}} \right]$$

$$Cs_7 = \frac{8\overline{Z}(t_s - \overline{Z})^2}{2Q^{0.5}} \left(\frac{e^{2LQ^{0.5}}}{1 + e^{2LQ^{0.5}}} \right)$$

$$C_{s8} = 4\overline{Z}(t_s - \overline{Z}) \left[\frac{2}{Q^{0.5}} \arctan(e^{LQ^{0.5}}) \right]$$

$$C_{s9} = \overline{Z}^3 L$$

Equations 5-55 and 5-56 can be simplified, or entered into a spreadsheet program, or both, and solved, but there are simpler and quicker methods for solving for the thermal deflections. We will illustrate easier methods in Examples 5-1 and 5-2 below.

A word about units: When using Eqs. 5-36 through 5-54, temperatures should be expressed in either °F or °C; however, in Eqs. 5-55 and 5-56, °F should be used, as the coefficient of thermal expansion used in the curve fits were to °F. All the reader using the metric system has to do is convert °C to °F and obtain thermal deflection in terms of inches and convert back to mm. Using absolute temperatures will result in an error, except in solving for the volume coefficient of expansion (β) in the Grashof number (N_{Gr}) Eq. 5-38. Sometimes some parameters, like thermal conductivity, are expressed as per °R versus °F. This fact has caused confusion among a few readers. The reader must remember that a parameter that varies per °F is the same as °R. For example, if thermal conductivity were expressed in terms of lb_m/ft-hr-°R, it would have the same magnitude as if it were expressed in terms of lb_m/ft-hr-°F. On the two temperature scales, a change in 1°F is the same as a 1°R change. Thus, if thermal conductivity was multiplied by temperature, the (lb_m/hr-ft-°R) ×°F would be equal to lb_m/hr-ft. The same is true for °K versus °C. Expressing thermal conductivity in terms of absolute temperature has been a good exam question for years, but our purpose here is not to give or pass exams but to apply principles to industrial applications. Clarity is what we are trying to achieve.

Example 5.1: Thermal Movements in a Vessel Skirt

A vessel was originally designed as shown in **Figure 5-23**. After the design was made and drawings were submitted, the process group discovered that they needed a skirt height of 20 ft for the net positive suction head (NPSH) for pumps connected to the vessel. Using the original design, calculate the temperature distribution down the length of a vessel skirt and the amount of thermal movement. Then recommend a fix to the height problem. The vessel contains a cold process fluid that varies in temperature because of cyclic process conditions. Three operating temperatures to be analyzed are $-200°F$, $-100°F$, and $-50°F$. The skirt is made of austenitic Type 304 stainless steel and is insulated on the inside and outside as shown in **Figure 5-23**.

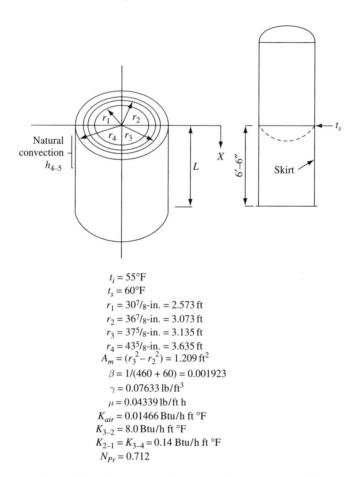

$$t_i = 55°F$$
$$t_s = 60°F$$
$$r_1 = 30^7/8\text{-in.} = 2.573\,\text{ft}$$
$$r_2 = 36^7/8\text{-in.} = 3.073\,\text{ft}$$
$$r_3 = 37^5/8\text{-in.} = 3.135\,\text{ft}$$
$$r_4 = 43^5/8\text{-in.} = 3.635\,\text{ft}$$
$$A_m = (r_3^2 - r_2^2) = 1.209\,\text{ft}^2$$
$$\beta = 1/(460 + 60) = 0.001923$$
$$\gamma = 0.07633\,\text{lb/ft}^3$$
$$\mu = 0.04339\,\text{lb/ft h}$$
$$K_{air} = 0.01466\,\text{Btu/h ft °F}$$
$$K_{3-2} = 8.0\,\text{Btu/h ft °F}$$
$$K_{2-1} = K_{3-4} = 0.14\,\text{Btu/h ft °F}$$
$$N_{Pr} = 0.712$$

Figure 5-23. Parameters for thermal profile down vessel skirt.

First, determine the natural convection film coefficient for the skirt. The temperature inside the skirt (t_i) is assumed to be 5°F lower than the ambient temperature (t_5).

$$U_4 = \left[\frac{r_1 \ln\left(\frac{r_2}{r_1}\right)}{k_{2-1}} + \frac{r_4 \ln\left(\frac{r_3}{r_2}\right)}{k_{3-2}} + \frac{r_4 \ln\left(\frac{r_4}{r_3}\right)}{k_{3-4}} + \frac{1}{h_{4-5}} \right]^{-1}$$

$$U_4 = \left[7.115 + \frac{1}{h_{4-5}} \right]^{-1}$$

Assume $h_{4-5} = 0.275$

$$U_4 = 0.093$$

$$t_4 = \left(\frac{U_4}{h_{4-5}} \right)(t_i - t_5) + t_5 = \left(\frac{0.093}{0.275} \right)(-5) + 60$$

$$= 58.31°F$$

$$t_4 - t_5 = 58.31 - 60 = -1.69°F$$

$$N_{Gr} = \frac{d^3 \gamma^2 g \beta(|\Delta t|)(3600)^2}{\mu^2}$$

$$= \frac{(7.27)^3 (0.07633)^2 (32.2)(0.001923)(1.69)(3600)^2}{(0.04339)^2}$$

$$= 1{,}613{,}720{,}723$$

where $d = 2r_4 = 2(3.635) = 7.27$ ft

$N_{Gr} N_{Pr} = (1{,}613{,}720{,}723)(0.712) = 1{,}148{,}969{,}155$

$N_{Gr} N_{Pr} > 10^9$

$$C = 0.129$$

$$m = 1/3$$

$$N_{Nu} = C(N_{Gr} N_{Pr})^m = 0.129 \, (1{,}148{,}969{,}155)^{1/3}$$

$$N_{Nu} = 135.11$$

$$h_{4-5} = \frac{(k_{air} N_{Nu})}{d} = \frac{(0.01466)(135.11)}{7.27} = 0.2725$$

$$t_4 = \left(\frac{0.093}{0.2725}\right)(-5) + 60 = 58.29°F$$

$$\Delta t_4' = 58.31 - 58.29 = 0.02$$

$$h_{4-5} = 0.275$$

For a cylinder with insulation on both sides,

$$Q = \left(\frac{2\pi k_{2-1}}{k_m A_m}\right)\left[\frac{1}{\ln\left(\dfrac{r_1}{r_2}\right)} + \frac{1}{\ln\left(\dfrac{r_4}{r_3}\right)}\right]$$

$$Q = \left[\frac{2\pi(0.14)}{(8)(1.209)}\right]\left[\frac{1}{\left|\ln\left(\dfrac{2.573}{3.073}\right)\right|} + \frac{1}{\ln\left(\dfrac{3.635}{3.135}\right)}\right]$$

$$Q = 1.1267 \text{ ft}^{-2}$$

$$Z = \left|\frac{2\pi}{k_m A_m}\left[r_4 h_{4-5}(t_4 - t_5) - k_{2-1}\left[\frac{t_4}{\ln\left(\dfrac{r_4}{r_3}\right)} - \frac{t_i}{\ln\left(\dfrac{r_1}{r_2}\right)}\right]\right]\right|$$

$$Z = \left|\frac{2\pi}{(8)(1.209)}\left[(3.635)(0.275)(-1.69) - 0.14\left[\frac{58.31}{\ln\left(\dfrac{3.635}{3.135}\right)} - \frac{55.00}{\ln\left(\dfrac{2.573}{3.073}\right)}\right]\right]\right|$$

$$Z = 65.102°F/ft^2$$

$$\bar{Z} = \frac{Z}{Q} = \frac{65.102}{1.1267} = 57.781°F$$

For $t_s = -200°F$,

$$t_x = \frac{2(t_s - \bar{Z})e^{xQ^{0.5}}}{1 + e^{2xQ^{0.5}}} + \bar{Z}$$

$$t_x = \frac{(-515.56)(2.89)^x}{1 + e^{2xQ^{0.5}}} + 57.781$$

Similarly for $t_s = -100°F$,

$$t_s = \frac{(-315.56)(2.89)^x}{1 + (2.89)^{2x}} + 57.781$$

and for $t_s = -50°F$,

$$t_x = \frac{(-215.56)(2.89)^x}{1 + (2.89)^{2x}} + 57.781$$

The temperature distribution curves are shown in **Figure 5-24**.

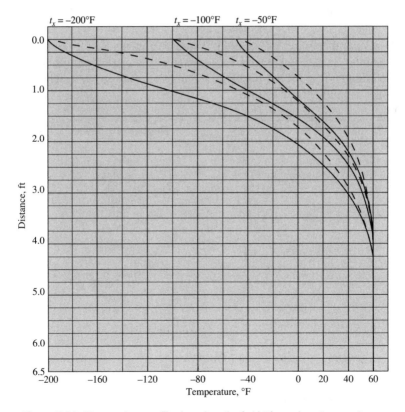

Figure 5-24. Temperature profile down length of skirt for various temperatures.

The thermal displacement of the skirt is checked using three different methods to illustrate the various ways of computing the deflection rather than using Eq. 5-56, which is rather cumbersome. The fastest but least accurate method involves taking the temperature at the skirt juncture, −200°F and viewing **Figure 5-24** to see where the temperature approaches ambient. In this case, it is 4 ft down the skirt from the skirt/vessel junction. Taking the average temperature between these points, we have

$$\bar{t} = \frac{(-200 + 60)}{2} = -70°F$$

Now the thermal displacement is

$$\delta = L\alpha\Delta t = (48)\,in.\,(9.0e^{-6})\,\frac{in.}{in.\text{-}°F}\,(-70 - 60)°F$$

$$\delta = -0.056\ in.$$

The second method uses numeric integration where the length of the 4 ft segment is divided arbitrarily into segments of temperatures of 25°F. The parameter x is solved for using the above equation for t_x. The table of the skirt segments is as follows:

$\alpha \times 10^{-6}$ in./in.-°F	t (°F)	x, length down the skirt, ft
8.47	−200	0
8.54	−175	0.435
8.60	−150	0.64
8.66	−125	0.83
8.75	−100	1.01
8.83	−75	1.21
8.90	−50	1.43
8.94	−25	1.70
8.98	0	2.05
9.03	25	2.59
9.07	50	3.95
9.08	57.7	6.50

We perform the numerical integration using

$$\delta = \sum_{t=-200}^{57.7} L(x)\alpha(x)t(x)$$

We use, for our purposes here, 70°F as the basis for the ambient temperature, on which the coefficients of thermal expansion are based.

δ_i	$\bar{\alpha} \times 10^{-6}$ in./in.-°F	Δx, ft	ΔT, °F	δ_i, ft
δ_1	8.505	0.435 to 0	−200 to 70	−0.000999
δ_2	8.570	0.64 to 0.435	−175 to 70	−0.000430
δ_3	8.630	0.83 to 0.64	−150 to 70	−0.000361
δ_4	8.705	1.01 to 0.83	−125 to 70	−0.000306
δ_5	8.790	1.21 to 1.01	−100 to 70	−0.000299
δ_6	8.865	1.43 to 1.21	−75 to 70	−0.000283
δ_7	8.920	1.70 to 1.43	−50 to 70	−0.000289
δ_8	8.960	2.05 to 1.70	−25 to 70	−0.000298
δ_9	9.000	2.59 to −2.05	0 to 70	−0.000340
δ_{10}	9.050	3.95 to −2.59	25 to 70	−0.000554
δ_{11}	9.075	6.50 to −3.95	50 to70	−0.000463
				$\Sigma \delta_i = -0.004620$

Note: $\bar{\alpha}$ is the average thermal coefficient of thermal expansion over each interval in length.

Thus the total thermal movement is $\delta = -0.004620$ ft $= -0.0555$ in.

The third method involves using finite element where a simple two-dimensional axisymmetric model is used, assuming a total adiabatic surface along the skirt, and with input of temperatures to the different segments. From the finite element assessment, the total thermal movement is

$$\delta = -0.0555 \text{ in.}$$

Thus, we have three methods that are in excellent agreement. Using these methods any material may be used in the analysis.

Using **Figure 5-19** we have

skirt thickness $= 3.135$ ft $- 3.073$ ft $= 0.062$ ft $= 0.744$ in.

$h = 6.5$ ft $=$ height of skirt

$T_a = 60°$F

Now $\Delta T = T - T_a = -200 - 60 = -260°$F

$\Delta T' = F(\Delta T)$

From **Figure 5-19**,

$$\frac{kh}{\sqrt{t}} = \frac{(1.0)(6.5)}{\sqrt{0.744}} = 7.54 \Rightarrow F = 0.29$$

Now

$$\Delta T' = (0.29)(-260) = -75.4°F$$

$$T' = \Delta T' + T_a = -75.4°F + 60°F = -15.4°F$$

Thus, with ΔT being the "average" temperature per **Figure 5-19**, the thermal contraction is as follows:

$$\delta = L\alpha\,(\Delta T) = (6.5)\,\text{ft}\left(\frac{12\ \text{in.}}{\text{ft}}\right)(8.96 \times 10^{-6})\,\frac{\text{in.}}{\text{in.-°F}}\,(-15.4 - 60)°F$$

$$\delta = -0.053\ \text{in.}$$

We obtain the approximately same answer for both methods for this *one* case. Now if the skirt is to be extended to 20 ft, as it was in this case, making the entire skirt of austenitic is cost prohibitive, especially if we are working on a lump sum project. So we select carbon steel as the material for the remaining section of the skirt. Now we are asked where to cut the austenitic skirt to weld the carbon steel portion. Using our analysis of Eq. 5-52 and **Figure 5-24**, we had the austenitic portion shortened to 4 ft. This was to give the skirt the same temperature for all three operating temperatures, $-200°F$, $-100°F$, and $-50°F$, to reduce the differential expansion between the austenitic stainless and the carbon steel component. We also want to have the carbon steel well outside the brittle fracture range. Thus, the austenitic stainless 304 portion was 4 ft long, and the low carbon steel portion was 16 ft long. Using **Figure 5-19** one would not be able to perform this analysis of the temperature profile, let alone calculate a safe length of the skirt.

The piping engineer needs to plan on thermal contraction of -0.0555 in. when the process fluid is $-200°F$. This method for thermal distribution has been verified in field measurements.

Note that in another example with a carbon steel skirt, the result obtained from using **Figure 5-19** was over 28% lower than that obtained by using Eq. 5-52 methodology. The skirt was made of SA-516-70 low carbon steel and was 10 ft long. The total deflection using **Figure 5-19** was 0.053 in., whereas using Eq. 5-52 methodology gave 0.074 in. These examples involving skirts will be discussed in Volume 2.

Thermal movements of vessel skirts are necessary for the piping engineer to calculate nozzle movements on the vessel. The vessel skirt thermal movement is added to that of the length from the shirt-shell tangent to the elevation of the nozzle in question. As mentioned, there are brittle fracture concerns, especially when using skirts with different metallurgies.

Example 5-2: Residual Temperatures in a Branch Pipe

A section of carbon steel process pipe is shown in **Figure 5-21C**. Three conditions will be analyzed for process fluid temperatures at the shut-off valve at 900°F, 600°F, and 300°F. The temperature of 900°F is outside the range acceptable for carbon steel, but the fluid is regeneration gas that reaches that temperature on a temporary basis such that graphitization is not a concern. The basic analysis is the same as in Example 5-1, beginning with the iteration procedure to find the natural convection film coefficient. Note that it is assumed that the temperature inside the empty pipe branch line (t_i) is 130°F and that the ambient temperature (t_o) is 60°F.

For a 6″ ϕ Schedule 40 pipe,

$$d_i = 6.0648 \text{ in.} = 0.5054 \text{ ft}$$

$$d_o = 6.625 \text{ in.} = 0.552 \text{ ft}$$

$$A_m = 5.58 \text{ in.}^2 = 0.03875 \text{ ft}^2$$

$$r_1 = \frac{0.5054 \text{ ft}}{2} = 0.2527 \text{ ft}$$

$$r_2 = \frac{0.552 \text{ ft}}{2} = 0.276 \text{ ft}$$

$$r_3 = 0.276 \text{ ft} + \frac{3.0}{12} \text{ ft} = 0.526 \text{ ft}$$

$$U_3 = \left[\frac{r_3 \ln\left(\frac{r_2}{r_1}\right)}{k_m} + \frac{r_3 \ln\left(\frac{r_3}{r_2}\right)}{k_1} + \frac{1}{h_o} \right]^{-1}$$

$$U_3 = \left[\frac{0.526 \ln\left(\frac{0.267}{0.2527}\right)}{25} + \frac{0.526 \ln\left(\frac{0.267}{0.2527}\right)}{25} + \frac{0.526 \ln\left(\frac{0.526}{0.276}\right)}{0.027} + \frac{1}{h_o} \right]^{-1}$$

$$U_3 = \left[12.565 + \frac{1}{h_o} \right]^{-1}$$

Let $h_o = 1.0$ Btu/hr-ft^2-°F

$$U_3 = 0.0737$$

$$t_3 = \left(\frac{U_3}{h_o}\right)(t_i - t_o) + t_o$$

$$t_3 = \left(\frac{0.0737}{1.0}\right)(130 - 60) + 60 = 65.16°F$$

$$\Delta t = t_3 - t_o = 5.16°F$$

$$N_{Gr} = \frac{d^3 \alpha^2 g \beta(|\Delta t|)(3600)^2}{\mu^2}$$

$$N_{Gr} = \frac{[(1.052)^3(0.07633)^2(0.001923)(32.2)(3600)^2(5.16)]}{(0.04339)^2}$$

$$N_{Gr} = 14,920,198.65$$

$$N_{Gr}N_{Pr} = (14,920,198.65)(0.712) = 10,623181.44$$

$$N_{Nu} = C(N_{Gr}N_{Pr})^m$$

where $C = 0.525$ and $m = \frac{1}{4}$

$$N_{Nu} = 0.525(10,623,181.44)^{1/4} = 29.97$$

$$h'_o = \left(\frac{k_{air}}{d}\right)N_{Nu} = \left(\frac{0.01466}{1.052}\right)29.57 = 0.4177$$

$$t'_3 = \left(\frac{U_3}{h'_o}\right)(t_i - t_o) + t_o = \left(\frac{0.0737}{0.4177}\right)(130 - 60) + 60$$

$$t'_3 = 72.35°F$$

$$\Delta t'_3 = t_3 - t'_3 = 65.16 - 72.35 = -7.19°F$$

This difference is too large; try another trial value.

Let $h_o = 0.49$ Btu/hr-ft-°F

$$U_3 = 1/(12.565 + 1/0.49) = 0.0687 \text{ Btu/hr-ft}^2\text{-°F}$$

$$t_3 = \left(\frac{0.06847}{0.49}\right)(70) + 60 = 69.781°F$$

$$\Delta t = 9.781°F$$

$$N_{Gr} = 28,279,559.99$$

$$N_{Gr}N_{Pr} = 20,135,046.71$$

$$N_{Nu} = 35.17$$

$$h'_o = 0.4901 \text{ Btu/hr-ft}^2\text{-}°F$$

$$t'_3 = (0.06847/0.4901)(70) + 60 = 69.779°F$$

$$\Delta t'_3 = 69.781 - 69.779 = 0.002°F < 0.1$$

$$h_o = 0.49 \text{ Btu/hr-ft}^2\text{-}°F$$

$$h_i = 25.0 \text{ Btu/hr-ft}^2\text{-}°F$$

$$Q = \frac{h_i \pi d_i}{k_m A_m} + \frac{2\pi K_i}{k_m A_m} \ln\left(\frac{r_3}{r_2}\right)$$

$$Q = \frac{(25)\pi(0.5054)}{(25)(0.0388)} + \frac{2\pi(0.027)}{(25)(0.0388)} \ln\left(\frac{0.526}{0.276}\right) = 41.03$$

$$Z = \left(\frac{\pi}{k_m A_m}\right)\left[\left[2r_3 h_o (t_3 - t_o) - \frac{2k_i t_3}{\ln\left(\frac{r_3}{r_2}\right)} - d_i h_i t_i\right]\right]$$

$$Z = \left(\frac{\pi}{(25)(0.0388)}\right)\left| 2(0.526)(0.49)(69.779 - 60) \right.$$

$$\left. - \frac{2(0.027)(69.779)}{\ln\left(\frac{0.526}{0.276}\right)} + (0.5054)(25)(130) \right|$$

$$Z = 5317.293$$

$$\overline{Z} = \frac{Z}{Q} = \frac{5317.293}{41.03} = 129.595$$

$$t_x = \frac{2(t_s - \overline{Z})}{1 + e^{2xQ^{0.5}}} + \overline{Z}$$

For $t_s = 900°F$,

$$t_x = \frac{2(900 - 129.595)e^{6.405x}}{1 + e^{12.81x}} + 129.595$$

For $t_s = 600°F$,

$$t_x = \frac{940.81e^{6.405x}}{1 + e^{12.81x}} + 129.595$$

For $t_s = 300°F$,

$$t_x = \frac{340.81e^{6.405x}}{1 + e^{12.81x}} + 129.595$$

Curves depicting t_x are shown in **Figure 5-25**. The closed-form solutions are compared to the finite element results. The closed-form solutions using the preceding formulations are shown with solid curves and the FE results are shown with dashed lines. The comparison of this problem and various others reveal that closed-form solution temperatures decrease slower than the FE temperatures. It is hard to make a general rule, but either method gives results that are accurate for most applications. In this example the fluid inside the pipe is air, a well defined substance. In situations where the film coefficient is not known, a low value can be used, but it will result in the temperature curve having a lower slope because the temperature is decreasing at a lower rate, giving a more conservative answer.

If one were performing a pipe stress assessment, the temperature profiles in **Figure 5-25** could be input to the pipe stress software to obtain the thermal considerations for the branch line.

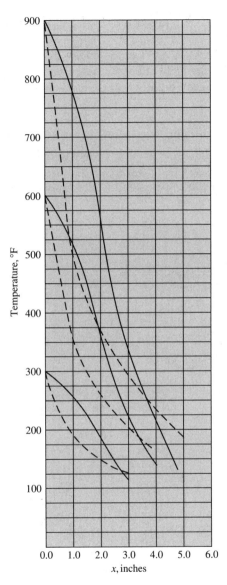

Figure 5-25. Temperature distribution of a branch pipe connected to a header. The configuration shown is in **Figure 5-21C**. The valve in the branch line next to the header is closed. The temperature curves originate from the valve. The solid lines show the results of the closed-form solution. The finite element solutions are shown in the curves with dashed lines.

Residual Heat Transfer Through Pipe Shoes

Heat transfer through plate surfaces is simpler than more complex surfaces because they can be handled with one-dimensional equations that are simple to apply. Based on **Figure 5-26**, we consider the heat balance down through the shoe as follows:

$$\left(\begin{array}{c}\text{Heat conducted through} \\ \text{shoe to base plate}\end{array}\right) = \left(\begin{array}{c}\text{Heat loss by convection from} \\ \text{shoe to outside air}\end{array}\right)$$

Writing in equation form, we have for one-dimensional steady state flow:

$$k_m A_m \left(\frac{\Delta t}{L}\right) = h_o A_p (\Delta t) \qquad \text{Eq. 5-57}$$

For the conduction process, $\Delta t = t_i - t_p$
For the convection process, $\Delta t = t_p - t_o$

Substituting these into Eq. 5-57, we have

$$k_m A_m \left(\frac{t_i - t_p}{L}\right) = h_o A_p (t_p - t_o)$$

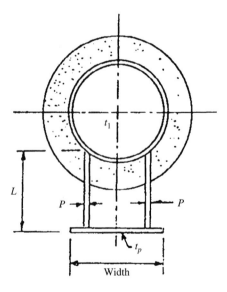

Figure 5-26. Heat transfer through pipe shoe.

Solving for t_p, we have

$$t_p = \frac{k_m A_m t_i + h_o A_p L t_o}{(k_m A_m + h_o A_p L)}, \text{°F} \qquad \text{Eq. 5-58}$$

where $A_m = (P \times \text{length of shoe}) \times 2$, in.^{2}
$A_p = \text{base width} \times \text{length of shoe}$, in.^{2}
$h_o = \text{free convection coefficient for shoe to air, Btu/hr-ft-°F}$
$K = \text{thermal conductivity of shoe material, Btu/hr-ft-°F}$
$L = \text{shoe height, in.}$

Like the analysis for cylinders, the free convection coefficient (h_o) can be substituted with a forced convection coefficient. However, most pipe shoes are normally protected by enough equipment and structures to prevent direct wind from blowing continuously on the shoe for any length of time. Of course, this depends on each individual case. **Figures 5-27a** and **5-27b** show thermal gradients for various simple pipe support configurations.

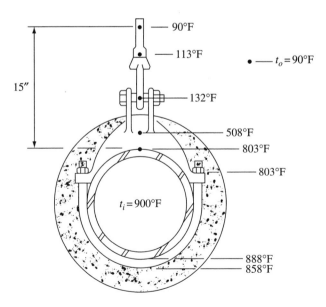

Figure 5-27a. Thermal gradients through a pipe clamp, clevis, and supporting rod.

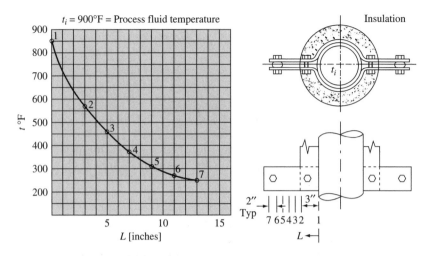

Figure 5-27b. Thermal gradient through pipe clamp support.

Example 5-3: Heat Transfer Through a Pipe Shoe

A 12 in. process header shown in **Figure 5-26** is supported by a shoe 14 in. long. The process fluid is at 750°F, and it is desired to determine the temperature of the bottom of the shoe base plate where Teflon is mounted to accommodate pipe movement. The Teflon cannot withstand temperature at or greater than 400°F. Referring to **Figure 5-28** and using Eq. 5-58, we have

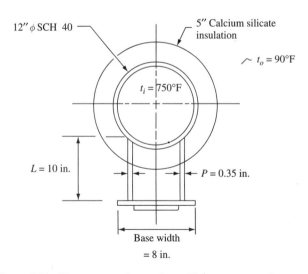

Figure 5-28. Pipe supported on a shoe with temperatures shown.

$$t_p = \frac{k_m A_m t_i + h_o A_p L t_o}{(k_m A_m + h_o A_p L)}, \, °F$$

where $h_o = 3.0$ Btu/hr-ft^2-°F for carbon steel in still air
$k_m = 26.0$ Btu/hr-ft-°F, thermal conductivity of pipe material
$L = 10.0$ in.
$A_m = (0.375)(14) = 5.25$ in.^{2}
$A_p = (8.0)(14) = 112$ in.^{2}
$t_o = 90$°F

$$t_p = \frac{(26.0)\dfrac{\text{Btu}}{\text{hr-ft-°F}}\left(\dfrac{5.25 \text{ in.}^2}{144 \text{ in.}^2}\right)\text{ft}^2 (750)°F + 3.0\dfrac{\text{Btu}}{\text{hr-ft}^2\text{-°F}}\left(\dfrac{112 \text{ in.}^2}{144 \text{ in.}^2}\right)\text{ft}^2\left(\dfrac{10}{12}\right)\text{ft } (90)°F}{\left[(26.0)\dfrac{\text{Btu}}{\text{hr-ft}^2\text{-°F}}\left(\dfrac{5.22 \text{ in.}^2}{144 \text{ in.}^2}\right)\text{ft}^2 + (3.0)\dfrac{\text{Btu}}{\text{hr-ft}^2\text{-°F}}\left(\dfrac{112 \text{ in.}^2}{114 \text{ in.}^2}\right)\text{ft}^2\left(\dfrac{10.0}{12}\right)\text{ft}\right]}$$

$$t_p = 306.303°F$$

Thus, the Teflon under the base plate is adequately protected. The amount of heat loss through the shoe base plate is

$$q = h_o A_p (t_p - t_o)$$

$$q = (3.0)\frac{\text{Btu}}{\text{hr-ft}^2\text{-°F}}\left(\frac{112 \text{ in.}^2}{144 \text{ in.}^2}\right)\text{ft}^2 (306.303 - 90) \, °F$$

$$q = 504.706 \frac{\text{Btu}}{\text{hr}}$$

Example 5-4: Emergency Constant Spring Replacement

The author was called out in the middle of the night when it was discovered that a constant spring hanger had bottomed out and become a rigid hanger on a hot superheated steam line. The engineering contractor representative discovered that the existing spring that bottomed out was designed only for 1.5 in. of travel, whereas 3.5 in. of travel is necessary, according to the pipe stress computer runs. The design temperature was 900°C, and the operating temperature was 780°C. The contractor placed a chain to support the pipe temporarily with operational monitoring. There were few options, and a chain support acting like a come-along could only be tolerated for a short time. Constant springs have a long delivery time half way around the world, so the option taken was to visit the junk yard and try to find a constant spring close to the required design.

The required spring was for a 1254 kg load on the super steam header. A constant spring was found in the junk yard rated for 1889 kg load with the same travel, 3.5 in. This spring and the one being replaced were the classical offset-slider crank mechanism design described previously, designed and fabricated by the same spring vendor. The spring was a size 28 shown in the spring table in **Figure 5-29**, rated for 4445 lb$_f$ (1889 kg).

When the spring was installed, its supporting force should have been in balance with the portion of the piping weight at which it was rated. The spring was preset to the cold position at the specified load. The spring had a turnbuckle, thus allowing for normal piping elevation adjustments. In our case, the actual piping load differed from the calculated load. The spring manufacturer claimed that an adjustment of 15 to 20% of the specified load could be made by turning the load adjustment bolt. Each division on the adjustment scale equals 5% of the rated load, according to the spring manufacturer's technical bulletin (catalog). Referring to **Figure 5-30**, we see the adjustment bolt setting has eight graduations.

Turning the bolt clockwise decreases the load 5% for each graduation; hence, four graduations represent a 20% decrease in the load. Similarly for a counterclockwise turn of the adjustment bolt, the load setting is increased. Thus, the load of 1889 kg reduced by 20% becomes

1889 kg(0.8) = 1511.2 kg

The acquired spring is adjusted by locking the spring assembly in the cold position and adjusting the load nut to match the required loading. The required loading is 20% over the operating load of the former spring. Since this load is 1254 kg, the adjusted load is

1254 × 1.2 = 1504.8 kg or 1505 kg

The 1505 kg is close enough to the 1511 kg reduction.

The load adjustment scale on the spring housing had ten graduations for the total adjustment range, as shown in **Figure 5-31**.

This is not to be confused with the scale on the load adjustment bolt. This scale indicates the amount of spring travel. Normally the actual is less than the total available travel to account for excursions. To translate the process temperatures and load requirements on the spring scale, we

Constant Support Size Selection Table

NOTE: The minimum load is the load for the preceding size

Hgr. Size 4000 4100 4200	2	2½	3	3½	4	4½	5	5½	6	6½	7	7½	8	8½	9	9½
1	90	75	60	55	45	40	35	35	30		25		20		20	
	130	100	85	70	65	55	50	45	40		35		30		30	
2	175	140	115	100	85	60	70	65	60		50		45		40	
3	235	190	160	135	120	105	95	85	80		70		60		55	
4	325	260	215	185	165	145	130	120	110		95		80		70	
5	445	360	295	255	225	200	180	160	150		130		110		100	
6	600	485	405	345	305	270	245	220	205		175		150		135	
7	755	605	505	435	380	335	305	275	255	235	215	205	190	180	170	160
8	905	725	605	520	455	405	365	330	305	280	260	245	230	215	205	190
9	1090	875	725	625	545	485	435	395	365	335	310	290	275	255	245	230
10	1310	1045	875	750	655	585	525	475	440	405	375	350	330	310	290	275
11	1465	1175	980	840	735	650	585	535	490	450	420	390	365	345	325	310
12	1620	1295	1080	925	810	720	650	590	540	500	465	435	405	385	350	345
13	1810	1450	1210	1035	905	805	725	660	605	560	520	485	455	430	405	385
14	2000	1605	1335	1145	1005	890	805	730	670	620	575	535	500	475	445	425
15	2240	1790	1495	1280	1120	995	900	815	750	690	640	600	560	530	500	475
16	2475	1980	1655	1415	1240	1100	990	900	825	765	710	660	620	585	550	525
17	2770	2215	1850	1585	1385	1230	1110	1010	925	855	790	740	695	655	615	585
18	3060	2445	2045	1750	1530	1365	1225	1115	1020	945	875	815	765	720	685	650
19	3400	2720	2265	1940	1670	1510	1360	1240	1135	1050	975	905	850	800	755	720
20	3735	2990	2490	2130	1870	1660	1495	1360	1245	1150	1070	995	935	880	830	790
21	4120	3295	2745	2350	2060	1830	1650	1500	1375	1270	1180	1100	1030	970	915	870
22	4500	3500	3000	2575	2250	2000	1800	1640	1500	1385	1285	1200	1125	1050	1000	950
23	4950	3650	3300	2830	2475	2200	1980	1800	1650	1525	1415	1320	1240	1165	1100	1045
24	5400	4320	3600	3085	2700	2400	2160	1965	1800	1665	1545	1440	1350	1270	1200	1140
25	5940	4755	3960	3395	2970	2640	2375	2160	1980	1830	1700	1585	1485	1400	1320	1255
26	6480	5185	4320	3710	3240	2880	2595	2355	2160	1995	1855	1730	1620	1525	1440	1370
27		5710	4755	4075	3565	3170	2855	2595	2380	2195	2040	1905	1785	1680	1585	1505
28		6230	5190	4445	3890	3460	3115	2830	2595	2395	2225	2075	1945	1830	1730	1645
29		6845	5705	4890	4280	3805	3425	3110	2855	2635	2445	2285	2140	2015	1905	1805
30		7645	6380	5465	4780	4250	3825	3475	3190	2945	2735	2555	2395	2250	2125	2020
31		8660	7215	6180	5405	4810	4330	3935	3600	3330	3090	2885	2705	2545	2405	2285
32		9480	7895	6760	5915	5265	4740	4305	3945	3640	3380	3150	2950	2780	2630	2500
33		10515	8770	7510	6575	5845	5260	4780	4380	4045	3750	3500	3285	3095	2920	2775
34		11700	9640	8265	7225	6425	5775	5250	4815	4450	4130	3845	3610	3405	3210	3050
35			10700	9160	8020	7130	6415	5830	5345	4940	4585	4220	4010	3775	3565	3385
36			11760	10060	8820	7840	7055	6415	5880	5430	5040	4600	4410	4150	3920	3720
37			13055	11180	9790	8700	7835	7120	6530	6030	5595	5170	4895	4605	4350	4130
38			14350	12305	10760	9565	8610	7830	7175	6625	6150	5745	5380	5060	4785	4545
39			15985	13700	11985	10650	9585	8725	7995	7375	6850	6390	5990	5640	5330	5060
40			17615	15095	13205	11740	10565	9610	8810	8125	7545	7040	6605	6215	5870	5575
41			19310	16545	14480	12875	11580	10530	9655	8910	8275	7720	7240	6815	6435	6115
42			21000	18000	15750	14005	12600	11455	10500	9690	9000	8400	7875	7415	7005	6650
43			22180	19140	16740	14880	13390	12170	11160	10300	9565	8925	8370	7875	7445	7070
44			24000	20280	17725	15750	14175	12885	11825	10910	10135	9450	8860	8335	7885	7485
45				21700	19050	16925	15235	13850	12705	11725	10890	10160	9525	8960	8470	8045
46				23250	20370	18100	16295	14815	13585	12535	11640	10870	10185	9585	9050	8600
47					22325	19845	17860	16240	14890	13740	12755	11910	11160	10510	9925	9520
48					24290	21590	19425	17660	16190	14945	13870	12960	12140	11435	10795	10260
49					26200	23290	20960	19055	17465	16125	14970	13980	13100	12330	11645	11065
50					28115	24990	22490	20450	18740	17300	16065	15005	14060	13230	12495	11870
51					30590	27195	24475	22250	20395	18830	17490	16320	15300	14390	13600	12935
52					33065	29400	26460	24045	22050	20360	18910	17640	16540	15550	14700	14000
53					35900	31920	28745	26120	23940	22110	20530	19145	17955	16885	15960	15180
	2	2½	3	3½	4	4½	5	5½	6	6½	7	7½	8	8½	9	9½

(Header for data columns: Total Travel (Inches) — Maximum Rated Load in Pounds For Each Size)

Figure 5-29. Spring support manufacturer's load table.

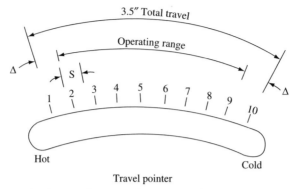

Figure 5-30. Spring travel scale on side of spring.

made the following calculations to see if the spring had enough travel to operate safely.

The 900°C represents 3.5 in. of total travel on the spring. This temperature is the very maximum that could be expected in an excursion. Thus for the operating case, which had a process temperature of 780°C, the travel incurred for this case is

$$\frac{900°C}{3.5 \text{ in.}} = \frac{780°C}{x} \Rightarrow x = 3.033 \text{ in.}$$

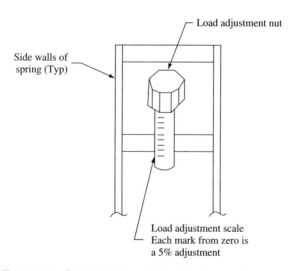

Figure 5-31. Elevation view—looking into the face of the spring.

On each side of the scale, there are 3.033/2 = 1.5165 in. Thus, the number of graduations on the spring scale on the housing is

$$\left(\frac{3.0}{3.5}\right) = 0.857$$

number of graduations = (1.0 − 0.857)(10 graduations)
= 1.429 graduations on scale

Δ = the amount of extra travel to prevent the spring from bottoming out

Thus,

$$\Delta = \frac{1.429}{2} = 0.71 \text{ graduation}$$

This represents the number of graduations on the spring scale between each side of the operating range and the design, or overall travel. Now the process temperature at the time of installation is 750°C. We calculate the difference between the installed position and the operating condition (δ). Thus, we have

$$\frac{900°C}{3.5 \text{ in.}} = \frac{750°C}{x} \Rightarrow x = 3.033 \text{ in.}$$

$$\delta = \frac{3.033 \text{ in.}}{2} = 1.5165 \text{ in.}$$

Now,

$$\left(\frac{3.0}{3.5}\right) = 0.857 \Rightarrow \Delta = \frac{1.5165 \text{ in.}}{2} = 0.758 \text{ in.}$$

The number of graduations on the scale is

(1.0 − 0.857)(10 graduations) = 1.429 graduations – difference
between 3.0 in. of travel and
3.5 in. of travel

Thus, the total adjustment for the installed temperature of 780°C is

$$\delta + \Delta = 1.5165'' + 0.758'' = 2.27 \text{ in. (Actual adjustment)}$$

Now we know the spring has enough travel at the desired load to operate safely. The spring was set and placed in use where it operated for ten months before the new spring arrived to replace it. The author monitored the spring during the entire period, and it performed very well. If a replacement spring had not been found in surplus, then the other alternatives would have been to shut down for months, which the plant management found unacceptable, or to fabricate a spring hanger from scratch.

Example 5-5: Pipe Header Simple Support

Problem: A 30 in. ϕ crude oil header is simply supported with the pipe resting on the steel. A company procedure calls for saddle-type supports with pads for piping 30 in. NPS (nominal pipe size) and larger. In case saddle pipe support cannot be installed, the line shall be analyzed for localized stresses. The question was: is the pipe resting on the steel overstressed? If it is overstressed, why has it not failed, as it has been in operation for thirteen years?

Solution: If the company procedure specifies a saddle support for lines 30 in. and larger, then this standard must be adhered to. Regarding its stress state, the pipe resting on the steel at a point is not a contact stress. Contact stresses are called Hertz stresses and are to applied solid bodies (e.g., meshing gears or ball bearings). They are not used for hollow surfaces (e.g., pipes or vessels). One classic example of this is the Zick analysis of horizontal drums supported on two saddles—contact stresses are not considered in the analysis.

In the case of the pipe resting on a steel support, the situation is like a ring. The cross section of the pipe resting on a solid surface represents a bulkhead or supporting ring in a pipe, supported at the bottom and carrying a total load (W) transferred by tangential shear (v) distributed as is shown in Roark Table 9.2 Case 20 [Reference 6]. The Roark model used in Case 20 is called a *tangential shear*, where the vertical reaction at the pipe support is applied to counter the pipe's weight, which includes the metal weight and content. The support reaction causes shear to flow around the pipe wall to replicate the pipe behaving like a beam. Rings are very common in structural design. This case is shown in **Figure 5-32**.

The Roark Table 9.2 Case 20 gives a bending moment due to the shear stress caused by the weight of the pipe and the internal fluid. This bending

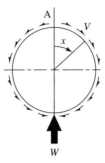

Figure 5-32. Roark Table 9.2 Case 20.

moment acts through the wall of the pipe. Being a bending moment, it has compression and tensile components through the cross section of the pipe. The internal pressure stress acts only in tension. Thus, when the pressure stress is combined with the shear bending moment stress, one side of the wall is added and the other side of the wall is subtracted from the tensile pressure stress. The sign convention on the shear bending moment is not relative, as the stress will be one value on each respective surface (we add tensile stresses and add and subtract tensile and compressive stresses). This is shown in **Figure 5-33**.

When solving for the shear bending moment (*I*), the Roark Table 9.2 Case 20, the resisting cross section acts in the plane of the paper. The effective width of this cross section is determined as shown in **Figure 5-34**.

Shear acts at a 45° angle; thus, the resisting cross section would be 45° diagonal lines drawn from the point of reaction to the axis of the pipe. As seen in **Figure 5-34**, the effective length is $2R_o$. Typically, the effective

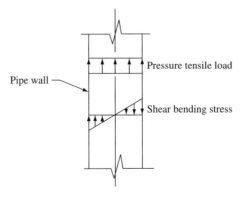

Figure 5-33. Hoop stress and shear moment bending stress in the pipe wall.

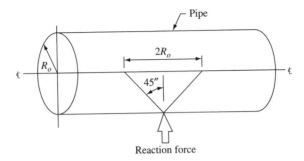

Figure 5-34. The effective length of pipe resisting the shear moment bending stress.

length of the cross section would be minimum of $2R_o$ or $40t_{nom}$, where t_{nom} is the nominal pipe wall thickness. The bending stress induced by the shear bending moment is calculated as follows:

$$be = \text{MAX}[40t_{nom}, 2R_o], \text{ see \textbf{Figure 5-34}} \qquad \text{Eq. 5-59}$$

$$S = \frac{(be)t_{nom}^2}{6} \qquad \text{Eq. 5-60}$$

where S = section modulus of the pipe, in.^{3}
$\quad t_{nom}$ = pipe nominal wall thickness, in.

This formula is derived from the relationship

$$S = \frac{I}{C} = \frac{\dfrac{bd^3}{12}}{\dfrac{d}{2}} = \frac{bd^2}{6}, \text{ for a rectangular cross-sectional area}$$

where I = moment of inertia, in.^{4}

In our case $be = 30$ in., as $40t_{nom} = 40(1.358)$ in. $= 54.32$ in. Thus, the shear moment bending stress is

$$\sigma = \frac{Mshear}{s} \qquad \text{Eq. 5-61}$$

The bending moments vary around the circumference of the pipe. Shown in **Figure 5-35** is the bending moment distribution for the pipe loaded with water.

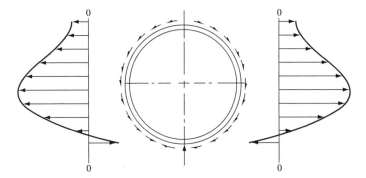

Figure 5-35. The shear bending moment across a pipe resting on a simple support.

The maximum bending moment occurs at 105.32° from the top point of the pipe. We loaded the pipe with crude oil using a specific gravity of 0.825 for the operational case. It was assumed that the pipe would not be operating at full pressure loaded with water. Combining the internal pressure stress and the shear moment bending stress for values of the angle from 0° to 180°, we have the following combined stresses for pipe loaded with crude oil:

Combined stresses for pipe loaded with crude oil

Angle x	Total Stress on Inner Wall, psi	Total Stress on Outer Wall, psi
0	16,924.6	14,452.1
15	17,194.2	14,182.4
30	17,960.3	13,416.3
45	19,908.7	12,278.0
60	20,417.2	10,959.5
75	21,676.8	9,699.9
90	22,617.8	8,758.8
105	22,989.8	8,366.8
120	22,580.6	8,796.0
135	21,244.7	10,132.0
150	18,923.6	12,453.0
165	15,661.2	15,715.5
180	11,607.7	19,769.0

The Section Modulus Conjecture: The criteria mentioned earlier concerning the effective area resisting the bending moment induced by shear in supporting the pipe has been a source of conjecture for many years. In the solution of a ring loaded in shear and inducting a moment into the plane of the pipe, the three-dimensional section resisting the bending moment is not considered because the ring is a two-dimensional problem. In **Figure 5-34** the effective length of the cross section resisting the bending moment induced by the shear is shown as $2R_o$. The effective cross section is the maximum of $40t_{nom}$ or $2R_o$; however, in structural applications, the minimum value of $40t_{nom}$ or $2R_o$ is often applied. In mechanical application, where we have internal pressure combined with the weight and thermal loads, the stresses can be significantly high—sometimes several times the ultimate strength of the pipe material. However, in reality, these pipes resting on simple supports do not fail, meaning that there is a fallacy in the equations—the theory does not match reality.

One author that addresses this phenomenon is Bednar [Reference 7], where in his brilliant work he discusses this problem on page 170. Bednar states that, to utilize the derived equation for the bending moment and bring the resulting stresses in the shell in agreement with actual measured stresses, we use what he describes as a "fictitious resisting width" of shell plate that is the smaller of $4R_o$ or $L/2$, where R_o is the shell radius and L is the length between supports. Of course, Bednar is discussing horizontal vessels supported on saddles, whereas here we are discussing continuous piping, like a continuous beam. Bednar's Figure 6.5 on page 168 is the same ring solution we use from Roark, but they are derived from different sources.

Another author, Troitsky [Reference 8] uses the same solution for the ring where the shear induces a bending moment from the load at the point of contact with the pipe support, with one exception. He is working with conveyor tube structures supported by ring girders. Troitsky uses the resisting section on pages 12–19 with the following expression:

$$c = 1.56\sqrt{rt} + b \qquad \text{Eq. 5-62}$$

Where r is the cylinder radius, t is the cylinder thickness, b is the width of the ring girder, and c is the effective width resisting the shear-induced bending moment.

Based on these criteria, we use the following to compute the effective width to obtain the section modulus, as follows:

$$beff = be + 1.56\sqrt{R(t_{nom} - CA)} \qquad \text{Eq. 5-63}$$

where R = mean radius, CA = corrosion allowance, and be is defined in Eq. 5-59. Thus, Eq. 5-60 becomes

$$S = \frac{beff\left(t_{nom}^2\right)}{6} \qquad \text{Eq. 5-64}$$

Now the stress criterion must be considered. In this case the pipe is subjected to primary and secondary stress. Thus, the maximum stress intensity σ_{max} is based on the primary or local membrane stresses plus primary bending stress plus the secondary stress (P_m or $P_L + P_b + Q$), using the nomenclature of the ASME Section VIII Division 2 Appendix 4. The value of σ_{max} cannot exceed $3S_m$, where all stresses may be computed under operating conditions. The following criterion must be met:

$$3S_m \leq 2(SMYS)$$

The value of $3S_m$ is defined as three times the average of the tabulated S_m values for the highest and lowest temperatures during the operation cycle. In the computation of the maximum primary-plus-secondary stress intensity range, it may be required to consider the superposition of cycles of various conditions that produce a total range greater than the range of any individual cycle. The value of $3S_m$ may vary with each cycle, or combination of cycles, being considered since the maximum temperatures may be different in each case. Thus, care must be used to ensure the applicable value of $3S_m$ for each cycle, and combination of cycles, is not exceeded except as permitted by ASME Section VIII Division 2 Appendix 4 Paragraph 4-136.4. This requirement is seldom required in piping systems, unless different services are used in the pipe with differing temperatures and pressures.

Since the pipe considered is operational, we can use the API 579 Fitness-for-Service recommended practice for the allowable stress, which is 20,000(1.2) = 24,000 psi from Figure B.1 of the API 579. For this case, the pipe wall thickness is large enough to accommodate the stress levels. The output of the computer run for the angular position of 105.23°, the point of maximum bending moment, is as follows.

Looking at **Figure 5-36**, we find the stresses in the 30 in. pipe are acceptable because, as mentioned previously, this is the angular position of maximum bending moment induced by the shear. However, if we take another pipe with a thinner wall, the stress magnitudes will differ. Consider a 24 in. line load with the same crude oil at the same facility. The pipe has a 3/8″ wall and is seamless, making the t_{nom}, less mill tolerance, 0.328 in. The results are as follows for the maximum bending moment at the same angular position of 105.23°.

We will now consider a 24″ ϕ pipe with a nominal wall thickness of 0.328 in. resting on a solid surface like we did for the 30″ ϕ pipe. The same methodology is used and the results are shown in **Figures 5-37a** and **b**.

The reader will notice that the outside wall is in compression and the inside wall is in tension because of the bending moment. Thus, only part of the wall is of significant stress, not the entire wall of the pipe. This being

Rules Sheet

Rules

---

if or(x $<$ 0, x $-$ π $>$ 1E-10) then caution $=$ 'Ang_Err else caution $=$ '_

call get_tab(matnum, matl, G, E, nu)

if ring $=$ 'thin then call get_tn_con(I, A, R, F, E, G; alpha, beta)

if ring $=$ 'thick then call get_tk_con(h, R, F, nu; alpha, beta)

k2 $=$ 1 $-$ alpha

k1 $=$ k2 $+$ beta

call case(W, R, I, x, k1, k2; v, LTM, LTN, LTV, M, N, V, MA, MC, NA, VA, DH, DV, delL, case)

plot $=$ given('plot, plot, 'y)

if and(solved(), plot $<>$ 'n) then call genplot(W)

Di $=$ Do $-$ 2 $\cdot$ tnom

$$Ro = \frac{Do}{2}$$

$$Ri = \frac{Di}{2}$$

$$Am = \left[\frac{\pi}{4}\right] \cdot \left[Do^2 - Di^2\right]$$

$$Wreaction = Am \cdot L \cdot 12 \cdot 0.283 + \left[\frac{\pi}{4}\right] \cdot \left[Di^2\right] \cdot \rho \cdot L \cdot 12$$

be $=$ MAX(40 $\cdot$ tnom, 2 $\cdot$ Ro)

$$beff = be + 1.56 \cdot \sqrt{R \cdot (tnom - CA)}$$

$$S = \frac{beff \cdot tnom^2}{6}$$

$$\sigma = \frac{M}{S}$$

$$\sigma p = \left[\frac{P}{Ej}\right] \cdot \left[\left[\frac{Do}{2 \cdot tnom}\right] - 0.4\right]$$

σinner $=$ σp $-$ σ

σouter $=$ σp $+$ σ

σallow $=$ 3.0 $\cdot$ Sm

If 3 $\cdot$ Sm $<$ 2 $\cdot$ SMYS then OK $=$ 'acceptable

If 3 $\cdot$ Sm $>$ 2 $\cdot$ SMYS then OK $=$ 'unaccpetable

Figure 5-36a. Equation sheet for 30 in. ϕ pipe.

Variables Sheet

Input	Name	Output	Unit	Comment
				Table 9.2: Roark's Formulas **Formulas for Circular Rings**
	case	'CASE_20		Reference Number
	plot	'y		Generate plots? 'n = no (Default = yes)
	caution	'_		Caution Message
17	matnum			Material Number (See Material Table)
	matl	"Steel - A.S.T.M. A7-61T"		Material name
	E	2.9E7	psi	Young's Modulus
	nu	0.27		Poisson's Ratio
'thin	ring			'thick or 'thin: thick/thin ring
7107.502	W		lbf	Applied total load
1943	I		in.^{4}	Area moment of inertia of X-section
11.813	R		in.	Mean radius of centroid of X-section
2	F			Shape Factor of X-section **THIN RINGS**:
27.83	A		in.^{2}	Area of X-section
1.087E7	G		psi	Shear Modulus of elasticity **THICK RINGS**:
	h		in.	Distance from Centroidal Axis to **Neutral Axis**
	DH	1.2938739E-4	in.	Change in horizontal diameter
	DV	−2.4236034E-4	in.	Change in vertical diameter
	v	184.791	lbf/in.	Tangential shear
	alpha	0.500310		Hoop-stress Deformation Factor
	beta	2.670776		Radial Shear Deformation Factor **AT SECTION**:
105.23	x		deg	Angular Position
	N	−2450.334	lbf	Internal Force
	V	3.129E-2	lbf	Radial Shear
	M	−15247.877	lbf-in.	Internal moment **AT A**:
	NA	1696.791	lbf	Internal Force
	VA	0	lbf	Radial Shear
	MA	−4.143	lbf-in.	Moment
	MC	13358.654	lbf-in.	Moment at C
	delL	−2.0957714E-4	in.	Increase in LOWER radius **LOAD TERMS**:

Figure 5-36b. Variable sheet showing results and answers for 30 in. ϕ pipe.

Input	Name	Output	Unit	Comment
	LTM	10065.9581776	lbf-in.	
	LTN	−2004.5965509	lbf	
	LTV	1637.2296434	lbf	
	k1	3.170466		Simplifying constants
	k2	0.499690		
	Di	27.284	in.	Inside diameter of pipe
30	Do		in.	Outside diameter of pipe
1.358	tnom		in.	Nominal pipe wall thickness
	Ro	15	in.	Outside radius of pipe
	Ri	13.642	in.	Inside radius of pipe
3.1416	π			Metal area of pipe x-section
	Wreaction	31272.705042	lbs	Combined of metal weight plus water
50	L		ft	Length between pipe supports
.03	ρ			Density of fluid, lbs/in.^{3}
	be	54.32	in.	Effective length of pipe wall resisting bending due to shear
	S	18.616281	in.^{3}	Section modulus of resisting pipe section to shear moment
	σ	−819.061411	psi	Bending stress due to shear moment
	Am	122.195158	in.^{2}	Cross sectional area of pipe
1400	P		psi	Internal pressure
1	Ej			Weld joint efficiency
	σinner	15722.978937	psi	Total stress in inner shell − pressure + shear bending
	σp	14903.917526	psi	Internal pressure stress
	σouter	14084.856114	psi	Total stress in outer shell − pressure + shear bending
	beff	60.568195		
0	CA			
	σallow	60000	psi	Allowable stress of pipe material at temperature
20000	Sm			Allowable stress for primary and secondary
35000	SMYS		psi	Specified Minimum Yield Strength of pipe material
	OK	'acceptable		

Figure 5-36b. cont'd.

the case, the high stress does not exist through the wall of the pipe, thus avoiding plastic deformation.

As mentioned previously, there are hundreds of pipes resting on flat solid surfaces that do not fail or exhibit plastic deformation. The 30 in. ϕ pipe mentioned was checked for cracks and deformation, and neither existed. The

Rules

---

if or(x < 0, x − π > 1E-10) then caution = 'Ang_Err else caution = '−

call get_tab(matnum, matl, G, E, nu)

if ring = 'thin then call get_tn_con(I, A, R, F, E, G; alpha, beta)

if ring = 'thick then call get_tk_con(h, R, F, nu; alpha, beta)

k2 = 1 − alpha

k1 = k2 + beta

call case(W, R, I, x, k1, k2; v, LTM, LTN, LTV, M, N, V, MA, MC, NA, VA, DH, DV, delL, case)

plot = given('plot, plot, 'y)

if and(solved(), plot <> 'n) then call genplot(W)

Di = Do − 2 · tnom

$$R_o = \frac{D_o}{2}$$

$$R_i = \frac{D_i}{2}$$

$$Am = \left[\frac{\pi}{4}\right] \cdot \left[Do^2 - Di^2\right]$$

$$Wreaction = Am \cdot L \cdot 12 \cdot 0.283 + \left[\frac{\pi}{4}\right] \cdot \left[Di^2\right] \cdot \rho \cdot L \cdot 12$$

be = MAX(40 · tnom, 2 · Ro)

$$beff = be + 1.56 \cdot \sqrt{R \cdot (tnom - CA)}$$

$$S = \frac{beff \cdot tnom^2}{6}$$

$$\sigma = \frac{M}{S}$$

$$\sigma p = \left[\frac{P}{Ej}\right] \cdot \left[\left[\frac{Do}{2 \cdot tnom}\right] - 0.4\right]$$

σinner = σp − σ

σouter = σp + σ

σallow = 3.0 · Sm

If 3 · Sm < 2 · SMYS then OK = 'acceptable

If 3 · Sm > 2 · SMYS then OK = 'unaccpetable

Figure 5-37a. Equation sheet for 24 in. ϕ pipe.

Variables sheet

Input	Name	Output	Unit	Comment
				Table 9.2: Roark's Formulas Formulas for Circular Rings
	case	'CASE_20		Reference Number
	plot	'y		Generate plots? 'n = no (Default = yes)
	caution	'_		Caution Message
17	matnum			Material Number (See Material Table)
	matl	"Steel-A.S.T.M. A7-61T"		Material name
	E	2.9E7	psi	Young's Modulus
	nu	0.27		Poisson's Ratio
'thin	ring			'thick or 'thin: thick/thin ring
7107.502	W		lbf	Applied total load
1943	I		in.^{4}	Area moment of inertia of X-section
11.813	R		in.	Mean radius of centroid of X-section
2	F			Shape Factor of X-section **THIN RINGS:**
27.83	A		in.^{2}	Area of X-section
1.087E7	G		psi	Shear Modulus of elasticity **THICK RINGS:**
	h		in.	Distance from Centroidal Axis to **Neutral Axis**
	DH	1.2938739E-4	in.	Change in horizontal diameter
	DV	−2.4236034E-4	in.	Change in vertical diameter
	v	184.791	lbf/in.	Tangential shear
	alpha	0.500310		Hoop-stress Deformation Factor
	beta	2.670776		Radial Shear Deformation Factor **AT SECTION:**
105.23	x		deg	Angular Position
	N	−2450.334	lbf	Internal Force
	V	3.129E-2	lbf	Radial Shear
	M	−15247.877	lbf-in.	Internal moment **AT A:**
	NA	1696.791	lbf	Internal Force
	VA	0	lbf	Radial Shear
	MA	−4.143	lbf-in.	Moment
	MC	13358.654	lbf-in.	Moment at C
	delL	−2.0957714E-4	in.	Increase in LOWER radius **LOAD TERMS:**

Figure 5-37b. Variable sheet for 24 in. ϕ pipe.

pipe has been in service for 13 years and has seen all possible operation cycles. The same is true for the 24 in. ϕ pipe and many others. It is concluded that this methodology adequately predicts the behavior of a pipe resting on a simple support, or flat surface. However, many companies consider

Input	Name	Output	Unit	Comment
	LTM	10065.9581776	lbf-in.	
	LTN	−2004.5965509	lbf	
	LTV	1637.2296434	lbf	
	k1	3.170466		Simplifying constants
	k2	0.499690		
	Di	23.344	in.	Inside diameter of pipe
24	Do		in.	Outside diameter of pipe
.328	tnom		in.	Nominal pipe wall thickness
	Ro	12	in.	Outside radius of pipe
	Ri	11.672	in.	Inside radius of pipe
3.1416	π			Metal area of pipe x-section
	Wreaction	7107.502462	lbs	Combined of metal weight plus water
30	L		ft	Length between pipe supports
.03	ρ			Density of fluid, lbs/in.^{3}
	be	24	in.	Effective length of pipe wall resisting bending due to shear
	S	0.485396	in.^{3}	Section modulus of resisting pipe section to shear moment
	σ	−31413.258210	psi	Bending stress due to shear moment
	Am	24.392689	in.^{2}	Cross sectional area of pipe
550	P		psi	Internal pressure
1	Ej			Weld joint efficiency
	σinner	51315.209430	psi	Total stress in inner shell − pressure + shear bending
	σp	19901.951220	psi	Internal pressure stress
	σouter	−11511.306991	psi	Total stress in outer shell − pressure + shear bending
	beff	27.070730		
0	CA			
	σallow	60000	psi	Allowable stress of pipe material at temperature
20000	Sm			Allowable stress for primary and secondary
35000	SMYS		psi	Specified Minimum Yield Strength of pipe material
	OK	'acceptable		

Figure 5-37b. cont'd.

it good engineering practice to place pipes 30 in. ϕ and larger on saddle supports. It is also noted that, for large diameter thin cylinders resting on the ground in stock yards and construction sites, angle beams 90° to one another are tack welded to the inside surface to prevent excessive ovaling.

In conclusion, a saddle support is required for the pipe if constructed new.

Note: The internal pressure stress was computed from the following formula, which is developed from Eq. 2-13(a), where $Y = 0.4$ and $D = R_o/2$:

$$t = \frac{PR_o}{\sigma EW + 0.4P}$$

where $W = 1$ for temperature less than 950°F, as in this case.

Solving for the stress we have the following:

$$\sigma = \frac{P}{E}\left(\frac{D_o}{2t} - 0.4\right)$$

where D_o = outside diameter
R_o = outside radius

References

1. A. Keith Escoe, *Mechanical Design of Process Systems*, Vol. 1, 2nd edition, 1994, Gulf Publishing Company, Houston, pp. 81–83.
2. A. Keith Escoe, *Mechanical Design of Process Systems*, Vol. 1, 2nd edition, 1994, Gulf Publishing Company, Houston, pp. 132–135.
3. J. P. Holman, *Heat Transfer*, 7th edition, 1990, McGraw Publishing Company, New York, pp. 335, 338.
4. Nicholas P. Cheremisinoff, *Heat Transfer Pocket Handbook*, 1984, Gulf Publishing Company, Houston, p. 106.
5. Alan E. Chapman, *Heat Transfer*, 3rd edition, 1974, Macmillan Publishing Company, New York.
6. Warren C. Young and Richard G. Budynas, *Roark's Formulas for Stress and Strain*, 7th edition, 2002, McGraw-Hill Publishing Company, New York.
7. Henry H. Bednar, *Pressure Vessel Design Handbook*, 2nd edition, 1986, Van Nostrand Reinhold Company, New York.
8. M. S. Troitsky, *Tubular Steel Structures, Theory and Design*, 2nd edition, 1990, The James Lincoln Arc Welding Foundation.

Chapter Six
Piping Maintenance and Repairs

Leaking Pipe Flanges and Hot Bolting

This chapter is oriented to in-plant piping maintenance and repairs. The methodology can be applied to pipelines, especially the sections on sleeve design. However, new developments are in the offing (e.g., the ASME B31.4 Committee, "Pipeline Transportation Systems For Liquid Hydrocarbons and Other Liquids," has passed a new ballot, and passed by the B31 Committee, that states that neither patches or half soles shall be installed on pipelines). These changes should appear in the ASME B31.4—2005 edition.

Hot bolting herein means to tighten the bolts on a flange while it is in service. Hot bolting has been used to stop small stable leaks or as a proactive measure in high temperature (over 800°F or 427°C) or cyclic services where there are temperature changes over 300°F (149°C) in less than 30 minutes. The vast majority of flanges bolted together properly should never have to be hot bolted, and most sites that have good maintenance do not routinely hot bolt flanges that are not leaking. However, process excursions or experience over time may require scheduled hot bolting. If hot bolting is planned, consideration should be given to coating the studs and nuts with electroless nickel to minimize corrosion-related binding.

There are safety precautions that need to be adhered to before proceeding with bolt torque. If there are hydrocarbons or other combustible gases in the atmosphere, a wrench can (and has) initiated a spark that can set off an explosion. Hot bolting is rarely done under the same conditions as cold bolting; thus, torque of the flange bolts is the most practical. The procedure for bolt torque regarding bolt torque increments and using the criss-cross method of bolt torque should be followed, and the reader is referred to the ASME PCC-1 [Reference 1] for further details.

The basic disadvantage to hot bolting is that the amount of preload remaining in the system is unknown. In addition, the torque values applied during initial joint assembly will no longer give the same stud stress due to the thread lubricity change with time at temperature and corrosion. Thus, hot bolting methods rely on limiting the amount of additional load applied to the joint.

Ideally if the line can be isolated the flange can be disconnected. We are assuming here that this is not possible. There are rare cases in systems that have temperatures close to ambient and are low in pressure where bolts are replaced individually. This is a very risky practice and should not be attempted in high-temperature and/or high-pressure service, or in lethal services.

Leak Sealing by Banding Flange or Wire Seal Peripheral Seal Repair

The simplest repair for a leaking flange is a band retaining ring or a barrier wire. The band retaining ring is shown in **Figure 6-1**, and the barrier wire configuration is shown in **Figure 6-2**. The band retaining ring comes with a band retaining cap with screws to grip and secure the space between the flanges. The wire wrap, which is perhaps more common, is simpler to install. These devices are recommended for low-pressure and relatively low-temperature services that are nonlethal. An example would be caustic soda at low pressure. Caustic soda tends to self-seal itself with the presence of salts that rapidly form a solid seal in itself. Another example would be low-temperature viscous fuel oil. These repairs are simple and quick to install.

Bolted Pipe Clamps

In operating process plants, it is often necessary to install temporary devices to keep the unit operating. If hot bolting does not stop the leak and isolation of the flange is impractical, leak sealing of the flange by banding, sealing, clamping, or boxing should be considered. Leaks do not always occur at flanges. They can occur at corroded or eroded locations in the parent pipe metal.

Typical barrier seal installation

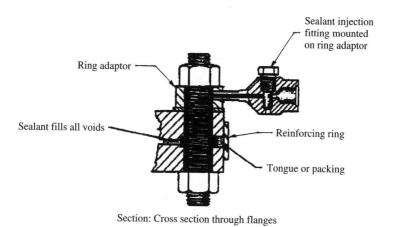

Reinforcing ring

Reinforcing ring
retaining cap screws

Tongue or packing

Sealant injection
fitting mounted
on ring adaptor

Sealant injection
fitting mounted
on ring adaptor

Ring adaptor

Sealant fills all voids

Reinforcing ring

Tongue or packing

Section: Cross section through flanges

Figure 6-1. Typical band sealing installation. (Courtesy of ExxonMobil, Inc.)

We will discuss bolted pipe clamps that are installed to seal off leaks. To accomplish this in a safe and efficient manner, the plant engineer must utilize various devices to prevent everyone's worst reality—an unscheduled shutdown. Unscheduled shutdowns not only interrupt the process facilities, and so halt the production of products, but also present a hardship on equipment. Particularly sudden trips in the system can result in thermal shock and equipment breakage. Bellows expansion joints are particularly sensitive to these unscheduled shutdowns.

This section deals with repairs requiring *non*-hot work. They are pre-assembled where if any welding or hot work is done, it is in the fabrication shop away from the process area where the leak exists. If necessary, the

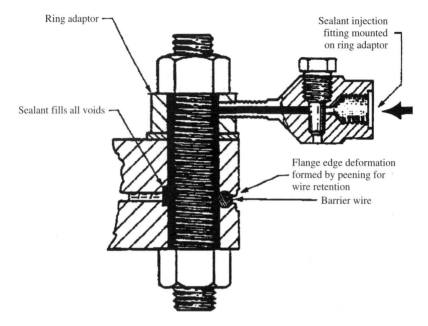

Ring adaptor

Sealant injection
fitting mounted
on ring adaptor

Sealant fills all voids

Flange edge deformation
formed by peening for
wire retention

Barrier wire

Figure 6-2. Typical barrier wire installation. Courtesy of ExxonMobil, Inc.

components are welded together to form a clamp or enclosure that are to be bolted to components in the field. This is very important because any welding or hot work performed on an operating line should fall under the guidelines of a hot tap. Components welded to the parent pipe become integral attachments, and thermal stresses can occur. Here we are concerned with bolted connections only. Bolts have a customary clearance between the bolt and bolt hole of 1/8 in. (3 mm), which can accommodate differential thermal expansion in most cases. End plates can have sharp shear edges that dig into the parent pipe, but that connection is not like fusion welding.

Various devices are used to keep a plant on stream. One such device is a clamp to seal off a leak. Areas of a unit that are prone to leaks are normally the weakest components in a piping system (e.g., flanges). Flanges are prone to leaks for many reasons, not the least of which is torquing the flange bolts out of sequence even using the proper bolt torque, using the correct gasket, or seating the gasket correctly. Maintenance crews vary across the world; some better than others. Installing flanges properly and torquing the bolts correctly cannot be overemphasized. A very helpful tool in this endeavor is the *Guidelines for Pressure Boundary Bolted Flange Joint Assembly, ASME PCC-1* [Reference 1]. Even following these helpful guidelines, sudden shutdowns can cause flanges to develop leaks. Another type of condition that can result in flange leaks is the thermal cycling of the piping. When a process fluid is cycled through various temperatures, the piping heats up and then cools down. After several of these cycles, depending on the temperature swings, the bolts eventually work themselves loose due to thermal ratcheting. Using jam nuts placed behind the hex nuts and spring-type washers will minimize this thermal ratcheting, but it will not completely prevent it. Usually in this type of piping systems the use of flanges is discouraged.

For our discussion here, we will assume that the leak has occurred. Such leaks can develop as the pipe corrodes, either internally or externally, and the process fluid exits the pipe wall into the atmosphere. This situation is highly undesirable, especially for hydrocarbons, which will ignite if exposed to a source of ignition. When a hydrocarbon leaks to the open atmosphere, a cloud can develop, and depending on the wind, a source of ignition may be present. If the gas cloud hits the source of ignition, a fire or explosion will happen. In the case of lethal substances (e.g., chlorine), even though an explosion is not possible with a source of ignition, the gas is lethal to personnel and animal life. There are countless lists of dangerous substances that, for many reasons, should be contained.

All bolted clamps should be treated as temporary (e.g., not for permanent) fixes. They are frowned upon in the pipeline industry because oftentimes a clamp is installed and simply forgotten about. In a long pipeline, a temporary repair can be installed out in the open country and left without proper inspection. In process plants, where clamp repairs are much more common, regular inspections are required to ensure their safety. Sometimes, in high-pressure systems, it is not uncommon to have the clamps re-injected with sealants to seal off the leak. This is not uncommon with high-pressure, high-temperature steam services. Generally, clamps probably are much more common in chemical plants than refineries. This is because chemical plants may contain highly corrosive substances that cause leaks to occur. The best proactive activity against leaks is to maintain an active inspection program to inspect the piping on regular intervals (e.g., a scheduled turnaround).

Vendors that manufacture and install clamps should be qualified as to their QA/QC (quality assurance/quality control) facilities, weld procedures, metallurgical capabilities, and shop capabilities. They should meet similar requirements as pressure vessel fabricators. There are many vendors throughout the world that do excellent work. It is the responsibility of the owner company of the operating unit to supervise the vendor's work, approve the vendor's facility, and, by all means, insist that calculations for the clamps be made and reviewed by the plant engineers. *Reviewing clamp calculations is an engineering function; inspection, maintenance, or operations do not fall under the purview of engineering.* An engineer needs to review and understand the methodology used for the clamp design. Clamps are very serious and should not be regarded lightly. Records should be kept of all calculations and materials, including sealants. These records should be filed by each operating unit superintendent and be scheduled for replacement during the next turnaround. Also a plant that requires many of these devices on a regular basis needs to reconsider its process conditions (process excursions) and maintenance procedures and then reconsider whether the correct metallurgy, gasket material, or type of flanges is being used. A typical situation is dealing with phosphoric acid when the pH level drops due to operational excursions and the acid literally eats itself out of the pipe wall. Another classic case is when the water in a cooling tower drops in pH and literally eats up tubes in heat exchangers downstream. This phenomenon is not so uncommon. If any of these factors are not right, clamps can become a very common experience. Usually no more than 5–10% of the flanges in a unit at most should have clamps. If clamps start appearing on the parent pipe material on a regular basis, then the pipe material probably should be considered for an upgrade. Some processes are naturally more leak-prone than others (e.g., chlorine-alkali units that separate chlorine and caustic soda from salt in chlorine cells). These plants generally tend to be more problematic than, say, ethylene plants in regard to corrosion.

In upgrading materials, it is a common mistake to assume that going from carbon steels to austenitic stainless steels is an "upgrade." Many times this is not true, especially if chlorides are present in the process stream. Another classic mistake is substituting carbon steel with titanium with powerful oxidizers (e.g., dry chlorine service). Although this mistake fortunately is not common, one must remember that chlorine reacts with titanium and the pipe can decompose. Going from carbon steel to a very exotic material such as titanium is a tremendous jump and should be flagged by everyone in the plant. This is not to say that it is necessarily wrong, but it needs to be looked at. One should try to avoid exotic materials because welding and decontaminating them can be very difficult. Using titanium because it is "high-tech" material is a dangerous mindset. The proper selection of any piping material needs the attention of a materials engineer before selection. If a small operating facility doesn't have a materials engineer, they should consider hiring a consultant to review material selection, especially if they do not have experience with a material being considered. One wants to stay with the simplest materials and work up to the exotics. Of course, this isn't always true; for example, caustic soda at high temperatures (depending on the concentration, normally above 210°F for 0% concentration down to 170°F for 100% concentration) normally requires nickel alloys without question.

For materials selection, the reader is referred to Hansen and Puyear [Reference 3 of Chapter 3]. If there is a leak at a flange in a caustic system, the caustic normally seals itself off by the salts that build up at the leak. This can happen before a clamp is installed.

Hydro test conditions are not routinely done in bolted clamp design, as ideally all bolted clamps discussed here are intended for temporary use. A permanent repair option should always be planned when installing a temporary repair. During a turnaround, all temporary repairs should be replaced with permanent repairs.

Flange Insert Clamps (Insert Ring or Tongue Clamps)

When leaks occur between two flanges, the single plane inset clamp is very common and cost-effective. There are simplistic devices that consist of two plates machined to fit between the two flanges. This configuration is shown in **Figure 6-3**.

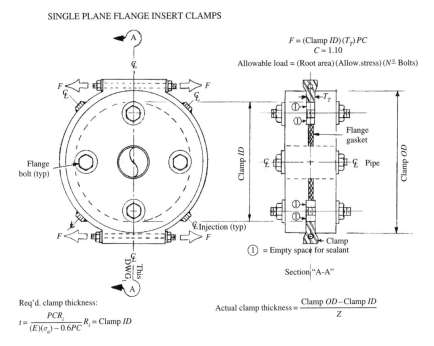

SINGLE PLANE FLANGE INSERT CLAMPS

$$F = (\text{Clamp } ID)(T_T)\,PC$$
$$C \approx 1.10$$

Allowable load = (Root area)(Allow.stress)($N^{\underline{o}}$ Bolts)

Req'd. clamp thickness:

$$t = \frac{PCR_i}{(E)(\sigma_a) - 0.6PC}\quad R_i = \text{Clamp } ID$$

$$\text{Actual clamp thickness} = \frac{\text{Clamp } OD - \text{Clamp } ID}{Z}$$

Figure 6-3. Flange insert clamp.

The two ring plates have holes tapped into the sides as shown. The sealant is injected at a minimum of 1.1 to 1.2 times the design pressure (some companies use a standard of 1.3 times). Note that, in any clamp, the bolts, not the sealant, hold the clamp together. The sealant acts to seal off the leak, acting much like a gasket. Sealants vary with clamp vendors, and they all are proprietary with each vendor. Typically, the type of sealant will vary with the type of process fluid being handled and the temperatures being generated in the process. This is very important because one does not want a sealant that will chemically react with the process fluid or become unstable at the process fluid temperature.

Figure 6-3 shows the two bolts that hold the assembly together. The thickness of the space between the flanges is T_T. Thus, the force acting on the bolts is

$$F = (Clamp\ ID)T_T PC_{inj},\ \text{lb}_\text{f}\ (\text{N})\qquad\text{Eq. 6-1}$$

where $C_{inj} = 1.1$
P = design pressure, psig (KPa$_g$)
T_T = distance between the flanges above the gasket, in. (mm)

The clamp ID will be specified by the clamp vendor, but it will meet the following criterion:

$$t_{reqd} = \frac{PC_{inj}(Clamp\ ID)}{E\sigma_A - 0.6PC}\qquad\text{Eq. 6-2}$$

where E = joint efficiency, usually 1.0, as the clamp is made of forged material or solid plate
σ_A = allowable stress of clamp material, psi (MPa)
C_{inj} = injection pressure factor, dimensionless

Typically clamps are designed to the ASME Section VIII Division 1 Boiler and Pressure Vessel code, as they are components, like flanges.

The allowable load on the bolts is as follows:

$$F_{allow} = A_B\sigma_B n\qquad\text{Eq. 6-3}$$

where A_B = the root area of each bolt, in.^{2} (mm^2)
σ_{BA} = allowable stress for the bolt material, psi (MPa)
n = number of bolts in clamp

One does not consider the flange bolts in the clamp bolt sizing. These insert clamps are quite common and are used ubiquitously because they are reliable and simplistic.

Simple Pipe Clamps with Single Plane Lug Plates

Simple pipe clamps with single plane lug plates are made from straight pieces of pipe, rolled plate, or piping elbows. They have lug or ear plates that are welded or formed on the sides in two pieces and then bolted together. Typically, the inside of the ear plate has grooves or channels formed with injection ports to inject the sealant. They are used on straight pipe or piping elbows. A straight pipe clamp is shown in **Figures 6-4a** and **6-4b**.

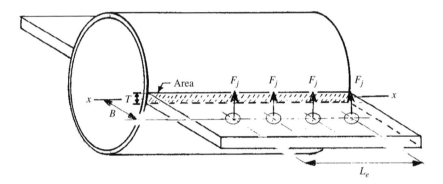

Figure 6-4a. Forces acting on clamp lug plate.

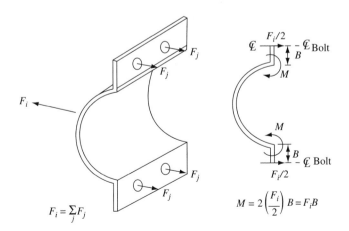

Figure 6-4b. Free body diagram of clamp lug plate.

The shaded region in **Figure 6-4a,** marked "AREA," represents the area of intersection between the lug and clamp, which is in two pieces. Since the curvature of the area is very small at the lug plate–clamp intersection, we can approximate it as a rectangular area as shown. The reaction forces acting at the bolts, marked F_j, resist the clamp from being separated. These reactions induce a bending moment about the *x-x* axis shown in the figure. The section modulus of the AREA is thus

$$Z_A = \frac{L_e T^2}{6} \qquad\qquad\qquad\text{Eq. 6-4}$$

where L_e = effective length, in. (mm)
$\quad\quad T$ = lug thickness, in. (mm)
$\quad\quad Z_A$ = section modulus, in.^{3} (mm^3)
$\quad\quad M$ = bending moment induced by injection pressure, where
$\quad\quad M = F_i B$, in.-lb$_f$ (J)
$\quad\quad P$ = design pressure $\times$ C, psig (KPa$_g$)
$\quad C_{inj}$ = injection pressure constant = 1.2

Solving for the stress, we have

$$\sigma = \frac{M}{Z_A} = \frac{F_i B}{\dfrac{L_e T}{6}} = \frac{6 F_i B}{L_e T^2} \qquad\qquad\text{Eq. 6-5}$$

where B = moment arm caused by the forces F_j, in.-lb$_f$ (J)

Referring to **Figure 6-5,** the reaction force (F_j) is derived as follows:

$$h = R \sin\theta$$
$$\text{AREA} = Gh = GR \sin\theta$$

$$F = 2PGR \int_0^{\frac{\pi}{2}} \sin\theta \, d\theta = -2PGR \left[\cos\left(\frac{\pi}{2}\right) - \cos(0) \right]$$

$$= -2PGR(0-1)$$

$$F = 2PGR = PGD$$

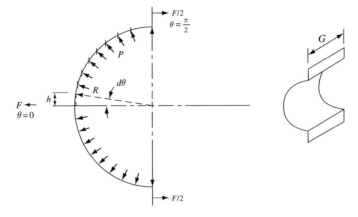

Figure 6-5. Loading from internal pressure on clamp wall.

$$F_j = \Sigma F_j = DGP, \text{ lb}_f \text{ (N)} \qquad\qquad \text{Eq. 6-6}$$

D = inside diameter of enclosure, in. (mm)
G = inside depth of enclosure, in. (mm)

Solving for the required lug plate thickness,

$$T = \sqrt{\frac{6F_iB}{L_e\sigma}} = 2.45\sqrt{\frac{F_iB}{L_e\sigma}} \qquad\qquad \text{Eq. 6-7}$$

where T = in. (mm)
L_e = length between bolt hole center lines on each end, in. (mm)

Now from **Figure 6-6** and the lug thickness/calculation data, we have

$B = W - 0.5g_1$
$L_e = 2L_i - nd_B$

where n = number of bolt holes
d_B = bolt hole diameter, in. (mm)
g_1 = sum of shell thickness and fillet weld length, in. (mm)
n = number of bolt holes
F_i = forces due to injection, lb$_f$ (N)
L_i = distance between outermost bolt centers, in. (mm)

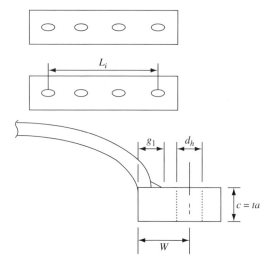

Figure 6-6. Dimensional parameters on clamp lug plate.

$2L_i$ = the distance between the outermost bolt center times two. The term L_i is multiplied by two because there are two flanges (lugs), as detailed in **Figure 6-6**. The term used in the ASME Section 8 Division 1 Appendix Y-9 is πC, where C is the bolt circle diameter for two circular flanges. The term πC is substituted for $2L_i$ for noncircular flanges (lugs).

σ_a = allowable stress at design temperature of lug material, psi (MPa)

Equation 6-7 becomes

$$T = 2.45\sqrt{\frac{F_i(W-0.5g_1)}{(2L_i-nd_B)\sigma_A}}$$
Eq. 6-8

Clamp Bolts

From the ASME Section 8 Division 1 Appendix Y Eq. 42, the required bolt area is

$$A'_B = \frac{\left[H + \dfrac{2M_p}{(A-C)}\right]}{\sigma_B}$$
Eq. 6-9

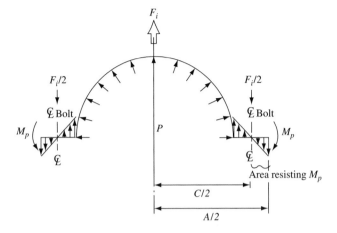

Figure 6-7. Bending moments at lug plate induced by internal pressure.

where $M_p = (F_i/2)(C/2)$, see **Figure 6-7**

H = hydrostatic end force on area inside of flange (same as F_i), lb$_f$ (N)

A = outside diameter of flange, in. (mm)

C = bolt circle diameter (for circular flange), in. (mm)

Referring to **Figure 6-7**, the part of the lug, or ear, that is in compression from the moment (M_p) is $(A - C)/2$. This is the effective area that is resisting the bending moment. Equation 42 in Appendix Y was written for circular flanges. Rewriting the equation, we solve for the total bolt area as

$$A_B' = \frac{\left[F_i + \dfrac{2M_p}{2L_i + (A-C)} \right]}{\sigma_B}$$ Eq. 6-10

The required bolt area for each bolt is

$$A_B = A_B'/n$$

where, again, n = number of bolts.

It is recommended when designing bolting connections that the plate minimum edge distance from the bolt center to the edge plate be per

AISC (American Society of Steel Construction, 9th edition), Table J3.5, p. 5-76 [Reference 2]. This table is shown as **Table 6-1**.
   Bolt dimensions and areas are shown in **Figure 6-8**.

Two Planar Clamps

Two planar clamps are clamps where lug plates or ears exist along two axes. They are used to fit around (secure) branch connections, tees, and valves. They are a mere extension of single plane clamps mentioned above. Shown in **Figure 6-9** is an example. The two diameters of the clamp are D_1 and D_2. The forces acting on the two planes are denoted as subscript H for the header and the subscript B for the branch. Thus, the internal pressure forces acting on the bolts along the header and branch are

$$F_{iH} = D_1 G_H P \text{ and } F_{iB} = D_2 G_B P \qquad \text{Eq. 6-11}$$

Let A_{bT} = total bolt area required, in.^{2} (mm^2)

The moments acting in each respective axis are as follows:

$$M_{PH} = \left(\frac{F_{iH}}{2}\right)\left(\frac{C_H}{2}\right) \qquad \text{Eq. 6-12}$$

Table 6-1
Minimum Edge Distance (Center of Standard Hole[1] to Edge of Connected Part)

Nominal Bolt or Rivet Diameter, in.	At Sheared Edges, in.	At Rolled Edges of Plates, Shapes or Bars, Gas Cut or Saw-cut Edges, in.[2]
1/2	7/8	3/4
5/8	1 1/8	7/8
3/4	1 1/4	1
7/8	1 1/2[3]	1 1/8
1	1 3/4	1 1/4
1 1/8	2	1 1/2
1 1/4	2 1/4	1 5/8
Over 1 1/4	1 3/4 × Dia.	1 1/4 × Dia.

Notes:
1. For oversized or slotted holes, see Table J3.6 in AISC.
2. All edge distances in this column may be reduced 1/8 in. when the hole is at a point where stress does not exceed 25% of the maximum design strength of the element.
3. These may be 1 1/4 in. at the ends of beam connection angles.

Bolt Size d	Standard Thread No. of Threads	Standard Thread Root Area	8-thread Series No. of Threads	8-thread Series Root Area	Bolt Spacing* Minimum B_s	Bolt Spacing* Preferred	Minimum Radial Distance R	Edge Distance E	Nut Dimension (across flats)	Maximum Fillet Radius r
$1/2''$	13	0.126	No. 8		$1 1/4''$	$3''$	$13/16''$	$5/8''$	$7/8''$	$1/4''$
$5/8''$	11	0.202	thread		$1 1/2''$	3	$15/16''$	$3/4$	$1 1/16$	$5/16$
$3/4''$	10	0.302	series		$1 3/4''$	3	$1 1/8$	$13/16$	$1 1/4$	$3/8$
$7/8''$	9	0.419	below $1''$		$2 1/16''$	3	$1 1/4$	$15/16$	$1 7/16$	$3/8$
$1''$	8	0.551	8	0.551	$2 1/4''$	3	$1 3/8$	$1 1/16$	$1 5/8''$	$7/16$
$1 1/8''$	7	0.693	8	0.728	$2 1/2''$	3	$1 1/2$	$1 1/8$	$1 13/16$	$7/16$
$1 1/4''$	7	0.890	8	0.929	$2 13/16''$	3	$1 3/4$	$1 1/4$	2	$9/16$
$1 3/8''$	6	1.054	8	1.155	$3 1/16''$		$1 7/8$	$1 3/8$	$2 3/16$	$9/16$
$1 1/2''$	6	1.294	8	1.405	$3 1/4''$		2	$1 1/2$	$2 3/8$	$5/8$
$1 5/8''$	$5 1/2$	1.515	8	1.680	$3 1/2''$		$2 1/8$	$1 5/8$	$2 9/16$	$5/8$
$1 3/4''$	5	1.741	8	1.980	$3 3/4''$		$2 1/4$	$1 3/4$	$2 3/4$	$5/8$
$1 7/8''$	5	2.049	8	2.304	4		$2 3/8$	$1 7/8$	$2 15/16$	$5/8$
$2''$	$4 1/2$	2.300	8	2.652	$4 1/4''$		$2 1/2$	2	$3 1/8$	$11/16$
$2 1/4''$	$4 1/2$	3.020	8	3.423	$4 3/4''$		$2 3/4$	$2 1/4$	$3 1/2$	$11/16$
$2 1/2''$	4	3.715	8	4.292	$5 1/4''$		$3 1/16$	$2 3/8$	$3 7/8$	$13/16$
$2 3/4''$	4	4.618	8	5.259	$5 3/4''$		$3 3/8$	$2 5/8$	$4 1/4$	$7/8$
$3''$	4	5.621	8	6.324	$6 1/4''$		$3 5/8$	$2 7/8$	$4 5/8$	$15/16$

* B_s = center-to-center distance between bolts, inches

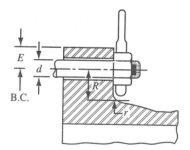

Figure 6-8. Bolt size parameters.

$$M_{PB} = \left(\frac{F_{iB}}{2}\right)\left(\frac{C_B}{2}\right) \qquad\qquad \text{Eq. 6-13}$$

The required bolt area for the header lug plate is

$$A_{bTH} = \frac{F_{iH} + \dfrac{2M_{PH}}{2L_{eH} + \dfrac{(A_H - C_H)}{2}}}{\sigma_{AB}}$$

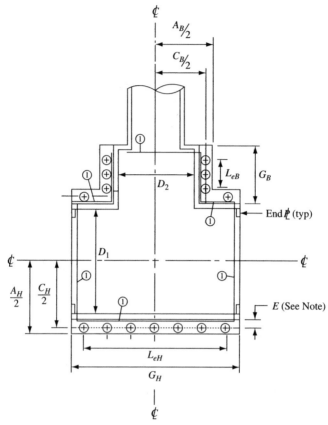

Note:

$$E_{min} = \begin{pmatrix} Bolt \\ size \end{pmatrix} + \begin{pmatrix} Injection\ channel \\ seal\ diameter \end{pmatrix}$$

$$E_{max} = \begin{pmatrix} Bolt \\ size \end{pmatrix} + \begin{pmatrix} Minimum\ edge\ distance \\ in\ AISC\ Table\ J3.5 \end{pmatrix}$$

① = Denotes sealant cavity

Figure 6-9. Two planar clamp.

Similarly, the required bolt area for the branch lug plate is

$$A_{bB} = \frac{F_{iB} + \dfrac{2M_{PB}}{2L_{eB} + \dfrac{(A_B - C_B)}{2}}}{\sigma_{AB}}$$

Adding the two required bolt areas for both the header and branch, we have

$$A_{bT} = \frac{F_{iH} + \dfrac{2M_{PH}}{2L_{eB} + \dfrac{(A_H - C_H)}{2}} + F_{iB} + \dfrac{2M_{PB}}{L_{eH} + \dfrac{(A_B - C_B)}{2}}}{\sigma_{AB}}$$

Eq. 6-14

where s_{AB} = allowable bolt stress, psi (MPa)

In this equation L_{eH} is not multiplied by 2 because there is no credit for bolts on the opposite end of the header.

The required thickness of the lug plates or ears is the greater of the following:

$$T_H = 2.45\sqrt{\frac{F_{iH}(W_H - 0.5g_1)}{(L_{eff} - nbd_B)\sigma_A}}$$

Eq. 6-15

$$T_B = 2.45\sqrt{\frac{F_{iB}(W_B - 0.5g_2)}{(L_{eff} - nd_B)\sigma_A}}$$

Eq. 6-16

where $L_{eff} = L_{eH} + 2L_{eB}$
$\quad\quad n$ = total number of bolts
$\quad\quad d_B$ = bolt hole diameter, in. (mm)
$\quad\quad s_A$ = allowable stress of lug plate (ear) material, psi (MPa)

The allowable stress of the lug plate should be the same as the clamp material. The clamp material should be identical, or as close as possible, to that of the pipe, or piping component, material.

The end plate thickness is per ASME Section VIII Division 1, Paragraph UG-34(c)(2); the minimum required thickness of flat unstayed circular heads, covers, and blind flanges shall be calculated by the following formula:

$$t_{EP} = d\sqrt{\frac{CP}{\sigma_A E}}$$

Eq. 6-17

where C = defined in Figure UG-34 (usually $C = 0.33$ for welded connec-
$\quad\quad\quad$ tions and $C = 0.75$ for threaded connections), dimensionless
$\quad\quad d$ = diameter of end plate (inside diameter of enclosure), in.

P = internal pressure, psi (MPa)

σ_A = allowable stress, same as lug plate (same material), psi (MPa)

E = weld joint efficiency (normally 0.7 without radiograph)

Elbow Clamps

Because of their geometry, piping elbows exhibit unequal pressure loadings at the intrados (inside of elbow) and at the extrados (outside of elbow). The internal pressure at the intrados is $1.28P$, with P being the internal pressure. The internal pressure at the extrados is $0.87P$. The lug plates containing the bolts are in the plane of the unequal pressure loading. This makes the bolts perpendicular to the equal pressure loading, which is P. There are two types—a 45° elbow clamp and a 90° clamp. Shown in **Figure 6-10** is the 45° elbow clamp.

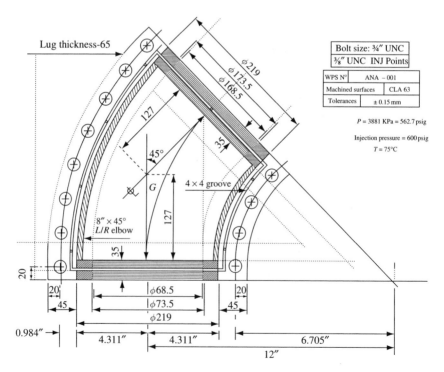

Figure 6-10. A 45° elbow clamp.

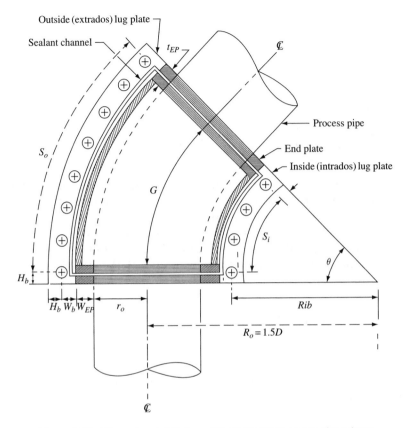

Figure 6-11. Dimension details for a 45° elbow clamp on top of a column.

Fabrication details are shown in **Figure 6-11**, showing the glands for the sealant for a 45° elbow clamp.

In **Figure 6-9**, the force resulting from internal pressure and sealant acting on the bolts is

$$F_i = P_i GD \qquad\qquad \text{Eq. 6-18}$$

where P_i = design pressure in process pipe times an injection factor

Most injection of the sealant is done at 10% over the design pressure, so

$$P_i = P_d(C_{inj})$$

where C_{inj} = 1.1 or whatever is used
P_d = design pressure of process pipe, psig (KPa$_g$)

$$G = 1.5D\left(\frac{\theta\pi}{180}\right) - 2t_{EP}$$ Eq. 6-19

where G = arc length of the centerline that is on the centerline of the
    elbow, in. (mm)
  D = nominal diameter of the pipe, in. (mm)
  t_{EP} = end plate thickness, in. (mm)
  θ = angle of elbow, usually 45° or 90°, degrees

The same equations for a straight clamp hold for the elbow clamp
because of the orientation of the lug plates. This is why elbow
clamps are fabricated with one lug plate on the extrados and the other
on the intrados. Equation 6-8 is calculated different from that of
two planar clamps. Referring to **Figure 6-11**, the outer arc connecting the centerlines of the bolts at each end of the lug plate on the
extrados is

$$S_o = \frac{(R_o + W_{EP} + W_b + r_o)\theta\pi}{180} - 2H_b$$ Eq. 6-20

The arc length connecting the centerlines at each end of the lug plate on
the intrados is

$$S_i = \frac{(Rib)\theta\pi}{180} - 2H_b$$ Eq. 6-21

where W_{EP} = width of the end plate, in. (mm)
  W_b = distance from centerline of bolt hole to inside surface of
    lug plate, in. (mm)
  r_o = OD of process pipe, in. (mm)
  R_o = bend radius of elbow clamp, in. (mm)
  H_b = distance from bolt centerline to edge of lug, in.
  θ = angle of bend (45° or 90°), degrees

The effective length of the bolts is now

$$L_{eff} = S_o + S_i$$ Eq. 6-22

The lug plate thickness is calculated in much the same way as Eqs. 6-14 and 6-15 as

$$T = 2.45\sqrt{\frac{F_i(W - 0.5g_1)}{(L_{eff} - nbd_B)\sigma_A}}$$ Eq. 6-23

where W and g_1 are defined in **Figure 6-6**, with W being the distance from inside the clamp enclosure to the bolt centerline.

The moment acting on the lug plate bolts is similar to that already discussed:

$$M_p = \frac{F_i C}{4}$$ Eq. 6-24

where C = distance from clamp centerline to lug bolt center, in. (mm).

The required area for the anchor bolts is calculated from Eq. 6-10. Note that installing a clamp on an elbow will eliminate any flexibility of the elbow. If the elbow is used in the piping system for thermal flexibility, new stress computations are in order because the elbow with a clamp becomes a rigid joint. This could affect the loadings on nozzles and pipe supports. Since the lug plates are in plane to thermal movement, any flexibility in this direction is lost. There will be some flexibility in the out-of-plane direction, but it would be safe and conservative to assume none. Placing a clamp on an elbow is like placing an angle valve at that point in terms of flexibility.

For large thin-walled elbows, the Bourdon effect can be pronounced. This effect occurs when the pressure increases and tends to straighten out (or "open up") the elbow, working just like Bourdon tubes in pressure gages. This effect will not exist because the lug plates make the elbow practically a rigid joint.

Mitered Elbow Clamps

Miter elbows are fabricated from pipe and consequently are not considered as fittings. Some companies forbid their use, but for large elbows they are the only option available. One must remember their limitations. Because of their stress concentrations at the miter joints, they should be restricted to relatively low pressure systems. Also the velocity heads are much higher than conventional flow because of the geometrical discontinuity at the miters. These factors must be considered in their use. The application of miter elbows is predominately restricted to in-plant piping. The ASME B31.3, Process Piping, gives criteria for miter bends in Paragraph 304.2.3. We will concentrate on multiple miter bends. The interested reader is referred to the code for single miter bends, in which the stress concentrations and velocity heads are more pronounced than for multiple miter bends. Paragraph 304.2.3 (a) for multiple miter bends gives two equations that give the maximum allowable internal pressure for the miter. The lesser of the two values calculated shall be the maximum allowable internal pressure. Referring to **Figure 6-12**, the equations are as follows:

$$P_m = \frac{SE(T - c)}{r_2}\left(\frac{T - c}{(T - c) + 0.643\tan\theta\sqrt{r_2(T - c)}}\right) \qquad \text{Eq. 6-25}$$

$$P_m = \frac{SE(T - c)}{r_2}\left(\frac{R_1 - r_2}{R_1 - 0.5r_2}\right) \qquad \text{Eq. 6-26}$$

Using the lesser pressure from these equations, the miter clamp will be designed like any other clamp. These pressure values are used for the clamp wall thickness. The remaining equations are the same as those discussed previously. A mitered elbow clamp is shown in **Figure 6-12b**.

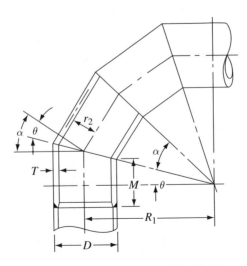

Figure 6-12a. Nomenclature for mitered elbows.

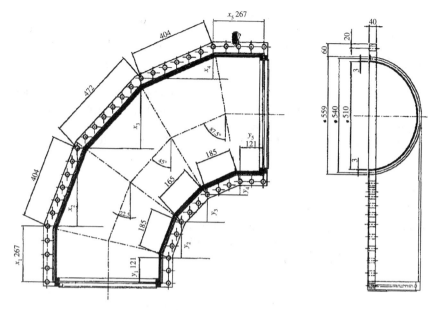

Figure 6-12b. A mitered elbow clamp for an overhead line on the top of a process column.

Clamps with Thrust Loads

Shear Pins and Serrated Teeth Connections

Sometimes a clamp may be subject to thrust loads that want to separate the clamp. In these circumstances, shear pins can be used to counter the thrust load in case the fluid escapes from the pipe and enters the clamp cavity. The thrust loads develop when there is a change in the cross section of the component being clamped (e.g., a pipe reducer, like a swage) or when the pipe being clamped becomes detached. Shear pins are typically used in high-pressure applications involving services like hydrogen and super-heated steam where the thrust loads can become high. Shear pins are a backup to the bolting on the clamp to ensure that the clamp has adequate strength. The bolting on the clamp should be the primary components to withstand all loadings. The shear pins are nonpressure components, so we can use the American Institute for Steel Construction criteria for shear.

For thrust loads where shear pins are not adequate, tie rods attached to lug plates or ears welded to the clamp may be necessary, but these are generally the exception.

In the case of the pipe reducer, the contained pipe diameter varies, making the magnitude of the thrust force the product of the difference in area between the two cross sections times the injection pressure of the sealant. This would be for a particular design of a box clamp that is rectangular in configuration and has an open cavity like that shown in **Figure 6-13**. The thrust force is developed by the unequal areas of the sealant cavity.

One must be careful in using the Type (a) construction because thermal expansion stresses must be considered. The Type (b) construction is much more common. In both types of construction, whether the process fluid will leak into the clamp chamber must be analyzed. The thrust force induced by the difference in diameter of the end plates is calculated as follows:

$$\Delta A = \frac{\pi}{4} (d_1^2 - d_2^2)$$

where d_1 = diameter of the larger end plate
d_2 = diameter of the smaller end plate
ΔA = differential area

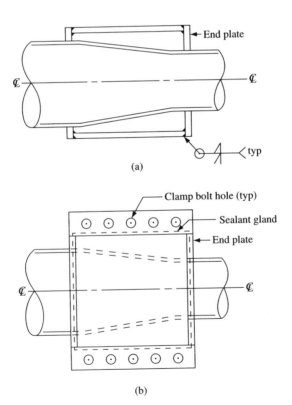

(a)

(b)

Figure 6-13. Type (a) is the welded version and Type (b) is the bolted version.

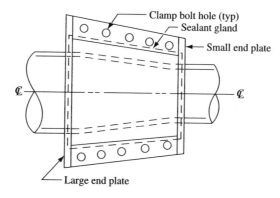

Figure 6-14. Tapered bolted clamp for reducer.

fluid force $= F_p = \Delta A(P)$, lb$_f$ (N)

where P = internal pressure, psig (KPa$_g$)

This can be avoided by using a configuration shown in **Figure 6-14**, where the clamp follows the outside surface of the reducer. The problem with the configuration in **Figure 6-14** is that the clamp is awkward to fabricate.

In cases where the contained pipe can become detached, the clamp bolts have to be checked for the thrust force. If the bolts' clamps are not adequate, then shear pins can be installed. Holes for the shear pins are typically drilled through the clamp wall. Tapped holes are drilled partially into the component (pipe, flange, fitting, etc.) wall for the pins to engage. The shear pin protrudes through the hole in the clamp wall and into the tapped hole in the component. The tapped hole in the component either is threaded and the shear pin is screwed in or has a tight fit and is hammered into place. In the latter case, the shear pin can be pointed and hammered into the component wall. One must use care to not damage the component.

In most applications, shear pins can be avoided. In many cases, a more practical approach is to machine serrated teeth into the end plates. These teeth are usually 2.5 mm long and are three to six in number. As the lug bolts are tightened, the teeth dig into the pipe and form a very tight connection. This approach has been proven to work in the field for many years and saves the time and effort of drilling holes into the clamp for shear pins. This type of arrangement is shown in **Figure 6-15** on an end plate.

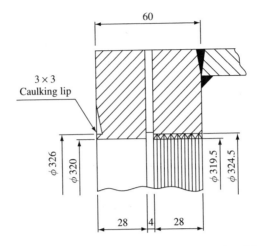

Figure 6-15. Serrated teeth at end of clamp to restrain axial forces.

The serrated teeth usually are formed at an angle of 60° and are made of a material with an SMYS equal to or greater than the parent pipe. The concept is similar to slip connections used on oil field equipment in retaining tubes. For applications such as a pipe welded to a flange that has a crack at the pipe-flange juncture for a high-pressure hydrogen line, shear pins are justified, but serrated teeth have proven experience in the field for many years.

Note that the enclosure-type clamps generally do not work on pipe (or pipelines) that have an oval shape, or out-of-roundness fabrication that is out of tolerance of applicable codes. Generally this type of fabrication error should be caught during the process of procurement of the pipe or pipeline.

Sealants

The kind of bolted clamps discussed earlier are accomplished by application of sealants. The sealant seals off the leak, but the bolts hold the assembly together. The sealant is injected into a cavity that runs along the inside perimeter of the clamps on the clamp periphery at usually 1.1 times the process design pressure. Normally, multiple injection ports are typically located equidistant around the clamp. A shutoff adapter is located at each injection point, and these are kept open during the sealant injection process. Sealant injection is initiated at a point farthest from the leak, continuing progressively on either side until only one injection point remains. Sealant is injected at one location until it is seen to extrude from an open adjacent adapter. Sealant injection then progresses to another point, and the process is repeated.

Sealants are injected into the sealant enclosure using a hand-operated hydraulic pump similar to a grease gun that extrudes the material into the cavity. A typical sealant injection pump can produce a pressure up to 2000–3000 psig (13,793–20,690 KPa$_g$) against the sealant, and higher pressure pumps are available. From a theoretical view, consideration should be given to limiting the differential pressure between the sealant pressure and the pipe internal pressure below the critical buckling pressure of the pipe. However, this is normally not a matter of concern (provided an experienced leak sealant sealing contractor is used and appropriate injection procedures are followed) for the following reasons:

- The actual pressure at the pipe outside diameter is less than the maximum pressure that the injection pump can produce due to the pressure drop in the hoses and the fittings.
- The technician should only apply the amount of pressure needed to get the sealant to flow and overcome the pipe operating pressure; hence, using 1.1 times the design pressure is normal. Some applications can go as high as 1.2 times the design pressure.
- As long as the shutoff adapters are kept open, they provide a path to relieve any pressure buildup at the pipe OD, keeping the external differential pressure low.

Before injecting a clamp or enclosure, the amount of sealant required to fill the cavity should be determined and monitored during injection to minimize the potential for injecting sealant into the line or for overloading the pipe component.

Sealant Material Considerations

Sealants fall into the specialty of nonmetallic materials. It is beyond the scope of this book to cover this subject, as many volumes could be written about it. Sealant materials are designed to flow plastically into the enclosure cavity when subjected to the injection gun pressure. The sealant compound cures at the pipe operating temperature and forms a solid, homogeneous molding as a barrier to the leak as shown in **Figure 6-16**.

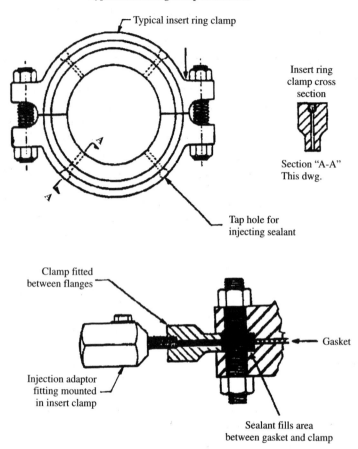

Figure 6-16. Cavity for sealant forms seal against leak of process fluid. Courtesy of ExxonMobil, Inc.

Sealants are proprietary with each contractor, although they generally come from two or three common sources. The sealants are rated to the application temperature, and a contractor normally has at least eight or nine sealants available for various temperature ranges.

Sealant compounds are made from a variety of materials, including thermosetting rubbers, fiber-reinforced rubbers, PTFE (Poly Tetra Fluoro Ethylene), resins, and fiber/graphite mixtures. Sealants are selected by the leak-sealing contractor for the specific application based on the design temperature and chemical compatibility with the process fluid. Factors influencing the selection of the appropriate sealant are as follows:

1. Sealants are selected to ensure they will cure and form a molding at the operating conditions. Note that some sealants are not designed to harden upon curing.
2. Sealants are selected so that there is no deterioration of the material at the design temperature. Compounds for use at temperatures up to 1100°F (600°C) are available and commonly used.
3. Sealants must be resistant to deterioration to chemical attack.
4. The effect that sealant trace impurities may have on the piping material should be assessed. In cases where sealant impurities create a risk for stress corrosion cracking of the piping or clamp, the use of high purity grade nuclear sealant compounds should be considered. Contractors can provide sealants with levels of chloride, sulfur, and fluoride below 0.100 ppm.

The shelf life of sealant compounds is limited by storage temperature. The procedures used by the leak-sealing contractor should be reviewed to assess whether a program to ensure sealant quality is available.

Sometimes when a sealant is injected, it can cure and become very hard, adhering to the clamp. In these cases, pneumatic hammers have to be used to remove the sealant. This is not always the case, but it can happen in some applications.

It is critical that the sealant contractor understand the process fluid inside the pipe and the most severe temperatures and pressures. At higher temperatures, it is not uncommon for the exposure of the sealant to heat to form an exothermic reaction, generating smoke. Smoke can exit from the clamp and/or bolt holes. Generally, this smoking can last from 3 days to a week. If this continues over a long period of time, then perhaps the wrong sealant was used. It is critical to make certain that the sealant used is compatible with the process fluid. Normally, with complex mixtures (e.g., diesel fumes mixed with hydrogen gas), selecting the correct sealant at high temperatures can be tricky. In these cases, it is best to consult with the sealant manufacturer—if time permits. Often these devices are installed under emergency conditions, and there is little time for consultation. For this reason, it is best to establish a contingency plan before an incident to plan for the correct type of sealant. Many technicians installing clamps know very little about the chemistry of the sealants, other than charts that give them guidance on which sealant to use for a particular service at various temperatures. Contingency plans are a must for complex mixtures.

Generally, for hot services, clamp vendors like to have sealant injection ports every 4 in. (102 mm) around the outside of the clamp. Depending on the viscosity of the sealant and the cure rate, this distance can be critical.

Re-injection of Leak Seal Repairs

In the event that a leak-sealed repair develops another leak in service, a sealant re-injection may be needed to reestablish a seal. Prior to re-injection, the piping integrity, as well as the clamp or enclosure integrity, should be assessed by inspection.

Sealant injection for resealing should be performed as previously described. Because these re-injections cause additional loading on the piping being sealed, injection pressure should be minimized to the extent needed so as not to damage the piping.

After the piping has been resealed, any further re-injections should be subject to an engineering assessment to confirm the integrity of the repair as well as to confirm that re-injection would not cause overloading of the piping or yielding of the studs.

Re-injection is not uncommon and should not be considered as a failure of the clamp. Sometimes when a sealant is exposed to heat it shrinks because it is curing, thus re-injection is required. Situations involving high-pressure superheated steam sometimes require re-injection, especially if the steam is leaking from the pipe component or flange. This happens in any service that involves high temperatures. Re-injecting also may be required if a clamp has been in service for an extended period of time and the sealant becomes brittle, forming pockets or voids. This is unusual, but it can happen. Most leak sealant contractors give a warranty period during which re-injection is done free of charge. If a bolted clamp is to be in for an extended period of time, temporary inspections are necessary to ensure that the sealant is functional. After curing, most sealants form into a very hard substance, becoming integral with the clamp. Most re-injections occur during the initial phases of the clamp installation while the sealant is curing.

Clamp Example 1

A 16″ ϕ gate valve in ethylene at 100% service developed a leak on the bonnet flange. Shutdown is not an option, making necessary a temporary repair. The internal pressure is 7600 KPa$_g$ with a temperature range from $-107°C$ to $46°C$. The material is 304 SS. A flange insert clamp is used, and the design calculations are as follows:

Required Clamp Thickness:

$P = 7600$; $KPa_g = 1102$ psig

The weld joint efficiency is 0.7, because there is no radiography. We use 1.1 times the design pressure $(C = 1.1)$ for injecting the sealant. The allowable stress of the 304 SS is 18,800 psi.

The pipe OD is the clamp ID. Thus using Eq. 6-2, we have for the internal pressure the required clamp wall thickness is, with $E = 0.7$ for no radiography,

$$t_r = \frac{PCR_o}{\sigma_A E - 0.6P} = \frac{(1102)(1.1)(8)}{(18800)(0.7) - (0.6)(1.1)(1102)} = 0.78 \text{ in.}$$

Eq. 6-27

Actual clamp thickness is 56 mm $= 2.20$ in. > 0.78 in. required.

Load on Clamp Bolts: From Eq. 6-1 and referring to **Figure 6-17**, the load on the flange bolts is using a tongue depth (TT) of 0.28 in. (7.0 mm)

$$F = (16)(0.28)(1102)(1.1) = 5430.7 \text{ lb}_f$$

For two $\frac{3}{4}''$ coarse thread bolts, referring to **Figure 6-8** above, the root area of each bolt is 0.302 in^2. The bolt was of SA-193 B8 (per the current ASME Section II Part D at the time). Using Eq. 6-3, the maximum allowable load on each bolt is

$$F_{allow} = (0.302)(18800)(2) = 11,355.2 \text{ lb}_f$$

Since the maximum allowable load is greater than the actual load, the clamp is acceptable. The sealant (proprietary to the clamp vendor) was injected, and the clamp held successfully until the next turnaround three years later. It was properly removed during the turnaround, and a new gasket was placed and properly installed. This was the permanent repair.

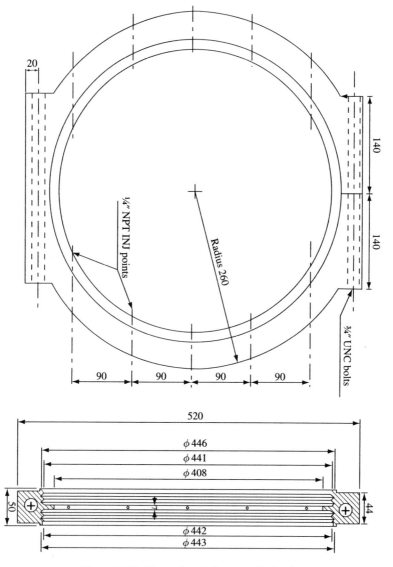

Figure 6-17. Flange insert clamp repair details.

Clamp Example 2

A valve bonnet flange connection on a 16″ ϕ gate valve is leaking. A clamp is proposed by a contractor with a clamp design. The design is shown in **Figure 6-18**. We are asked to assess the clamp.

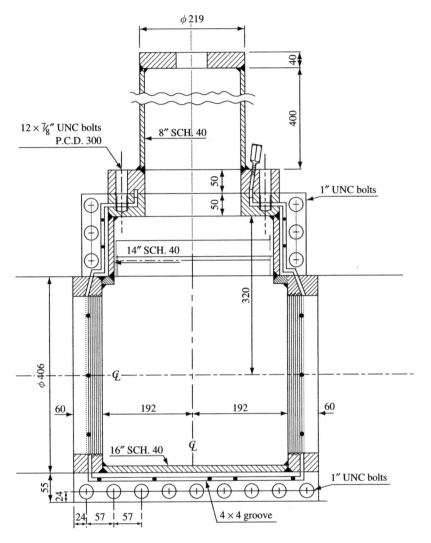

Figure 6-18a. Valve main body repair details (units shown are in mm).

The sealant injection pressure shall be 10% over design pressure inside the valve. The internal diameters of the 16″ ϕ and 14″ ϕ portions are as follows:

$D_1 = 16.0 - 2(0.5) = 15.0$ in.
$D_2 = 14.0 - 2(0.437) = 13.126$ in.

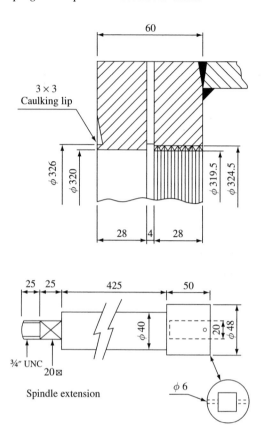

Figure 6-18b. Valve clamp extension component detail (units shown are mm).

Now,

$$G_H = (2)(24) + 8(57) = 504 \text{ mm} = 19.842 \text{ in.}$$
$$G_B = (320/25.4) - 16.0/2 + (50/25.4) = 6.567 \text{ in.}$$
$$P = 300(1.1) = 330 \text{ psig} = \text{injection pressure}$$
$$C_H/2 = (16.0/2) + (55 - 24)/25.4 = 9.22 \text{ in.}$$
$$C_B/2 = (14.0/2) + (55 - 24)/25.4 = 8.22 \text{ in.}$$

With the same size lug plates and bolts on the branch and header planes,

$$W_H = W_B = (55 - 24)/25.4 = 1.22 \text{ in.}$$

$$L_{eH} = 8(57/25.4) = 17.952 \text{ in.}$$

$$L_{eB} = 2(57/25.4) = 4.488 \text{ in.}$$

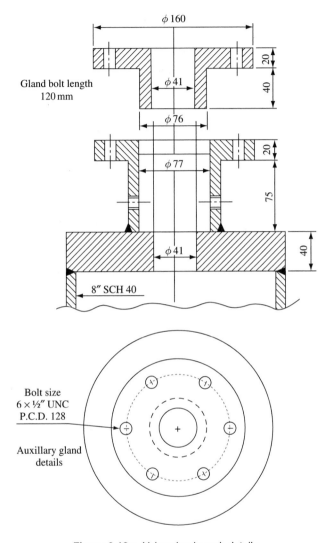

Figure 6-18c. Valve gland repair details.

Because there are bolts only on one side of the clamp in the horizontal plane,

$$L_{eff} = L_{eH} + 2L_{eB} = 17.952 + 2(4.488) = 26.928 \text{ in.}$$

$$F_{iH} = D_1 G_H P = (15.0)(19.842)(330) = 98,217.9 \text{ lb}_f$$

$$F_{iB} = D_2 G_B P = (13.126)(6.567)(330) = 28,445.5 \text{ lb}_f$$

$$M_{PH} = \left(\frac{F_{iH}}{2}\right)\left(\frac{C_H}{2}\right) = \left(\frac{98,217.9}{2}\right)(9.22) = 452,784.5 \text{ in.-lb}_f$$

$$M_{PB} = \left(\frac{F_{iB}}{2}\right)\left(\frac{C_B}{2}\right) = \left(\frac{28,445.5}{2}\right)(8.22) = 116,911.0 \text{ in.-lb}_f$$

Now using Eq. 6-14, we have the required bolt area as

$$A_{bT} = \frac{98,217.9 \text{ lb}_f + \dfrac{2(452784.5) \text{ in.-lb}_f}{2(4.488) \text{ in.} + \dfrac{(20.33-18.44) \text{ in.}}{2}} + 28,445.5 \text{ lb}_f + \dfrac{2(116,911.0) \text{ in.-lb}_f}{17.952 \text{ in.} + \dfrac{(18.33 - 16.44) \text{ in.}}{2}}}{(40,000)\left(\dfrac{\text{lb}_f}{\text{in.}^2}\right)}$$

$$A_{bT} = 5.75 \text{ in.}^2$$

where N = number of bolts = 15
  A_{bR} = root area of bolts, in.^{2}
  A_{brq} = required root area of each bolt, in.^{2}
  $A_{brq} = \dfrac{5.75 \text{ in.}^2}{15} = 0.383 \text{ in.}^2 < 0.551 \text{ in.}^2$ root area for a 1 in. coarse
    thread bolt

Thus, the bolts are satisfactory.

  $g_1 = 0.50 + 0.25 = 0.75$ in.
  $g_2 = 0.437 + 0.25 = 0.687$ in.
  d_B = diameter of bolt hole, in.
  $E = 0.7$ for no radiography

For a 1 in. bolt, $d_B = 1.125$ in., the required lug thickness on the horizontal plane is

$$T = 8.625\sqrt{\frac{(0.33)(330)}{(20000)(0.7)}}$$
$$= 0.760 \text{ in.} < 50 \text{ mm } (1.9685 \text{ in.}) \text{ plate used}$$

For the vertical lug plate,

$$T_H = 2.45\sqrt{\frac{(28,445.5)[1.22 - 0.5(0.687)]}{[26.928 - (15)(1.125)](20000)}}$$

$$= 0.863 \text{ in.} < 50 \text{ mm (1.9685 in.) plate used}$$

Using Eq. 6-20, the required top end plate thickness is

$$T = 8.625\sqrt{\frac{(0.33)(330)}{(20000)(0.7)}}$$

$$= 0.760 \text{ in.} < 40 \text{ mm (1.575 in.) plate used}$$

Checking the Pipe Wall of the Clamp Components: Piping is specified by outside diameter. For this reason we use the following equation for finding the required wall thickness:

$$t = \frac{PR_o}{\sigma_A E + 0.4P} = \frac{P(OD)}{2(\sigma_A E + 0.4P)} \qquad \text{Eq. 6-28}$$

$$t = \frac{(330)(8.625)}{2[(20000)(0.7) + 0.4(330)]} = 0.101$$

Clamps are generally made of rolled plate, especially for large pipe. Seamless pipe can be difficult with the mill tolerance, as it may be thicker on one side and may be hard to fit around the component. The clamp contractor did not specify seamless or rolled plate, so we will use seamless pipe.

For the 16″ ϕ schedule 40 pipe,

$$t = \frac{(330)(16.0)}{2[(20000)(0.7) + 0.4(330)]} = 0.187 \text{ in.}$$

Let cm = mill tolerance of 12.5% of the nominal wall as allowed by ASME B31.3. Thus the required thickness is

$$t = 0.187 \text{ in.} + 0.5(0.125) \text{ in.} = 0.25 \text{ in.} < 0.5 \text{ in. nominal wall}$$

The 16 in. ϕ Sch 40 pipe is acceptable.

Similarly for the 14" ϕ Sch 40 pipe,

$$t = \frac{(330)(14.0)}{2[(20000)(0.7) + 0.4(330)]} = 0.163 \text{ in.}$$

Now $t = 0.163$ in. $+ 0.437(0.125)$ in. $= 0.218$ in. < 0.437 in. nominal wall

The 14 in. ϕ Sch 40 pipe is acceptable.
   For the 8 in. Sch 40 pipe,

$$t = \frac{(330)(8.625)}{2[(20000)(0.7) + 0.4(330)]} = 0.101 \text{ in.}$$

Now $t = 0.101 + 0.322(0.125) = 0.141$ in. < 0.322 in. nominal wall

The 8 in. ϕ Sch 40 pipe is acceptable.
   The clamp design has been checked and is satisfactory. This clamp was put into service with 48 months left before the next turnaround. There was no shutdown in that period, and the clamp performed satisfactorily with no leaks or further maintenance required.

Clamp Example 3

A 6" ϕ process line containing 12% caustic soda has a leak at a 45° elbow. The line design pressure is 3762 KPa$_g$ (545.5 psig) at 52°C (125.6°F). Check the clamp vendor's design to ensure that it is satisfactory.
   Referring to **Figure 6-10**, the following are computed:

$$G = \frac{R_o \theta \pi}{180} - 2t_{EP} = \frac{(12)(45)\pi}{180} - 2(1.378) = 6.6688 \text{ in.}$$

where $P_{inj} = c(P) = (1.1)(545.45) = 600.0$ psig
$\qquad F = P_{inj}(G)(D) = (600)(6.6688)(8) = 32{,}009.97 \text{ lb}_f$

Computing the effective length of the lug plate bolts:

$$S_o = \frac{[12.0 + 3.415 + 0.984 + 3.3125](45)\pi}{180} - 2(0.7874)$$

$$= 13.907 \text{ in.}$$

$$S_i = \frac{(6.705)(45)\pi}{180} - 2(0.7874) = 3.691 \text{ in.}$$

$$L_{eff} = S_o + S_i = 13.907 \text{ in.} + 3.691 \text{ in.} = 17.598 \text{ in.}$$

Calculating the required lug plate thickness,

$$T = 2.45 \sqrt{\frac{(32{,}009.973) \text{ lb}_f [1.31 \text{ in.} - 0.5(0.7) \text{ in.}]}{[17.598 \text{ in.} - (12)(0.875) \text{ in.}](20{,}000) \dfrac{\text{lb}_f}{\text{in.}^2}}}$$

$$= 1.140 \text{ in. is approx. } 29 \text{ mm}$$

The 29 mm is less than the 30 mm plate actually used. The lug plate is satisfactory.

Calculating the required bolt area, the moment on the bolts is

$$M_p = \frac{(32{,}009.973 \text{ lb}_f)(5.295 \text{ in.})}{4} = 43{,}373.20 \text{ in.-lb}_f$$

The required bolt area is calculated on a 40,000 psi stress allowed to be placed on the bolts, per ASME PCC-1 [Reference 1], which allows up to 50,000 psi stress, using SA-193 B7 bolts. Note that the bolt material in this application was SA-193-B7, which has a specified yield strength of 105 ksi at ambient. The ASME Section II Part D code listed the allowable stress at 25 ksi. This stress is not only overly conservative, but if bolts were torqued to 25 ksi, many bolted connections would leak. The stress 40 ksi has proven to be a good base for the design of bolted clamps. Thus,

$$A_{bTOT} = \frac{32{,}009.973 \text{ lb}_f + \dfrac{2(42{,}373.20) \text{ in.-lb}_f}{\left(17.598 \text{ in.} + \dfrac{(6.083 \text{ in.} - 5.295 \text{ in.})}{2}\right)}}{40{,}000 \dfrac{\text{lb}_f}{\text{in.}^2}}$$

$$= 0.918 \text{ in.}^2$$

For 12 bolts, the required area for each bolt is

$$A_b = \frac{0.918}{12} = 0.076 \text{ in.}^2$$

During the time this clamp was designed, an allowable stress on the bolts was 25,000 psi, which gave a required total bolt area of

$$A_b = 1.4688 \text{ in.}^2$$

Or a required area for each bolt as

$$A_b = 0.1224 \text{ in.}^2 < 0.302 \text{ in.}^2 \text{ for a } \tfrac{3}{4}'' \text{ coarse thread bolt}$$
which was used

Thus, the $\tfrac{3}{4}''$ bolt is more than adequate.

Now computing the required thickness for the end plate, we have

$$T = (8.625) \text{ in.} \sqrt{\dfrac{(0.33)(600)\,\dfrac{\text{lb}_f}{\text{in.}^2}}{(20,000)\,\dfrac{\text{lb}_f}{\text{in.}^2}(0.7)}} = 1.026 \text{ in.}$$

$$= 26 \text{ mm} < 35 \text{ mm actual plate used}$$

Thus, the end plates are satisfactory.

Serrated Teeth Connection: The end plates have six teeth of 2.5 mm depth. As the clamp bolt torque is applied, the teeth will dig into the pipe material. The theoretical allowable thrust load is calculated as follows:

D_{c1} = diameter of the teeth at the root = 173.5 mm = 6.831 in.
D_{c2} = diameter of the teeth at the sharp edge = 168.5 mm = 6.63 in.
A_{con} = metal contact area
$$A_{con} = \dfrac{\pi[(D_{c1})^2 - (D_{c2})^2]}{4} = 2.125 \text{ in.}^2$$
A_{Tcon} = total contact area of teeth = 6(2.125) in.^{2} = 12.75 in.^{2}
F_{TA} = maximum allowable thrust load
$$F_{TA} = \sigma_A A_{Tcon} = (20,000)\,\dfrac{\text{lb}_f}{\text{in.}^2}\,(12.75)\text{ in.}^2 = 255,000 \text{ lb}_f$$

This thrust load is well beyond any design pressure for the pipe. So the teeth are acceptable.

Usually this calculation for the serrated teeth is not performed, as they have considerable strength. The engineer should be the judge on when such a calculation is necessary.

This clamp was installed successfully and stayed in place until the next turnaround 18 months later.

Repairs Involving Hot Work

Hitherto we discussed repairs involving bolt-on clamps that are injected with sealant material to stop the leak. Now we will focus on all-welded enclosures. When geometry considerations make a bolt-on clamp impractical, weld-on repairs may be used. Before a weld repair is made over a leak, one should determine if such a repair can be made safely. This may involve consulting with maintenance and operations personnel. Inspection plays a critical role because UT (ultrasonic) measurements must be made on the component to be welded to ensure there is enough wall left for welding to be practical.

Welding and other hot work repairs are normally used where leaks develop (or could develop as a consequence of local wall thinning detected by on-stream inspection) in lines or components that are too large or have complex geometries that make bolt-on clamps impractical.

Lap Patches

Lap patches are attached to the pipe by fillet welds, which introduce design limitations. Fillet welds have a lower joint efficiency. Because of the geometry of the lap patch, stress concentrations occur because the patch is eccentric in the path of hoop stress, and bending stresses the welds. This fact is aggravated if the leaking component has lost significant wall thickness in the area of the leak. Significant loss of wall can cause crack propagation with the loss in strength in the pipe wall. Thermal stresses caused by differential temperature between the lap patch and the pipe wall are minimal as the lap patches are welded directly on the pipe, and most piping is 1 in. or less in wall thickness, making the thermal gradient through the pipe wall insignificant. This assumption is not valid for thick-walled pipe, which is used for high pressures where lap patches would not be desirable. A typical configuration of a lap patch is shown in **Figure 6-19**.

The design procedure for a lap patch is as follows: The lap patch repair is mentioned in API 570 Appendix D. The first step is to select the material of the lap patch. The material should either be pipe or plate material—the choice depends on material availability. The material of the patch should have the same generic chemistry and be at least equal in strength to the pipe material. It is most preferable for the patch to be the same as the pipe material.

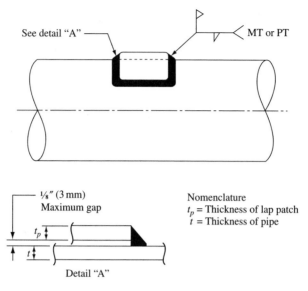

Figure 6-19. Typical lap patch repair.

The next step is to calculate the required minimum thickness of the lap patch from the following:

$$t_{LPr} = \frac{\left(\dfrac{PD}{2\sigma_A E} + CA\right)}{0.707}$$ Eq. 6-29

where t_p = minimum required thickness of the lap patch, in.
CA = corrosion allowance, in. (mm)
P = pipe design pressure, psig
D = pipe outside diameter plus 1 (assume that lap patch is $\frac{1}{2}$ in. (12.7 mm) thick as initial estimate and iterate as required, in. (mm)
σ_A = allowable stress of the patch material at the design temperature per the applicable code (e.g., ASME B31.3), psi (MPa)
E = fillet weld joint efficiency = 0.45

The minimum required thickness of the lap patch must include the 0.707 factor to account for the throat thickness of the fillet weld.

Bending stresses induced by pressure are generally not considered for lap patches on piping systems, as the subtended angle (angle made through the arc length of the patch plate) is significant enough such that there is minimal bending stress transferred to the lap patch plate.

The eccentricity of the hoop stress path should be assessed because this stress is always present for lap joints. The following equation is recommended to calculate the resultant stress at the edge of the patch plate:

$$\sigma = \frac{PD}{2(t_{LP} - CA)} + \frac{3PDe}{(t_{LP} - CA)^2} \qquad \text{Eq. 6-30}$$

where e = distance from the centerline of the patch plate to the centerline of the pipe, or

$$e = \frac{(t_p + t_{LP})}{2}$$

where t_p = thickness of process pipe, in. (mm)
t_{LP} = thickness of lap patch, in. (mm)

The calculated stress is compared to $1.5\sigma_A$. The value of t_{LP} should be iterated to arrive at a stress that is equal to or less than $1.5\sigma_A$. One should use full-thickness welds to attach the lap patch to the pipe. The patch should be formed to match the curvature of the pipe.

Note that the research performed by the PRCI [Reference 3] and the ASME B31.4 in the 2005 edition recommend against the use of lap patches. The PRCI document qualifies the lap patch as customarily manufactured to cover half the circumference of the pipe and may be up to 3 m (10 feet) in length. Past research has shown [Reference 3] that these types of repairs are very sensitive to fabrication techniques and should not be used in high-pressure pipelines. The ASME B3.4 2005 edition prohibits their use entirely; however, this rule so far does not apply to in-plant piping.

Example of a Lap Patch

A 12″ ϕ Sch 40 pipe made of API 5L Gr B pipe operates at 500°F (260°C) at 150 psig (1034.5 KPa$_g$). The allowable stress at this temperature is 18,900 psi (130.3 MPa). The corrosion allowance is 1/16″ (1.59 mm). The pipe is in a utility unit that cannot be shut down because it must provide steam to the other units. It is decided to place a lap patch over a leak in the line, as a bolted clamp is temporary. The lap patch is designed as follows:

We will try a patch 3/8" (9.5 mm) thick.

$$D = OD + 1 = 13.625$$

$$t_{LPr} = \cfrac{\cfrac{(150)\,\dfrac{\text{lb}_f}{\text{in.}^2}\,(13.625)\,\text{in.}}{2(18,900)\,\dfrac{\text{lb}_f}{\text{in.}^2}\,(0.45)} + 0.0625}{0.707} = 0.258\,\text{in.}$$

$$e = \frac{(0.406 + 0.375)\,\text{in.}}{2} = 0.3905\,\text{in.}$$

$$\sigma = \frac{(150)\dfrac{\text{lb}_f}{\text{in}^2}(13.625)\,\text{in.}}{2(0.375 - 0.0625)\,\text{in.}} + \frac{3(150)\dfrac{\text{lb}_f}{\text{in.}^2}(13.625)\,\text{in.}\,(0.3905)\,\text{in.}}{(0.375 - 0.0625)^2\,\text{in.}^2}$$

$$= 27,787.15\,\text{psi}$$

$$\sigma_{AT} = 1.5(18,900)\,\text{psi} = 28,350\,\text{psi}$$

Since $\sigma < \sigma_{AT}$, *the lap patch is acceptable.*

Welding Caps

Weld caps have been used with good success for leaks on large diameter pipes. A pipe cap of appropriate size is welded to the pipe with an attached valve to allow the escape of weld gases. This configuration is shown in **Figure 6-20**.

The wall thickness of the cap should be close to that of the pipe wall. This allows adequate thickness for pressure and distributes the dissipation of heat during the welding process such that the heat is balanced. Note that one must perform UT analysis of the pipe to make sure there is adequate pipe wall to be welded; otherwise, burn-through of the weld

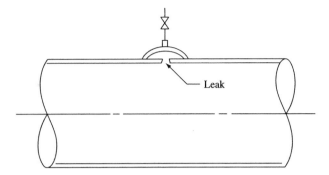

Figure 6-20. Typical welded cap (small leak) repair.

will result in a worse leak and potential pipe burst and, depending on the process fluid, an explosion. Any welding on an operating pipe must meet hot tap criteria.

Welded-on Nozzle

If the size of the leak is too great to permit a weld cap or if a larger valve is required to vent the leak during installation of the cap, it may be possible to weld a nozzle with a full-size flanged gate valve. After the nozzle is welded on, the end of the gate valve will have a blind flange installed to close off the leak. This technique is common with boiler feed water service and can be used for other applications.

The ratings of the valve and flanges are based on the design pressure and temperature of the leaking pipe. The nozzle thickness and the requirement for additional reinforcement would be according to the applicable code, such as ASME B31.3. Like any reinforcement computations, the loss of original pipe material enclosed by the nozzle must be considered. The nozzle material should be of the same material as the pipe or of comparable metallurgy.

It is best that the nozzle neck be kept as short as practical to minimize the potential for dead leg corrosion. A realistic corrosion allowance should be applied, usually equal to that of the pipe with the contained fluid. For such a repair, regular inspection monitoring is necessary to ensure that the corrosion rates do not increase. This configuration is shown in **Figure 6-21**. Any welding on an operating pipe must meet hot tap criteria.

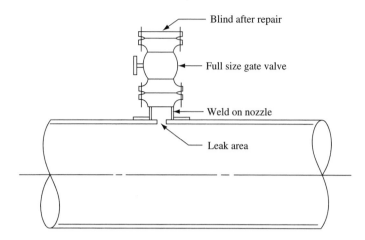

Figure 6-21. Welded-on nozzle repair (large leak).

Full Encirclement Sleeves

Full encirclement sleeves are very common in pipelines because the temperature differential between the pipeline and sleeve is normally minimal. This occurs because pipelines operate at temperatures relatively close to ambient. For in-plant process piping, this is often not the case. Much in-plant process piping operates at temperatures well above or below ambient, and differential thermal expansion (or contraction) can induce high thermal stresses. For this reason, full encirclement sleeves are outlawed by some companies for in-plant process piping.

When a full encirclement welded sleeve is considered, it is usually done in situations where the pipe damage is more extensive than can be attenuated by a lap patch. Full encirclement sleeves are also used in situations where the pipe wall thickness is low. The required thickness of a complete encirclement sleeve is less than a lap patch, but it is normally more than that of the contained pipe. This is because the weld joint efficiency of the longitudinal seams considered in the hoop stress computations are only 0.6 (or 0.65 if a backing strip is used), since these welds cannot be radiographed. Even though the hoop stress in the sleeve normally governs the required sleeve thickness, a check must be made to ensure that the longitudinal stress induced by pressure and eccentricity of the hoop stress path does not exceed $1.5\sigma_A$, where σ_A is the allowable stress in tension of the full encirclement sleeve at design temperature. Guaranteeing this requirement may necessitate a thicker sleeve.

A full encirclement pressure-containing sleeve may be considered permanent if the following criteria are met:

- The sleeve is designed to contain full design pressure.
- All longitudinal seams in the sleeve are full penetration butt welds with the weld joint efficiency and associated inspections consistent with the applicable code, such as ASME B31.3.
- The circumferential fillet welds attaching the sleeve to the pipe wall are designed to transfer the full longitudinal load in the pipe wall, using a joint efficiency of 0.45, without relying on the integrity of the original piping material covered by the sleeve. Where significant, the eccentricity effects of the sleeve relative to the original pipe wall shall be considered in sizing the sleeve attachment welds.
- Fatigue of the attachment welds caused by differential expansion of the sleeve relative to the pipe shall be considered.
- The sleeve material must be suitable for contact with the contained fluid at the design conditions, and appropriate corrosion allowance must be accounted for in the sleeve. Ideally, the sleeve material should be the same as the pipe material.

A design more complex than a full encirclement sleeve may be required to clear certain obstructions on the pipe (e.g., epoxy wrap, temporary clamping devices, bleeder valves, pipe flanges). For such situations full encirclement sleeves with end plates may be required.

The length of the sleeve normally is at least 4 inches and is also long enough to extend at least 2 inches beyond both ends of the defect. Even though there is no upper limit on sleeve length, one should consider the sleeve's weight, the snugness of fit, the impact of carrier pipe curvature (if any), the impact of girth weld reinforcement, and any high-low condition. Also, the ability of a repair crew to install the sleeve is a consideration.

Full Encirclement Welded Sleeve Without End Plates

Shown in **Figure 6-22** is the welded full encirclement sleeve. Refer to the API 570 Appendix D and **Figure 6-22** for typical design details and installation procedures. Again the sleeve material should be the same generic material as the pipe, and preferably the same material as the pipe.

The sleeve material must be compatible with the process fluid and have the same coefficient of thermal expansion as the pipe material.

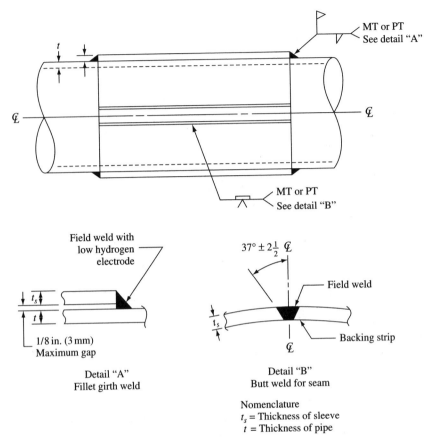

Figure 6-22. Typical welded full encirclement sleeve.

First, calculate the minimum required thickness of the sleeve from the following equation:

$$t_s = \frac{P(OD)}{2\sigma_A E} + CA \qquad\qquad \text{Eq. 6-31}$$

where t_s = minimum thickness of the sleeve, in.
  P = design pressure of the pipe, psig (KPa$_g$)
  OD = outside diameter of the pipe being sleeved + 1 (assume that the sleeve thickness is $\frac{1}{2}$ in. or 13 mm as an initial estimate and iterate as required), in. (mm)

σ_A = allowable stress of the sleeve material at the pipe design temperature per the applicable code (e.g., ASME B31.3), psi (MPa)

E = weld joint efficiency of the longitudinal welds (0.6 if no backing strip is used, 0.65 if a backing strip is used)

Next the eccentricity of the hoop stress path at the circumferential fillet welds must be considered, since this stress is always present for lap joints. The following equation is used to calculate the resultant stress at the edge of the sleeve, using t_s computed from Eq. 6-31:

$$\sigma = \frac{P(OD)}{4(t_s - CA)} + \frac{3P(OD)e}{2(t_s - CA)^2}$$ Eq. 6-32

where e = distance from the centerline of the sleeve to the centerline of the pipe (process pipe thickness + sleeve thickness)/2. See **Figure 6-23**.

Use the value of t_s calculated in Eq. 6-31. The calculated stress is then compared to $1.5\sigma_A$. The value of t_s should be iterated upon to arrive at a calculated stress below $1.5\sigma_A$. Thermal stresses must be considered for welded full encirclement sleeves. However, as long as the sleeve is preheated to the approximate temperature of the run pipe, thermal stresses should be minimal. Thermal stresses are discussed later in this chapter.

Use full-thickness longitudinal butt welds to attach sleeve halves. Use full-thickness circumferential fillet welds between the sleeve ends and the pipe. Thermal stresses must be considered for welded full encirclement sleeves. However, as long as the sleeve is preheated to the appropriate temperature of the process pipe, thermal stresses should be minimal.

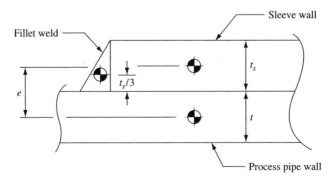

Figure 6-23. Weld detail of full encirclement sleeve.

Full Encirclement Repair with End Plates on Straight Pipe Section

A methodology for the design of a full encirclement repair with end plates on a straight pipe section is as follows:

1. Select the sleeve material. The material ideally should be the same as the pipe or plate material. The selection depends on enclosure diameter and material availability. The sleeve material should be of the same generic chemistry and be equal in strength to the contained pipe.
2. Use the same corrosion allowance as that of the contained pipe. Normally, a welded containment sleeve is considered a permanent repair because hot work is being performed. If the repair is a temporary repair, a bolted clamp as described previously should be considered.
3. Determine the diameter of the enclosure. This diameter should be as small as possible to minimize the amount of welding and the required thickness of the end plates. The sleeve diameter must be large enough to clear any obstructions as mentioned earlier.
4. The diameter of the sleeve must be large enough to permit the appropriate welding details and, preferably, a standard pipe size. If a standard pipe size is selected, be careful in selecting the seamless pipe because the mill tolerance can result in uneven wall thickness around the pipe, which could cause the sleeve not to clear obstructions.
5. Determine the required length of the sleeve. The sleeve must be long enough to enclose the area of concern completely and to allow welding to sound metal.
6. Calculate the minimum required thickness of the cylindrical portion of the sleeve from the following:

$$t = \frac{PD}{2\sigma_A E} + CA + cm \qquad\qquad \text{Eq. 6-33}$$

where P = design pressure of the enclosed pipe, psig (KPa_g)
D = outside diameter of the sleeve, in. (mm)
σ_A = allowable stress of the sleeve material at the contained pipe design temperature, psi (MPa). This is obtained from the applicable code.

E = weld joint efficiency of the longitudinal weld seams of the enclosure (0.6 if no backing strip is used and 0.65 if a backing strip is used)

CA = corrosion allowance of the contained pipe, in. (mm)

cm = mill tolerance for seamless pipe, which is 12.5% of nominal pipe wall, $cm = 0$ for rolled plate

Full Encirclement Repair with End Plates at an Elbow

Unlike the bolted elbow clamp discussed previously, this repair involves a welded enclosure with two flat end plates enclosing an elbow. Instead of serrated edges on the end plates or the use of shear pins along the clamp body, the end plates are fusion welded to the enclosed pipe. Depending on the size and actual location of the leak, thickness of the elbow, and other parameters, the enclosure may be sized to enclose the entire elbow or only a portion thereof. The following procedure is followed in designing such an enclosure:

1. Select the material, diameter, and corrosion allowance in the same manner as for a straight section of pipe. Note that if a short radius elbow is used for the enclosure, the diameter may have to be somewhat larger than would otherwise be required, as noted later.
2. Calculate the required thickness for the elbow following the requirements of ASME B31.3 Paragraph 304.2.1 for Pipe Bends. The equation for the required wall thickness is as follows:

$$t_{rell} = \frac{PD}{2\left[\left(\dfrac{SE}{I}\right) + PY\right]} \qquad \text{Eq. 6-34}$$

where, P, D, E, and Y have already been defined.

For the intrados (inside bend radius),

$$I = \frac{4\left(\dfrac{R_1}{D}\right) - 1}{4\left(\dfrac{R_1}{D}\right) - 2}$$

For the extrados (outside bend radius),

$$I = \frac{4\left(\dfrac{R_1}{D}\right) + 1}{4\left(\dfrac{R_1}{D}\right) + 2}$$

where I = 1.0 at the sidewall on the bend centerline radius
R_1 = bend radius of welding elbow or pipe bend, in. (mm)

3. Determine if a short radius elbow may be used to enclose a leaking long radius elbow. This may be possible if the enclosure diameter is made large enough and the amount of welding required is decreased.

   Note that if a short radius elbow is used, the distance between the OD of the pipe elbow and the ID of the enclosure elbow is not constant along their perimeters. This should be a consideration in sizing and positioning the enclosure, especially if there is an obstruction on the elbow, such as epoxy wrap.
4. Find where the enclosure must terminate to ensure that it encloses the area of concern (the leak), permits welding to sound metal, and is an adequate distance from the pipe-to-elbow welds.

Full Encirclement Repair with End Plates at a Branch Connection

These enclosures that are installed at branch connections are "T" shaped. The following procedure is used for such a repair:

1. The material selection and geometric parameters (e.g., diameters and wall thicknesses) are performed in the same manner as before.
2. The branch connection reinforcement calculations of the enclosure should be in accordance with that in the ASME B31.3 code.
3. If additional reinforcement is required, it is preferred practice to increase the enclosure wall thickness as necessary based on the area replacement calculations per ASME B31.3. Reinforcing pads are an alternative if the pipe is not leaking and the pad is welded on after the enclosure has been installed. Note that in this process one must be careful of adding significant weight with spring supports in the system, particularly constant springs. Constant springs are very

sensitive to additional weight added to piping systems. Subject to the particular situation, a piping stress analysis may be in order to check the springs. Variable springs are more "forgiving" than constant springs, but it is good practice to check the springs if significant weight is added.
4. Find where the enclosure sections must terminate to ensure that it encloses the area of the leak or area of concern (thin wall). Also UT measurements should be made to ensure that there is enough wall available for welding to sound metal and that there is adequate distance from the pipe-to-branch welds.

Required End Plate Thickness Without Pressure Thrust Load

The ASME Section VIII Division 1 Appendix 9 gives a procedure to calculate the minimum required thickness of flat end plates for an enclosure. This procedure is appropriate when the pressure thrust load in the pipe is not a design consideration. The pressure thrust load does not need to be considered if there is no concern of the following:

- Deteriorated pipe enclosed by the enclosure may completely separate during operation.
- Flange bolts may fail when exposed to the process fluid, for example, if the enclosure completely encloses flanged joints.

If these issues are not a concern, the following procedure can be applied:

1. The minimum required thickness of the flat end plates shall be the greater of the following:

$$t_{rep} = 2t_{rj} = CA \text{ in. (mm)} \qquad \text{Eq. 6-35}$$

$$t_{rep} = 0.707j \sqrt{\frac{P}{\sigma_A}} + CA \text{ in. (mm)} \qquad \text{Eq. 6-36}$$

where t_{rep} = minimum required thickness of flat end plate, in. (mm)
t_{rj} = minimum required thickness of cylindrical enclosure excluding CA (i.e., $t_{rj} = t - CA$)
j = distance between the OD of the process pipe and the ID of the enclosure box, in. (mm). The maximum

value for this distance is calculated using the *CA* and mill tolerance.

P = design pressure of the process pipe, psig (KPa$_g$)

σ_A = allowable stress of flat plate material at the design temperature of the process pipe per the applicable code (e.g., ASME B31.3), psi (MPag)

CA = enclosure allowance of the enclosure—should be the same as for the process pipe being contained, assuming comparable metallurgy (strongly recommended)

2. All well sizes should meet the requirements of the ASME Section VIII Division 1 Appendix 9.
3. The end plate maximum thickness should be limited to two to three times the thickness of the process pipe being contained, with a maximum practical limit of approximately 1 in. (25 mm) considering the field welding case. This is normally not a concern when the pressure thrust load is not a design parameter.

Required End Plate Thickness Considering Pressure Thrust Load

The following procedure is derived from the methodology of the ASME Section VIII Division 1 Appendix 2. This procedure is applied if there is a concern that the contained pipe sections may separate inside the enclosure. This phenomenon could be caused by severe circumferential corrosion or failed bolts in enclosed flanges.

1. Calculate the acting loads and moments using the following equations:

$$W = \frac{\pi B^2 P}{4} \qquad\qquad \text{Eq. 6-37}$$

$$H_D = \frac{\pi B^2 P}{4} \qquad\qquad \text{Eq. 6-38}$$

$$H_T = W - H_D \qquad\qquad \text{Eq. 6-39}$$

$$H_D = h_D H_D \qquad\qquad \text{Eq. 6-40}$$

$$M_T = h_T H_T \qquad\qquad\qquad \text{Eq. 6-41}$$

$$M_o = M_D + M_T \qquad\qquad\qquad \text{Eq. 6-42}$$

where B = outside diameter of the pipe being enclosed, in. (mm)

D_{ic} = inside diameter of the enclosure (considering mill tolerance and corrosion allowance), in. (mm). If enclosure is made from pipe, the following should be used: D_{ic} = pipe OD $- 2(0.875 t_{nom} - CA)$.

h_D = radial distance from OD of the pipe being enclosed to the average diameter of the enclosure (D_{ave}), in. (mm) (e.g., if enclosure is made from pipe, D_{ave} = pipe OD $- t_{nom}$)

$h_T = 0.5 h_D$, in. (mm)

P = design pressure of process pipe being contained, psig KPa$_g$)

2. The minimum required thickness of the flat end plates (t_{EP}) is calculated as

$$t_{EP} = \sqrt{\frac{M_o Y}{\sigma_A B}} + CA \qquad\qquad\qquad \text{Eq. 6-43}$$

where σ_A = allowable stress of end plate material at design temperature of the process pipe being contained per the applicable code (e.g., ASME B31.3), psi (MPa)

CA = corrosion allowance of enclosure, in. (mm)

Y = value obtained from Figure 2-7.1 of Appendix 2 in the ASME Section VIII Division 1 Appendix 2. This value can be alternatively computed from the following:

$$Y = \frac{1}{K-1}\left[0.66845 + 5.7169\left(\frac{K^2 \log_{10} K}{K^2 - 1} \right) \right]$$

where $K = \dfrac{\text{(Enclosure OD)}}{\text{(Pipe OD)}}$

3. Limit the maximum end plate thickness to two to three times the process pipe thickness, with a maximum practical limit of 1 in. (25 mm).

4. If a flat end plate would be too thick, then a pipe cap, reducer, or conical transition piece may be used for the enclosure. The thickness of the pipe cap or reducer would normally be the same as that of the enclosure cylinder. However, for pipe caps that have had holes cut in them to accommodate installation around thinned wall pipe, branch reinforcement calculations should be made. Alternatively, as previously discussed, flat end plates could be designated as an end plate without pressure thrust load, or a "strong-back" system with tie rods could be designed to withstand the pressure thrust load, or serrated end plates or shear pins could be used.

Thermal Stress Criteria in Welded Enclosure Designs

Welded enclosures are commonly used with pipelines because they normally operate at temperatures relatively close to ambient temperatures. For this reason thermal stresses are not a major concern for using welded enclosures, such as split tees and sleeves, in pipelines. This fact is not true for in-plant piping, where often piping systems routinely operate at temperatures far above or below ambient temperatures. Thus, many companies forbid welded enclosures for in-plant process piping. The term "ambient" temperature can vary considerably depending where on earth the facility is located. It can range from many times below zero on the northern slope of Alaska or in Siberia, Russia, to the highest official recorded temperature on earth, 57.8°C (136°F) in Al Azizyah, Libya (south of Tripoli). Ambient is assumed to be 21°C (70°F) in the United States and most parts of Europe. Thermal stresses need to be considered for welded enclosures if the design temperature of the process pipe is more than approximately 350°F (177°C).

A classic case of welded attachment failure was a plate ring welded all around a 20 in. superheated steam line beneath a furnace. The welded ring plate was used to attach rods for two constant spring hangers beneath the steam header at several locations. The entire header and support ring plates were insulated to minimize the differential temperature between the support ring and steam header. The problem came at shutdown when saturated steam was shot through the steam header to cool it down. As the steam header cooled down, the ring plate remained hot and the differential temperature between the support ring and the steam header resulted in massive cracking in the welds between the support ring and steam header. This cracking became worse with each shutdown until the cracking progressed into the header parent metal, setting up

leaks. This phenomenon has to be considered when using welded enclosures for in-plant piping.

The enclosure will typically be cooler than the process pipe when
installed. If there are no leaks in the process pipe before or after enclosure installation, thermal stresses will occur when the line shuts down
and the process pipe cools to ambient temperature, as previously
described.

If the enclosure is installed while the process pipe is leaking or if a
leak occurs in the process pipe after the enclosure is installed, thermal
stresses will result during operation when the enclosure is heated to the
temperature of the process pipe by the leaking fluid. After the enclosure
and the process pipe reach thermal equilibrium, thermal stress will be
negligible. Thermal stresses can be minimized by preheating the enclosure to the approximate temperature of the process pipe during the installation welding process.

The methodology of assessing thermal stresses is as follows:

1. Calculate the differential thermal movement (Δ) between the
   enclosed pipe and the enclosure using the following equation:

$$\Delta = \Delta_p - \Delta_e = \alpha_p(T_p - T_a)L_p - \alpha_e(T_e - T_a)L_e \qquad \text{Eq. 6-44}$$

where Δ_p = thermal expansion/contraction of enclosed pipe, in. (mm)
Δ_e = thermal expansion/contraction of enclosure, in. (mm)
α_p = coefficient of thermal expansion of enclosed pipe,
in./in./°F (mm/mm/°C)
α_e = coefficient of thermal expansion of enclosure, in./in./°F
(mm/mm/°C)
T_p = temperature of enclosed pipe when enclosure is
installed, °F (°C)
T_e = temperature of enclosure when it is installed, °F (°C)
T_a = ambient temperature normally used for the pipe stress
analysis at the particular plant site, °F (°C)
L_p = length of pipe being enclosed, in. (mm)
L_e = Length of enclosure, in. (equal to L_p)

We will now consider several operating cases.

Case 1: The enclosure is installed before a leak occurs or after
the leak has been eliminated by some other method (e.g., a
bolted clamp described earlier in the chapter). The leak does
not reoccur after the enclosure is installed. As the process line
shuts down, we should examine the thermal stresses. Calculate

the differential thermal movement (D) that occurs during a line shutdown using Eq. 6-44.

Case 2: A leak occurs after installation of the enclosure even though the pipe being enclosed is not leaking when the enclosure is installed. The thermal stresses that develop as the enclosure temperature reaches equilibrium with the temperature of the pipe being enclosed need to be examined. In this case, compute the differential thermal movement (Δ) using the following equation:

$$\Delta = \Delta_e = \alpha_e(T_e - T_a)L_e \qquad \text{Eq. 6-45}$$

where T_e = operating temperature of enclosed pipe, °F (°C)
T_a = temperature of enclosure when installed, °F (°C)

Once both the enclosure and the enclosed pipe reach equilibrium, thermal stresses also develop when the line is shut down. However, this effect is minimal because both the enclosure and the pipe will be cooling down (or heating up) at approximately the same rate.

Case 3: The pipe is leaking when the enclosure is being installed. Again the thermal stresses that develop as the enclosure reaches equilibrium with the pipe need to be examined. Calculate the thermal movement (D), using Eq. 6-45, with

T_e = operating temperature of the enclosed pipe, °F (°C)
T_a = average temperature of the enclosed pipe and ambient temperature when the enclosure is installed, °F (°C)

For conservatism, use Case 2 to determine the larger thermal displacement. For all three cases, preheating the enclosure minimizes the differential thermal movement.

2. For conservatism, assume that the flat end plates are infinitely rigid. Then calculate the axial force (F) in the pipe and enclosure using the following formula:

$$F = \frac{\Delta E}{L\left(\dfrac{1}{A_p} + \dfrac{1}{A_e}\right)} = \left(\frac{\Delta}{\dfrac{1}{k_p} + \dfrac{1}{k_e}}\right) \qquad \text{Eq. 6-46}$$

where = modulus of elasticity at ambient temperature, psi (MPa)

D = differential thermal movement calculated above, in. (mm)

A_p = metal cross-sectional area of pipe using nominal thickness, in.^{2} (mm^2)

A_e = metal cross-sectional area of enclosure using nominal thickness, in.^{2} (mm^2)

$L = L_e = L_p$, in. (mm)

k_p = axial spring rate of enclosed pipe, lb/in.

$$= \frac{A_p E}{L}$$

k_e = axial spring rate of the enclosure, lb/in.

$$= \frac{A_e E}{L}$$

3. Calculating the axial spring stress in the enclosed pipe (σ_p), we have

$$\sigma_p = \frac{P(OD)}{4t_p} + \frac{F}{A_p}, \text{ psi (MPa)} \qquad \text{Eq. 6-47}$$

where OD = outside diameter of pipe, in.

t_p = thickness of pipe (use actual thickness if UT data are available), in. (mm)

4. If the pipe is leaking, the axial stress in the leak box (σ_e) is calculated as follows:

$$\sigma_e = \frac{pD_e}{4t_e} - \frac{F}{A_e}, \text{ psi (MPa)} \qquad \text{Eq. 6-48}$$

where D_e = outside diameter of enclosure, in. (mm)

t_e = nominal thickness of enclosure, in. (mm)

If the enclosed pipe is not leaking, then only the second term (F/A_e) applies for calculating σ_e.

5. The design of the configuration is acceptable if both σ_p and s_e do not exceed three times the ASME B31.3 allowable tensile stress (σ_a) for each respective component. If buckling of the enclosure is a concern, the resultant compressive stress in the enclosure or the pipe should be within the allowable compressive stress (σ_c).

Note that if the design cannot be verified using the simplified approach already discussed, then the flexibility of the flat end plates may be considered using the following methodology:

6. Compute the spring rate of each flat end plate (k_{ep}) as follows:

$$k_{ep} = \frac{4Et_{ep}^3}{CD_e^2}, \text{ lb/in. (N/mm)} \qquad \text{Eq. 6-49}$$

where t_{ep} = end plate thickness, in. (mm). Use the uncorroded thickness because this gives more conservative results.

C = a function of $D_e/(OD)$ based on **Table 6-2**. In this table linear interpolation may be used.

7. Compute the axial force in the pipe and enclosure (F) in much the same way as was done earlier as follows:

$$F = \left(\frac{\Delta}{\dfrac{1}{k_p} + \dfrac{2}{k_{ep}} + \dfrac{1}{k_e}} \right), \text{ lb (N)} \qquad \text{Eq. 6-50}$$

8. Compute σ_p and σ_e as previously, using F calculated from Eq. 6-50. Use the acceptance criteria specified previously for these stresses.

9. Compute the stress in the flat end plate (σ_{ep}) as follows:

$$\sigma_{ep} = \frac{\beta_1 P(OD)^2}{4t_e^2} + \frac{\beta_2 F}{t_e^2}, \text{ psi (MPa)} \qquad \text{Eq. 6-51}$$

where **Table 6-3** is used.

The stress criterion for σ_{ep} is three times the ASME B31.3 allowable tensile stress. If this criterion cannot be met, then a more detailed analysis may be made, such as finite element using the elastic-plastic analysis approach presented in the ASME Section VIII Division 2 Appendix 4.

			Table 6-2			
			C versus $D_e/(OD)$			
$D_e/(OD)$	1.25	1.5	2	3	4	5
C	0.0013	0.0064	0.0237	0.062	0.092	0.114

Table 6-3
Flexible End Plate Parameters

$D_e/(OD)$	1.25	1.5	2	3	4	5
β_1	0.0412	0.114	0.245	0.422	0.520	0.579
β_2	0.115	0.220	0.405	0.703	0.933	1.130

This would be used for highly critical situations. For most practical purposes, another approach could be considered depending upon the situation. For example, if the situation needed to have a repair to last until the next shutdown or turnaround, the bolted clamp assembly presented earlier in the chapter may be an option.

Full penetration welds should be made between all the components that attach to each other—the pipe, the end plates, and the enclosure—to minimize stress concentrations. However, full penetration welds may not be practical under field conditions when the end plates are larger than approximately ½ in. (13 mm) thick because field welding becomes increasingly difficult.

Pipe caps, reducers, or conical transitions may provide more flexibility than flat plates for absorbing differential thermal expansion. However, no simplified formulas or algorithms are available for evaluating the flexibility of these geometries. Therefore, an axisymmetric finite element model can be developed if this approach is considered for an enclosure.

Welded Full Encirclement Sleeve on Straight Section of Pipe with End Plates

A straight portion of a 16 in. NPS pipe developed a pinhole leak in the 6 o'clock position in a thinned area. The leak was plugged with a gasket material with steel banding to keep it in place. The location is in an isolated desert and the clamp contractor cannot access the site for a day. Thus, we will consider a welded sleeve design. The following information is for the design case:

design temperature = 316°C = 600°F
design pressure = 862.07 KPa$_g$ = 125 psig
pipe material: API 5L Gr B
pipe weld joint efficiency = E = 1.0
pipe size = 16″ ϕ Std. wall = 0.375 in.
service = heat transfer oil
future corrosion allowance = FCA = 1.6 mm = 0.0625 in.
banding projects beyond the pipe OD by approximately 32 mm = 1.25 in.

The Inspector UT measurements confirm that the corroded length is approximately one meter (39.37 in.) in length in the 6 o'clock (bottom) position. There is no significant corrosion found in the pipe. The ambient temperature before installation is 38°C (100°F).

With the banding being 1.25 in. thick, the inside clearance for the sleeve pipe is a minimum of 18.5 in. A 20 in. pipe is the next size commercially available pipe for the sleeve to allow for sufficient clearance between the jacket ID and the banding. The sleeve pipe is also API 5L Gr B.

Since the corrosion is along a straight line for a meter, the jacket pipe does not need to be designed for pressure thrust loads. With this presumption, flat end plates are satisfactory. However, the design temperature is relatively high; thus, thermal stresses should be assessed. There was no preheat indicated in the records to indicate that the thermal stresses will be minimized.

To install the 20 in. jacket, it must be cut longitudinally and fit around the leaking pipe and rewelded. This rewelding has no radiography; thus, the joint efficiency is $E = 0.6$. Calculating the required jacket thickness using Eq. 6-31,

$$t = \frac{P(OD)}{2\sigma_A E} + CA = \frac{(125.0)(20.0)}{2(17,300)(0.6)} + 0.0625 = 0.183 \text{ in.}$$

Considering a mill tolerance of 12.5% for the seamless pipe, the adjusted required thickness is

$$t = \frac{0.183 \text{ in.}}{0.875} = 0.209 \text{ in., which confirms that the } 0.375 \text{ in.}$$
$$20'' \phi \text{ pipe is sufficient}$$

The sleeve is 1.2 m (48.0 in.) long and extends beyond any corroded region. Still welding on an operating line requires a hot tap permit, and the hot tap procedures need to be followed. An extensive UT scan is made of the pipe, and the flow velocity in the line is sufficient to dissipate the weld heat. We will discuss this later, but the heat transfer oil has a high flash point, and we will assume that the hot tap procedures are adequate.

We now calculate the required thickness of the flat end plates.

$$ID_j = \text{ID of jacket pipe} = 20.0 \text{ in.} - 2[(0.375)(0.875) - 0.0625] \text{ in.}$$
$$= 19.469 \text{ in.}$$

Now from Eqs. 6-35 and 6-36,

$$j = \frac{ID_j - Pipe\ OD}{2} = \frac{19.469 - 16}{2} = 1.734\ \text{in.}$$

For the end plates, we select SA-516-Gr 70. The allowable stress is 19,400 psi at 600°F.

$\sigma_A = 19{,}400$ psi

$t_{rc1} = 2t_{rj} + CA$

$t_{rj} = t_r - CA = 0.183 - 0.0625 = 0.1205$ in.

$t_{rc1} = 2(0.1205) + 0.0625 = 0.304$ in.

$$t_{rc2} = 0.707j\sqrt{\frac{P}{\sigma_A}} + CA = 0.707\,(1.734)\sqrt{\frac{125}{19{,}400}} + 0.0625$$

$$= 0.161\ \text{in.}$$

The end plates's minimum required thickness is 0.304 in. We select 3/8″ plate (9.5 mm).

To calculate the axial force on the end plates, we use Eq. 6-44 in the following:

$T_p = 600°F =$ temperature of hot oil pipe
$T_e = 115°F =$ temperature when enclosure is installed
$T_a = 100°F =$ ambient temperature

For conservatism and calculation ease, we will use the ASME B31.3 tables for thermal expansion based on 70°F up to the indicated temperature. Thus,

$\alpha_p = 7.23 \times 10^{-6}$ in./in.-°F
$\alpha_e = 6.13 \times 10^{-6}$ in./in.-°F
$\Delta_p = \alpha_p(T_p - T_a)L_e = (7.23 \times 10^{-6})(600 - 100)(48) = 0.173$ in.
$\Delta_e = \alpha_e(T_e - T_a)L_e = (6.07 \times 10^{-6})(115 - 100)(48) = 0.0044$ in.
$\Delta = \Delta_p - \Delta_e = 0.173 - 0.0044 = 0.1686$ in.
$E = 29.3 \times 10^6$ psi for the modulus of elasticity at ambient temperature (100°F)

$A_p = 18.41$ in.^{2} = the metal cross-sectional area for the 16″ ϕ STD WGT pipe

$$k_p = \frac{A_p E}{L} = \frac{(18.41)\ (\text{in.}^2)(29.3 \times 10^6)\left(\dfrac{\text{lb}}{\text{in.}^2}\right)}{48\ \text{in.}} = 11.239 \times 10^6 \frac{\text{lb}}{\text{in.}}$$

$A_e = 23.12$ in.^{2}, metal cross-sectional area for 20″ ϕ NPS STD WGT pipe

$$k_e = \frac{A_p E}{L} = \frac{(23.12)\ (\text{in.}^2)(29.3 \times 10^6)\left(\dfrac{\text{lb}}{\text{in.}^2}\right)}{48\ \text{in.}} = 14.11 \times 10^6 \frac{\text{lb}}{\text{in.}}$$

$$F = \frac{\Delta}{\left(\dfrac{1}{k_p} + \dfrac{1}{k_e}\right)} = \frac{0.1686}{\left(\dfrac{1}{11.239 \times 10^6} + \dfrac{1}{14.11 \times 10^6}\right)}$$

$$= 1.05 \times 10^6\ \text{lb}$$

Calculating the axial stress in the pipe using Eq. 6-45:

$$\sigma_p = \frac{P(OD)}{4t_p} + \frac{F}{A_p} = \frac{(125)(16)}{4(0.375)} + \frac{1.05 \times 10^6}{18.41} = 58{,}368\ \text{psi}$$

Stress criterion $= 3\sigma_A = 3(17{,}300) = 51{,}900$ psi

Since $\sigma_p > 3\sigma_A$ using the rigid plate assumption, we adjust by considering the flexibility of the end plates as follows:
Using Eq. 6-47, we have

$t_{ep} = 0.375$ in.

With

$$\frac{D_e}{OD} = \frac{20}{16} = 1.25 \rightarrow C = 0.0013 \text{ from Table 6-2}$$

$$E = 29.3 \times 106 \text{ psi}$$

$$k_{ep} = \frac{4Et_{ep}^3}{CD_e^2} = \frac{4(29.3 \times 10^6)(0.375)^3}{(0.0013)(20)^2} = 11.89 \times 10^6 \frac{\text{lb}}{\text{in.}}$$

$$F = \frac{0.1686}{\left(\dfrac{1}{11.239 \times 10^6} + \dfrac{1}{11.89 \times 10^6} + \dfrac{1}{14.11 \times 10^6} \right)}$$

$$F = 691,120 \text{ lb}$$

Revising the calculation for σ_p,

$$\sigma_p = \frac{(125)(16)}{4(0.375)} + \frac{691,120}{18.41} = 38,874 \text{ psi}$$

Since $\sigma_p < 3\sigma_A = 51,900$ psi, the system is satisfactory.

Calculating the Axial Stress in the Enclosure (Jacket):

$$\sigma_e = -\frac{F}{A_e} = -\frac{1.05 \times 10^6}{23.12} = -45,415 \text{ psi} < 3\sigma_{al} = 51,900 \text{ psi}$$

Thus, the axial stress in the enclosure is acceptable.

Because of the relatively large thermal displacement, the buckling of the enclosure should be checked. The allowable longitudinal compressive stress (s_c), using the ASME Boiler and Pressure Vessel Code, Section VIII Division 1, Paragraph UG-23, is as follows:

$$t_{ep} = 0.375 \text{ in.}$$

$$R_o = OD/2 = 10.0 \text{ in.}$$

$$E = 29.3 \times 10^6 \text{ psi at } 100°F$$

$$A = \frac{0.125}{\left(R_o/t \right)} = \frac{0.125}{\dfrac{10.0}{0.375}} = 0.0047$$

In calculating the stiffness of the end plates, we used the modulus of elasticity at ambient temperature. Thus, the axial force is also calculated using the modulus of elasticity at ambient temperature, so from Figure CS-2 of the ASME Code Section II Part D, the allowable compressive stress (B) is

$$B = 17,000 \text{ psi}$$

Thus,

$$s_c = 17,000 \text{ psi}$$

Since $-51,900 \text{ psi} \gg -17000 \text{ psi}$, the buckling condition is not satisfactory.

Using the result by accounting for the flexibility of the end plates,

$$\sigma_e = \frac{-691,120}{23.12} = -29,893 \text{ psi}$$

Since $\sigma_e = -29,893 \text{ psi} > -17,000 \text{ psi}$, the configuration is still not satisfactory.

Use of Figure CS-2 of the ASME Code Section II Part D is an older more conventional and conservative approach to computing the allowable compressive stress. Applying the API 579 (*Fitness-for-Service*) Appendix B.4.4.1(d), the allowable axial compressive membrane stress of a cylinder subject to an axial compressive load acting alone (F_{xa}) is computed as follows:

$$\lambda_c = \frac{KL_u}{\pi r_g} \left(\frac{F_{xa}(FS)}{E_y} \right)^{0.5} \qquad \text{Eq. 6-52}$$

where K = coefficient based on end conditions of a member subject to axial compression, with $K = 0.65$ for a member with both ends fixed

L_u = unbraced length of a member in compression, in. (mm)

$r_g = 0.25\sqrt{D_o^2 - D_i^2} = 0.25\sqrt{(20)^2 - (19.25)^2} = 3.99$

FS = defined in Paragraph B.4.4.4 when the predicted buckling stress is equal to the SMYS, $FS = 1.667$

$E_y = 29.3 \times 10^6 \text{ psi}$ = modulus of elasticity

The parameter F_{xa} is determined from the following criteria:

$$F_{xa} = MIN[F_{xa1}, F_{xa2}]$$

$$t_c = t - LOSS - FCA = 0.375\text{-}0 - 0.0625 = 0.3125 \text{ in.}$$

$$\frac{D_o}{t_c} = \frac{20.0}{0.3125} = 64 < 135$$

$$F_{xa1} = \frac{\sigma_{ys}}{FS} = \frac{35,000 \text{ psi}}{1.667} = 20,995.8 \text{ psi}$$

Now,

$$\lambda = \frac{(0.65)(48)}{\pi(3.99)} \sqrt{\frac{(20,995.8)(1.667)}{29.3 \times 10^6}} = 0.086 < 0.15$$

– Local buckling criterion

Computing the parameter F_{xa2},

$$C_x = MIN[\Psi, 0.9]$$

$$\Psi = \frac{(409)\bar{c}}{\left(389 + \dfrac{D_o}{t_c}\right)}$$

$$M_x = \frac{L}{\sqrt{0.5 \, D_o t_c}} = \frac{48}{\sqrt{(0.5)(20)(0.3125)}} = 27.152$$

$$\bar{c} = 1.0 \text{ for } M_x \geq 15$$

Now,

$$\Psi = \frac{(409)(1.0)}{(389 + 64)} = 0.903$$

Now $C_x = 0.9$

$$F_{xe} = \frac{C_x E_y t_c}{D_o} = \frac{(0.9)(29.3 \times 10^6)(0.3125)}{20} = 412,031.25$$

$$F_{xa2} = \frac{F_{xe}}{FS} = \frac{412,031.25}{1.667} = 247,169.32 \text{ psi}$$

Now,

$$F_{xa} = MIN[F_{xa1}, F_{xa2}] = MIN[20,995.8, 247,169.2] = 20,995.8 \text{ psi}$$

Now the allowable compressive stress is 20,995.8 psi. The actual stress, 29,893 psi, still exceeds this stress. This compressive stress of 29,893 psi may be reduced by applying preheat to the enclosure to reduce the thermal displacement, or a larger enclosure pipe may be used to provide extra flexibility in the end plates. An axisymmetric finite element model may be used to permit pipe caps, reducers, or conical transitions for the end plates.

Note that, as seen here, some companies prohibit the use of rigid welded-on jackets or enclosures (sleeves) for in-plant process piping for this very reason. A bolted clamp design presented earlier in the chapter would have been more satisfactory and required less design calculations.

Continuing on, we consider the next case for this example.

$T_e = 600°F$

$T_a = 100°F$ for enclosure

$\Delta = \alpha_e(T_e - T_a)L = (7.23 \times 10^{-6})(600 - 100)(48) = 0.174$ in.

This differential expansion is slightly more than in Case 1; however, the jacket or enclosure is exposed to the design pressure and will actually act to decrease the calculated compressive stress in the jacket. As a result, Case 1 governs the design.

It is recommended in this case to consider the bolted clamp connection design as discussed earlier in the chapter.

Welded Partial Leak Containment Box

Referring to **Figure 6-24** the following procedure may be used to design a welded partial leak containment box.

This design is similar to that with a bolted clamp connection, the significant difference being that to weld on an operating line a hot tap procedure must be approved. This involves doing a thorough UT scan of the remaining wall in the vicinity of the leak where the containment box is to be welded.

The material of the welded box should be the same as that of the process pipe or have the same generic chemistry. The leak box may be made from pipe or plate material, depending on the box diameter, material availability, and whether a formed cover plate will be used.

The corrosion allowance (CA) for the box should be the same as that of the process pipe. Determine the required height of the inside surface (h). This dimension should be kept as small as possible to

Welded partial containment box

Note: End plates not shown for clarity

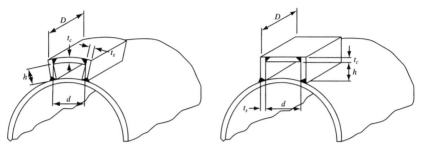

Curved cover plate (preferred construction)
Full penetration attachment welds to pipe

Flat plate (less desirable construction)
Fillet weld attachments to pipe

Figure 6-24. Welded partial containment box. Courtesy of ExxonMobil, Inc.

limit the amount of welding and the required thickness of the closure plates. However, the height (*h*) must be large enough to clear any obstructions on the pipe and permit the use of appropriate welding details.

Next determine the length of the box that will completely enclose the area of concern and permit welding to sound metal.

The minimum required thickness of the cover plate portion of the box (*t_c*) is found from the following:

$$t_c = \frac{P(D_o)}{2\sigma_A E} + CA$$

where D_o = [OD of process pipe + 2(h + t_c)] = outside diameter of the box enclosure, in. (mm). Assume t_c = ½ in. (13 mm) as an initial estimate, then adjust if later required.

E = 1 for a seamless cover plate
P = design pressure of the process pipe, psig (KPa_g)
CA = Corrosion allowance of the leak box, in. (mm)
σ_A = allowable stress of box material at the process pipe design temperature per the applicable code (e.g., ASME B31.3)

If the cover plate is made from seamless plate, the value of t_c must be divided by 0.875 to adjust for the 12.5% mill tolerance to find the required thickness of the cover plate.

For a Flat Cover Plate:

$$t_c = \Omega \sqrt{\frac{ZCP}{2\sigma_A E}} + CA$$

where $\Omega = MIN(d, D)$, in. (mm)

d = inside width of the box, in. (mm)

D = box length, in. (mm)

$Z = 3.4 - \dfrac{2.4d}{D}$, where the value of Z will not exceed 2.5

P, σ_A, E, and CA are previously defined

Next we calculate the minimum required thickness of the side plates as follows:

$$t_s = \Phi \sqrt{\frac{ZCP}{2\sigma_A E}} + CA$$

where $\Phi = MIN(h, D)$, in. (mm)

$C = 0.33$

Z = same as earlier

P, σ_A, E, and CA are previously defined

Next we calculate the minimum required thickness of the end plates (t_{ep}) as

$$t_{ep} = \Phi \sqrt{\frac{ZCP}{2\sigma_A E}} + CA$$

where all the parameters are as defined earlier.

It is desirable to make the side and end plates the same thickness. If the two have different thicknesses, the side plates would be the thicker component. Full penetration welds are desirable so that no fillet weld calculations are required.

Consideration should be given to the possible need for reinforcement. If the thinned/leaking area possibly grows—due to increased localized corrosion once the leak box is installed—or if the leaking area is presently quite large, adequate reinforcement should be provided by means of a reinforcement pad. Reinforcement calculations should be based on the applicable code (e.g., ASME B31.3).

Equipment Isolation Repairs—Stoppling

Stoppling involves the installation of a plug on each side of the repair area to cut off flow. Shown in **Figure 6-25** is a typical scheme. Normally a hot tap is performed to install a bypass line around the repair area. This bypass line is to avoid interruption of service during repairs. Hot taps are normally used because nozzle connections are seldom available to provide for the stopple and bypass connections. Also, a pressure bleed-off equalization connection is installed between the stopples to serve as a pressure equalization fitting to permit the stopple fitting to be withdrawn under line pressure. This is required, after completion of repairs. We will discuss hot taps in the next chapter. During the

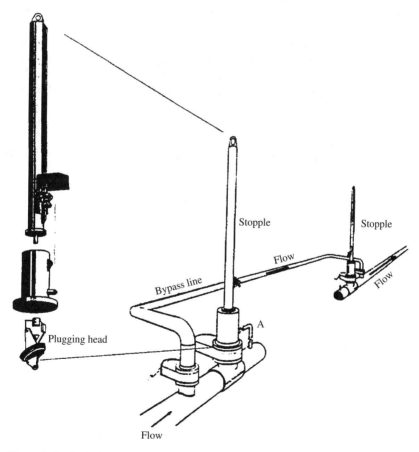

Figure 6-25. Typical line stopple with separate bypass tap. Courtesy of ExxonMobil, Inc.

stoppling operation, a plug is installed through a tee branch connection on a special carrier that permits the plug to be rotated into the line by means of a hydraulic jack.

Proper alignment of the hot tap nozzle is very important because the diameter of the cutter used for stoppling is approximately the same as the inside diameter of the line being isolated. The use of a full encirclement split tee for reinforcement as an added precaution is recommended in case the cutter cuts into the pipe wall.

The stopple device includes an elastomeric stopple sealing element. This elastomeric sealing element must be suitable for the temperature of the line being isolated and must be compatible with the chemicals to which it is being exposed. Solid metal plugs have been used for stoppling in smaller sizes. The pressure rating of the stopple fitting should be carefully checked if the application involves line sizes or pressures higher than previously experienced by the stoppling contractor.

To bypass a section of piping, a special lateral tee may be used to eliminate the requirement for the two additional hot taps for the bypass connections. When these are used, only two lateral tees need to be attached. The bypass piping also needs to be attached to the laterals as shown in **Figure 6-26**.

The attachment can be either welded or flanged. Following normal hot tap procedures, two hot tap valves should be attached and leak tested. A hot tap cutting machine should be attached. Once the cutting is

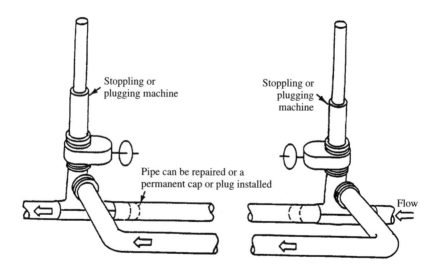

Figure 6-26. Line stopples with special lateral tee for bypass. Courtesy of ExxonMobil, Inc.

initiated, the bypass line will be pressurized. The bypass line needs to be supported adequately, and, if necessary, pipe stress computer runs may have to be performed with very cold or hot lines.

After both hot taps are completed, line stoppling machines may be placed through each lateral tee and the flow stopped in the line being repaired. Repair to the pipe being bypassed can be performed or the section can be capped off with a pipe cap. Once repairs or capping are complete, the valves may be blinded to prevent accidental leaks.

A summary of items that need to be considered when a stoppling operation is being assessed follows:

1. The plug fitting should be suitable for the line pressure.
2. The elastomeric sealing cup should be with the line contents and temperature.
3. The flow velocity in the line should not exceed 10 ft/sec (3.5 m/sec).
4. During operation metal shavings may cut the sealing element or carry downstream to rotating equipment. Many companies prohibit hot taps upstream of rotating equipment, even when strainers are used. One cannot depend on strainers to collect all the metal shavings, and the consequences can be disastrous.
5. Stoppling of lines in the vertical position should be avoided as the possibility exists that the stopple head may hang up when being retracted (when the flow is upward, the risk is less). The reason for this is that one doesn't want the coupon falling down onto a valve or piece of machinery.
6. Possible heavy scale on the inside of the line to be stoppled could interfere with the effectiveness of the seal.
7. After the installation of the stopple, there should be adequate pressure bleed down in the isolated section of the line to ensure that the stopple is effective before initiating repairs or removing the pipe section.
8. There should be adequate access to the pipe for the installation of the hot tapping and stoppling equipment.

Shown in **Figure 6-27** is a typical situation where stoppling can be desirable. A clamp temporary repair was made previously to seal a pin-hole leak. Since the installation of the temporary repair was installed, the isolation valve started leaking and cannot be completely closed (i.e., was passing). A clamp was in the proximity of the valve. Placing one clamp over another clamp is not recommended and consumes much space. Here stoppling can be used to isolate the line and repair the valve.

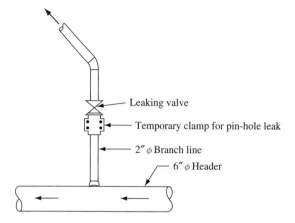

Leaking valve

Temporary clamp for pin-hole leak

2″ φ Branch line

6″ φ Header

Figure 6-27. Piping situation where stoppling was effective in repairing valve. Arrows indicate direction of process fluid flow.

Equipment Isolation Repairs by Freeze Sealing

Freeze sealing is a common practice that can be applied without the work intensity of stoppling. However, not all process fluids can be conveniently frozen. One classic example is a solution of brine in water. Brine has a very low freezing temperature, and because of its heterogeneous composition, freezing is not done with this substance. Also freeze sealing of gas lines is out of the question, as most gases have a very low freezing temperature.

Freeze sealing is used for isolating equipment for performing both permanent and temporary repairs. It is mentioned again in Chapter 7 as a possible alternative to stoppling and hot taps. If an area can be isolated where there is no fluid at pressure, a tap can be installed without going through the procedures with an operating line with pressurized fluid at temperature. The reader is referred to Chapter 7, particularly Figure 7-11, in making decisions to apply freezing for plant repairs.

Freeze sealing has been used successfully for equipment isolation by freezing the process liquid into a solid mass, creating a plug of solid material formed by application of a coolant to the exterior of the pipe. Most process fluids are water or hydrocarbons. The coolant is typically liquid nitrogen at $-320°F$ ($-196°C$). Sometimes dry ice can be used as a coolant. Dry ice is frozen carbon dioxide (CO_2) that has a temperature of $-78.5°C$ ($-109.3°F$).

A chamber, enclosure, or jacket forming an annular space is bolted onto the outside of the process pipe. The liquid nitrogen is injected into the jacket. For water service, the freezing can take 30 minutes for a 4 in. line and up to 10 hours for a 24 in. line. An experienced freeze-sealing contractor should know how long the freezing should take for the fluid being frozen.

Freeze sealing has been used for a wide variety of applications, from sealing leaking valves to isolating a section of pipe for hydrostatic testing. Because of the risks involved, the use of freeze sealing with hydrostatic testing should be done with utmost care. Stoppling would be preferred for a hydrostatic test. Freeze sealing is considered by many companies as a last resort when other methods cannot be used, especially in a process unit.

The main concern about freeze sealing is brittle fracture. With the temperatures used for this technique, typical carbon steels and low chrome piping is subjected to conditions where the risk of brittle fracture can be high. This could result from the propagation of defects that would be safe at normal temperatures but become unstable or self-propagating at the temperature used for freeze sealing; consequently, they could lead to a sudden, brittle fracture. Thermal stresses can result from the low temperatures involved, and any shock or sudden loading can cause the piping to be sensitive to the shock loading that can happen. One such scenario would be to drop a heavy wrench on the frozen area and trigger a brittle fracture (it has happened).

Because of the foregoing concerns, each application of freeze sealing should be approached with caution. All aspects should be reviewed with safety personnel and include consultation with the appropriate engineers associated with fitness-for-service. The reader is referred to Chapter 4 for details about brittle fracture in piping systems.

Finally, freeze sealing using a hydrocarbon plug to block hydrocarbon flow for the purpose of opening equipment is not recommended because of the potential unreliability of this method. Normally water is introduced into the line, and freeze sealing is achieved by creating a plug of ice.

Guidelines for freeze sealing are provided as follows:

1. Risk analysis should be performed for each application of freeze sealing for plugging an operating line, considering both the probability of brittle fracture and the consequences of failure. See Chapter 4 for details.
2. The area to be exposed to freezing should be located in a section of pipe free of circumferential or longitudinal welds and any restraints, dead ends, or side connections to minimize the possibility of a defect being present because of welding defects and the possibility of pressure buildup from freezing.

3. The area to be freeze sealed should be inspected for cracks introduced by environmental factors such as stress corrosion cracking.
   If HICs (see Chapter 3) exist in the piping, freeze sealing should
   not be considered.
4. The freeze plug should be remote from the area where mechanical
   work will be performed. The pipe should be well supported
   between the two areas.
5. If the pipe is to be cut, the ends need to be supported to minimize
   the bending stress at the plugs.
6. By all means minimize vibration of the pipe. Vibration may be
   induced from the work in cutting the pipe for installation of new
   equipment, from the travel of vehicles on roadways near the pipe,
   or from operation of machinery in the vicinity.
7. As mentioned before, consider using dry ice rather than liquid
   nitrogen. Although the temperature of dry ice is below the
   accepted engineering limit for most carbon steels, it is significantly higher than that of liquid nitrogen. Thus, there will be
   slightly more ductility in the pipe at dry ice temperatures than at
   liquid nitrogen temperatures.
8. The freeze-sealing operation should be planned to minimize the
   potential for impact loadings on the pipe area that is frozen and
   several pipe diameters from the plug.
9. The consequences of failure should be addressed when planning
   for freeze sealing. When freeze sealing is to be done to hydrostatic
   test a section of the piping system that contains only water, the
   consequence of failure should be minimal in terms of a hazard to
   the surrounding neighborhood.
10. If freeze sealing is intended to plug hydrocarbon flow in the pipe,
    other considerations should be considered. If the area of the pipe
    that is freeze sealed fractures, the hydrocarbon will be released to
    the environment. The possibility of pipe fracture during freeze
    sealing is high, so the likely consequence of a hydrocarbon release
    should be assessed.
11. Provide adequate monitoring and sensing devices to reliably con-
    firm the existence of a plug.

Safety Considerations of Freeze Sealing

There are safety considerations that result from the low temperature of
the freezing coolant and the release of large quantities of nitrogen gas in
the scenario of a nitrogen coolant release. Nitrogen expands to 700 times
its liquid volume when vaporized; thus, the risk of asphyxiation can be

high unless adequate precautions are taken. This is especially true in confined areas such as pits. Oxygen monitors are usually used to ensure safe conditions.

Another concern is if the liquid nitrogen comes in contact with human tissue, resulting in "cold burns." Gloves, boots, overalls, and safety glasses should always be worn.

The potential for oxygen enrichment is another concern when liquid nitrogen is used. Bare metallic surfaces cooled by liquid nitrogen may cause the surrounding air to condense. This can result in an increased fire hazard because of the resulting oxygen-enriched atmosphere.

Failure Experiences with Freeze Sealing

1. A workman dropped a heavy wrench onto the freeze area of a carbon steel line causing a shock loading that instantly resulted in a brittle fracture. Liquid nitrogen was being used as the coolant. The pipe, containing pressurized hydrocarbon, opened up, and the hydrocarbon was exposed to atmospheric conditions. The hydrocarbon flashed, resulting in a vapor cloud. The vapor cloud hit a source of ignition and exploded.
2. The valve bonnet bolts on a pump suction valve failed during an attempt to plug freeze the valve body to allow removal of the pump for repairs. In this case, dry ice was being used for the freezing in a box enclosing both the bonnet and the body of the valve. The expansion of the liquid as it froze in the valve overstressed the bolts, causing them to fail.
3. A hydrocarbon release happened when a hydrocarbon plug did not effectively seal a line that was being freeze sealed. When the "plug" was thought to have effectively sealed the plug, hydrocarbon gushed from the pipe, engulfing a vehicle operating nearby, causing a fire.

Note that, except for fairly heavy hydrocarbons, water can be introduced into hydrocarbon lines for effective freeze sealing.

Closure—Threaded Connections

Bolts are integral to many component designs. They are what hold the components together, namely clamps. Of critical importance is how much thread is required to be inserted into a bolt to result in a fully structural connection. Paragraph 335.2.3 in the ASME B31.3 states: "Bolt Length. Bolts should extend completely through the length of the nuts they pass through. Any which fail to do so are considered acceptably engaged if the lack of complete engagement is not more than one thread."

Example of a Bolt-up Problem in a Plant

An 84 in. valve was to be bolted in place on the nozzle of a process column. The valve was new and was larger than the space provided. As a result, not all the threads would engage through the nuts. There was not enough space to thread 7 of 64 bolts on an 84 in. flange next to the vessel head. The thread engagement was as follows:

0 position 13.76 out of 16 threads engaged the nut
1 position 14.56 out of 16 threads engaged the nut
2 position 14.56 out of 16 threads engaged the nut
3 position 15.36 out of 16 threads engaged the nut

For 2 in. heavy hex nuts, the height is approximately 2.0 in. Thus, there are 8 in./thread on the external thread of the bolt and internal thread of the nut.

Let Esmin = minimum pitch diameter of the external thread for the class of thread specified, in.
Knmax = maximum minor diameter of the internal thread, in.
At = tensile stress area of screw thread, in.^{2}

From the *Machinery Handbook* [Reference 4], p. 1168, and Bickford [Reference 5], p. 28, the formula for Le, the length of thread engagement required to develop full strength, is as follows:

Esmin = 1.9087; Knmax = 1.890; At = 2.77

$$Le = \frac{2(At)}{\left[\pi(Kn\text{max})[0.5 + 0.577359(Es\text{min} - Kn\text{max})]\right]}$$ Eq. 6-53

Le = 1.827 in.

The most deficient bolt is 13.76 threads engaged out of 16 threads. The required length for threads to have full engagement is denoted as *Thrq*. Thus,

$$Thrq = \left(\frac{1.827}{2.0}\right)16 = 14.616 \text{ in.}$$

The percent of engaged bolts at the "0" position is as follows:

$$Thr\% = \left(\frac{13.76}{14.616}\right)100 = 94.143\%$$

The percent engaged (*Thr%*) for bolts at "1" and "2" positions is

$$Thr\% = \left(\frac{14.56}{14.616}\right)100 = 99.617\%$$

These calculations are based on the nut and bolt having comparable strengths. For $2\frac{1}{4}''$ UNC threads, we have the following:

Esmin $= 1.8433$; Knmax $= 1.795$; $At = 2.302$

$$Le = \frac{2(At)}{\left[\pi Kn\text{max}[0.5 + 0.577359(Es\text{min} - Kn\text{max})]\right]}$$

$Le = 1.547$ in.

Thus, the required length for the threads to have full engagement is as follows:

$$Thrq = \left(\frac{1.547}{2.0}\right)4.5 = 3.481 \text{ in.}$$

Thus, we need 3.481 in. (approximately 3.5 in.) of threads to be engaged to take the full load.

Now the worst case given is the stud bolt being 7 mm (0.276 in.) short of the outside of the nut. Now the length of the stud inside the nut becomes

$$L = 2.0 - 0.276 = 1.724 \text{ in.}$$

Since $1.724 > 1.547$ in. required, the flange assembly is acceptable to take the full load.

As a result of this analysis, an expensive alteration was avoided, where the nozzle would have been required to be cut off, the inside lining in the column removed, and an ASME "R" stamp issued for the vessel. The plant fired up and operated successfully for many years.

Example of Clamp Design Using Shear Pins for Thrust Forces

A 4″ ϕ hydrogen quench gas line, made of 316 austenitic stainless steel, is connected to a 20″ ϕ quench header with a 4″ 1500 lb flange top of a hydrocracker in a refinery. The flange is made of 321 austenitic stainless steel. The operating temperature of the hydrogen gas is 850°F (454°C) at 2600 psig (17,931 KPa$_g$). The leak location on the existing makeshift flange is shown in **Figure 6-28a**, and the location is shown in **Figure 6-28b**.

The flange was made from a blind flange with a hole cut in the center for the 4″ ϕ hydrogen quench line, with no hub or reinforcement. The ASME/ANSI B16.5 Table 7 (Reducing Threaded and Slip-On Flanges for 150 to 2500 lb) limits the allowable center hole size cut into a blind flange to 1.5 in. Any hole size over 1.5 in. must have a hub. Since there are no 4″ 1500 lb slip-on flanges specified in the B16.5, the blind flange is an undesirable design. The design was made and installed by a contractor. To complicate matters, the flange developed a leak. To shut down the quench line would mean to shut down the entire refinery, a prospect that is highly undesirable. The only other timely option was to install a bolted clamp to allow time to last until the next scheduled shutdown in a month and a half.

The clamp design is shown in **Figure 6-29**. One portion of the clamp fits directly over the top of the flange with a connecting pipe extending

Figure 6-28a. Leak location on top of flange.

Figure 6-28b. Location of leaking flange on top of vessel.

upward. The top connecting pipe is joined together with a flange. In this vertical section of the pipe are clamp bolts. In the case of the flange rupturing, it is desired to install shear pins to hold the clamp in place, preventing the pipe from separating.

The problem is complicated by the fact that the 20″ φ header to which the 4″ φ hydrogen quench line is connected is supported by constant spring hangers. This limits the weight of the bolted clamp because constant spring hangers cannot tolerate much overloading.

A. Calculation for the Clamp Sleeve (Section above Section B-B in Figure 6-29a): Calculating the minimum required wall thickness for the sleeve, we have the following:

$$t_{sr} = \frac{PR}{(S_a E - 0.6P)}$$

with P = 2600 psig,
   R = inside radius of enclosure = 2.7 in.
   E = weld joint efficiency = 0.7
   S_a = sleeve material allowable stress at temperature = 15,900 psi
   t_{sr} = 0.734 in.

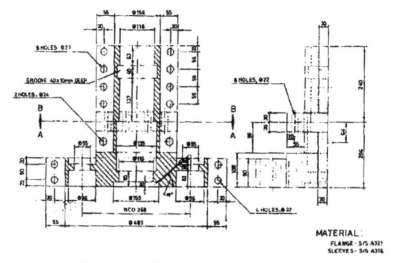

Figure 6-29a. Elevation view of quench line clamp.

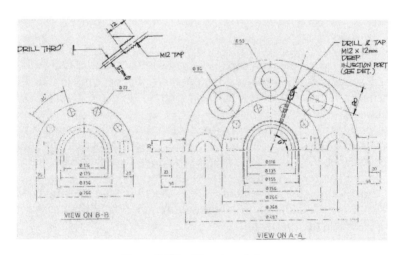

Figure 6-29b. Plan view of clamp.

Calculation for the Forces Due to Line Pressure on Bottom Flange Plate:

FLP = force due to line pressure, lb
D = inside diameter of pipe, in.
G = inside length of enclosure (plate thickness), in.
P = design pressure

DO NOT SCALE

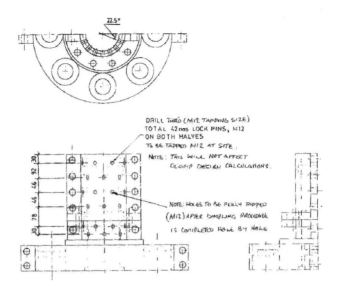

Figure 6-29c. Detail showing shear pins.

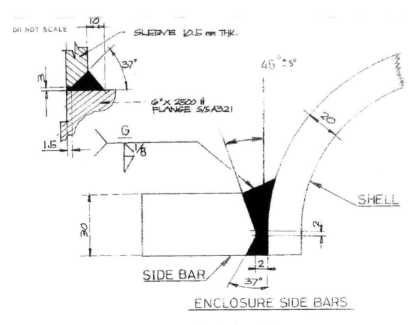

Figure 6-29d. Welding detail of clamp.

ϕ = injection factor
$FLP = DGP$ = (5.4 in.)(0.472 in.)(2600 psig)
FLP = 6226.88 lb$_f$

With an injection pressure $FL = 2600(1 + \phi) = 2600(1.2) = 7952.2$ lb$_f$

Calculation for the Lug Thickness:

T = lug plate thickness, in.
f_i = force due to injection, lb$_f$ = FL
g_1 = sum of shell thickness and fillet weld length, in. = 1.19 in.
z = distance from ID clamp to center of bolt hole, in. = 2.16 in.
n = number of bolts = 12
d_B = diameter of bolt holes, in. = 0.866 in.
σ_A = allowable design stress, psi = 15,900 psi
li = distance between outermost bolt centers, in = 14.17 in.

$$T = 2.45\sqrt{\frac{F_i(W - 0.5g_1)}{(2L_i - nd_B)\sigma_A}}$$ Eq. 6-8

T = 0.511 in.

Calculation for the Stud Load:

Let AR = total minimum root area of stud required, in.^{2}
       S_b = allowable stress of bolts, psi = 21,000 psi

$$AR = \frac{FL}{S_b} = \frac{7952.2}{21000} = 0.379 \text{ in.}^2$$

Let ARU = root area of studs used = 0.302 in.^{2}
       $ATOT$ = total bolt area = $n(ARU)$ = (12)(0.302) = 3.624 in.^{2}

Since $ATOT \gg AR$, stud bolts are satisfactory.

Calculation for Clamp End Plate Thickness (T): From the ASME
Section VIII Division 1 Figure UG-34, the design of an end plate using
Equation 6-17, which contains the following parameters:

d = diameter of enclosure, in. = 5.4 in.
C = factor on method of attachment = 0.33
P = design pressure, psig = 2600 psig

σ_A = maximum allowable stress for end plate material on Section B-B in **Figure 6-29a**, psi = 15,900 psi

E = weld joint efficiency = 0.7

$$t_{EP} = d\sqrt{\frac{CP}{\sigma_A E}} = (5.4)\sqrt{\frac{(.33)(2600)}{(15900)(0.7)}} = 1.499 \text{ in.} \qquad \text{Eq. 6-17}$$

B. Calculations for Bottom Flange Assembly (Section A-A in Figure 6-29b): Calculate the minimum required wall thickness (t_r)

t = minimum required shell thickness, in.
P = design pressure, psig
R = inside radius of enclosure, in.
S = maximum allowable stress, psi
E = weld joint efficiency

$$t = \frac{PR}{(SE - 0.6P)} = \frac{(2600)\frac{\text{lb}}{\text{in.}^2}(2.73)\text{ in.}}{\left[(15,900)(0.7) - 0.6(2600)\right]\frac{\text{lb}}{\text{in.}^2}}$$

$t = 0.742$ in.

Calculations for Forces Due to Line Pressure:

FLP = force due to line pressure, lb
D = inside diameter of enclosure, in.
G = inside length of enclosure, in.
P = design pressure, psi
FLP = (5.48 in.)(0.472 in.)(2600 psi) = 6725 lb
FL = $FLP(1 + \phi)$ = (6725)(1.2) = 8070 lb

The lug thickness is not applicable to the bottom flange assembly.

Calculations for the Stud Load:

Let AR = total minimum root area of stud required, in.^{2}
S_b = allowable stress of studs = 21,000 psi

$$AR = \frac{FL}{S_b} = \frac{8070}{21000} = 0.384 \text{ in.}^2$$

Size of studs used = 2.00 in.
ARU = root area of studs = 2.30 in.^{2}

n = number of studs used = 8
$ATOT$ = total root area of bolt studs, in.^{2}
$ATOT = n(ARU) = (8)(2.30) = 18.4$ in.$^2 \gg AR$

Thus, the stud area is satisfactory.

Calculation for the Clamp End Plate Thickness (T_{EP}):

d = inside diameter of enclosure, in. = 5.46 in.
C = factor based on ASME Section VIII Division 1 Figure UG-34 = 0.33
P = design pressure, psig = 2600 psig
σ_A = maximum allowable stress value for end plate material on Section A-A in **Figure 6-29a**, psi = 12,400 psi

$$t_{EP} = d\sqrt{\frac{CP}{\sigma_A E}} = (5.46)\sqrt{\frac{(0.33)(2600)}{(12400)(0.7)}} = 1.717 \text{ in.} \qquad \text{Eq. 6-17}$$

C. Calculation for the Thrust Load for the Shear Pins: Using twelve $\frac{1}{2}''$ lock pins to hold the assembly in place, let

$APIN$ = total pin area, in.^{2}

Use the diameter of the pin as the root area, since there are no threads. Thus,

$$APIN = \pi\left(\frac{0.5^2}{4}\right)12 = 2.356 \text{ in.}^2$$

where $SPIN$ = allowable stress for each pin = 23,500 psi (from ASME Section II Part D Table 3, for SA-193 B-16 at 850°F)
OD = 4.5 in.
$ID = OD - 2(0.673) = 3.154$ in

$$AP = \pi\left(\frac{ID^2}{4}\right) = 7.813 \text{ in.}^2$$

Let FT = thrust force in case weld fails and pipe dislodges from flange, lb$_f$
$FT = (AP)P = (7.813 \text{ in.}^2)(2600 \text{ lb}_f/\text{in.}^2) = 20,313 \text{ lb}_f$

The stress in the shear pins is the thrust force divided by the total number of available pin area.
Let $SPINA$ = actual stress in pins, psi

$$SPINA = \frac{FT}{APIN} = \frac{20,313.8 \text{ lb}}{2.356 \text{ in.}^2} = 8,622 \text{ psi}$$

The shear pins are nonpressure components. Thus, we use the AISC criteria for shear, accounting for the pin being at the design temperature of the pipe. Per the *AISC Manual of Steel Construction*, "Bolts, Threaded Parts and Rivets," Table 1-D, pp. 4–5 [Reference 2], the allowable stress for shear of nonthreaded bolts, threaded components, and rivets is $0.22Fu$, where Fu is the tensile stress at 850°F. Thus the tensile stress (Fu) at 850°F is calculated with the allowable stress being 0.25 of the tensile (below the creep range). In this case, the pins are shear pins are below the creep range. Now,

$$Fu = \left(\frac{S}{0.25}\right) = \left(\frac{15,900}{0.25}\right) = 63,600 \,\text{psi}$$

Fv = allowable shear stress, psi
$Fv = 0.22Fu = 13,990$ psi

With 8621 psi actual shear stress on the pins being less than 13,990 psi allowable shear stress, the 12 shear pins are adequate for the maximum thrust load. It was learned later that 42 shear pins were used in the actual clamp.

D. Assessment of Existing Makeshift Flange Bolting Joints: Computing the effective length of the flange bolted joint per Bickford [Reference 5], we have

Le = effective length
H = height of 2″ heavy hex nut
TFB = thickness of makeshift flange
TFN = thickness of nozzle flange

$TFB = 3.0$ in.; $TFN = 4.5$ in.; $H = 2.0$ in.

$$Le = \left(\frac{H}{2}\right) + TFB + TFN + \frac{H}{2}$$

$Le = 9.5$ in.
D_b = nominal diameter of bolt, in. = 2.0 in.

$$AB = \left(\frac{\pi}{4}\right)D_b^2 = 3.142 \,\text{in.}^2$$

ΔL_c = approximate change in length of bolt under load, in.
As = root area of 2″ ϕ threaded stud = 2.30 in.^{2}

$$Lbe = \left(\frac{H}{2}\right) + \left(\frac{H}{2}\right)$$

$Lse = TFB + TFN$

$E = 29 \times 10^6$ psi

$Fp = $ force exerted by bolt on flange, lb_f

$P = $ bolt stress, psi $= 5000$ psi

$Fp = P(As) = 115,000$ lb_f

$$\Delta Lc = F_P\left[\left(\frac{Lbe}{E(Ab)}\right) + \left(\frac{Lse}{E(As)}\right)\right] = 0.015 \text{ in.}$$

$Kj = $ stiffness of bolted joint, lb_f/in.

$$Kj = \frac{F_P}{\Delta Lc} = 7.441 \times 10^6 \frac{lb_f}{\text{in.}}$$

E. Calculation for the Force of Injection:

$A = $ area of injection $= 0.8$ in. wide $\times$ 0.6 in. high

$D1 = OD$ of 4" ϕ pipe

$D2 = $ Outer diameter of injected area $= 4.50 + 2(0.8) = 6.1$ in.

$$A_i = \left(\frac{\pi}{4}\right)(D2^2 - D1^2) = 13.32 \text{ in.}^2$$

$Pi = $ injection pressure, psig

$Fi = $ injection force, lb_f

$Pi = 1.2(2600) = 3120$ psig

$Fi = Pi(A_i) = 41,560$ lb_f

There are 2" bolts on the retaining plate above the existing bolts.

Let $BS = $ bolts torqued to 20,000 psi

$ROOT\ AREA = 2.30$ in.^{2}

$FR = $ required force from bolt torque, lb_f

$FR = BS(As) = 46,000$ lb_f

Since $FR = 46,000$ $lb_f > Fi = 41,560$ lb_f injection pressure, a bolt tension producing a stress of 20,000 psi is adequate to secure the plate.

F. Synopsis of Results: The clamp was designed, fabricated, and installed. Sealant was injected to seal off any leak. The clamp stayed until the proper materials were purchased and delivered. The clamp stayed in service until the next available shutdown 2 months later. The undesirable blind flange was replaced with a 4" ϕ 1500 lb weld neck flange and reentered into service, where it has been successfully operating for several years. Actual photos of the clamp are shown in **Figure 6-30**.

Figure 6-30a. Photo of fabricated clamp.

Figure 6-30b. Photo of clamp dismantled. The shear pins were installed in the holes.

Figure 6-30c. Photo of clamp installed and injecting sealant.

Figure 6-30d. Arrows indicate location of shear pins. The shear pins were installed in each hole.

References

1. ASME PCC-1–2000, *Guidelines for Pressure Boundary Bolted Flange Joint Assembly*, an American National Standard, 2000, American Society of Mechanical Engineers, New York.
2. *AISC Manual of Steel Construction*, 2003, American Institute of Steel Construction, Chicago.
3. *Pipeline Repair Manual, Contract PR-218-9307*, December 31, 1994, Pipeline Research Council International, p. 49.
4. Paul B. Schubert, Editor, *Machinery Handbook*, 20th edition, 1978, Industrial Press, New York.
5. John H. Bickford, *Introduction to the Design of Behavior of Bolted Joints*, 2nd edition, 1990, Marcel Dekker, New York.

Chapter Seven

Hot Tapping (Pressure Tapping) and Freezing

This chapter deals with the selection, location, design, and construction of hot taps in piping systems. The emphasis is on hot taps using welded branch fittings, but there is a discussion on bolted-on fittings. This discussion is not intended to be a full treatise on hot tapping. Such a work would be extensive and would fill several volumes. This chapter is a mere introduction to hot tapping and how it relates to piping and pipelines.

The reader is encouraged to obtain a copy of the API RP 2201 *Safe Hot Tapping Practices in the Petroleum & Petrochemical Industries* [Reference 1].

Hot tapping is the technique of creating an opening in an operating pipe by drilling or cutting a portion of the pipe with an attached fitting. The attached fitting can be a mechanical (bolt-on) or welded branch fitting to the operating piping.

The purpose of hot taps is to add connections to piping without depressurization or disruption of normal process operations. Hot taps may also be used to make connections into piping in circumstances where it would be impractical to use hot work. Hot tapping is also used to isolate sections of piping for maintenance by plugging or stoppling, as discussed in the previous chapter, or to connect a new piping system to an existing one. Oftentimes during expansion projects, hot tapping into existing lines is necessary to avoid shutting down the existing operating facilities.

Hot taps are used only where it is impractical to take the system out of service. Specifying hot taps requires inspection, design, and testing to ensure that this operation is accomplished in a safe and reliable manner. Thus, hot tapping should be considered only when other options are evaluated and rejected. Each hot tap should be properly designed, the hot tap location should be inspected, and the installation procedures should be reviewed. Relevant design conditions and safety procedures should be assessed accordingly. A typical hot tap installation is shown in **Figure 7-1**.

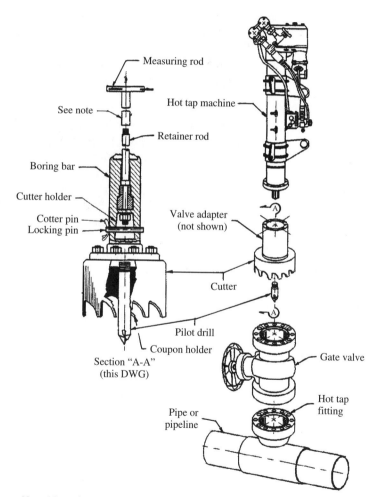

Note: Measuring rod slides into feed screw to engage end of retainer rod

Figure 7-1. Typical hot tap installation. Courtesy of ExxonMobil, Inc.

Shown in this configuration are stopples used to isolate a portion of an existing line while providing a bypass to continue operation. For readers not familiar with hot tapping, **Figure 7-1** illustrates how sections of piping can be isolated and repaired, utilizing hot tap machines and stopples while not interrupting the process flow. In this case, a valve that is passing is replaced with a new valve. In a similar manner, sections of pipe can be isolated to repair damaged areas.

Hot tapping requires special tools and knowledge. The following is a list of common terms associated with hot tapping:

- *Hot tap (or pressure tap)*—Any connection made by drilling or cutting a pipeline or pipe that has not been purged and cleared for conventional construction methods.
- *Burn-through*—An event that happens when the metal beneath the weld pool melts or no longer has the strength to contain the internal pressure of the pipe. A rupture occurs, allowing the internals to escape.
- *Combustible liquid*—A liquid with a flash point at or above 100°F (38°C) and handled at more than 15°F (8°C) below its flash point.
- *Flammable liquid*—Liquid with a flash point below 100°F (38°C).
- *Flammable material*—Flammable liquid, hydrocarbon vapors, and other vapors, such as hydrogen and carbon disulfide, that are readily ignitable when they are released to the atmosphere and hit a source of ignition, such as a spark or flame.
- *Flammable mixture*—Mixture in which a flammable liquid and air (oxygen) exist in the correct quantities to sustain combustion. Only a source of ignition is lacking.
- *Minimum allowable temperature*—See Chapter 4, page 208.
- *Plugging or stoppling*—A procedure used to isolate a section of pipe for repair or alteration without depressuring or clearing the entire line. This procedure requires a hot tapped connection or any suitable size-on-size branch connection and valve arrangement.

The Hot Tap Process

A typical hot tap process is graphically illustrated in **Figure 7-2**, where there is a passing valve in a piping system. It is desired to isolate the valve and replace it without shutting down the line.

As shown in **Figure 7-2a**, split tee fittings are welded into place with 2 in. sockolets for equalizing the pressure. The hot tap machine, containing a sandwich valve designed to allow the cutter to pass, is mounted on the split tee. In **Figure 7-2b**, after coupons are drilled from the pipe, a branch connection is connected to make a bypass line. In **Figure 7-2c**, stopple machines are mounted on the two inner split tees to isolate the passing valve with the pressure equalization lines. In **Figure 7-2d**, the old valve is replaced with the new valve, and the branch line is removed. Lock-o-ring plugs are mounted after the stopple machine is removed. In **Figure 7-2e**, the blind flanges are installed for all the branches, and the line is operating with the new valve. The steps are indicated in the figure in more detail.

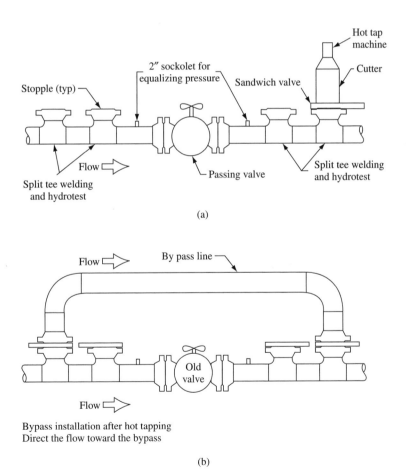

(a)

(b)

Figure 7-2. A typical application of a hot tap to isolate a passing valve and replace it with a new one.

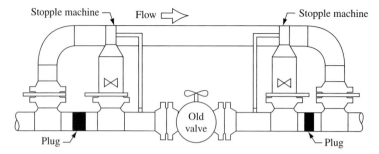

Stopple machines – insert the stopple to plug the line
Bypass installation after hot tapping
Divert the flow toward the bypass
Remove the defective valve & replace with new valve
Remove the stopple plug and insert the lock-o-ring plug

(c)

Place blind flanges above the lock-o-ring plugs
Remove the bypass
Install the stopple machine

(d)

Install blind flanges for all the branches
The line is in service

(e)

Figure 7-2. cont'd.

Assessing the Feasibility of Hot Tapping

Before one proceeds to a hot tap, the question as to whether a hot tap can be performed should be addressed. Many hot taps are performed without difficulty; however, some require extreme caution. These problematic situations should be assessed on the particular service, pipe material, and location of the hot tap. Conditions where hot taps are not recommended are as follows:

- Piping containing a flammable mixture at any pressure. Certain flammable mixtures may contain low flash points that could easily result in an ignition. Hydrogen is one such example; it is very volatile and cannot be seen when burning.
- Piping containing caustic (sodium hydroxide—NaOH). Caustic becomes more corrosive with an increase in temperature, and welding onto a caustic line would initiate accelerated corrosion in the localized region of a hot tap.
- Piping containing strong oxidizers such as pure oxygen or chlorine.
- Air lines where the presence of hydrocarbons cannot be assured.
- Monel piping containing sulfur compounds.
- Austenitic stainless steel piping containing catacarb solution.
- Piping containing chlorides, acids, peroxides, or other chemicals likely to decompose or become hazardous from heat of welding. One example is sulfuric acid or acetylene.

Engineering considerations for a hot tap are as follows:

- Review the feasibility of whether a hot tap can be performed, based on the design and service conditions, and addressing any concerns about the pipe material.
- Review previous pipe stress (flexibility) calculations or computer runs and perform new ones as required.
- Perform pressure stress and branch reinforcement calculations as required for a hot tap connection.
- Advise the operations staff of any maximum operating restrictions during the hot tap, special inspection requirements, and test pressure restrictions.
- Review the appropriate engineering standards and procedures for the pipe being hot tapped for any additional requirements.

Special Considerations for Welded-on (Hot Work) Hot Taps

Piping with certain services can be hot tapped with extra precaution. The physical parameters involved are as follows:

1. Liquid systems with no flow should be open to another system or equipment.
2. There should be confirmation that no hydrogen attack has occurred in the piping in hydrogen service. The Nelson Curve limits of API RP 941, *Steels for Hydrogen Service at Elevated Temperatures and Pressures in Petroleum Refineries and Petrochemical Plants* [Reference 2], are sufficient to satisfy this requirement as an alternative to inspection. Metals of C-$^1/_4$ Mo and C-$^1/_2$ Mo should have been operated within the limits of carbon steel to be considered for hot taps.
3. For piping containing flammable liquids operating below atmospheric pressure, provisions should be made to prevent air from being exposed to the process since air contains oxygen and an air leak would create a flammable mixture.
4. H_2S or other toxic chemicals in specific services require special safety precautions. Air lines that are free of hydrocarbons with greater than 23.5% oxygen–nitrogen purging may be required during hot tapping. With butadiene, the hot tap area must be inspected. Butadiene peroxide can exothermically react. With ethylene, the minimum flow rate should be approximately 10 ft/sec (3 m/sec), with the maximum kept as close to this value as possible. Welding procedures should minimize heat input by using low-hydrogen weld rods and by using minimum diameter electrodes and low current settings. Wet H_2S is a very dangerous service that requires extreme caution. Generally hot tapping to this service is discouraged.
5. Piping operating at temperatures above 750°F (400°C) is more susceptible to creep cracking because of high local strains around the hot tap nozzle and reinforcement attachment welds.
6. The piping needs to be checked for brittle fracture using the guidelines in Chapter 4. Special welding precautions are necessary to prevent brittle fracture.
7. Austenitic stainless steel subject to chloride stress corrosion cracking should be inspected for existing cracking. Care must be made in referring to the generic term "stainless steel." Martensitic stainless steel is not subjected to this phenomenon, nor is duplex stainless steel.
8. Hot taps made at low temperatures (e.g., carbon steel below 40°F or 4°C—see Chapter 4) may require special preheat, gas purge (e.g.,

dry air, CO_2, etc.), and welding procedures to prevent moisture accumulation, high weld quench rates, and potential weld cracking. In this regard, chilled water service can be very temperamental with condensate forming. Such condensate can make the welds porous.

9. For services that may result in carburization, nitriding, or other forms of embrittlement of the material to be welded, special evaluation of the welding procedure with respect to the process side metal temperature should be performed to determine the potential for embrittlement during hot tapping. The thickness of the unembrittled material should be determined and verified to be satisfactory for the design and conditions.

10. The pipe material should be confirmed appropriate by the proper NDE method before commencing of welding for services in which aqueous or room temperature hydrogen fissuring may occur.

11. Piping requiring post-weld heat treatment should be assessed on a case by case basis. Some examples are as follows:

   a. For ferritic steel piping that requires PWHT for residual tensile stresses, it may be possible to develop and make weld bead sequences that produce compressive stresses on the pipe inside surface. This approach should be applied after a thorough review of the proposed hot tap is performed by an experienced welding, mechanical, or materials engineer.

   b. For ferritic steels that require PWHT for hardness reasons, it may be possible to demonstrate the necessary hardness limit by using a mock-up. In such situations, hot taps may be performed provided the demonstrated test conditions and procedures are followed.

   c. For air hardenable steels where PWHT is required (e.g., $1^1/_4$ Cr-$^1/_2$ Mo), $2^1/_4$ Cr-1 Mo, 5 Cr-$^1/_2$ Mo, hot tapping may be permitted if the pipe metal temperature is maintained at or above the preheat temperatures listed in ASME B31.3, Table 330.1.1. In such cases, this minimum pipe metal temperature must be maintained from the time of the hot tap until the next planned shutdown, at which time the connection must again be PWHT to recommission the line.

12. Piping with internal linings, cladding, or weld overlay should be assessed for lining damage. Possible subsequent corrosion should be considered because if a hot tap is performed on lined pipe, the immediate area of the hot tap will become unprotected. This may be acceptable for short periods of operation in few circumstances. Generally hot tapping is prohibited on most internally lined, internally coated or internally insulated lines because there is no way to effect repairs to the internals with the equipment in service.

13. Hot taps made on such piping may result in lining detachment, and subsequent flow blockage, hot spots (in refractory lined pipe), or erosion must be considered. Concrete and refractory lined pipe is subject to damage from hot taps.
14. Hot tapping on reinforced concrete piping can be performed using bolt-on tapping fitting with special tapping equipment and techniques.
15. Care must be taken to avoid welding on casings. Hot taps should be performed on piping, but not on casings.
16. Reinforcement of the coupon to prevent it from becoming flat and binding up the cutter should be considered.
17. Hot tapping using a welded fitting is prohibited on cast-iron lines. Bolt-on connections are used for hot tapping cast-iron equipment. Bolt-on connections should not be used where the results of a leak or failure of the seal material could cause damage to equipment or injury to personnel.
18. Hot tapping is prohibited on piping that is susceptible to brittle fracture failure.

Hot Tap Design Considerations

The hot tap should conform to the applicable code for the installation. The area to be hot tapped should be inspected prior to initiating the design. This inspection should determine the materials of construction, wall thickness, and freedom from laminations.

One of the key parameters in hot tap design is flow rate. Adequate flow rate of the process fluid is necessary to transfer heat away from the area of welding, or the weld source. As discussed earlier, the type of process fluid is critical to hot tapping. For hot tapping, there is a minimum and maximum range of flow velocities. For thin pipe, it is necessary to maintain a minimum flow to dissipate welding heat. It is also necessary to require a limit on the maximum flow rate so as not to quench the weld and contribute to the cracking of hard welds and fusion problems. In gas services, the minimum flow rate should be typically 1.3 ft/sec (0.4 m/sec). Hot taps on piping without flow may be possible and should be assessed on an individual case basis. The maximum flow rate should be generally limited to 10 ft/sec (3 m/sec) to minimize the possibility of spinning the cut coupon and having it drop into the pipe. It is possible to have higher flow rates if appropriately designed pilot bits are used to dissipate the heat or avoid major operational disruptions because of flow rate adjustments. Some suppliers of hot tapping equipment recommend a maximum flow rate of 30 ft/sec (9 m/sec) when using their standard nonpositive retention pilot. They also have positive retention pilot bits that can retain coupons for gas velocities up to 100 ft/sec (30 m/sec) and solid drills that do not create a coupon but that result in more chips.

For liquid services, the flow rate should typically be in the range of 1.3 to 4 ft/sec (0.4 to 1.3 m/sec). Hot taps on piping without flow (e.g., gas service) should be evaluated on an individual case basis. It may also be possible to increase the maximum flow velocity if required in certain cases, depending on the liquid and metallurgy involved. Wet steam should be considered a liquid.

Certain suppliers of hot tap equipment recommend a maximum flow rate of 8 ft/sec (2.4 m/sec) when using their standard nonpositive retention pilot. There are also positive retention pilot bits that can retain coupons for liquid velocities up to 50 ft/sec (15 m/sec) and solid drills that do not create a coupon but that result in more chips.

Flow can be discontinued during cutting if the heat generated by cutting can be adequately dissipated without flow. This could typically be done for branch connections that are less than 50% of the header diameter. This is normally not the case with stainless steel piping. The cutting process generates enough heat to put austenitic stainless steel in the blue color range after the cutting is complete. When cutting austenitic stainless steel piping, the cutters should be used at low speed and a low feed rate. C-5 carbide cutter teeth are normally not used since the carbide tip may chip off and make cutting the coupon more difficult.

Two-phase flow systems should be treated as either gas or liquid, depending on the flow regime (i.e., annular, stratified, or slug).

Flow rates should be adjusted as required to be within the flow velocities discussed. In cases involving relatively small diameter thin wall connections into heavy wall pipe, it may be possible to tolerate lower flow rates if necessary for operational reasons because the lower welding heat magnitude can be dissipated by the heavy wall pipe. Mock-up tests may be required to determine the weld procedure details and required flow rates.

Systems containing flammable mixtures may require the injection of continuous nitrogen, steam, or hydrocarbon flow to achieve adequate safety.

Most non–air hardenable materials normally fabricated by welding can be hot tapped provided the correct conditions of pressure and temperature exist.

A realistic maximum metal temperature for performing a hot tap is 700°F (370°C). The specific temperature is usually set by the lower of the following:

1. The design temperature of the hot tap machine. The primary concern is normally the packing or the boring bar material.
2. The safe temperature in which adequate protective equipment for the welder can be provided.

The minimum design temperature for a hot tap operation is determined by the material to be welded, the hot tap equipment, and the welding conditions. The reader is referred to the discussion of brittle fracture assessments in Chapter 4 for guidance in determining the minimum temperature permitted during hot tap. Temperatures below the dew point can result in welding problems because of the moisture or frost on the metal surfaces. This can be compensated by using special procedures such as heating the area or blowing the area with dry air.

Wall Thickness of Header Pipe

LMT Approach for Process Piping

The wall thickness of the header pipe in the area of the hot tap has to have enough remaining wall thickness to be acceptable for welding. UT (ultrasonic) wall thickness measurements should be made within the area within the bounds of the limits of branch reinforcement and the nozzle outside diameter. The remaining wall thickness of the header pipe within this area should be adequate for the following:

1. Containing the pressure and preventing burn-through when welding on the operating line.
2. Adequate for the pipe design conditions based on the applicable code.
3. Withstanding the increased metal temperature generated during the hot tap welding and cutting operations.
4. Pressure testing of the nozzle assembly.

When a minimum thickness for a line is determined, this thickness should be checked in the field by UT or radiography to verify the pipe wall is of sufficient thickness. To minimize the risk of burn-through, the minimum wall thickness should be 3/16 in. (5 mm) if a set-on type nozzle is used for the hot tap connection. If a split tee or full encirclement type reinforcing pad is used, then the minimum required wall thickness may be reduced to 5/32 in. (4 mm). Thinner pipe walls may be considered subject to verification by use of a mock-up. However, the minimum thickness of the piping and equipment should be reviewed on an individual basis.

The maximum material thickness that can be hot tapped using a welded-on nozzle should be below the thickness at which the PWHT becomes mandatory per the applicable code.

Per item 4, the maximum allowable pressure for hot tapping should be calculated. The applicable code formula should be used to calculate the maximum permitted internal pressure while welding the hot tap nozzle. The allowable stress, thickness, and diameter to be used are established as follows:

1. If the welding is in an existing HAZ region, assume the weld HAZ extends to 60% of the actual measured wall thickness. Therefore, the wall thickness available to contain the internal pressure is

$$t = 0.4t_{mm}$$

   where t_{mm} = minimum measured wall thickness in hot tap region, in. (mm)

2. Adjust the outside diameter to be used in the calculations as follows:

$$D = \text{Pipe OD} - \beta t_m$$

   where $\beta = 1.2$ = usable thickness factor

3. Let T_1 = service temperature.
4. Assume the temperature in the HAZ is T_2 and equal to 1380°F (750°C).
5. Calculate the logarithmic mean temperature (LMT):

$$LMT = \frac{T_2 - T_1}{\ln\left(\dfrac{T_2}{T_1}\right)} \qquad \text{Eq. 7-1}$$

6. Determine the allowable stress as $S_a = 0.9$ (code allowable stress at LMT). The maximum allowable pressure for in-plant piping is calculated as follows:

$$P = \frac{2tS_aE}{(D-2yt)} \qquad \text{Eq. 7-2}$$

   where t is calculated in Step 1.

Less conservative assumptions for HAZ temperature and penetration may be used. This can only be verified by past experience or mock-up tests on the actual pipe size and thickness, material, and flow conditions involved in the particular hot tap situation under consideration.

Maximum Allowable Pressure for Pipelines

The LMT approach is not practical for pipelines because the allowable stress is a direct function of the SMYS. Being that pipelines operate at temperatures somewhat below much process in-plant piping, ASME B31.4 in Table 402.3.1(a) has temperatures ranging from $-20°F$ to $250°F$ ($-30°C$ to $120°C$), and ASME B31.8 in Table 841.116A has the temperature derating factor (T) for steel pipe for temperatures at $250°F$ and less up to $450°F$. Thus, the SMYS is not tabulated versus temperature in the same way that process piping is as in the ASME B31.3 or ASME B31.1. In pipelines, the concern is the minimum thickness necessary to prevent a blow-through during welding. The minimum thickness during a hot tap to prevent a blow-through during welding has been cited in various industry standards based on tests performed by the Battelle Memorial Institute. This thickness should be that required by the applicable code or standard plus 2.4 mm (3/32 in. = 0.094 in.) for weld penetration into the parent metal. To prevent blow-through during welding of the hot tap nozzle (t_{mm}), the minimum measured wall thickness in the hot tap region should not be less than the larger of (1) 3/16 in. (4.8 mm) or (2) the required thickness by code plus 3/32 in. (2.5 mm).

The maximum pressure in the pipeline during a hot tap operation is as follows:

$$P = \frac{2(SMYS)(t_{mm} - 0.094)(F)(T)}{OD} \qquad \text{Eq. 7-3}$$

where $SMYS$ = specified yield strength of the pipeline material, psi
$\quad\quad\quad F$ = design factor per code
$\quad\quad\quad T$ = temperature derating factor per ASME B31.8 Table 841.116A; for the ASME B31.4, $T = 1$

According to the API 2201 [Reference 1] Paragraph 3.2, to minimize burn-through, the first weld pass to the piping of less than $1/4$ (6.4 mm) thick should be made with a 3/32 in. (2.4 mm) or smaller diameter welding electrode to limit the heat input. Subsequent passes should be made with a 1/8 in. (3.2 mm) diameter electrode, or smaller if the pipe wall thickness does not exceed $1/2$ in. (12.7 mm). Note that the use of low heat input levels can increase the risk of cracking in high carbon equivalent materials. For piping wall thicknesses greater than $1/2$ in. (12.7 mm), where burn-through is not a major concern, larger diameter electrodes may be used. The use of low hydrogen rods may be desirable to reduce the risk of burn-through and cracking when welding on high carbon-equivalency components in many situations.

The LMT method for process piping and Eq. 7-3 for pipelines have been successfully used for many years in industry.

Example: Calculating the Maximum Allowable Pressure for Hot Tapping

An 8 in. Sch 40 pipe made of API 5L Gr B operates at 400°F (204.4°C) at 200 psig (1379.3 KPa$_g$). The minimum measured wall thickness in the hot tap region is 7.5 mm (0.295 in.). Calculate the maximum allowable pressure for hot tapping, assuming that the welding is done in an existing HAZ region in proximity to another weld.

From the ASME B31.3 code, the allowable stress for the API 5L Gr B pipe at 400°F is 20,000 psi. Thus, the minimum required thickness is

$$t_{min} = \frac{PD_o}{2(S_aE + 0.4P)} = \frac{(200)(8.625)}{2[(20,000)(1.0) + 0.4(200)]} = 0.043 \text{ in.}$$

FCA = future corrosion allowance = 0.0625 in. (1.6 mm)
Now, $t = 0.4(0.295) = 0.118$ in.
$\quad\quad D = 8.625 - 1.2(0.322) = 8.239$ in.

$$LMT = \frac{T_2 - T_1}{\ln\left(\dfrac{T_2}{T_1}\right)} = \frac{(1380 - 400)}{\ln\left(\dfrac{1380}{400}\right)} = 791.3°F$$

From the ASME B31.3, the allowable stress for API 5L Gr B at 791.3°F is $S_a = 10,400$ psi. Now $S_aE = 0.9(10,400)$ psi $= 9360$ psi. The maximum allowable pressure for hot tapping is as follows:

$$P = \frac{2t_{mm}S_aE}{(D-2yt)} = \frac{2(0.118)(9360)}{[8.239-2(0.4)(0.118)]} = 271.2 \text{ psig} > 200 \text{ psig}$$

Test Pressure and Temperature

Per the API RP 2201, conduct a hydrostatic test at a pressure at least equal to the operating pressure of the piping to be tapped, but not exceeding the present internal pressure by more than approximately 10% in order to avoid possible internal collapse of the pipe wall. If there exist conditions that could cause collapse of the pipe wall, the test pressure can be reduced. If a hydrostatic test is not practical, then a pneumatic test may be performed, using the common precautions. Note that header walls can collapse during testing if insufficient internal operating pressure or excessive external hydrostatic pressure is applied. This is especially true for large diameter piping and pipelines.

The test pressure can be limited if necessary to prevent shell buckling because of differential external pressure between the outside and inside pipe wall being hot tapped. This is accomplished by either a reduction in the test pressure from that calculated using the applicable code rules or an increase in the pipe internal pressure.

Shown in **Figure 7-3** is a hot tap nozzle welded to pipe with external load (hydrostatic test pressure). It shows the two different configurations for the nozzle connection—a welded fitting or a saddle and a full encirclement sleeve. Shown in the weld configurations, the hydrostatic test pressure is contained within the confines of the inside of the nozzle wall. If the hydrostatic test water leaks through one of the welds, then that would be a hydrostatic test failure. A more graphic detail of the typical 90° nozzle connection is shown in **Figure 7-4**.

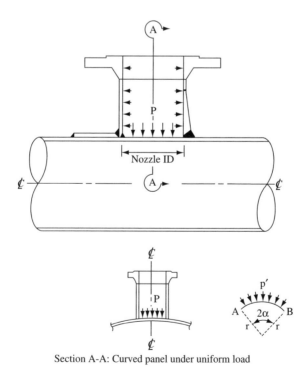

Section A-A: Curved panel under uniform load

Figure 7-3. Hot tap nozzle welded to a pipe with external load on the pipe wall during hydrostatic test.

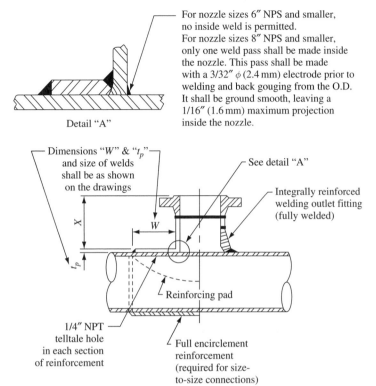

For nozzle sizes 6″ NPS and smaller, no inside weld is permitted.
For nozzle sizes 8″ NPS and smaller, only one weld pass shall be made inside the nozzle. This pass shall be made with a 3/32″ ϕ (2.4 mm) electrode prior to welding and back gouging from the O.D. It shall be ground smooth, leaving a 1/16″ (1.6 mm) maximum projection inside the nozzle.

Detail "A"

Dimensions "W" & "t_p" and size of welds shall be as shown on the drawings

See detail "A"

Integrally reinforced welding outlet fitting (fully welded)

W

Reinforcing pad

1/4″ NPT telltale hole in each section of reinforcement

Full encirclement reinforcement (required for size-to-size connections)

Figure 7-4. Typical 90° nozzle connection.

The maximum pressure required to buckle the shell wall consists of a curved plate clamped at the edges. This situation is for a saddle or reinforcing pad assembly that does not encompass the entire circumference of the pipe. This problem of elastic stability was first solved by E. I. Nicolai in St. Petersburg in 1918, cited in Timoshenko and Gere, *Theory of Elastic Stability* [Reference 3]. The buckling pressure is in the following form:

$$p' = \frac{Et^3 \, (k^2 - 1)}{12r^3(1 - v^2)} \qquad \text{Eq. 7-4}$$

where r = radius of curvature (radius of header pipe), in. (mm)
v = Poisson ratio
E = modulus of elasticity, psi (MPa)
t = Thickness of header wall, in. (mm)
k = tan($k\alpha$)/tan(α)

The relationship between α and k is as follows:

α	15°	30°	60°	90°	120°	150°	180°
k	17.2	8.62	4.37	3.0	2.36	2.07	2.0

This particular case is in Roark's 7th edition, Table 15.2 Case 21 [Reference 4].

Curved panels under uniform loading were a topic of great interest in Russia, where it snows heavily during the blizzard winters. As the story goes, whenever snow loads would build up, a curved roof sometimes would collapse. Consequently, interest to solve the problem stimulated a formal analytical solution, shown above.

As seen in **Figure 7-3**, the hot tap connection can be either a saddle or full encirclement sleeve. The latter type is shown in more detail in **Figure 7-5**.

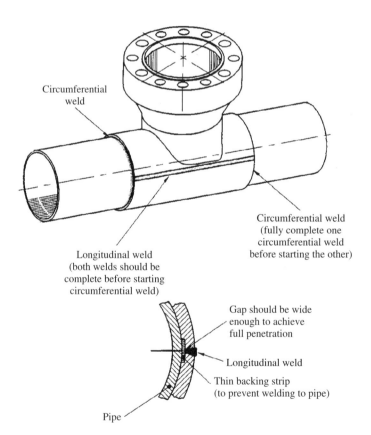

Figure 7-5. A full encirclement sleeve for hot tap installation. Courtesy of ExxonMobil, Inc.

The hydrostatic test pressure is assessed as in the ASME Section VIII Division 1 code for external pressure on a cylindrical shell. The assessment uses the A and B values in the ASME Section II Part D curves for the material considered. Before any component is welded, be it a saddle, reinforcing pad, or a full encirclement sleeve, an NDE such as UT needs to be performed to certify that the remaining wall thickness on the pipe is substantial enough to have hot work performed. If the pipe has an LTA (see Chapter 3) and a sleeve is to be welded on, the pipe is no longer of uniform thickness, and the assessment in the ASME Section VIII Division 1 for external pressure is not valid. For a pipe with an LTA, a buckling assessment using finite element is required to find the critical buckling pressure. Normally, it is accepted practice not to perform hot taps close to corroded regions to avoid this situation; however, this event cannot always be avoided.

Typical nonperpendicular nozzle hot tap connections used for pipe sizes 3 in. NPS and smaller are shown in **Figure 7-6**.

To facilitate drilling and cutting, guide plates are required for nozzles—both flanged and threaded—when attached to elbows or when installed at angles other than in the perpendicular direction. Typical

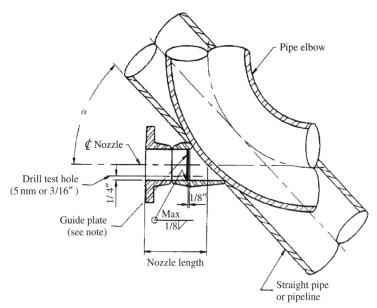

Note: Guide plate should be same metallurgy as pipe or pipeline

α = angle as shown on drawings

Figure 7-6. Typical hot tap installation for a nonperpendicular installation (e.g., an elbow). Courtesy of ExxonMobil, Inc.

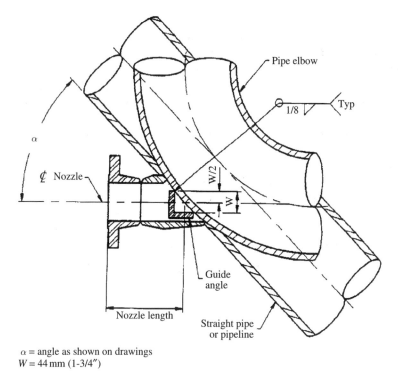

α = angle as shown on drawings
W = 44 mm (1-3/4″)

Figure 7-7. Typical nonperpendicular nozzle hot tap connections for pipe sizes 4 in. NPS and larger. The angle beam shown is to allow the drill with the cutter to maintain a common line of drilling to ensure a proper connection. Courtesy of ExxonMobil, Inc.

nonperpendicular nozzle hot tap connections for pipe sizes 4 in. NPS and larger are shown in **Figure 7-7**.

The guide angle sizes shown in **Figure 7-7** are standard AISC (American Society of Steel Construction—see Chapter 6) structural shapes. If required to make angle surface perpendicular to the axis of the pilot drill, one leg of the angle may have to be trimmed. Guide angles are not installed for cutter or drill sizes less than 2 in. OD. These guide angles provide the means to drill straight into an elbow. Without them, a worker could drill at an angle resulting in an improper fit-up.

Bolted-on fittings should be used in services where bolted-on fittings should be considered (e.g., caustic or piping requiring PWHT). When heat is applied to caustic, it becomes much more corrosive. Caustic becomes highly corrosive at high temperatures and can either cause severe corrosion damage or even eat through the pipe. Also, bolted-on fittings are used where the material is non-weldable (e.g., concrete) or difficult to weld (e.g., cast iron).

The use of bolted-on fittings is limited by design to piping normally less than 12 in. (400 mm) in diameter. These mechanical clamps are normally fabricated of carbon steel and use a rubber compression joint to seal against the pressure. This type of hot tap fitting is shown in **Figure 7-8**. Various other connections acceptable for hot taps are shown in **Figure 7-9**.

Hot tap connections are normally summarized on a computer spread sheet, as shown in **Figure 7-10**. The Type 1 is the full encirclement saddle, which is permitted in all cases. This type is preferred if vibration is possible. The Type 2, the full encirclement sleeve, is required when vibration will occur and is used when a Type 1 connection is not available; otherwise, it is permitted in all cases. The Type 3 is the split tee, which is used for hot tap installation with a standard flange or lock-o-ring flange. This type is more expensive than the Type 1 or 2. The Type 4 is the welding outlet. It is permitted only when supplied by a reputable manufacturer. The Type 5 is the circular reinforcing pad and is permitted in all cases except when there are large amounts of vibration. The full encirclement sleeve is preferred when the branch is greater than 70% of the header size. The Type 6 connection is the saddle, which has the same applications as the Type 5 except that it is preferred for high levels of vibration services with small branch connections. The Type 7 is the contour insert, which has the same application as the Type 6 connection except that it is preferred and recommended with 100% radiography. A typical hot tap calculation is shown in spreadsheet form in **Figure 7-10**.

Summary Procedures

The engineer of the proponent organization fills out the spreadsheet, and it is checked by a unit engineer as well as inspection and operations personnel before issued to the contractor. The steps required for implementing a hot tap and stoppling vary with each company. The reader will notice that there are three solutions to the hydro test of the hot tap connection. If a saddle or reinforcing pad is utilized, the Nicolai solution of Eq. 7-1 (Roark's 7th edition, Table 15.2 Case 21 [Reference 4]) is applied. If the hot tap connection is a full encirclement sleeve, then the ASME Section VIII Division 1 rules for external pressure are applied. However, as mentioned previously, if the full encirclement sleeve covers an LTA, then a buckling assessment should be performed. This assessment can be done with a linear elastic finite element model for the differential pressure between the pipe internal pressure and the applied external test pressure. Some companies avoid the finite element

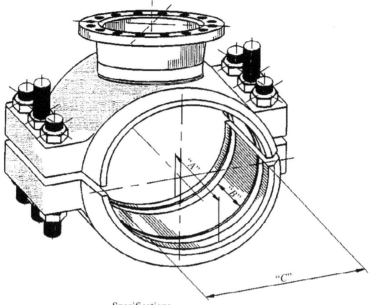

Specifications

Main API pipe size	Branch pipe size	Approx. overall length "A"	Sealing area between seals "B"	Inside diameter "C"
1 1/2	Thru 1 1/2	8 1/2	5 1/4	2 3/8
2	Thru 2	9	5 1/4	3 1/8
3	Thru 3	8 1/2	5 1/4	4
4	Thru 3	8 1/2	5 1/4	5
4	4	18	12	5
6	Thru 3	9	5 1/4	7 1/8
6	4 Thru 6	18	12	7 1/8
8	Thru 3	10	5 1/4	9 1/8
8	4 Thru 8	18	12	9 1/8
10	Thru 3	10 1/2	5 1/2	11 1/4
10	4 Thru 8	18	12	11 1/4
10	10	24	18	11 1/4
12	Thru 3	10 1/2	5 1/2	13 1/4
12	4 Thru 8	18	12	13 1/4
12	10 Thru 12	24	18	13 1/4

Above dimensions are in inches

Figure 7-8. Typical bolt-on hot tap fitting. Courtesy of ExxonMobil, Inc.

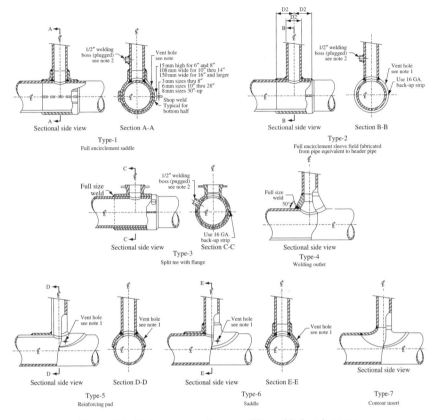

Figure 7-9. Various connections used for welded-on hot taps.

assessment by not allowing a hot tap or stopple close to an LTA; however, this event cannot always be avoided.

During welding the inner temperature of the pipe wall can rise to 19,000°F (10,400°C). This temperature can vary, depending on the wall thickness of the pipe, welding amperage, and welding technique. Temperatures of this magnitude can result in metallurgical changes in steels. Also the contents inside the pipe can be affected by such temperatures. Materials that become unstable with heat should not be subject to hot tapping. Oxidizers (e.g., oxygen and chlorine) can cause explosions with mixtures of air and fuel. Hydrogen, hydrogen mixtures, and caustic can result in cracking of the pipe in the weld metal or heat-affected zone.

Hot tapping on high purity ethylene can result in exposure of the chemical to high temperatures, and violent decomposition can occur. Tests have been performed that show that for clean systems, pressures as high

HOT TAP CALCULATION FORM

Hot tap number	11A-007	
Engineer	Yagi Siyan	
Phone number	007-3-878-1007	

PLANT NAME	Styrene 1
AREA	11
Header Type	Trunk line

Piping code: ASME B31.4

1.

HEADER DATA

Outside diameter (OD) - inches	30
Nominal wall thickness (tn) - inches	0.255
Measured minimum wall thickness (tmm) - inches	0.255
SMYS - psi	52000
Design factor per B31.4 or B31.8	0.5
Flange class	300
Flange material group	1.1
MAOP/Design pressure - psig	433
Estimated operating pressure during welding and cutting (OP)	210 psig
Expected operating pressure during branch hydrotest (Oph) - psig	210
Operating temperature - deg. F	110
Temperature derating factor, T, (B31.8) (1.0 for B31.4)	1

BRANCH DATA

Outside diameter (Db) - inches	30
Nominal wall thickness - tnb - inches	0.65
Length of split tee-inches (if applicable)	60
SMYS - psi	60000
Design factor per B31.4 or B31.8	0.5
Flange class	600
Flange material group	1
Valve material group	1.7

2. **ALLOWABLE HEADER PRESSURE DURING WELDING**

Pmax = 2*(SMYS)*(tmm – 0.094)*F*T/OD   279.0667

3. DURING WELDING

Flow velocity for gas in pipeline = 1.3–10 ft/sec
Flow velocity for liquid in pipeline = 1.3–4 ft/sec

Fluid =	Liquid	ft/sec
Velocity =	3	ft/sec

DURING CUTTING

Flow velocity (max = 15 ft/sec)   Velocity =   2   ft/sec

4. **HEADER PRESSURE DURING BRANCH HYDROTEST**   210
5. **MAXIMUM ALLOWABLE EXTERNAL PRESSURE (DIFFERENTIAL) ON HEADER**

E.I. Nicolai solution:

$p^r = E*t^3*(k^3 – 1)/(12*r^3*(1 – nu^2)) =$ ____ psig

ASME SOLUTION FOR FULL ENCIRCLEMENT

L/Do =	2
Do/MIN(tn, tnb) =	117.6
Factor A =	0.0005
Factor B =	7620

Maximum external pressure (1.25*4*Factor B)/(3*Do*MIN(tn, tmm)) =   108   psig
Maximum external pressure from FE buckling analysis for LTA =

6. **TOTAL OF 4 AND 5 =**   318   psig
7. **BRANCH VALVE SEAT PRESSURE LIMIT =**   1650
8. **BRANCH FLANGE/WALL TEST PRESSURE LIMIT**

Branch flange test pressure limit =	2250
Branch valve test pressure limit =	2250

9. Branch wall test pressure limit (0.9*2^tb*Syb/Lo)=   2340

HEADER FLANGE/WALL TEST PRESSURE LIMIT

Header flange test pressure limit =   1125

Header wall test pressure limit

10.   (0.9*MIN(tn, tmm)*Syh/Do =   ____   Lower of 6, 7, or 13
11. **VALVE SEAT TEST PRESSURE**   318   psig   Lower of 6, 8, or 13
12. **VALVE BODY/FITTING TEST PRESSURE =**   318   psig
13. **MAOP × 1.25 (Refer to B31.4 or B31.8) =**   541   psig   Lower of 6, 8, or 13
14. **BRANCH (STICKER) HYDROTEST PRESSURE =**   318   psig   Lower of 6, 8, or 13
15. **HOT TAP MACHINE TIGHTNESS TEST PRESSURE =**   318   psig   Lower of (5+OP), 8, or MAOP of header or machine
16. **REQUIRED CUTTER SIZE FOR HOT TAP =**   27   in. (mm)

Figure 7-10. Typical hot tap calculation on a spreadsheet.

as 1200 psig (8.0 MPag) can be tolerated without decomposition. However, experience indicates that pressures of the magnitude of 300 psig (2.0 MPag) are more reliable as a safe limit for operating equipment.

Piping that contains pure acetylene should not be hot tapped. The limiting pressure for decomposition depends on the temperature of the acetylene. However, with temperatures experienced during welding, pressures as low as 15 psig can be sufficient for decomposition. Vinyl acetylene has been shown to decompose at 10 psig pressure at moderate temperature.

Butadiene is normally more stable than ethylene; applying the same restrictions for ethylene to butadiene can avoid explosive decomposition. Butadiene in the presence of oxygen reacts to form a peroxide polymer that can decompose explosively. One must prevent the forming of butadiene peroxide, even in small quantities, because of its highly unstable nature. During hot tapping, the cutting machine must be purged of all air to prevent the formation of butadiene peroxide in the hot tapping equipment. Any line that contains butadiene peroxide should not be hot tapped.

When hot tapping piping contains hydrogen, hydrogen attack can occur. Hydrogen attack is a function of the hydrogen partial pressure, temperature, time, and material of construction. It can take the form of internal decarburization and fissuring, hydrogen blistering, and dissolved hydrogen leading to embrittlement. Allowable hydrogen partial pressures are based on the API RP 941, *Steels for Hydrogen Service at Elevated Temperatures and Pressures in Petroleum Refineries and Petrochemical Plants* [Reference 2]. For hot tapping, the piping should be operating at at least 100 psi (0.7 MPag) below the appropriate Nelson curve. Typically, low hydrogen welding electrodes are used for hydrogen service. The area hot tapped should be inspected by magnetic particle or liquid penetrant approximately 2 days after welding. To help distribute the residual stress in the weld connection, full encirclement fittings are recommended for hydrogen service.

Shown in **Table 7-1** are typical problematic processes for hot tapping. This table is a general guideline for hot tapping piping where one should use caution.

Table 7-1
Hot Tapping Selected Process Fluids

Service	Comments
MEA or DEA	No welded hot taps should be made for ferritic steel containing these services. Use a bolt-on hot tap saddle for hot taps. Ferrite steel cracks under weld heat in amine service.
Caustic	Under conditions requiring PWHT, use caution. The hot tap package should be reviewed by a materials and welding engineer, unless a bolt-on hot tap saddle is used.
Combustible or flammable mixture at any pressure	Hot tapping is not recommended

(Table continued on next page)

Table 7-1—cont'd
Hot Tapping Selected Process Fluids

Service	Comments
Acids, chlorides, peroxides, or other chemicals likely to decompose or become hazardous from weld heat	Hot tapping is not recommended. Examples are sulfuric acid, acetylene, high purity ethylene if air content is above 1000 ppm or the pressure is over 300 psig (2 MPag), butadiene, or elemental sulfur.
Ethylene[1]	The flow should be as high as practical, but not less than 10 ft/sec (3 m/sec). Weld procedures should minimize heat input to the extent practical by using low hydrogen weld electrodes and low current settings. Special precautions are required on ethylene lines at any pressure if the stream contains 10% inerts such as CO_2, N_2, or steam. If the ethylene content in a hydrocarbon stream is greater than 50%, special precautions are required.
Air where absence of hydrocarbons cannot be assured	Hot tapping is not recommended
Pure oxygen, or air containing more than 50% oxygen, chlorine, or liquefied gases	Hot tapping is not recommended
Catacarb solution	Hot tapping is not recommended if line is austenitic stainless steel.
Sulfur compounds	Hot tapping is not recommended if line is monel.
Hydrogen	Hot tapping is permissible provided the equipment has not operated above the Nelson curve limits.
Sour service	Hot tapping is permitted.
Chilled water[2]	This service is difficult because the water condensate can cause weld porosity. However, hot tapping lines in this service are done successfully on a routine basis.

Notes:
1. If piping containing ethylene, butadiene, or acetylene is to be hot tapped, special precautions must be taken to maintain circulation and prevent overheating and thermal decomposition (with possible explosion) of the contents.
2. In general, hot tapping is not allowed on piping that is cold enough to be below the ambient dew point temperature. This is because the moisture formed on the outside of the pipe will make the welding impractical, as already mentioned.

The Hot Tap Package

When it is decided to perform a hot tap, a package is compiled with the pertinent data and responsible agencies. Each hot tap has the relevant documents assembled together in a package before the task is performed. The typical documents in a hot tap package are as follows:

1. Justification for stopple explains why a hot tap is required and lists all the other options, noting why they are not acceptable.
2. Hot tap approval form includes a list of signatures of responsible agencies.
3. The tie-in list provides a list of tie-ins approved to be hot tapped.
4. The line list is the list of piping lines indicating the material, temperature, design and operating pressures, line sizes, wall thickness, line number, and flange rating. On this list is also shown the location of where the pipe originates and to what location it terminates and the piping specification that indicates the flange rating and process service.
5. The tie-in isometric drawing is an isometric of the piping to be hot tapped shown on an isometric drawing
6. The bill of materials includes all the materials required for the hot tap operation.
7. The safety review form is a checklist of safety items to be reviewed and checked off before the hot tap operation commences.
8. The mechanical flow diagram is a flow diagram of the piping, the flow rates, and the location of equipment. If rotating equipment is downstream of the hot tap, then a hot tap is not approved because the shavings and cut pieces could enter and damage the machinery (e.g., a pump or compressor).
9. A mill certificate details the metal to be worked on.
10. NDE forms include the UT thickness measurements of the hot tap region showing the location of the measurements and the Liquid Penetrant Examination Report, which is a check for defects in the piping (e.g., cracks).
11. Welder Procedure Qualification Card (PQR) (ASME Section IX) and the Welding Procedure Specifications (WPS), which include the weld process to be used, the weld electrode(s) to be used for the root pass and subsequent passes, electrode sizes for each pass, travel speeds for the welding, a sketch of the weld, and the type of weld. The amperage and voltage ranges are also listed. A Weld Procedure Qualification Record is included in the package to document that the WPS was qualified.
12. The stress analysis of the applicable calculations made for the hot tap includes the reinforcement calculations if a nonreinforced fitting

is used and any other calculations relevant to the hot tap, as mentioned previously.

13. The welding fittings to be used in the hot tap are described.
14. A certification record of the hot tap equipment to be used in the hot tap, the calibration tests, and the manufacturer's certification of the equipment.
15. A detailed tie-in procedure that lists the steps to be performed during the hot tap.
16. P&IDs (piping and instrumentation diagrams) show all the piping and connecting equipment, along with the line numbers and equipment numbers. The P&IDs also indicate where in the facility the hot tap is to be made, and that location is shown on the P&ID with the assigned hot tap number. All hot taps have assigned reference numbers for all documents associated with the hot tap.

Freeze Sealing

Freeze sealing is a technique that has been successfully used for many years in isolating equipment. A jacket is bolted over the area where the process fluid is to be frozen, and liquid nitrogen at $-320°F$ ($-196°C$) is injected into the annulus. The process liquid freezes to form a plug, thereby isolating the desired area. Normally, liquid nitrogen is used; however, dry ice has been used in certain applications. With liquid nitrogen, the time to freeze the process liquid can range from a half an hour for a $4''$ line to ten hours for a $24''$ ϕ line. The freezing time depends on the process liquid being frozen.

For substances such as brine, freezing is impractical, because the high salt content makes freezing virtually impossible. Water with a high salt content is also very difficult to freeze. Thus, there are applications where freezing is not practical.

When the coolant is injected into the chamber, if there is a defect in the piping, brittle fracture is a possibility, particularly when the pipe is made of carbon or low chrome steel. Also the thermal shock of the coolant entering the annulus can result in high stresses. This concern of brittle fracture makes freezing a last resort versus hot tapping. NDE should be performed to help ensure that there are no defects in the pipe or pipeline. One must bear in mind that if defects exist in a pipe or pipeline, they may be acceptable at normal temperatures, but they become critical at low temperatures. The reader is referred to Chapter 4 for further details regarding brittle fracture.

Another concern about freezing is applying it in hydrocarbon service where an attempt is made to form a hydrocarbon plug to block hydrocarbon flow. This is not recommended because of the unreliability of the technique. The standard practice is to introduce water into the line and then to freeze the water to form a plug of ice. There was an incident where a hydrocarbon line was attempted to be frozen to form a "hydrocarbon" plug to isolate a passing valve. It was found that the solidification of the hydrocarbon was not sufficient, resulting in a hydrocarbon leak.

Other failures include a case where dry ice was used to attempt to isolate a pump suction valve. The dry ice was applied to enclose both the valve bonnet and body of the valve. The expansion of the liquid as it froze in the valve overstressed the bolts on the valve, causing them to fail. Another case involved a 6 in. pipeline that ruptured during a freezing operation at the start of a pressure test. This failure was caused by a crack propagating at a dent that developed during an excavation and was not noticed. The freeze zone extended halfway over the dent, the crack became self-propagating, and the pipeline ruptured.

When freeze sealing is applied, the area to be frozen should be free of circumferential welds and any restraints, dead ends, or side constraints. This will lessen the possibility of a defect being present as a result of welding and buildup of pressure during freezing. The area to be frozen should be inspected for cracks and defect mechanisms (see Chapter 3 for details on defect mechanisms). The freeze plug should be remote from the area of mechanical work to be performed, and the pipe should be well supported between the two areas. Vibration in the piping should be at a minimum. Vibration response can result from work cutting the pipe for equipment installation, travel of vehicles on roadways near the pipe, or operation of rotating machinery near the pipe.

It is safer to use dry ice rather than liquid nitrogen. Albeit the temperature of the dry ice is below the accepted material limit for most carbon steels, it is significantly higher than that of liquid nitrogen. Thus, the pipe will be somewhat more ductile with dry ice.

Impact loadings should never happen during freezing. Any impact loading should be several diameters away from the frozen plug. The consequence of a failure should be assessed for freeze sealing a pipe or pipeline. When freeze sealing for a hydrostatic test where a section of the pipe or pipeline contains only water, the consequence of a failure is minimal in relation to the hazard to the surrounding area.

If freeze sealing is selected for hydrocarbon service, the consequences of a line rupture or hydrocarbon leak should be assessed. Hydrocarbons released to the atmosphere are very dangerous, especially if they come in contact with a source of ignition.

During freezing adequate monitoring and instrumentation should be used to confirm reliably the existence of a solidified plug.

In the application of freeze sealing, one must be cognizant of the fact that nitrogen expands seven hundred times its liquid volume when vaporized. The risk of asphyxiation can be very probable unless adequate safety precautions are made. This is especially true in confined spaces, in pits, and inside equipment. Oxygen monitors should always be used for confined spaces.

Also if liquid nitrogen comes in contact with exposed skin, blisters can develop. Appropriate safety attire should be used during a freezing operation.

Finally bare metallic surfaces cooled by liquid nitrogen may result in the surrounding air to condense. This oxygen-enriched atmosphere can be a fire hazard.

Figure 7-11 is a flow chart for decision making for hot tapping and freezing, as well as repairs. When using the figure, the reader is referred to Chapter 6 for details concerning temporary repairs. It is recommended that the reader refer to Chapter 6 for more details about the technique of freezing. In the figure where welding is not permitted, the use of bolt-on clamps is an example of an acceptable temporary repair.

Note in **Figure 7-11** that the option of hot bolting needs to be performed with caution. If bolts are loose and a combustible substance is exposed to the atmosphere, applying torque to the bolts could set off a spark for ignition. Several workers were killed in a refinery in the Middle East when workers attempted to torque bolts on a vessel in service. The wrench on one of the bolts set off a spark, which caused an explosion that resulted in several fatalities. For safety reasons, a sample of the air taken in the area of hot bolting needs to be checked for combustibles.

Example 7-1: Area Replacement Calculation for a Hot Tap

A hot tap is to be made for a connection where a non-self-reinforced connection is made. For this reason, an area reinforcement calculation needs to be performed to find whether a reinforcing pad is required. The piping was designed per the ASME B31.3. The preferred units of the ASME B31.3 2004 edition are the metric SI units. The location is outside the United States where the metric SI units are official. The U.S. readers will find this a handy reference for using the metric SI units, and the readers that use these units will find it a welcomed relief. Because round-offs are required to convert to plate sizes in the English system, the interested reader may want to perform the computations in the English system.

The process fluid in the pipe to be hot tapped has a design pressure of 4.138 MPa. The header pipe is 219.08 mm in diameter and the branch

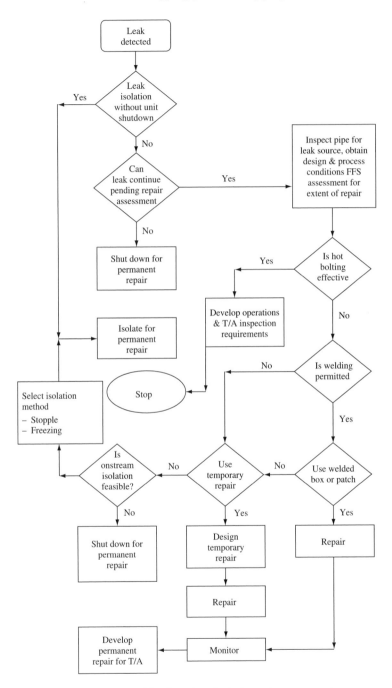

Figure 7-11. Decision flow chart for leak repair assessment.

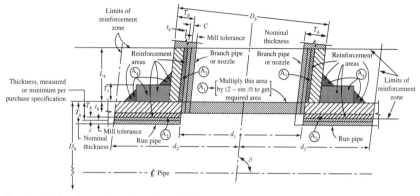

Figure 7-12. Nozzle opening configuration nomenclature.

connection is 114.3 mm. The header pipe is 8.2 mm (Sch 40) thick and
the branch pipe is 8.56 mm (Sch 80) thick. The ASME B31.3 nozzle
opening configuration (NOC) nomenclature is shown in **Figure 7-12**
(Figure 304.3.3 of the ASME B31.3 2004). The branch is to attach to the
header at 90°. Find the size of the reinforcement pad if one is necessary.

The solution follows in computer spreadsheet format. In these calculations, the weld joint factor (W) is equal to 1 because the design
temperature is under 510°C (950°F). Following the equations is the
variable spreadsheet, which explains the parameters.

$$th = \frac{P \cdot Dh}{2 \cdot (\sigma ah \cdot Eh + 0.4 \cdot P)}$$

$$tb = \frac{P \cdot Db}{2 \cdot (\sigma ab \cdot Eb + 0.4 \cdot P)}$$

$$Ths = Thbar \cdot 0.875$$

$$Th = Ths$$

$$Tb = Tbranch \cdot 0.875$$

$$VLIM1 = 2.5 \cdot (Th - c)$$

VLIM2 = 2.5 · (Tb − c) + Tr

If VLIM1 < VLIM2 Then L4 = 2.5 · (Th − c)

If VLIM1 > VLIM2 Then L4 = 2.5 · (Tb − c) + Tr

$$d1 = \frac{Db − 2 \cdot (Tb − c)}{SIND(\beta)}$$

HLIM1 = d1

$$HLIM2 = (Tb − c) + (Th − c) + \frac{d1}{2}$$

If HLIM1 > HLIM2 Then HLIM = d1

If HLIM1 < HLIM2 Then HLIM = HLIM2

If HLIM > Dh then HLIM = Dh

If HLIM < Dh then HLIM = HLIM

d2 = HLIM

A1 = th · d1 · (2 − SIND(β))

A2 = (2 · d2 − d1) · (Th − th − c)

$$A3 = \frac{2 \cdot L4 \cdot (Tb − tb − c)}{SIND(\beta)}$$

WON = LW1^2

WOP = LW2^2

DOP = 2*(d2 − LW2)

AP = Tr*(DOP − Db)

A4 = WON + WOP + AP

A5 = A2 + A3 + A4

AREA = A5 − A1

APA = Tr*(PODA − Db)

If PODA > DOP then PODA = DOP

A4A = WON + WOP + APA

A5A = A2 + A3 + A4A

AREACT = A5A − A1

Variable Sheet: The solutions to the foregoing equations are shown in spreadsheet form in **Figure 7-13**. The equations and variables are self-explanatory. Thus, a reinforcing pad of 199 mm outside diameter with a thickness of 6.35 mm is adequate.

Each hot tap package should have all supporting calculations when needed for branch reinforcement if a nonreinforcing connection is installed.

Variables Sheet

Input	Name	Output	Unit	Comment
	th	3.812388		Design required wall thickness of header, mm
4.138	P			Design pressure, MPa
219.08	Dh			Outside diameter of header, mm
137.93	σah			Maximum alowable stress of header material, MPa
.85	Eh			Quality factor for header weld
	tb	1.694210		Design required wall thickness of branch, mm
114.3	Db			Outside diameter of branch pipe, mm
137.93	σab			Maximum alowable stress of branch material, MPa
1	Eb			Quality factor of branch pipe weld – if required
	Ths	7.175		Wall thickness of header less mill tolerance, mm

Figure 7-13. NOC algorithm solution for hot tap connection.

Input	Name	Output	Unit	Comment
8.2	Thbar			Header nominal wall thickness, mm
	Th	7.175		Header wall thickness, mm
	VLIM1	11.5875		Parameter for vertical reinforcement limit, mm
2.54	c			Corrosion allowance, mm
	VLIM2	18.725		Parameter for vertical reinforcement limit, mm
	Tb	7.49		Wall thickness of branch less mill tolerance, mm
8.56	Tbranch			Nominal wall thickness of branch, mm
6.35	Tr			Thickness of repad, mm
	L4	11.5875		Vertical limit of reinforcement, mm
	d1	104.4		Effective length removed from pipe at branch
90	β		deg	Angle branch makes with header
	HLIM1	104.4		Horizontal limit parameter each side of branch, mm
	HLIM2	61.785		Horizontal limit parameter each side of branch, mm
	HLIM	104.4		Limit of horizontal reinforcement each side of branch, mm
	d2	104.4		Outside radius of horizontal limit, mm
	A1	398.013290		Required area of reinforcement, mm^2
	A2	85.880710		Available area of reinforcement in header, mm^2
	A3	75.452924		Available area of reinforcement in branch, sq. mm
	WON	22.6576		Area of weld between header and branch, mm^2
4.76	LW1			Length of weld at header branch connection, mm
	WOP	22.6576		Area of weld at pad header connection, mm^2
4.76	LW2			Length of weld at pad header connection, mm
	DOP	199.28		Diameter of repad, mm^2
	AP	539.623		Area of repad, mm^2
	A4	584.9382		Area of welds and repad, mm^2
	A5	746.271834		Total available reinforcing area, mm^2
	AREA	348.258544		Net area – positive means there is enough reinforcement, mm^2
	APA	537.845		Actual pad area, mm^2
199	PODA			Actual pad outside diameter, mm
	A4A	583.1602		Actual area of welds and repad, mm^2
	A5A	744.493834		Actual total available reinforcing area, mm^2
	AREACT	346.480544		Net area – positive means there is enough reinforcement, mm^2

Figure 7-13. cont'd.

References

1. API RP 2201, *Safe Hot Tapping Practices in the Petroleum & Petrochemical Industries*, 5th edition, July 2003, American Petroleum Institute, Washington, DC.
2. API RP 941, *Steels for Hydrogen Service at Elevated Temperatures and Pressures in Petroleum Refineries and Petrochemical Plants*, 6th edition, March 2004, American Petroleum Institute, Washington, DC.
3. Stephen P. Timoshenko and James M. Gere, *Theory of Elastic Stability*, 2nd edition, 1961, McGraw-Hill Publishing Company, New York.
4. Warren C. Young and Richard G. Budynas, *Roark's Formulas for Stress and Strain*, 7th edition, 2002, McGraw-Hill Publishing Company, New York.

Chapter Eight
Pipeline Fitness-for-Service, Repair, and Maintenance—Selected Topics

The empirical formulations developed by the PRCI over the years to produce the recursive software that predicts the burst pressure for carbon steel pipelines were discussed in Chapter 1. In Chapter 3, we reviewed the API 579 Fitness-for-Service methodology as applied to in-plant piping, namely ASME B31.1, "Power Piping," and ASME B31.3, "Process Piping." Now we focus our attention on pipelines.

Pipelines have many similarities to in-plant piping, but they also have significant differences. Most pipelines extend much further than in-plant piping and operate at temperatures closer to ambient conditions. Pipelines are maintained differently; many have pigging (called "scrapers" in the Middle East) capability. Pigs are very useful in detecting corrosion and cleaning the pipelines. Pigging technology has increased tremendously during the past decade, with different design configurations developed and tailored for specific applications.

Our purpose here is to assess the damage defects that exist in pipelines and understand how to extend service life and avoid unnecessary shutdowns. To accomplish this end, we need to address the basic concepts of fitness-for-service.

Useful RSF Equations Using API 579 Methodologies

As of this writing, API 579 [Reference 1] does not cover ASME B31.4, "Pipeline Transportation Systems For Liquid Hydrocarbons and Other Liquids," [Reference 3] or ASME B31.8, "Gas Transmission and

Distribution Piping Systems" [Reference 4]. The following discussion is an extrapolation of the API 579 rules to pipelines and thus is not an official part of the API 579. In Chapter 1, we discussed many aspects about pipeline fitness-for-service, and in Chapter 3 we discussed the subject in more detail. This chapter is a supplement to Chapter 3.

Since we are on the topic of RSF, it may be helpful to delve into how this parameter is connected to other problem parameters. The RSF concept is very useful, especially for LTAs. In FFS, one may know if a component is "acceptable" or "not acceptable," and API 579 is very clear about acceptance criteria.

What is not evident is how long a component will remain acceptable once it is found to be acceptable. This question is fair and inevitably should be asked by the organization—inspectors and managers alike. To answer this we will "jump through some hoops" and get ahead of ourselves since we are talking about RSF and its relevance.

The API 579 mentions MAWP—maximum allowable working pressure. This term is sanctioned for use with piping by the API 570, "Piping Inspection Code." It is defined in Paragraph 3.21 as "The maximum internal pressure permitted in the piping system for continued operation at the most severe condition of coincident internal or external pressure and temperature (minimum or maximum) expected during service. It is the same as the design pressure, as defined in ASME B31.3 and other code sections, and is subject to the same rules relating to allowances for variations of pressure or temperature or both." The following is solving for the MAWP in terms of the RSFa and other parameters.

The remaining thickness ratio (R_t) is defined in API 579 (Eq. 4.2) as follows:

$$R_t = \left(\frac{t_{mm} - FCA}{t_{\min}} \right)$$
Eq. 8-1

where FCA = future corrosion allowance, in. (mm)
$t_{\min}$ = minimum required wall thickness, in. (mm)
t_{mm} = minimum measured thickness, in. (mm)

The RSFa is the *allowable* remaining strength factor. Now we are placing a limit on the RSF factor. The parameter (λ) is defined as

$$\lambda = \frac{1.285(s)}{\sqrt{Dt_{\min}}}$$
Eq. 8-2

where s = meridional (axial) dimension of the LTA, in. (mm)

D = ID (inside diameter) of the shell, in. (mm)

API 579 places the value of R_t in Eqs. 5.61, 5.62, and 5.63 as follows:

$$R_t = 0.2 \text{ for } \lambda \leq 0.3475 \qquad \text{Eq. 8-3}$$

$$R_t = \frac{\left(RSFa - \dfrac{RSFa}{M_t}\right)}{\left(1.0 - \dfrac{RSFa}{M_t}\right)} \text{ for } 0.3475 < \lambda < 10.0 \qquad \text{Eq. 8-4}$$

$$R_t = 0.885 \text{ for } \lambda \geq 10.0 \qquad \text{Eq. 8-5}$$

where $M_t = \sqrt{1 + 0.48\lambda^2}$, defined in Eq. 5.12 of API 579

The term M_t is called the Folias factor. The API 579, like most standards, sets the procedures and rules, but is not very didactic. In other words, to many readers the Folias factor is a pure abstraction. It does have physical significance, however. We dealt with it in Chapter 1 with the Keifner et al. algorithms. What it represents is the "bulging effect" of an LTA to internal pressure. Suppose we had a hypothetical pipe containing an LTA that is tissue thin compared to the surrounding cylinder. By applying internal pressure inside the cylinder, we would notice the LTA bulging outward as we increased the pressure. This bulging, or balloon, effect causes bending moments at the edge of the LTA junctions with the surrounding shell. For shallow LTAs (i.e., where the LTA, or corroded region, has shallow boundaries or the remaining wall is slightly less than the surrounding wall), these bending moments can be marginal. However, if the LTA has a remaining wall that is significantly less than that of the surrounding wall, these bending stresses can become significant. The Folias factor takes this phenomenon into account, which is explained later in this chapter.

Now in Appendix A of API 579 basic parameters are defined. The pipeline codes are based on the Barlow equation, $PD/2t$. The Barlow equation is more conservative than the Boardman equation of the ASME B31.1 and ASME B31.3, which is based on a smaller pipe, except when $Y = 0$. Consequently, we must rewrite the equations for the MAWP for pipelines. The minimum required wall thickness in the longitudinal direction (axial) is as follows:

$$t = \frac{P_i D}{2SFE} \text{ for ASME B31.4} \qquad \text{Eq. 8-6}$$

$$t = \frac{PD}{2S(F)(E)(T)} \text{ for ASME B31.8}$$

Eq. 8-7

where P = internal pressure, psig (KPa_g)
$\quad\quad S$ = allowable stress, psi (MPa)
$\quad\quad F$ = design factor, based on the location classification
$\quad\quad E$ = longitudinal joint factor per ASME B31.4, ASME B31.8
$\quad\quad T$ = temperature de-rating factor, see ASME B31.8
$\quad\quad D$ = nominal outside diameter of pipe that includes $LOSS + FCA$, in. (mm)
$\quad\quad R$ = $ID/2$, in. (mm)
$\quad LOSS$ = metal loss in shell prior to the assessment equal to the nominal (or furnished thickness if available) minus the measured minimum thickness at the time of inspection, in. (mm)
$\quad FCA$ = future corrosion allowance as mentioned in Paragraph A.2.7, in. (mm)

The FCA is the expected or anticipated corrosion or erosion that will occur. From Eq. A.1,

$$t_{req} = t_{min} + FCA \text{ (in., mm)}$$

Eq. 8-8

The parameter t_{req} is the required thickness for future operation.
From Paragraph 4.4.2.1.f.1,

$$t_{am} - FCA > t_{min}$$

Eq. 8-9

where t_{am} = average measured wall thickness for general (or uniform) metal loss, in. (mm). This is accounted for in the term UML (uniform metal loss).

General metal loss in pipelines is normally not common, especially for internal corrosion. Computing the t_{mm} we have

$$t_{mm} = t_{nom} - UML$$

From Paragraph 4.4.2.1.f.2, one must make the following check:

$$t_{mm} - FCA > MAX\ [0.5t_{min}, 2.5 \text{ mm } (0.10 \text{ in.})]$$

Now we set $R_t = R_{ta}$ and set Eq. 8-1 equal to Eq. 8-4, obtaining

$$R_{ta} = \left(\frac{t_{mm} - FCA}{t_{min}}\right) = \frac{\left(RSFa - \dfrac{RSFa}{M_t}\right)}{\left(1.0 - \dfrac{RFa}{M_t}\right)} \qquad \text{Eq. 8-10}$$

Now,

$$Do = 2R = 2[R_i + (LOSS + FCA)]$$

where R_i = inside radius of pipeline, in. (mm)

From Eq. 8-7, we have

$$t_{min} = \frac{PR}{S(F)(E)(T)} \qquad \text{Eq. 8-11}$$

Now substituting $MAWP = P$, $S_a = S = SMYS$, and $R_c = R_i + t_n - (LOSS + FCA)$, we have

$$MAWP = \frac{(t_{mm} - FCA)\left(1.0 - \dfrac{RSFa}{M_t}\right)S_a(F)(E)(T)}{\left(RSFa - \dfrac{RSFa}{M_t}\right)R_c} \qquad \text{Eq. 8-12}$$

The terms F, E, and T are defined as previously. For ASME B31.4, use $T = 1$.

As in Chapter 3, when $\lambda \leq 0.3475$, $R_t = 0.2$, and when $\lambda \geq 10.0$, $R_t = 0.885$. The provisions in the API 579 call for $R_t = 0.2$ if $\lambda \leq 0.3475$ and $R_t = 0.885$ for $\lambda \geq 10.0$ (see Figure 5.6 in the API 579). If $\lambda > 10.0$, making $R_t = 0.885$ is a penalty paid for a long defect. Long defects are not uncommon in pipelines; in pressure vessels, the above criterion is a safer approach. Letting $a = 0.2$ or 0.885, we derive the relationship for the MAWP as follows:

$$MAWP = \frac{2S(F)(E)(T)(t_{min} - FCA)}{a(D)} \qquad \text{Eq. 8-12a}$$

The terms F, E, and T are defined as previously. For ASME B31.4, use $T = 1$.

Now we have obtained a working relationship for the MAWP in terms of the RSF, and we can predict the remaining life. Here we see that the MAWP is inversely proportional to the RSF. This equation results in the lower in value the RSF the higher the value of the MAWP. In **Figure 8-1**, Eq. 8-12 is applied for a specific case, where $R_c = 12.0$ in., $t_{mm} = 0.150$, $M_t = 1.38$, $FCA = 0.0$, $UML = 0.0$, $S_a = 35000$ psi, $T = 1$, and $E = 1$. **Figure 8-1** illustrates the relationship between the MAWP and the RSFa.

Equation 8-12a has proven impractical for pipelines. When using $R_t = 0.885$, the MAWP value is lowered, providing a more conservative result, which is less consistent with the burst test results. For this reason, we use Eq. 8-12 for pipelines. See **Table 8-2b** in the discussion that follows below under Kiefner Case 68.

The term M_t is called the Folias factor. The API 579, like most standards, sets the procedures and rules, but it is not very didactic. In other words, to many readers the Folias factor is a pure abstraction. It does have physical significance, however. We dealt with it in Chapter 1 when we discussed the Kiefner et al. algorithms. It represents the "bulging effect" of an LTA to internal pressure. Suppose we had a hypothetical pipe containing an LTA that is tissue thin compared to the surrounding cylinder. By applying internal pressure inside the cylinder, we would notice that the LTA bulges outward as the pressure is increased. This bulging, or balloon, effect causes bending moments at the edge of the LTA junctions with the surrounding shell. For shallow LTAs (i.e., where the LTA, or corroded region, has shallow boundaries, or remaining wall slightly less than the surrounding wall), these bending moments can be

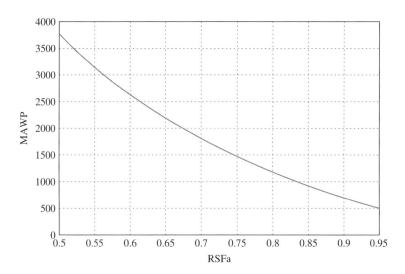

Figure 8-1. MAWP versus RSFa for the specific pipeline.

marginal. However, if the LTA has a remaining wall significantly less than that of the surrounding wall, these bending stresses can become significant. The Folias factor takes this phenomenon into account. Shown in **Figure 8-2** is what the Folias effect means in practical terms.

Now that we have obtained a working relationship for the MAWP in terms of the RSF, we can predict the remaining life. Here we see that the MAWP is inversely proportional to the RSF. This equation results in the lower in value the RSF the higher the value of the MAWP. **Figure 8-1** shows the application of Eq. 8-12 for a specific case, where $R_c = 12.0$ in., $t_n = 0.5$, $t_{mm} = 0.15$, $M_t = 1.1$, $FCA = 0.0$, $UML = 0.0$, $SMYS = 35000$ psi, $F = 0.72$, $T = 1$, and $E = 1$. This figure illustrates the relationship between the MAWP and the RSFa.

API 579 CRITERIA
Level 1 (Par. 5.4.2.2)

(1) $R_t \geq 0.20$ Eq. 8-13

(2) $t_{mm} - FCA \geq 2.5$ mm (0.10 in.)   Eq. 8-14

(3) $L_{msd} \geq 1.8\sqrt{Dt_{min}}$ Eq. 8-15

Level 2 (Par. 4.4.3.2)

(4) $\lambda \leq 5.0$ Eq. 8-16

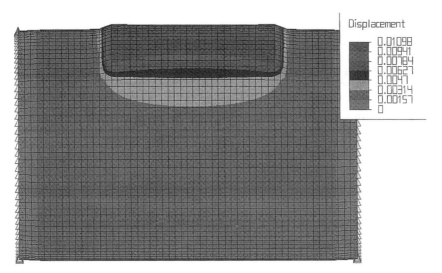

Figure 8-2. Exaggerated view of LTA displacement relative to thicker pipe (also refer to Figure 3-2 for a three-dimensional view).

These parameters will be defined in the examples that follow. Criterion (2) is invalid in pipelines because of the *Kiefner 20%–80% Rule*. This rule states that if the depth of a defect is 20% or less than the nominal wall thickness, the surface can be recoated and reentered into service. If the depth of the defect falls between 20% and less than 80% of the nominal wall, then an FFS assessment is required. If the depth of the defect exceeds 80% of the nominal wall thickness, then the pipeline component must be replaced. Thus, criterion (2) for pipelines would be $0.20t_n$, where t_n is the nominal wall thickness. Criterion (3) is very conservative. L_{msd} is the distance from the edge of the defect (LTA) to the nearest structural discontinuity. A better criterion would be by revising Eq. 8-15 as follows:

$$L_{msd} \geq 7t_n \qquad \qquad \text{Eq. 8-15 revised}$$

Criterion (5) was intended for pressure vessels. It is not realistic in pipelines, where defects can run very long lengths. This is supported by the Kiefner 20%–80% Rule, which states that if the depth of a defect is less than 20% of the nominal wall, it can have an indefinite length. It further states that if a defect falls between 20% and 80% of the nominal wall, the defect can run indefinitely. Besides, in the Kiefner Case 68, which is discussed later in this chapter, it was validated by a burst test.

API 579 Criteria Modified to Pipelines

Example: Pipeline LTA Assessment

A 48" ϕ wet crude oil pipeline in the Middle East operates at 250 psig at 160°F and contains a local thin area on the inside of the pipeline, determined by UT readings. The pipeline operation unit wishes to know if the pipeline is safe to operate with this LTA. The pipeline material is API5L Gr B (*SMYS* = 35,000 psi) and has a nominal wall thickness of 0.5 in. (12.7 mm). The company procedure requires that a pipeline running through a plant facility have a design factor of 0.5, as an extra safety requirement to protect personnel. The LTA is shown in **Figure 8-3**. The corresponding tabulated values of the pit scale readings for **Figure 8-3** are shown in **Table 8-1**. The "M" stands for the meridian direction and the "C" designates the circumferential direction. The C values (e.g., C17, C18) follow along the axial direction of the pipeline; the M values are along the girth or circumferential direction of the pipeline. Given the UT readings, is the pipeline currently acceptable? The new pumps installed

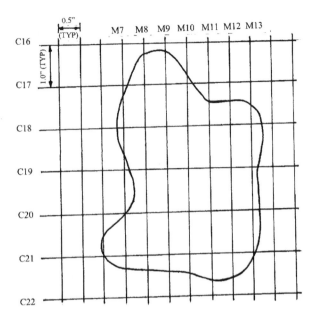

Figure 8-3. LTA shown at location 3 and designated as LTA 3.

Table 8-1
Tabulated Values of the Pit Scale Readings

(mm)	C17	C18	C19	C20	C21	C22
M7	12.1	12.4	12.5	12.3	12.5	12.5
M8	10.8	10.6	12.0	11.7	11.9	11.9
M9	9.0	9.2	9.4	11.8	9.5	11.4
M10	10.0	6.8	7.8	11.5	8.3	11.8
M11	12.8	7.9	7.6	11.0	7.3	10.6
M12	12.3	11.9	11.0	11.3	7.8	12.5
M13	11.6	11.3	11.4	11.8	10.9	12.8
MIN (mm)	9.0	6.8	7.6	11.0	7.3	10.6
MIN (in.)	0.354	0.268	0.299	0.433	0.287	0.417

can cause a temporary surge in pressure to 289 psig. What is the stress at this pressure?

Per API 579, we need to calculate the length of the LTA in the longitudinal direction, known as *s*. The circumferential direction is in the vast majority of cases not a concern in pipelines unless there are high axial loads (e.g., thrust loads). **Figure 8-4** shows how to calculate the parameter *s*.

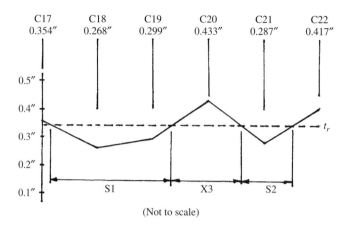

Figure 8-4. Solving for the longitudinal defect length *s*.

Using the formulation described earlier, the LTA is assessed as follows:

$$tr = \frac{P \cdot D}{2 \cdot S \cdot F \cdot E}$$

$$s1 = n \cdot i + \left[\frac{tr - tc18}{tc17 - tc18} \right] \cdot i + \left[\frac{tr - tc19}{tc20 - tc19} \right] \cdot i$$

$$X3 = \left[\frac{tr - tc19}{tc20 - tc19} \right] \cdot i + \left[\frac{tr - tc21}{tc20 - tc21} \right] \cdot i$$

$$s2 = \left[\frac{tr - tc21}{tc20 - tc21} \right] \cdot i + \left[\frac{tr - tc21}{tc22 - tc21} \right] \cdot i$$

$$s = s1 + s2$$

$$\lambda = \frac{1.285 \cdot s}{\sqrt{D \cdot tr}}$$

$$Rt = \frac{tmm - FCA}{tr}$$

$$\gamma = tmm - FCA$$

If $\gamma > 0.10$ then $OK1 = 1$

$$Lmsd = 1.8 \cdot \sqrt{D \cdot tr}$$

If $Lmsd < 10$ then $OK2 = 1$

$$Mt = \sqrt{1 + 0.48 \cdot \lambda^2}$$

$$RSF = \frac{Rt}{1 - \left[\dfrac{1}{Mt}\right] \cdot (1 - Rt)}$$

$$MAWP = \frac{(tmm - FCA) \cdot \left[1.0 - \dfrac{RSFa}{Mt}\right] \cdot S \cdot F \cdot E}{\left[RSF - \dfrac{RSFa}{Mt}\right] \cdot R}$$

The variable sheet showing the solutions to these equations is shown in **Figure 8-5**.

From the preceding algorithm, one can see that for a pressure of 250 psig the calculated $RSF = 0.9558$. According to API 579, the computed RSF should not be lower than 0.90. From the preceding assessment, the pipeline is satisfactory to operate at 250 psig assuming no uniform metal loss (UML) or future corrosion allowance (FCA). Since the operations group had no corrosion data, the pipeline was found satisfactory in the present condition, but one cannot calculate the remaining life without corrosion rate data. Hence, it is important to maintain corrosion data in pipeline service. A linear-elastic finite element run got a stress (Tresca) approximately 22,000 psi for a surge pressure of 289 psi.

One must remember that API 579 is based on the yield strength of the component material. The software RSTRENG, mentioned previously, is based on the burst pressure or ultimate strength of the pipeline material. Consequently, the Level 2 of Section 5 of API 579 is more conservative than RSTRENG because it is based on the yield strength rather than the ultimate strength of the pipeline steel. A Level 3 of API 579 will give results closer to RSTRENG, depending on the plastic offset applied in the analysis. Because the allowable plastic deformation used in a Level 3 is far less than the ultimate failure point, the Level 3 is also more conservative than the ultimate strength approach. We ran the algorithm for Eq. 8-12 to see what modifications had to be made to approximate an RSTRENG assessment. Summarized in **Table 8-2** are results of varying the computed RSF to agree with RSTRENG. The

460 *Escoe: Piping and Pipeline Assessment Guide*

Variables Sheet

Input	Name	Output	Unit	Comment
	tr	0.342857		Minimum required thickness per B31.4, in.
250	P			Internal pressure of pipeline, psig
48	D			OD of pipeline, in.
35000	S			SMYS, psi
.5	F			Design factor for pipeline inside plant
1	E			Weld joint factor
	s1	2.197724		Length of first segment of LTA with remain thickness < tr
1	n			Number of inspection intervals
1	i			Length of inspection interval, in.
.268	tc18			Remaining wall thickness at Point C18, in.
.354	tc17			Remaining wall thickness at Point C17, in.
.299	tc19			Remaining wall thickness at Point C19, in.
.433	tc20			Remaining wall thickness at Point C20, in.
	X3	0.709875		Segment between S1 & S2 with remain, thickness > tr
.287	tc21			Remaining wall thickness at Point C21, in.
	s2	0.812253		Length of second segment of LTA with remain, thickness < tr
.417	tc22			Remaining wall thickness at Point C22, in.
	s	3.009978		Length of LTA in longitudinal direction, in.
	λ	0.953431		Shell parameter
	Rt	0.781667		Remaining thickness ratio
.268	tmm			Minimum measured wall thickness, in.
0	FCA			Future corrosion allowance, in.
	γ	.268		Computed minimum measured wall thickness less FCA, in.
	OK1	1		Gamma parameter > 0.10, then okay (OK1 = 1)
	Lmsd	7.302133		Length from structural discontinuity, in.
	OK2	1		Length to nearest structural discontinuity is satisfactory
	Mt	1.198472		Folias factor
	RSF	0.955789		Computed remaining strength factor
	MAWP	215.239208		Maximum allowed working pressure (MAOP), psi
.96	RSFa			Allowed remaining strength factor
24	R			Outside radius of pipeline, in.

Figure 8-5. Pipeline FFS example of LTA 3.

Kiefner RSTRENG results are from [Reference 2], where he conducted 215 burst tests.

In developing **Tables 8-2** and **8-2a**, we took the safe maximum pressure (SMP) computed from RSTRENG and then used that pressure as the MAWP in the API 579 run. To compute the MAWP in the API 579, we found the SMP with the defect parameters that would yield the desired MAWP. The term "MAWP" is an API 579 term, and we use it here to be consistent with that document. Actually the computed MAWP is the MAOP in pipelines, although the term "MAWP" has a more

Table 8-2
Tabulation of Kiefner RSTRENG Results and Wet Crude Line with API 579 Method for Pipelines

RSTRENG Kiefner Case No.	OD of Pipe Line, in.	SMYS, psi	t_m in.	API 579 RSF*	API 579 RSFa	RSTRENG Safe Maximum Pressure, psig	API 579 Safe Maximum Pressure, psig	API 579 MAWP, psig
177	12.75	54,100	0.253	0.803	0.96	1373	1245	1373
86	22.0	60,967	0.198	0.564	0.96	608	442	605.6
194	20.00	35,000	0.593	0.715	0.96	1432	1323	1431
193	20.00	35,000	0.593	0.68	0.96	1287	1208	1287
68	30.00	59,400	0.372	0.736	0.96	979	968	979
90	36.00	73,440	0.400	0.947	0.96	1175	917.5	1175
48″ Wet Crude Line (LTA3)	48.00	35,000	0.500	0.956	0.96	363.8	250.0	251.2
48″ Wet Crude Line (LTA3) Revised Pressure	48.00	35,000	0.500	0.827	0.96	363.8	313.0	363.8

*RSF is the computed value.

Table 8-2a
Tabulation of API 579 Parameters for Each Case in Table 8-3

RSTRENG Kiefner Case No.	OD of Pipe Line, in.	Defect Length, in.	t_m in.	Rt	t_{mm}* Allow.	t_{mm}	λ	Design Factor (F)
177	12.75	2.82	0.253	0.652	0.05	0.133	2.25	0.72
86	22.0	6.00	0.198	0.451	0.04	0.05	3.68	0.72
194	20.00	10.00	0.593	0.623	0.12	0.327	3.97	0.72
193	20.00	13.00	0.593	0.611	0.12	0.293	5.395	0.72
68	30.00	36.00	0.372	0.713	0.07	0.242	13.01	0.72
90	36.00	1.60	0.400	0.416	0.08	0.13	0.419	0.72
48″ Wet Crude Line (LTA3)	48.00	7.00	0.500	0.782	0.10	0.268	0.953	0.50
48″ Wet Crude Line (LTA3) Revised Pressure	48.00	7.00	0.500	0.624	0.10	0.268	1.675	0.50

*Min. wall for a defect for Kiefner 20%–80% Rule.

Table 8-2b
Ratio of the API 579 SMP to the MAWP

Case	SMP/MAWP
Case 177	0.91
Case 86	0.73
Case 194	0.92
Case 193	0.94
Case 68	0.99
Case 90	0.781
48″ Wet Crude Line	0.995
48″ Wet Crude Line Revised Pressure	0.860

comprehensive meaning with pressure vessels than the MAOP with pipelines. The SMP is the predicted failure pressure multiplied by the design factor. The ratio of the SMP to the MAWP using API 579 methodology is shown in **Table 8-2b**. As illustrated in the table, there is no constant ratio between the SMP and the MAWP because of the defect parameters—length of LTA, RSFa, λ, M_t, γ, RSF. Also, there was no uniform metal loss or future corrosion allowance pertaining to the remaining life in these cases. The FCA will be shown later in an actual problem of a seawater injection pipeline.

Now we can see that the computed value of the RSF varies to match the burst test results. If one uses $RSF > 0.90$, as in the preceding example, then the results can be very conservative. Thus, for pipeline, the computed RSF can vary from approximately 0.56 to 0.96 to give results compatible with the ultimate strength assessment. This is done to offset the yield strength assessment from the ultimate strength assessment. From **Figure 8-1**, one can see that as the RSF decreases the MAWP (MAOP) will increase. The criteria for API 579 are originally intended for pressure vessels, heat exchangers, API storage tanks, and in-plant piping where the codes are based on yield strength rather than ultimate strength.

Note that even though the computed RSF is stipulated to be greater than the RSFa of 0.90 in Example 5.11.1 of the API RP 579, if the computed RSF is less than the RSFa of 0.9, then the MAWP is re-rated by multiplying the design pressure by the ratio of the RSF to the RSFa. The preceding benchmark test using the previously mentioned PRCI burst tests by Kiefner show this to be a conservative criterion. The RSF is more likely to vary as a function of the RSFa and other parameters than to be a constant, like 0.90, or the proposed 0.96.

Now re-running the example of the LTA 3 in the wet crude line using ultimate strength for an assessment, we have the following as shown in **Figure 8-6**.

```
FILENAME: D:\RSTRENG\G26LTA3 .RST

Specimen = 48 in G26 LTA 3        Date = 11-22-2004
Diameter = 48.00 in.              Thickness = 0.500 in.
Yield Str. = 35,000 psi.          Comment = LTA 3 ON SAFANI
 !--------------------------------------------------------!
       0              2              4              6
0.00
 *--------------------------------------------------------*
0.04    -
0.08    -                               *              *
0.12    -
0.16    -       *
0.20    -                       *               *
0.24    -           *
0.28    -
0.32    -
0.36    -
0.40    -

CASE 1 MINIMUM      CASE 1 MINIMUM       72% MINIMUM
Failure Stress      Failure Pressure     FAILURE PRESSURE
psi.                psi.                  psi.
40,047              834                   601

(*) NOTE: NO SAFETY FACTOR APPLIED TO CASE 1,2, or 3
SMYS = 35,000 psi.  (NOTE: No Safety Factors Applied to
CASE 1, 2 or 3)
100% 72%

CASE 2 MODIFIED METHOD USING AREA = 0.85 dL : 769 psi.
554psi.

CASE 3 EXISTING B31G METHOD USING AREA = 2/3 dL : 684 psi.
493psi.

--- PIT DEPTH MEASUREMENTS (MILS) --- (MAX. Pit Depth =
0.232 inch) ------

0.00    0
1.00  146
2.00  232
3.00  201
4.00   67
5.00  213
6.00   83
7.00    0
```

Figure 8-6. Pipeline FFS example with refined parameters to match RSTRENG.

$$tr = \frac{P \cdot D}{2 \cdot S \cdot F \cdot E}$$

$$s1 = n \cdot i + \left[\frac{tr - tc18}{tc17 - tc18}\right] \cdot i + \left[\frac{tr - tc19}{tc20 - tc19}\right] \cdot i$$

$$X3 = \left[\frac{tr - tc19}{tc20 - tc19}\right] \cdot i + \left[\frac{tr - tc21}{tc20 - tc21}\right] \cdot i$$

$$s2 = \left[\frac{tr - tc21}{tc20 - tc21}\right] \cdot i + \left[\frac{tr - tc21}{tc22 - tc21}\right] \cdot i$$

$$s = s1 + s2$$

$$\lambda = \frac{1.285 \cdot s}{\sqrt{D \cdot tr}}$$

$$Rt = \frac{tmm - FCA}{tr}$$

$$\gamma = tmm - FCA$$

If $\gamma > 0.10$ then $OK1 = 1$

$$Lmsd = 1.8 \cdot \sqrt{D \cdot tr}$$

If $Lmsd < 10$ then $OK2 = 1$

$$Mt = \sqrt{1 + 0.48 \cdot \lambda^2}$$

$$RSF = \frac{Rt}{1 - \left[\dfrac{1}{Mt}\right] \cdot (1 - Rt)}$$

$$MAWP = \frac{(tmm - FCA) \cdot \left[1.0 - \dfrac{RSFa}{Mt}\right] \cdot S \cdot F \cdot E}{\left[RSF - \dfrac{RSFa}{Mt}\right] \cdot R}$$

The variable solution sheets are as shown in **Figure 8-7**.

Variable Sheet

Input	Name	Output	Unit	Comment
	tr	0.462857		Minimum required thickness per B31.4, in.
337.5	P			Internal pressure of pipeline, psig
48	D			OD of pipeline, in.
35000	S			SMYS, psi
.5	F			Design factor for pipeline inside plant
1	E			Weld joint factor
	s1	4.488595		Length of first segment of LTA with remain, thickness < tr
1	n			Number of inspection intervals
1	i			Length of inspection interval, in.
.268	tc18			Remaining wall thickness at Point C18, in.
.354	tc17			Remaining wall thickness at Point C17, in.
.299	tc19			Remaining wall thickness at Point C19, in.
.433	tc20			Remaining wall thickness at Point C20, in.
	X3	2.427315		Segment between S1 & S2 with remain, thickness > tr
.287	tc21			Remaining wall thickness at Point C21, in.
	s2	2.557248		Length of second segment of LTA with remain thickness < tr
.417	tc22			Remaining wall thickness at Point C22, in.
	s	7.045843		Length of LTA in longitudinal direction, in.
	λ	1.920844		Shell parameter
	Rt	0.579012		Remaining thickness ratio
.268	tmm			Minimum measured wall thickness, in.
0	FCA			Future corrosion allowance, in.
	γ	.268		Computed minimum measured wall thickness less FCA, in.
	OK1	1		Gamma parameter > 0.10, then okay (OK1 = 1)
	Lmsd	8.484312		Length from structural discontinuity, in.
	OK2	1		Length to nearest structural discontinuity is satisfactory
	Mt	1.664640		Folias factor
	RSF	0.775013		Computed remaining strength factor
	MAWP	417.118529		Maximum allowed working pressure (MAOP), psi
.96	RSFa			Allowed remaining strength factor
24	R			Outside radius of pipeline, in.

Figure 8-7. The MAWP per API 579 Level 2 versus the computed RSF.

The safe maximum failure pressure from the RSTRENG run is computed as follows:

$$SMP = \left(\frac{35{,}000}{40{,}047}\right)(0.5)(834) = 364.4 \text{ psig}$$

We now make the MAOP equal to 364.4 psig. The corresponding MAOP from the API 579 algorithm yields a computed RSF of 0.827. If we run the API 579 algorithm to obtain an RSF equal to the RSFa, the operating pressure is 250 psig with an MAOP of 251.2 psig, considerably below the 364.4 psig. This line with the MAOP of 364.4 psig has been operating successfully for 5 years.

As mentioned previously, the current API 579 recommended practice also does not include the ASME B31.4 or B31.8 pipeline codes, so there is flexibility in interpreting the application of the methodology to pipelines.

The other software package, KAPA (Kiefner and Associates), is an Excel spreadsheet that accomplishes the same thing as RSTRENG, except that RSTRENG is an older version and is DOS based. The KAPA output is shown in **Figure 8-8**.

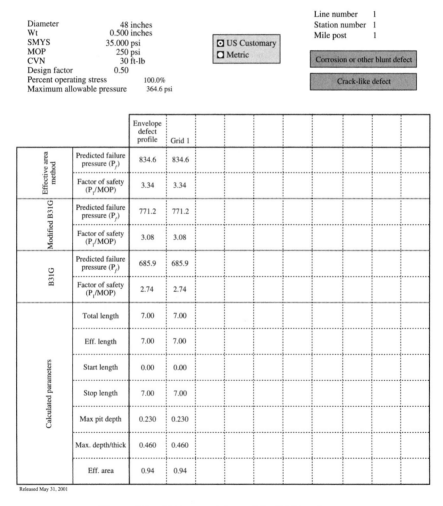

Figure 8-8a. Main spreadsheet of the KAPA software.

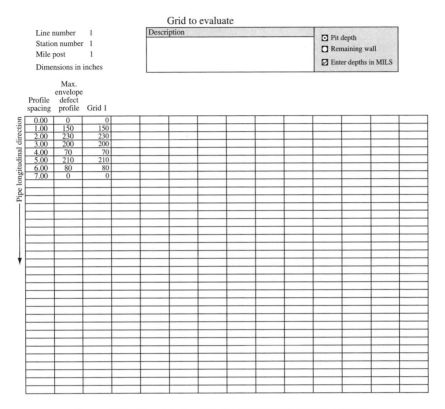

Profile spacing	Max. envelope defect profile	Grid 1
0.00	0	0
1.00	150	150
2.00	230	230
3.00	200	200
4.00	70	70
5.00	210	210
6.00	80	80
7.00	0	0

Line number   1
Station number   1
Mile post   1
Dimensions in inches

Grid to evaluate

Description

☑ Pit depth
☐ Remaining wall
☑ Enter depths in MILS

Figure 8-8b. The data input for the KAPA spreadsheet software.

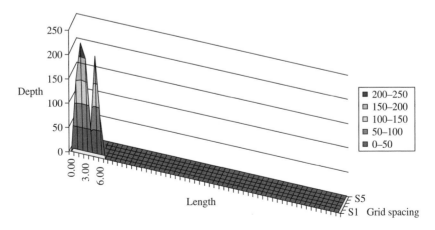

Figure 8-8c. Graphic output of the defect geometry in the KAPA software.

Thus, the API 579 algorithm can be used to back-calculate a safe maximum operating pressure from the MAOP predicted by RSTRENG DOS version. The API 579 methodology, which includes the algorithms and finite element, also is used where RSTRENG is not valid (see limitations of RSTRENG listed later in this chapter).

Notice that the factor of safety is computed differently than that of RSTRENG. KAPA takes the predicted burst pressure and divides by the entered value of the MOP (Maximum Operating Pressure) to determine the factor of safety. As stated in the KAPA User Manual on page 5, the program does not compute a safe operating pressure (SOP). It also states on page 5 that the safe operating pressure should be determined by an engineer who is familiar with pipeline integrity issues. In the RSTRENG DOS version, the MAOP was found by multiplying the minimum predicted burst pressure by the design factor.

Another observation of the Kiefner burst tests is that in cases 177, 86, 194, 193, and 68, the pipe burst at a pressure lower than the yield strength. This is significant because when a defect becomes self-propagating, failure will occur. The tabulated results of the minimum failure pressure to the yield strengths for the various cases are shown in **Table 8-2c**.

The defect assessed in the KAPA software for LTAs is for blunt edges. It also gives the capability to assess cracks and that is where the toughness of the material, or Charpy impact notch value (CVN) is used.

Table 8-2c
Comparisons of Yield Strengths to Minimum Failure Stresses

Kiefner Case No.	Yield Strength, psi	Minimum Failure Stress, psi
177	54,100	48,116
86	60,967	46,926
194	35,000	33,538
193	35,000	30,160
68	59,400	54,846
90	73,440	80,028

The value of the CVN has no effect on the results of a blunt defect such as an LTA.

For the sake of inquiry, we can see what the stress level would be in a pipe with no defects, using the same design factor of 0.5 and the weld joint factor (*E*) of 1.0:

$$S = \frac{P_i D}{2Ft} = \frac{(417.0)(20.0)}{2(0.5)(0.5)} = 16{,}680 \text{ psi}$$

So we can see what effect an LTA has on the stress level in a pipeline.

Limitations of RSTRENG

RSTRENG is an iterative algorithm based on burst tests. Most, if not all, of the piping subject to burst tests was 5/8″ (16 mm) thick. All pipe tested was carbon steel. The limitations of RSTRENG are as follows:

1. It is generally accepted that RSTRENG can be used for up to ¾″ pipe, but over that wall thickness it is not valid. Some use it for up to a 1″ wall, but no known burst test specimens had a 1″ wall thickness.
2. Also if a pipeline material is not carbon steel or is of a different grade than the specimens tested, RSTRENG is not valid. (Most grades of carbon steel tested were API 5L GrB through X70.)
3. For large LTAs where the LTA extends significantly in the circumferential direction, or where there is corrosion around girth welds, RSTRENG is not valid.
4. RSTRENG is not valid for the assessment of elbows and branch connections.
5. RSTRENG is not valid for situations where brittle fracture may be a possibility.
6. RSTRENG is not valid in cases where environmental cracking or hydrogen-induced cracking is present.

For these exceptions, the API 579 methodology or finite element can be used. One typical application where RSTRENG is not valid is with seawater injection pipelines, where seawater is injected into oil reservoirs. These pipelines have a nominal wall that is larger than 1″, normally between 1⅛″ (28.6 mm) to 1.5″ (38 mm). These pipelines normally operate between 2500 and 3000 psig.

Another Actual Field Example

A 22.0 in. seawater injection line to pump seawater into oil reservoirs is constructed of API 5L X60 and is a 1.188 in. nominal wall. The operating pressure is 3110 psig. In this case, RSTRENG does not apply because of the wall thickness of the pipe. We use the API 579 methodology to assess this pipeline with an LTA at point UT-12 on the south flank header. The LTA is measured 33 in. in the longitudinal direction and 15 in. in the circumferential direction. The UT measurements indicate that the measure wall varies across the LTA by 2 mm. The UT measured value of t_{mm} is 0.984 in. The FCA is computed for a 3 year period with the corrosion rate being 0.0079 in./yr. The design factor of the pipeline is 0.5, as per company specifications; it is in the proximity of people and plant equipment and assets. The following equations show the output of the API 579 algorithm:

$$tr = \frac{P \cdot D}{2 \cdot S \cdot F \cdot E}$$

$$\lambda = \frac{1.285 \cdot s}{\sqrt{D \cdot tr}}$$

$$Rt = \frac{tmm - FCA}{tr}$$

$$\gamma = tmm - FCA$$

If $\gamma > 0.10$ then $OK1 = 1$

$$Lmsd = 1.8 \cdot \sqrt{D \cdot tr}$$

If $Lmsd < 10$ then $OK2 = 1$

$$Mt = \sqrt{1 + 0.48 \cdot \lambda^2}$$

$$RSF = \frac{Rt}{1 - \left[\dfrac{1}{Mt}\right] \cdot (1 - Rt)}$$

$$MAWP = \frac{(tmm - FCA) \cdot \left[1.0 - \dfrac{RSFa}{Mt} \right] \cdot S \cdot F \cdot E}{\left[RSF - \dfrac{RSFa}{Mt} \right] \cdot R}$$

The variable solution sheet to the preceding equations is shown in **Figure 8-9**.

Note that the computed RSF must be less than 1.0, by definition. Also the operating pressure is approaching the MAWP which it would with LTA with the UT readings indicating defects with small readings. The required wall thickness (t_r) is close to the nominal wall with the high internal pressure.

To verify the results, a linear-elastic finite element model of the pipeline containing the LTA was constructed. The results were that the maximum Tresca stress was 29,519.5 psi, which is right below the 30,000 psi allowable. Tresca stresses were used because ASME codes are based on them. With the FE model providing verification the line was

Variables Sheet

Input	Name	Output	Unit	Comment
	tr	1.140333		Minimum required thickness per B31.4, in.
3110	P			Internal pressure of pipeline, psig
22	D			OD of pipeline, in.
60000	S			SMYS, psi
.5	F			Design factor for pipeline inside plant
1	E			Weld joint factor
33	s			Length of LTA in longitudinal direction, in.
	λ	8.466225		Shell parameter
	Rt	0.841859		Remaining thickness ratio
.984	tmm			Minimum measured wall thickness, in.
.024	FCA			Future corrosion allowance, in.
	γ	.96		Computed minimum measured wall thickness less FCA, in.
	OK1	1		Gamma parameter > 0.10, then okay (OK1 = 1)
	Lmsd	9.015706		Length from structural discontinuity, in.
	OK2	1		Length to nearest structural discontinuity is satisfactory
	Mt	5.950205		Follas factor
	RSF	0.864844		Computed remaining strength factor
	MAWP	3121.179940		Maximum allowed working pressure (MAOP), psi
.96	RSFa			Allowed remaining strength factor
11	R			Outside radius of pipeline, in.

Figure 8-9. Algorithm of seawater injection pipeline with LTA.

found acceptable, re-coated, and entered back into service. (RSTRENG was run for this case and given a minimum failure pressure of 6374 psig, which would result in a safe operating pressure of $(0.5)(6374) = 3187$ psig, which is greater than the acceptable level.)

For a problem where RSTRENG does not hold, a Level 1 or Level 2 per API 579 is difficult without some verification (e.g., finite element or from field experience). The FE model can be a linear-elastic model that will predict realistic stresses inside the elastic range. This result can be used in the API 579 methodology (Eq. 18) for pipelines.

Grooves, Plain Dents, and Dents with Gouges and Crack-like Defects

The reader is referred to Chapter 3 for these topics.

Pipeline Protection

The subject of pipeline protection is beyond the scope of this book, but some mention is in order. The areas of corrosion control are not the subject of this book, as corrosion engineering is a different discipline.

Coatings are used to provide a barrier between the pipeline steel and the surroundings. Pipelines are sometimes coated on either the inside or outside to resist corrosion. Coatings should exhibit properties of adhesion, chemical resistance, electrical resistance, compatibility with cathodic protection, resistance to abrasion, flexibility, resistance to impact loads and penetration, and resistance to soil and weather. There are many types and forms of coatings, and the interested reader is referred to [Reference 5].

Cathodic Protection

Cathodic protection (CP) is also a subject beyond the scope of this book, but it deserves mention. Cathodic protection is the first line of defense against corrosion on pipelines. The technique is an application of an electric current to flow through the pipeline from an external current source to prevent corrosion. Corrosion can exist only if all of the following four conditions are met:

1. There must be an anode and cathode.
2. There must be an electrical potential difference between the anode and cathode.

3. The anode and cathode must have a metallic connection between them.
4. The anode and cathode must be immersed in a mutual electrically conductive medium—an electrolyte (e.g., soil).

Cathodic protection works by negating one of the four conditions above. The electrical potential difference between the cathode and anode is eliminated by making the entire pipeline cathodic. The two types are 1) Impressed Current System and 2) Galvanic Current System. The former consists of an external AC power source that converts the AC current to DC current, the rectifier, a common layer of anodes with a metallic connection from the anodes to the pipeline, and a common environment (e.g., the soil). The material components allow a large amount of electric current to flow through them in proportion to their own corrosion rate. The external power source enables the anodes to function. The power source sends the electric current to the rectifier and then flows through the anode header cable to the anodes. As the current exits the anodes, it enters the pipeline at voids where it travels through a cathode cable. The electric current is completed by the current flowing from the pipeline back to the rectifier, where the current and voltage are monitored.

A typical CP survey graph is shown in **Figure 8-10**. The upper line shows the pipeline with electric current applied. It should be above the lower line, which indicates the minimum criteria level. At location 37.0 kilometer, there is a blank space, which indicates a reading was not taken. A technician needs to be sent out to take a reading to fill the void. Overall, the CP group decided that the pipeline had no areas of concern. Such surveys can vary with each different company, but it gives a quick look at the situation of a pipeline many kilometers long. If there is a concern, then the area or areas that are seen to be problematic need to be investigated by a site visit. Such surveys need to be assessed by qualified CP personnel.

Pigging Technology

The use of pigs in pipelines has been in wide use for many years. The term "pig" is ubiquitous in use; however, in the Middle Eastern countries the term "scraper" is used instead of "pig" for cultural reasons. Even though the term "scraper" may be more descriptive, we will use the term "pig" because it is more widely used.

Pigging technology is evolving everyday. This book is by no means intended to be an exhaustive discussion of pigging technology, but it does

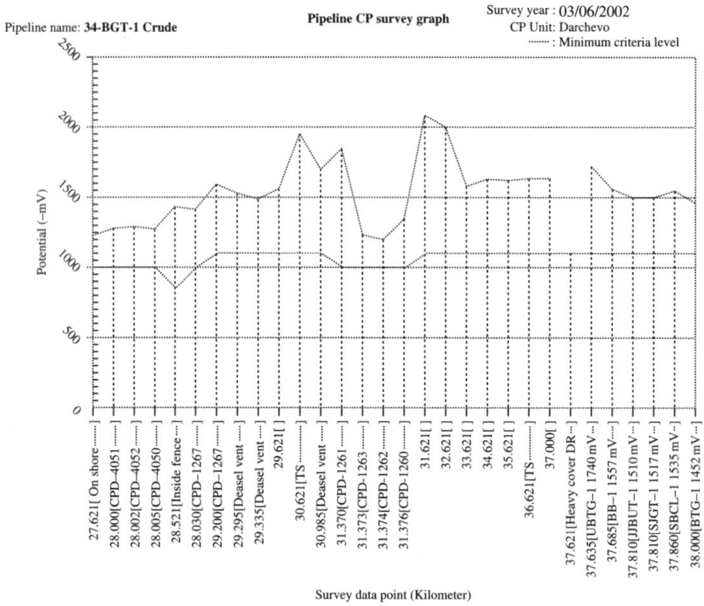

Figure 8-10. A typical CP survey graph.

provide a brief discussion about its function and some problems encountered in practice. For a more detailed discussion of the subject, the interested reader is referred to [References 5, 6 and 7].

Pigs act like free-moving pistons inside pipelines, sealing against the inside wall with a number of sealing elements. The pigs are used for the following:

- Cleaning debris from the wall
- Flooding lines for hydrostatic testing
- Dewatering and drying lines
- Applying protective coatings to the inside wall
- Gauging the internal bore
- Carrying inspection tools
- Separating differing liquids and gases in the line
- Isolating the pipeline for repair and/or maintenance work

Pipelines with pigs include, but are not limited to, such components and isolation equipment as:

- Closures
- Traps

- Foam, metal–bodied, and solid polyurethane pigs
- Pig signalers
- Piggable tees
- High- and low-pressure isolation pigs
- Flange and joint testers

Piggable tees usually are metal bars welded over tees to form a mesh or grid that allows both flow and the pig to pass through the tee without being an obstruction. Some of the various types of tees are described in this section.

Pigs are custom designed for each application, process fluid, and pipeline. They vary considerably in types, but some typical designs are shown here.

The cleaning pig is a short-bodied three-cup design that allows negotiation of bends down to $1.5D$ radius, where D is the inside diameter. The driving cups are made of polyurethane, and the pig has a polyurethane protector nose. It is a general-purpose cleaning pig suitable for traveling long distances in cross-country pipelines. See **Figure 8-11**.

The gauging pig is made of two polyurethane driving pigs with a polyurethane protector nose. It is fitted with a gauging plate that is made of carbon steel or aluminum, which is normally machined to 90 to 95% of the line bore, depending on the client's specification. It is used for precommissioning duties and is used for proving roundness of the constructor's pipe, removing debris, and identifying excessive weld penetration. This design has been used for many years by pipeline operators. See **Figure 8-12**.

The separation pig is made of four driving polyurethane cups with a polyurethane protector nose. It will traverse bends to $1.5D$ radius for sizes 8″ to 14″. It is used for product separation, batching, displacement,

Cleaning Pig

16″- 48″ CLEANING PIG

Figure 8-11. Cleaning pig. Courtesy of Pipeline Products Limited.

Figure 8-12. Gauging pig. Courtesy of Pipeline Products Limited.

swabbing, and line clearance duties. It can be used for gauging when fitted with an optional gauging plate. See **Figure 8-13**.

The foam pig is used for pipelines that may or may not have been
designed to run conventional pigs or spheres. It is able to traverse any
bend mitre, ball, gate, or check valve. It can traverse full 90° barred tees
and easily pass through reduced pipe diameters. See **Figure 8-14**.

Figure 8-13. Separation pig. Courtesy of Pipeline Products Limited.

Figure 8-14. Foam pig. Courtesy of Pipeline Products Limited.

Launching and Retrieving Pigs

Pigs are launched in a pipeline by a launching trap for a gas line and retrieved by a receiving trap. The horizontal pig trap has proven more practical for general service in both suction and discharge locations. It should be mounted at a convenient location above ground. A typical pig launcher for a trap in a gas line is shown in **Figure 8-15**.

This particular pig trap launcher is for a natural gas pipeline. The launching barrel is usually one to two times larger than the line pipe. The barrel length should be at least 1.5 times the longest pig to be launched. The barrel is equipped with a quick-opening end closure. A bypass line $\frac{1}{4}$ to $\frac{1}{3}$ the pipe diameter enters the trap at a point near the end closure of the launcher such that the flow will enter the trap behind the rear cup of the pig.

A typical pig receiver trap for a natural gas line is shown in **Figure 8-16**. The receiver barrel is typically one to two sizes larger than the line pipe. The barrel lengths can vary considerably and depend on the number of pigs and amount of debris received before the receiver is opened. The minimum barrel length should be at least 2.5 times the length of the longest pig to be received.

A typical pig launcher for a liquid line is shown in **Figure 8-17**. For the liquid pig launcher, the launcher barrel is typically one to two times the pipeline size. The barrel length should be at least 1.5 times the length of the longest pig to be launched. An exception to this criterion would be a product line where it would be necessary to launch two or more pigs in succession to separate a buffer batch. The pigs are loaded into the barrel

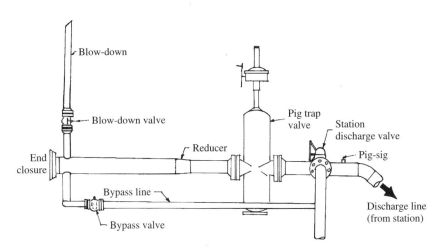

Figure 8-15. Typical pig launcher trap for gas lines. Courtesy of T. D. Williamson, Inc.

and contained in place by mechanical rams that are activated at each interface. The barrel is equipped with a quick-opening end closure. A bypass line ¼ to ⅓ the line pipe size enters the trap at a point close to the end closure of the launcher such that the flow will enter the trap behind the cup of the pig. A drain line at least 2″ (50 mm) in diameter is connected to the bypass line to facilitate drainage of the trap. A vent

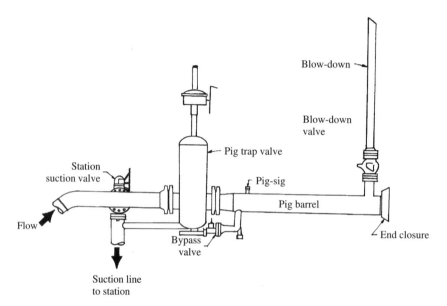

Figure 8-16. Typical pig receiving trap gas line. Courtesy of T. D. Williamson, Inc.

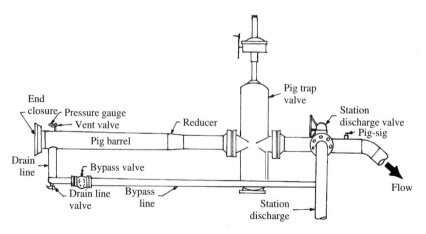

Figure 8-17. A typical pig launcher for a liquid pipeline. Courtesy of T. D. Williamson, Inc.

valve is mounted above the barrel and should be the same size as the drain line. A pig signal indicator is installed downstream (not shown) from the launcher to indicate the passage of the pig into the pipeline.

A typical pig receiver for liquid lines is shown in **Figure 8-18**. Length can vary considerably, depending on the number of pigs launched and amount of debris recovered. The minimum barrel length should be 2.5 times the length of the longest pig. The barrel is equipped with a quick-opening end closure. A bypass line $\frac{1}{4}$ to $\frac{1}{3}$ of the pipeline size connects the trap at a point near the reducer or valve end of the barrel. This location causes a decrease in the flow behind the pig, which reduces the speed at which the pig enters the pipeline. A drain line 2″ (50 mm) in diameter is connected to the bypass line to facilitate complete drainage of the trap. A vent valve is installed on top of the barrel and normally is the same size as the drain line. The vent valve should be piped so that the product may be caught in a bucket during purging procedure. A pig signal indicator (not shown) is installed on the receiver barrel to signal the arrival of the pig.

Pig launchers and receivers are normally fitted with various manufacturers' patented designs. Each one has its own design for applications of many process services.

Generally, there is potentially less damage to a pig in a liquid versus a gas pipeline. This is because a liquid pipeline has residual liquid around the inner walls of the pipe that can lubricate the cups on the pig. The velocity of pig travel can affect the abrasion and wear of the pig. A typical average travel speed is 1.0 to 1.5 m/sec (3.28 to 4.92 ft/sec),

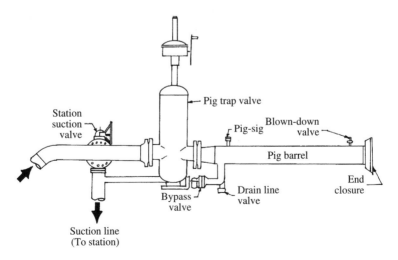

Figure 8-18. Typical pig receiver for a liquid pipeline. Courtesy of T. D. Williamson, Inc.

preferably 1.0 m/sec. Pigs traveling at 3.0 m/sec are too fast and can cause damage to the cups on the outer part of the pig.

The weight of the pig is of utmost concern. Pigs are constructed of polymers (e.g. polyurethane). They normally have a steel body construction with Standard Omnithane®, Hyper Omnithane®, and Super Omnithane®. These materials have proved to be extremely wear-resistant. The particular material used depends on the particular pipeline and process fluid. Pig suppliers have engineering teams that design pigs for use in sour service, corrosive duties such as sulfuric acid, using the latest codes and standards.

Pigs can be obstructed by "nose diving," which occurs when the tip is too heavy or an obstruction is in the pipeline. Attempting to rotate pigs does not work, particularly in large pipelines.

Large diameter pipelines, particularly those 48 in. and larger, can present unique challenges. Damage in these pipelines can vary from minor wear to total damage, but it can also be caused by having pigs stuck in the pipeline. Most incidences of damage happen when the process service is changed (e.g., when a pipeline that has been transporting crude oil is converted to dry gas). The larger the pipeline is, the heavier the pig will be, and this mass affects the rigidity and strength of the cups that scrape along the inside of the pipeline. For this reason, some vendors use aluminum instead of steel to construct the pig body. The quality of the cups or disks is a significant factor that contributes to the performance of the pig. Low-quality cups or disks will accelerate their damage and reduce their effectiveness to clean, gauge, or patch the pipeline. The lower quality of the disk or cups acts in combination with the shelf life, low-quality manufacturing, inadequate material, or improper application.

To augment the pigging process, chemical gels are used when more than one pig is launched into a pipeline. The gel is placed between the pigs. A gel is a high-viscosity liquid. Gels are also sometimes used to assist in cleaning the pipelines, but there will also be gel deposited inside the pipeline. Thus, one must be able to remove the residual gel left over in the pipeline.

Thus, the design and fabrication of the pig is of utmost importance in ensuring proper performance. Pigs perform better when they are supported by wheels. Also the foam-type pig passes with minimum damage because it has less mass, which is evenly distributed over the entire body of the pig.

The length of the pipeline is a major parameter in the longevity of the pig. The pig must be able to withstand the travel throughout the length of the pipeline before wearing out or becoming damaged. The other factor is the expected amount of debris that accumulates in front of the pig and that may hinder its travel and its effectiveness. This aspect is based on experience and types of transported fluids. Some debris collected at the end can form black powder, which is nothing more than oxidized metal that has been removed from the pipelines.

Generally, the pipeline is cleaned by a cleaning pig before an intelligent pig is run inside the pipeline. Intelligent pigs carry equipment that scans for corrosion damage. Some advanced intelligent pigs can feed data back to a computer that can perform RSTRENG (see Chapter 1) or another burst pressure algorithm to determine the fitness-for-service of the pipeline throughout the length of the pig run. Intelligent pigs are invaluable for reliability purposes and are the second line of defense after CP.

After intelligent pigs find corroded regions, more detailed inspection of damaged areas may be necessary for remediation.

Pigging can be a challenge for smaller diameter pipelines as well as for larger ones. The information from [Reference 8] reports chronic problems in smaller pipelines regarding the waxing of pipelines with the paraffins found in crude oil or condensate that restricts flow through the pipe. Wax buildup becomes a major problem in subsea pipelines when the oil cools and wax is deposited by molecular diffusion. The waxing mechanism is described as wax precipitating in a concentrated gradient between the dissolved wax in the turbulent core and the wax remaining in solution on the pipe wall. The wax buildup can seal off the pipeline or block a pig from passing if either the wrong pig is used or pigging is not performed often enough. Directly cleaning or scraping the pipeline is the preferred method of removing the wax. The desired plan is to manage the wax buildup and remove all of it, called wax management, versus merely keeping the line open, called bore management.

Pigging presents risks. A pig can be immersed in the wax, forming a plug and blocking the pipeline. As the pig moves along the pipeline, the wax will be scraped off the pipe wall and accumulate in front of the pig; a force will be applied to the rear of the wax buildup. The pressure gradient over the wax causes oil to be squeezed out, forming a harder surface in front of the pig. This harder wax and the wax buildup in front of the pig will increase the amount of friction required to move the pig. Thus, a plug will be formed in the pipeline. To correct such a problem, the pipeline must be cut open and the pig removed, resulting in expensive down time and operation losses.

To avoid this scenario, a bypass pig is used. With a bypass pig, the fluid behind the pig is passed through it to break up the wax. The amount of bypass through the pig is significant because it is a function of the expected volume of wax accumulation in the line. If the bypass flow is too small, wax buildup can occur and block the pig. Such a pig is shown in the following schematic in **Figure 8-19**.

Shown in **Figure 8-20** is a theoretical differential pressure required to push a volume of wax in pipelines of various diameters. The figure shows the differential pressure (P_R) in MPa.

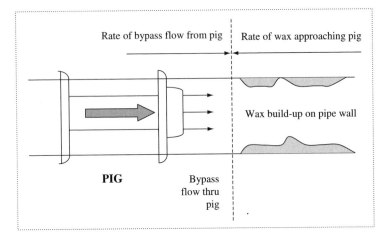

Figure 8-19. Schematic of the bypass flow pig. The arrow shown in the center of the pig is the throughput fluid flow. This flow exiting the end of the pig is the flow to break up the wax.

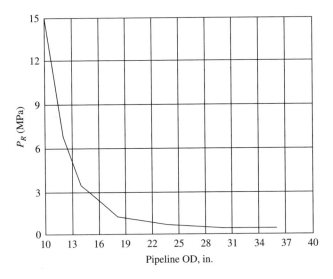

Figure 8-20. The effect of the pressure to move the pig and plug versus pipe diameter.

Because larger pipelines have greater area and consequently a higher pressure thrust, the risk of plugging is much less. The problem with larger pipelines is the quantity of wax at the receiver. There are normally strainers and filters that can easily become plugged. If this happens, then the backup pressure can rapidly build up.

Repair Options for Pipelines

Repair options depend on the type and severity of the defect in the pipeline. The new ASME B31.4 gives a detailed description of repair types and when they can be used. There are two major types of sleeves—metallic and nonmetallic, or composite.

Metal Sleeves

Metallic sleeves have proven to be acceptable repair options for pipelines that have flaws. They are very practical for leaks and in situations where the flaw depth equals or exceeds eighty percent of the nominal wall of the pipeline. Typical configurations are given in the API RP 1107 [Reference 9].

There are two types of metallic sleeves—pressure containing and non-pressure containing. A nonpressure-containing sleeve is defined in the new revised ASME B31.4 as a Type A sleeve, as shown in **Figure 8-21**. This sleeve is used to reinforce a pipeline that has nonleaking flaws. Nonpressure-containing sleeves should not be used for circumferentially oriented flaws.

The pressure-containing sleeve is used for leaks and flaws that equal or exceed 80% of the nominal wall. They are referred to in the new revised ASME B31.4 code as Type B sleeves. A typical pressure-containing sleeve is shown in **Figure 8-22**. Type B sleeves are best for cracks—especially ERW cracks. They can be used for leaking or nonleaking defects including circumferentially oriented defects.

Advantages of Metal Sleeves

- True pressure containment can be accomplished only with a metal pressure-containing sleeve construction.
- Metal sleeves are time proven and economical.
- Metal sleeves can be considered permanent repairs.
- Pressure-containing full-encirclement sleeves can be used where the external or internal corrosion exceeds nominal wall thickness of the pipeline.

Disadvantages of Metal Sleeves

- Metal sleeves require hot work (welding).
- Operating pressure of the pipeline may have to be lowered to weld the sleeve.

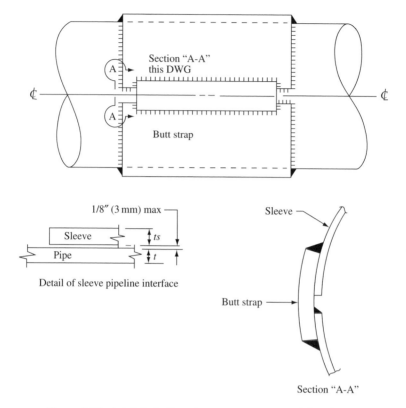

Figure 8-21. A detail of a Type A or nonpressure-containing sleeve.

- Hot work requires a hot tap procedure and guidelines.
- An area of sleeve must be protected.

Note that when placing metal sleeves for a hot tap application (a pressure-containing sleeve), it is important to perform pit scale readings for external corrosion or UT for internal corrosion in the area where the sleeve is to be welded. If an LTA exists within the sleeve area, a buckling assessment must be made to determine if the hydrostatic test pressure will result in buckling of the shell. The easiest manner to accomplish this is to use the finite element method. A defect is the raison d'etre for using a sleeve; thus, using proper caution is in order for using sleeves.

Pipelines have collapsed under sleeves. This phenomenon is well known in gas services. One scenario is reported by [Reference 7] where a through-the-wall leak can result in the gas inside the pipeline filling the gap between the sleeve and the outside surface of the pipeline. When sections of a pipeline are separated by block valves, rapidly closing the

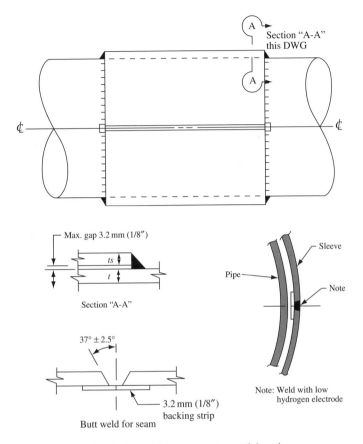

Figure 8-22. A Type B or pressure-containing sleeve.

valves can result in pressure waves. The magnitude of the pressure variations can cause the section of the pipeline under the sleeves to implode on the inside. Also rapidly venting the pipeline can result in the same phenomenon.

Typically, in sour service, a pipeline can implode inside a sleeve. This is a different phenomenon from that mentioned earlier, where hydrogen permeates into the metal wall of the pipeline and recombines in the pipe wall to form a defect in the form of a HIC or blister. One solution has been to drill a hole 25 mm (1 in.) in diameter in the sleeve. The hole is threaded and a plug is screwed into the hole. As hydrogen builds inside the interstice between the sleeve and pipeline, it can escape through the threaded connection of the plug. Another technique is mentioned in [Reference 10], which suggests that injecting a sealant in the gap between the sleeve and pipeline would be a very practical solution to keep gas out of the void between the pipe and sleeve. To accomplish this task the reader is referred to Chapter 6 where bolted clamps are discussed along with sealant injection systems. As discussed in Chapter 6, a sealant would have to be injected at various ports around the sleeve to seal the gap. This technique would satisfy the problem of residual gas in the annulus between the sleeve and pipeline; however, one would have to be cognizant of the injection pressure and any defects (e.g., an LTA that would result in an implosion).

Composite or Nonmetallic Sleeves

Composite sleeves have become more common and more widely used in recent years. Composite structures first got started in the aerospace industry and are used mainly to save on weight. In the petrochemical, refining, and process industries, composite sleeves should not be used for the following:

- For leak repair
- For pipes where the metal loss has a depth greater than 80% of the nominal wall thickness
- For circumferentially oriented defects
- For crack-like defects

Composite sleeves can be used for the following:

- To repair defects that have been removed by grinding
- To reinforce a pipe that is not leaking and the defect depth does not exceed 80% of the nominal wall of the pipe

A composite sleeve must be tested for compatibility with cathodic protection. The composite must retain its essential properties in a moist environment at temperatures within the operational temperature of the pipe. The load-carrying capacity of the remaining pipe and composite sleeve should be a minimum equal to that of the pipe. Composite sleeves should be marked and/or documented to be evident that a repair was made.

Other Types of Repairs

In the new revised ASME B31.4 to be released, lap patches (half soles), are illegal. Does this mean that all the existing ones need to be replaced?

The answer is certainly not; the code is not retroactive. A summary of repair options is given in the new ASME B31.4 in **Table 8-3**.

A table of acceptable repair options for dents, buckles, ripples, wrinkles, leaking couplings, and defective prior repairs from the new revised ASME B31.4 is listed in **Table 8-4**.

Also in the upcoming ASME B31.4, temporary repairs are allowed and the rule is worded as follows:

> Temporary repairs may be necessitated for operating purposes. Such temporary repairs shall be made in a safe manner and in accord with sound engineering principles. Temporary repairs shall be made permanent or replaced in a permanent manner as soon as practical in accordance with this code.

Grit Blasting of Operating Pipelines

Often it is necessary to grit blast pipelines to perform coating operations without shutting down the pipeline. Investigations by the author among various shops around the world revealed that metal loss during grit blast operations can range from 3 to 5 mil (0.003 to 0.005 in.). The maximum allowable operating pressure during these operations takes the form of Eq. 7-4 for hot tap operations. The equation is as follows:

$$P = \frac{2(SMYS)(t_{mm} - a)F}{OD} \qquad \text{Eq. 8-17}$$

where $a = 0.005$ in.

$SMYS =$ specified minimum yield strength of pipeline material

$t_{mm} =$ minimum measured wall thickness, in. (see Chapter 3)

$OD =$ outside diameter of pipeline, in.

$F =$ design factor

Typical Example of Grit Blast (Abrasive Blast) of an Operating Pipeline

A pipeline containing disposal water operates between 1500 and 2000 psig with a design factor of $F = 0.72$. The pipeline is API5L X60 and is 8″. Ultrasonic thickness inspection showed that the wall thickness varied from 11.0 to 15 mm (0.433 to 0.59 in.). At one location, the minimum measured wall thickness was 8 mm (0.315 in.). Two test holes were found in bad condition because of coating damage, and they needed grit

Table 8-3
ASME B31.4 Acceptable Pipeline Repair Methods (Revised Edition)
Table 451.6.2(b)-1 Acceptable Pipeline Repair Methods (non-indented, non-wrinkled and non-buckled pipe)

Repair Methods

Type of defect	1 Replace as Cylinder	2 Removal by Grinding	3 Deposition of Weld Metal	4a Reinforcing Full Encirclement Sleeve (Type A)	4b Pressure Containing Full Encirclement Sleeve (Type B)	5 Composite Sleeve	6 Mechanical Bolt-on Clamps	7 Hot Tap	8 Fittings
External Corrosion ≤80% t (excluding grooving, selective, or preferential corrosion of ERW, EFW seams)	Yes (a)	No	Limited (b)	Limited (e)	Yes	Yes (e)	Yes	Limited (c)	Limited (h)
External Corrosion >80% t	Yes (a)	No	No	No	Yes	No	Yes	Limited (c)	Limited (h)
Internal Corrosion ≤80% t	Yes (a)	No	No	Limited (d)	Yes	Limited (d)	Yes	Limited (c)	No
Internal Corrosion >80% t	Yes (a)	No	No	No	Yes	No	Yes	Limited (c)	No

Type of Defect								
Grooving, Selective or Preferential Corrosion of ERW, EFW Seam	Yes (a)	No	No	Yes	No	Yes	Limited (c)	No
Gouge, Groove or Arc Burn	Yes (a)	Limited (g)	Limited (e) (f)	Yes	Limited (e) (f)	Yes	Limited (c)	Limited (f) (h)
Crack	Yes (a)	Limited (g)	Limited (g)	Yes	Limited (g)	Yes	Limited (c)	No
Hard Spot	Yes (a)	No	Limited (e)	Yes	No	Yes	Limited (c)	No
Blisters	Yes (a)	No	No	Yes	No	Yes	Limited (c)	No
Defective Girth Weld	Yes (a)	No	Limited (b)	Yes	No	Yes	No	No
Lamination	Yes (a)	No	No	Yes	No	Yes	No	No

(a) Replacement pipe should have a minimum length of one-half of its diameter or 3 inches (76.2 mm), whichever is greater, and shall meet or exceed the same design requirements as those of the carrier pipe.

(b) The welding-procedure specification shall define minimum remaining wall thickness in the area to be repaired and maximum level of internal pressure during repair. Low-hydrogen welding process must be used.

(c) Defect must be contained entirely within the area of the largest possible coupon of material that can be removed through the hot-tap fitting.

(d) May be used only if internal corrosion is successfully mitigated.

(e) Tight-fitting sleeve at area of defect must be assured or a hardenable filler such as epoxy or polyester resin shall be used to fill the void or annular space between the pipe and the repair sleeve.

(f) May be used only if gouge, groove, arc burn or crack is entirely removed and removal is verified by visual and magnetic particle or dye-penetrant inspection (plus etchant in the case of arc burns).

(g) Gouge, groove, arc burn or crack must be entirely removed without penetrating more than 40% of the wall thickness. The allowable length of metal removal is to be determined by para 451.6.2(a)(2). Removal of gouge, groove, arc burn or crack must be verified by visual and magnetic-particle or dye-penetrant inspection (plus etchant in the case of arc burns).

(h) The defect shall be contained entirely within the fitting and the fitting size shall not exceed NPS3.

Courtesy of the American Society of Mechanical Engineers.

Table 8-4
Acceptable Pipeline Repair Methods for Other Types of Defects

Table 451.6.2(b)-2 Acceptable Pipeline Repair Methods for Dents, Buckles, Ripples, Wrinkles, Leaking Couplings, and Defective Prior Repairs

	Repair Methods					
Type of defect	1 Replace as Cylinder	2 Removal by Grinding	4a Reinforcing Type Full Encirclement Sleeve (Type A)	4b Pressure Containing Full Encirclement Sleeve (Type B)	5 Composite Sleeve	6 Mechanical Bolt-on Clamps
Dents ≤6% of the Diameter of the Pipe Containing Seam or Girth Weld	Yes (a)	No	Limited (b)	Yes	Limited (b)	Yes
Dents ≤6% of the Diameter of the Pipe Containing Gouge, Groove or Crack	Yes (a)	Limited (d)	Limited (b) (c)	Yes	Limited (b) (c)	Yes
Dents ≤6% of the Diameter of the Pipe Containing External Corrosion with Depth Exceeding 12.1/2% of Wall Thickness	Yes (a)	No	Limited (b)	Yes	Limited (b)	Yes

Dent Exceeding 6% of the Diameter of Pipe	Yes (a)	No	Limited (b)	Yes	Limited (b) (c)	Yes
Buckles, Ripples, or Wrinkles	Yes (a)	No	Limited (b)	Yes	No	Yes
Leaking Coupling	Yes (a)	No	No	Yes	No	Yes
Defective Sleeve from Prior Repair	Yes (a)	No	No	Yes	No	Yes

(a) Replacement pipe should have a minimum length of one-half of its diameter or 3 inches (76.2 mm), whichever is greater, and shall meet the same design requirements as those of the carrier pipe.

(b) A hardenable filler such as epoxy or polyester resin shall be used to fill the void between the pipe and the repair sleeve.

(c) May be used only if gouge, groove, arc burn or crack is entirely removed and removal is verified by visual and magnetic-particle or dye-penetrant inspection (plus etchant in the case of arc burns).

(d) May be used only if the crack, stress riser, or other defect is entirely removed, removal is verified by visual and magnetic-particle or dye penetrant inspection (plus etchant in the case of arc burns), and the remaining wall thickness is not less than 87.5% of the nominal wall thickness of the pipe.

Courtesy of the American Society of Mechanical Engineers.

blasting and reconditioning. The customer wishes to know if it is possible to grit blast the pipeline while it is pressurized or must they depressurize the line to zero pressure, which is very undesirable, since the line cannot be isolated due to isolation valves are passing.

The UT exam show that $t_{mm} = 0.315$ in. Thus, using Eq. 8-17, we have the following:

$$P = \frac{2(60,000)\left(\frac{lb_f}{in.^2}\right)0.315 - 0.005 \text{ in. } (0.72)}{8.625 \text{ in.}}$$

$P = 3105.4$ psig

Thus, the pipeline is safe to grit blast while being operated at 2000 psig. The pipeline was successfully grit blasted in service, and the coating repairs were performed.

Hydrogen Attack

Hydrogen blisters are much more common in pressure vessels than in pipelines in sour service. However, hydrogen atoms diffuse into pipeline steels, causing blisters or filling the air space between a sleeve and the outside pipeline wall in sour service. As hydrogen gas builds in the annulus, the pipeline wall, especially if it is thinned by defects, can implode inward, obstructing a pig from passing. Typically, in these situations, tap holes no larger than 2 in. are drilled into the sleeve wall. The preferable diameter would be a $\frac{3}{4}''$ to $1''$ diameter. This practice is done while installing the sleeve. After drilling, the hole is threaded, and a tap screw is screwed into place. If hydrogen gas accumulates inside the sleeve, it will escape through the tap threads. This practice has been very successful in avoiding sleeved pipelines from imploding inside because of buildup of hydrogen gas. The problem is aggravated if a sleeve is placed over an LTA or another type of defect, which is not recommended.

Soil-Structure Interaction Abnormality of Pipe Bowing

In some instances, buried pipelines cannot expand in the manner in which they were designed. This can occur if they are installed in winter conditions; during warmer months, they may expand as a function of the temperature differential between the seasons. This phenomenon is widely seen in arid desert climates where the temperature differences between the winter and summer seasons can vary significantly. Another reason for pipeline bowing is accumulative friction forces at the supports. Frequently, the pipeline moves away from the designed offsets that were installed for thermal displacement. Also changes in the operating temperature can result in the pipeline bowing to relieve the thermal stresses. An example of pipeline bowing is shown in **Figure 8-23**.

The section of the trunkline shown in **Figure 8-23** had bowed as a result of inadequate sand cover; the nature of the land profile had shifted, making it easier for the pipeline to bow upward. This phenomenon was aggravated with the variation in the operating temperature. Nondestructive examinations (UT was used in this case) were performed on the line, and no damage or flaws were revealed.

In another case, a pipeline bowed in the desert environment shown in **Figure 8-24**. The bowed pipeline was restored to its original state by using the following steps:

1. Depressurize the line to zero pressure.
2. Dig up the soil around the pipe down to the level of the plate anchor.
3. Inspect the pipe for any defects, using NDE (e.g., UT).
4. Reset the pipe to its original position.
5. Place external coating on the pipe.
6. Place soil around the pipeline as indicated in the figure.

The steps were implemented and the pipeline continued in successful service. Pipeline bowing normally happens after being in service for a long time. If it is located in soft soil (e.g., sand), it is more likely to

Figure 8-23a. Top view of bowing of a trunkline in the desert.

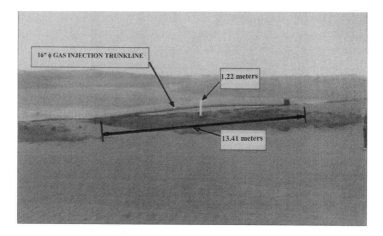

Figure 8-23b. Side view of trunkline bowing in desert, showing deformation.

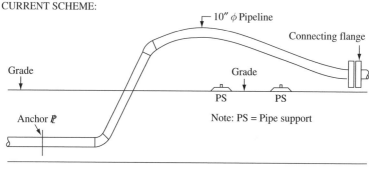

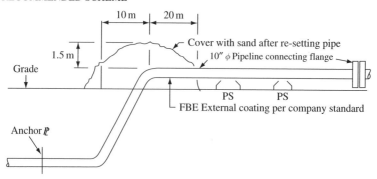

Figure 8-24. The bowing of a pipeline, showing the original bowed shape and the solution.

occur, especially when the operating temperature is increased beyond the design parameters.

Over-bends, or pipeline elbows that extend in the vertical direction, such as that shown in **Figure 8-24**, must be held down by the weight of the pipe and soil to counteract the uplift component of the axial restraining force in the hot operating condition. The length of the pipe, which includes the length of the elbow, contributes to resisting the uplift and is the critical length for an axially loaded column. The axial force in the pipeline tends to initiate arching of the pipeline and increase the amount of vertical displacement, lifting it off the bottom of the ditch. If there is a correct amount of soil cover, the soil loads over the pipeline should cause a sag in the same length of pipeline, which exceeds the amount of vertical displacement, keeping the pipeline below grade. In addition to the combined weight of the soil, pipeline, and fluid internals, one must take credit for soil friction in the vertical planes on each side of the pipe.

Generally, pipelines with smaller diameters (e.g., 8″ ϕ and smaller) need relatively more cover than large diameter pipelines for small angle bends. However, it is reasonable to accept a lower safety factor for small pipelines. A general guideline is to use 0.6 m (2 ft) for oil and water pipelines and 0.9 m (3 ft) for gas and NGL pipelines containing bends up to 2°. Generally, only a small percentage of bends need to be larger than 2°.

Sag-bends in pipelines are subject to uplift when the line operates below the tie-in temperature and is depressurized. The uplift force is a function of the axial tension and the bend radius.

For small sag-bends, it is assumed that the length of pipe that is effective is the same as that critical length for over-bends. The maximum temperature drop will normally not exceed 25°C (77°F), albeit higher temperature drops must be anticipated in NGL pipelines that are subject to blowdown. The minimum soil cover will usually be adequate for all sag-bends except when a large temperature drop can occur or when a smaller than standard bend radius is used. In the hot operating condition, sag-bends experience a downward load that could double the soil pressure under the pipe. This condition is not considered unsafe.

For lateral bends, the net passive soil resistance against the lateral load exerted by the pipeline must be conservatively estimated because often the cover of the pipe is mounded with a width of only two pipe diameters. The three following conditions must be considered:

1. Maximum width of the berm
2. Soil cover measured to original grade
3. Whether the ditch is in marl or rocky soil

Tie-in Temperatures

When segments of pipelines are welded together, the welding is performed at a certain ambient temperature. This ambient temperature is what is referred to as the tie-in temperature. The actual tie-in temperature is the average of two readings, one at the top and one at the bottom of the pipe. If a contact measuring device is used, it should be shielded from direct sunlight to prevent inaccurate reading. The tie-in weld is a weld that connects one of the following:

- Two pipeline strings together
- An existing pipeline to a pipeline under construction
- A pipeline to an anchor
- Pipeline sections to an intermediate mainline valve that is being inserted

The construction engineer shall determine the highest practical tie-in temperature for each weld, depending on the area of the world in which the construction is being performed. In hot climates, it is desirable to have the tie-in temperature higher to be close to average ambient conditions. Different companies have varying standards as to the length of pipeline segments being welded and how they are restrained. This is a construction issue and is not within the scope of this book because we are concerned with assessments rather than design and construction. The tie-in temperature is an important parameter used in pipeline assessments as discussed later.

Thermal Expansion of Buried Pipelines

To consider pipeline bowing adequately, the thermal expansion of the buried pipe must be addressed first. Buried pipelines will, in most cases, enter and exit the ground. These points of entry and exit will experience thermal expansion. The amount of thermal expansion of buried pipe can be approximated by the use of certain assumptions developed from tests data acquired over several years. The inconsistency of the data, particularly regarding the resistance of the soil, prevents accurate determination of the displacements of buried pipe induced by thermal expansion. An example of such inconsistencies is published in [References 11 and 12], indicating soil restraint values varying from 47 pounds per square foot to 1080 pounds per square foot of pipe surface area. These values were found in the same general area and on the same pipe size. Because the problems related to thermal expansion are proportional, one must apply conservative methodology. However, conservatism can be expensive, so the approach here is to be moderately conservative.

The resistance of the soil to thermal expansion depends on the contact between the pipe and the soil. The frictional resistance depends on the surface roughness of the pipe and the type and condition of the soil. Values for the coefficient of friction vary between 0.1 and 0.8. Average values of 0.4 to 0.5 may be assumed.

Because the soil around the pipe is in a state of compression, it tends to exert pressure on the walls of the pipe. This pressure is referred to as passive pressure. Values of passive pressure have been found by Karl Terzaghi (the Father of Soil Mechanics) to vary from 0.2 to 0.8 times the weight of the soil above the centerline of the pipe. On the conservative side, the lighter load must be implied, and a passive pressure factor of 0.2 will be used for finding soil resistance to thermal expansion.

Because buried pipe is in intimate contact with the soil in which it is embedded, the behavior of the pipe during thermal expansion is a direct function of the soil properties. These soil properties and their interaction with the pipe are difficult, and expensive, to assess and can be subject to controversy. These soil properties will vary considerably over the length of the line or with time or the moisture content of the soil, to mention just a few factors.

It should be understood that assumptions that are conservative when calculating the amount of thermal expansion—values used for the coefficient of friction, passive pressure and fill loads—are nonconservative when calculating the unbalanced frictional forces on anchors. For this reason, engineering judgment should be used when evaluating the soil conditions and their effects on the response of the pipe.

Soil Resistance Equations.   The frictional resistance (force) of the pipe (F_r) to thermal growth is given by the following equation:

$$F_r = \mu(P_p + F_L + W_p + W_c)\frac{\text{lb}}{\text{ft}} \qquad \text{Eq. 8-18}$$

where μ = coefficient of friction between the soil and pipe (varies from 0.1 to 0.8)

F_L = fill load = W_{cm} = CWBD, for calculation of thermal expansion, lb/ft

$C = 1.25$

W = specific weight of the soil (such as 120 lb/ft^3)

B = bed width, ft
D = OD of pipe, ft
W_p = weight of pipe per linear foot, lb/ft
W_c = weight of the fluid inside the pipe per linear foot, lb/ft

Note to the reader about units: In pipeline and piping computations it is common to combine force loads, such as thermal forces (lb_f) with mass loads (lb_m). When loads in lb_m mass are used with force loads, lb_f, the mass load in lb_m must be multiplied by g/g_c (or divided by g_c/g). The constant g is the average acceleration due to gravity, written as follows:

$$g = 32.174 \, \frac{ft}{sec^2}.$$

The constant g_c is the acceleration of gravity at sea level at 45 degrees latitude. The constant g_c is written as follows:

$$g_c = \frac{32.174 \text{ ft-}lb_m}{sec^2\text{-}lb_f}$$

The term lb_m is considered by some to be obsolete, but is still used and exists in older documents and is still widely used, namely by the U.S. government. It would be difficult to say the use of a unit is obsolete if it is in wide use and sanctioned by the U.S. government. The interested reader can refer to *http://www.epa.gov/eogapti1/module1/units/ units.htm* for additional information about U.S. government use of lb_f and lb_m mass. This source is an official U.S. government website that addresses the issue here. The pound as a unit of force is written as lb_f. When encountering lb_m, such as a liquid in a pipeline (lb_m) or the mass of the pipeline material, the following equations show how the two (lb_m and lb_f) are combined together:

From Newton's second law of motion, Force = (Mass)(Acceleration). Thus

$$\text{Force} = \frac{\text{Mass } (lb_m) \text{ Accleration}\left(g = 32.174\dfrac{ft}{sec^2}\right)}{\left(g_c = \dfrac{32.174 \, lb_m\text{-ft}}{lb_f\text{-sec}^2}\right)} = lb_f$$

Hence

$$\text{Mass} = \frac{\text{Force}\,(\text{lb}_f)\left(g_c = \dfrac{32.174\,\text{lb}_m\text{-ft}}{\text{lb}_f\text{-sec}^2}\right)}{\text{Acceleration}\left(g = \dfrac{32.174\,\text{ft}}{\text{sec}^2}\right)} = \text{lb}_m$$

As mentioned before, we are using two systems of units in this book—the U.S. system of measurements and the metric SI system. We have referred to the U.S. system of units as the English, or Imperial, system, but it is officially referred to as the American Engineering System in the U.S. (AES) (see source above). There are slight differences between the British Imperial system and the AES. Engineers prefer to work in a gravitational system where mass is a derived unit and force is a primary unit, such as in the AES. By lb_m we can convert to force, or, rounding 32.174 to 32.2, we have

$$\text{Force} = \frac{M\,(\text{lb}_m)\left(32.2\dfrac{\text{ft}}{\text{sec}^2}\right)}{32.2\dfrac{\text{lb}_m\text{-ft}}{\text{lb}_f\text{-sec}^2}} = \text{lb}_f$$

Thus, we use the terms lb_f and lb_m interchangeably. In the AES system of units, lb_m and lb_f have the same value (magnitude). **Pound (mass), lb_m, and pound (force) have identical numerical values**. Thus one pound mass is equal in magnitude to one pound force, hence we use the term lb (pound) in this section on thermal expansion of buried pipelines. **Force is not mass and this is hard to understand using the AES system of units**, where the same term is used for both mass and force—pounds. The word "pound" is used for both mass and force even though these are clearly not the same concept. The pound mass is the older definition of the pound. It is also the primary definition of the pound. We list both units because many people are used to using one or the other and like to see them listed separately. In our case here, weight is a force in computation of loadings and stresses and is added to other forces, such as thermal forces in piping and pipeline flexibility (stress) computations. To simplify the discussion, we apply just lb or lbs in the equations **only** in this section on thermal expansion of buried pipelines, using lb_f as lbs interchangeably. In the metric SI system this problem does not exist, as mass is expressed in kilograms and force in newtons. In the metric SI system mass and force do not have the same name and are not numerically equal, but the conversion is similar, with the gravitational constant, g_c, being 1.0 Kg-m/N-sec^2 and g, the average acceleration due to gravity, being 9.807 m/sec^2. In the metric SI system, mass is a primary unit and force is a derived unit. Consequently, for many there is less confusion between mass and force.

P_r = passive pressure, where

$$P_r = C_s WD\left[H + \frac{D}{2}\right], \text{lbs/ft} \qquad \text{Eq. 8-19}$$

where C_s = soil coefficient (varies from 0.2 to 0.8)
$\quad\quad W$ = the specific weight of the soil—defined earlier
$\quad\quad D$ = OD of the pipe, ft
$\quad\quad H$ = height of the soil fill above the top of the pipe, ft.

The value of C_s varies with the soil conditions, the degree of compaction, and the condition of the trench, among other factors.

Forces and Stresses Induced in Buried Pipe

Long runs of pipelines tend to anchor themselves and place the pipes into compression. This self-anchoring phenomenon is referred to as a virtual anchor, and the following discussion may be used to find the locations of these virtual anchors and the forces developed on them by the tendency of the pipe to expand. The force on the anchor caused by thermal compressive stress (F_A) is

$$F_A = E\alpha(\Delta T)A_m \qquad \text{Eq. 8-20}$$

And the thermal compressive stress in the pipe wall (Sc) is

$$Sc = E\alpha\Delta T \qquad \text{Eq. 8-21}$$

where E = modulus of elasticity, psi (MPa)
$\quad\quad \alpha$ = mean coefficient of thermal expansion, in./in.-°F (mm/mm-°C)
$\quad\quad \Delta T$ = temperature differential between installation (tie-in) to operating temperature, °F (°C)
$\quad\quad A_m$ = pipe cross-sectional area, in.^{2} (mm^2)

Note that the amount of axial strain developed from the coefficient of thermal expansion and the temperature differential is found by

$$\varepsilon = \frac{\Delta L}{L} \frac{\text{in.}}{\text{in.}} \left(\frac{\text{mm}}{\text{mm}} \right) \qquad \text{Eq. 8-22}$$

where $\Delta L = L\alpha(\Delta T)$

L = length of pipeline between anchors, be it physical or virtual anchors

Allowable Stress for Buried Pipelines.   Referring to Eq. 8-6, the allowable stress for pipelines containing liquids, ASME B31.4, is as follows:

$$S_A = 0.72E_j(SMYS)$$

We differentiate the E_j from E, the elastic modulus, to prevent confusion of terms. The E_j in the previous equation is the weld joint factor in the ASME B31.4.

Finding the Location of the Virtual Anchor.   The length of pipeline from the virtual anchor to the point of entry or exit from the ground is defined as

$$L_v \overset{\Delta}{=} \frac{\text{Anchor force}}{\text{Frictional resistance}}$$

Thus,

$$L_v = \frac{F_A}{F_r} \qquad \text{Eq. 8-23}$$

where F_A is defined in Eq. 8-20 and F_r is defined in Eq. 8-18.

The resultant thermal displacement from the point where the pipe exits or enters the ground is

$$\Delta L_v = \text{Thermal displacement} = \frac{12L_v e}{R_f} \qquad \text{Eq. 8-24}$$

where L_v is defined in Eq. 8-23

e = linear coefficient of thermal expansion at a given temperature, in./in. (mm/mm)

R_f = soil resistance factor = 2.0

Note that the resultant thermal displacement is one half of the free thermal expansion. This is because of the restraint of the soil.

In cases where the pipeline is connected to critical equipment, the calculated thermal displacement should be multiplied by a safety factor to account for thermal differences in the pipeline from one end to the other. Tests have indicated that the end of the pipe at the origin of flow will not displace the same amount as the terminal point of the flow. These tests indicate that the terminal end may displace as much as twice that of the calculated displacement. During start-up, the initial end of the system will be hot and the terminal end will be cold. This condition causes the differences of displacement at the opposite ends. Another fact is that during shutdown, the line does not totally return to its original length. In the past, practice has been to incorporate a safety factor of 2 with the calculation of thermal displacements [Reference 13].

Example Problem of Buried Pipe

Referring to **Figure 8-25**, we wish to locate the virtual anchor on a pipeline exiting and entering the ground. The pipe is $12'' \phi$ STD WGT made of A53 Gr B material. The tie-in temperature was 70°F, and the operating temperature is 140°F. The process fluid in the pipeline is crude oil with a specific gravity of 0.72. The soil is sand with a specific weight of 120 lb/ft^3.

The solution is shown in spreadsheet format in **Figure 8-26**.

Restraining Bowing of Pipelines

Because buried pipe is virtually anchored in more than one location on straight runs, analysis must be performed with regard to column buckling and the possibility of the piping bowing out of the ground. The following method is based on the consultations of the Whitegate Refinery product line failures with Professor J. P. Den Hartog of MIT back in the 1970s.

The approach is to enable the engineer to specify the necessary installation conditions to prevent the pipeline from bowing out of the ground due to an increase in the temperature of the internal fluid. The methodology is as follows:

1. Find the length of pipe between anchors (virtual or real) from the previous section.
2. Find the critical buckling length of the pipeline as a function of temperature.
3. Compare the results of steps 1 and 2. If the length of pipeline between anchors is greater than the critical buckling length of the pipeline, an analysis for bowing and depth of burial is necessary.

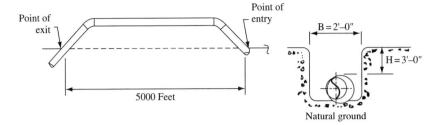

Figure 8-25. Pipeline layout scheme. Refer to Figure 8-24 on an actual pipe configuration exiting the ground.

Example of Thermal Growth of Buried Pipeline

Rules Sheet

Rules

$\Delta T = Toper - Ttiein$

$FA = E \cdot \alpha \cdot \Delta T \cdot Am$

$Sc = E \cdot \alpha \cdot \Delta T$

$Pp = Cs \cdot W \cdot D \cdot \left[H + \dfrac{D}{2} \right]$

$\Delta T = Toper - Ttiein$

$FA = E \cdot \alpha \cdot \Delta T \cdot Am$

$Sc = E \cdot \alpha \cdot \Delta T$

$Pp = Cs \cdot W \cdot D \cdot \left[H + \dfrac{D}{2} \right]$

$Wcm = C \cdot W \cdot B \cdot D$

$WT = Wp + \gamma \cdot Ww$

$Fr = \mu \cdot (Pp + Wcm + WT)$

$Lv = \dfrac{FA}{Fr}$

$\Delta Lv = \dfrac{12 \cdot Lv \cdot e}{Rf}$

$\Delta Lvadj = SF \cdot \Delta Lv$

Figure 8-26a. Equations used for assessment procedure.

Variables Sheet

Input	Name	Output	Unit	Comment
140	Toper			Maximum operating temperature, deg F
70	Ttiein			Tie-in temperature, deg F
	ΔT	70		Temperature differential, deg F
	FA	177625.42812	lbs	Anchor force
27900000	E		psi	Modulus of elasticity of pipeline material
0.000006	α			Coefficient of thermal expansion of pipe, in./in.-F
14.58	Am			Metal area of pipe cross section, in.^{2}
	Sc	12182.814	psi	Compressive stress in pipeline
	Pp	84		Passive pressure, lb/ft
.2	Cs			Soil coefficient
120	W			Specific weight of soil, lb/ft^3
1	D		ft	OD of pipe
3	H		ft	Height of soil above top of pipeline
	Wcm	300		Fill load, lb/ft
1.25	C			Fill load constant
2	B		ft	Width of bed
	WT	84.84		Total weight of pipe and internal liquid lb/ft
49.56	Wp			Weight of pipe per foot, lb/ft
.72	γ			Specific gravity of lquid in pipe
49	Ww			Weight of water in pipe per foot, lb/ft
	Fr	187.536		Frictional force, lb/ft
.4	μ			Coefficient of friction between soil and pipe
	Lv	947.153763	ft	Distance to virtual anchor
.00045	e			Coefficient of thermal expansion at given temperature, in./in.
2	Rf			Factor of safety for soil friction
	ΔLv	2.557315	in.	Thermal growth at ground exit
	ΔLvadj	5.114630	in.	Adjusted thermal growth at ground exit
2	SF			Factor of safety for thermal growth

Figure 8-26b. Variable sheet showing solution to example problem.

4. Find the theoretical height of bowing that the critical buckling length of the pipeline would experience, caused by an increase in temperature.

5. Find the deflection of the critical buckling length of the pipeline, supported with fixed ends, caused by the weight of the pipeline and the contained fluid and earth cover.

6. Compare the results of steps 4 and 5. If the deflection caused by dead weight (from step 5) is greater than the theoretical height of bowing (from step 4), the pipeline will remain in the ground.

Note that all computations based on thermal increases are made considering an axially loaded column with pinned ends.

The anchor force caused by thermal compressive stress is

$$F_A = E\alpha(\Delta T)A_m \qquad\qquad\qquad \text{Eq. 8-20}$$

The critical buckling length is

$$L_c = \pi\sqrt{\frac{4EI}{F_A}} \text{ in. (mm)} \qquad\qquad\qquad \text{Eq. 8-25}$$

The theoretical height of bowing caused by thermal expansion is

$$\delta_b = \sqrt{\frac{4\alpha\Delta TL_c^2}{\pi^2} - 2R^2} \qquad\qquad\qquad \text{Eq. 8-26}$$

The bowing deflection due to weight

$$\delta_w = \frac{W_g L_c^4}{384EI}, \text{ in. (mm)} \qquad\qquad\qquad \text{Eq. 8-27}$$

The critical temperature based on the critical buckling length is

$$T_{cr} = \left(\frac{\pi^2}{2\alpha}\right)\left(\frac{R^2}{L_c^2}\right), \text{ in. (mm)} \qquad\qquad\qquad \text{Eq. 8-28}$$

Example of Pipeline Bowing

Consider the preceding example, which dealt with the thermal expansion of a buried pipeline in **Figure 8-25**. It would be helpful to refer to **Figure 8-24** to refer to a similar configuration in the example. Using the same example, we will assess the possibility of the pipeline bowing. The equations for pipeline bowing are solved in spreadsheet form, which is shown in **Figure 8-27**.

As seen in the solved algorithm in **Figure 8-27** there is enough soil cover to prevent the pipeline from bowing. If $L_c <$ 5000 ft, the length of buried pipeline, is greater than the critical buckling length, then a bowing calculation must be made. If the height of the bowing (*db*) is less than the deflection due to sag of the critical buckling length, 108 ft, then the cover is sufficient. As seen from above, the configuration is acceptable.

Permissible Bending of Pipelines

ASME B31.4 does not require that bending stress in buried pipelines be included in the calculation of equivalent tensile stress. However, to

Rules Sheet

Rules

$$LC = \pi \cdot \sqrt{\frac{4 \cdot I}{e \cdot Am}}$$

If $\dfrac{LC}{12} < 5000$ then $Lc1 = \text{°bowing}$

$$R = \frac{D}{2}$$

$$\delta b = \sqrt{\left[\frac{4 \cdot e \cdot Lc^2}{\pi^2}\right] \cdot 2 \cdot R^2}$$

$Wg = Wt + Wc + Wpr$

$$\delta w = \frac{Wg \cdot Lc^4}{384 \cdot E \cdot I}$$

If $\delta b < \delta w$ then def = 'acceptable

Variables Sheet

Input	Name	Output	Unit	Comment
	Lc	1295.680631	in.	Critical buckling length
3.1416	π			
279	I		in.^{4}	Moment of inertia of pipe cross-section, in.^{4}
.00045	e			Coefficient of thermal expansion, in./in.
14.58	Am		in.^{2}	Pipeline metal area of cross section
	Lc1	'bowing		Calculated value of Lc to test if it is bowing is possible
	δb	14.996386	in.	Amount of bowing due to thermal expansion
	R	6.375	in.	Outside radius of pipe
	Wg	467.34		Total weight of pipe plus liquid plus fill, lb/ft
49.56	Wt			Weight of pipe per foot, lb/ft
35.28	Wc			Weight of internal liquid per foot, lb/ft
382.5	Wpr			Weight of fill per foot, lb/ft
27900000	E		psi	Modulus of elasticity of pipe material at temperature
	δw	440.641290	in.	Deflection of pipeline due to sag
	def	'acceptable		Is deflection of pipeline acceptable?
12.75	D		in.	Outside diameter of pipeline

Figure 8-27. Algorithm of pipeline bowing.

ensure proper field repairs and bowing of pipelines, as previously discussed, we will discuss the subject. ASME B31.4, Paragraph 419.6.4(a), stipulates that beam-bending stresses shall be included in the longitudinal stress for those portions of the pipeline that are supported *above* the ground.

The equation for the maximum bend a cold pipe can take without being overstressed is derived in [Reference 14], pp. 46–47. The equation reads as follows:

$$R = \frac{3EI}{2ZS_A} \qquad \text{Eq. 8-29}$$

where E = modulus of elasticity of the pipe at temperature, psi
$\quad\quad I$ = moment of inertia of the pipe cross section, in.^{4}
$\quad\quad Z$ = section modulus of the pipe cross section, in.^{3}
$\quad\quad S_A$ = allowable stress of the pipeline at temperature

Since $Z = \left(\dfrac{2}{D_o}\right)I$, Eq. 8-29 becomes

$$R = \frac{3ED_o}{4S_o} = \frac{3ER_o}{2S_A} \qquad \text{Eq. 8-30}$$

Equation 8-30 is the bend radius that a cold pipe can take without being overstressed.

References

1. John F. Kiefner, Patrick H. Vieth, and Itta Roytman, Line PRCI, Line Pipe Research Supervisory Committee, Pipeline Research Committee of PRCI, Contract No. PR218-9304, Dec. 20, 1996.
2. *API RP 579, Fitness-for-Service*, 1st edition, January 2000, American Petroleum Institute, Washington, DC.
3. American Society of Mechanical Engineers, ASME B31.4-2002, *Pipeline Transportation Systems for Liquid Hydrocarbons and Other Liquids,* New York, NY.
4. American Society of Mechanical Engineers, ASME B31.8-2002, *Gas Transmission and Distribution Piping Systems*, New York, NY.
5. M. Mohitpour, H. Golshan, and A. Murray, *Pipeline Design and Construction*, 2nd edition, 2003, American Society of Mechanical Engineers, New York.
6. *Guide to Pigging*, 1976, T. D. Williamson, Inc., Tulsa, OK.
7. J. N. H. Tiratsoo, *Pipeline Pigging Technology*, 2nd edition, 1999, Gulf Professional Publishing, imprint of Butterworth-Heinemann, Houston, TX.

8. Dr. Aidan O'Donoghue, "Pigging as a Flow Assurance Solution: Estimating Pigging Frequency for Dewaxing," *The Journal of Pipeline Integrity*, Quarter 4, 2003, pp. 251–258.

9. API RP 1107, *Recommended Pipeline Maintenance Welding Practices*, 1991, American Petroleum Institute, Washington, DC.

10. J. L. Otegui, Rivas A. Urquiza, and A. Trunzo, "Local Collapse of Gas Pipelines Under Sleeve Repairs," *International Journal of Pressure Vessels and Piping*, Vol. 77, 2000, pp. 555–566.

11. Fritz Karge, *The Petroleum Engineer*, October 1952.

12. Fritz Karge, *Oil and Gas Journal*, March 22, 1951.

13. F. C. Benedict, AER-1304 Rev. 1, *Pipeline Movements During Start Up*, April 27, 1973, Aramco Overseas Co., The Hague, The Netherlands.

14. M. W. McAllister, editor, *Pipeline Rules of Thumb Handbook*, 5th edition, 2002, Gulf Professional Publishing, Woburn, MA.

Appendix A
Properties of Pipe

Properties of Pipe

The following formulas are used in the computation of the values shown in the table:

† weight of pipe per foot (pounds) $= 10.6802t(D - t)$
weight of water per foot (pounds) $= 0.3405d^2$
square feet outside surface per foot $= 0.2618D$
square feet inside surface per foot $= 0.2618d$
inside area (square inches) $= 0.785d^2$
area of metal (square inches) $= 0.785(D^2 - d^2)$
moment of inertia (inches^4) $= 0.0491(D^4 - d^4)$
$= A_m R_g^2$

section modulus (inches^3) $= \dfrac{0.0982\,(D^4 - d^4)}{D}$

radius of gyration (inches) $= 0.25\sqrt{D^2 + d^2}$

$A_m =$ area of metal (square inches)
$d \ =$ inside diameter (inches)
$D \ =$ outside diameter (inches)
$R_g =$ radius of gyration (inches)
$t \ \ =$ pipe wall thickness (inches)

† The ferritic steels may be about 5% less, and the austenitic stainless steels about 2% greater than the values shown in this table which are based on weights for carbon steel.

*schedule numbers

Standard weight pipe and schedule 40 are the same in all sizes through 10-inch; from 12-inch through 24-inch, standard weight pipe has a wall thickness of 3/8-inch.

Extra strong weight pipe and schedule 80 are the same in all sizes through 8-inch; from 8-inch through 24-inch, extra strong weight pipe has a wall thickness of 1/2-inch.

Double extra strong weight pipe has no corresponding schedule number.

a: ANSI B36.10 steel pipe schedule numbers

b: ANSI B36.10 steel pipe nominal wall thickness designation

c: ANSI B36.19 stainless steel pipe schedule numbers

nominal pipe size outside diameter, in.	schedule number* a	b	c	wall thickness, in.	inside diameter, in.	inside area. sq. in.	metal area. sq. in.	sq ft outside surface, per ft	sq ft inside surface, per ft	weight per ft, lb †	weight of water per ft, lb	moment of inertia, in.⁴	section modulus, in.³	radius gyration, in.
1/8 0.405	–	–	10S	0.049	0.307	0.0740	0.0548	0.106	0.0804	0.186	0.0321	0.00088	0.00437	0.1271
	40	Std	40S	0.068	0.269	0.0568	0.0720	0.106	0.0705	0.245	0.0246	0.00106	0.00525	0.1215
	80	XS	80S	0.095	0.215	0.0364	0.0925	0.106	0.0563	0.315	0.0157	0.00122	0.00600	0.1146
1/4 0.540	–	–	10S	0.065	0.410	0.1320	0.0970	0.141	0.1073	0.330	0.0572	0.00279	0.01032	0.1694
	40	Std	40S	0.088	0.364	0.1041	0.1250	0.141	0.0955	0.425	0.0451	0.00331	0.01230	0.1628
	80	XS	80S	0.119	0.302	0.0716	0.1574	0.141	0.0794	0.535	0.0310	0.00378	0.01395	0.1547
3/8 0.675	–	–	SS	0.065	0.710	0.396	0.1582	0.220	0.1859	0.538	0.1716	0.01197	0.0285	0.2750
	–	–	10S	0.065	0.545	0.2333	0.1246	0.177	0.1427	0.423	0.1011	0.00586	0.01737	0.2169
	40	Std	40S	0.091	0.493	0.1910	0.1670	0.177	0.1295	0.568	0.0827	0.00730	0.02160	0.2090
	80	XS	80S	0.126	0.423	0.1405	0.2173	0.177	0.1106	0.739	0.0609	0.00862	0.02554	0.1991
1/2 0.840	–	–	5S	0.065	0.710	0.3959	0.1583	0.220	0.1859	0.538	0.171	0.0120	0.0285	0.2750
	–	–	10S	0.083	0.674	0.357	0.1974	0.220	0.1765	0.671	0.1547	0.01431	0.0341	0.2692
	40	Std	40S	0.109	0.622	0.304	0.2503	0.220	0.1628	0.851	0.1316	0.01710	0.0407	0.2613
	80	XS	80S	0.147	0.546	0.2340	0.320	0.220	0.1433	1.088	0.1013	0.02010	0.0478	0.2505
	160	–	–	0.187	0.466	0.1706	0.383	0.220	0.1220	1.304	0.0740	0.02213	0.0527	0.2402
	–	XXS	–	0.294	0.252	0.0499	0.504	0.220	0.0660	1.714	0.0216	0.02425	0.0577	0.2192
3/4 1.050	–	–	5S	0.065	0.920	0.665	0.2011	0.275	0.2409	0.684	0.2882	0.02451	0.0467	0.349
	–	–	10S	0.083	0.884	0.614	0.2521	0.275	0.2314	0.857	0.2661	0.02970	0.0566	0.343
	40	Std	40S	0.113	0.824	0.533	0.333	0.275	0.2157	1.131	0.2301	0.0370	0.0706	0.334
	80	XS	80S	0.154	0.742	0.432	0.435	0.275	0.1943	1.474	0.1875	0.0448	0.0853	0.321
	160	–	–	0.218	0.614	0.2961	0.570	0.275	0.1607	1.937	0.1284	0.0527	0.1004	0.304
	–	XXS	–	0.308	0.434	0.1479	0.718	0.275	0.1137	2.441	0.0641	0.0579	0.1104	0.2840
1 1.315	–	–	5S	0.065	1.185	1.103	0.2553	0.344	0.310	0.868	0.478	0.0500	0.0760	0.443
	–	–	10S	0.109	1.097	0.945	0.413	0.344	0.2872	1.404	0.409	0.0757	0.1151	0.428
	40	Std	40S	0.133	1.049	0.864	0.494	0.344	0.2746	1.679	0.374	0.0874	0.1329	0.421
	80	XS	80S	0.179	0.957	0.719	0.639	0.344	0.2520	2.172	0.311	0.1056	0.1606	0.407
	160	–	–	0.250	0.815	0.522	0.836	0.344	0.2134	2.844	0.2261	0.1252	0.1903	0.387
	–	XXS	–	0.358	0.599	0.2818	1.076	0.344	0.1570	3.659	0.1221	0.1405	0.2137	0.361
1 1/4 1.660	–	–	5S	0.065	1.530	1.839	0.326	0.434	0.401	1.107	0.797	0.1038	0.1250	0.564
	–	–	10S	0.109	1.442	1.633	0.531	0.434	0.378	1.805	0.707	0.1805	0.1934	0.550
	40	Std	40S	0.140	1.380	1.496	0.669	0.434	0.361	2.273	0.648	0.1948	0.2346	0.540
	80	XS	80S	0.191	1.278	1.283	0.881	0.434	0.335	2.997	0.555	0.2418	0.2913	0.524
	160	–	–	0.250	1.160	1.057	1.107	0.434	0.304	3.765	0.458	0.2839	0.342	0.506
	–	XXS	–	0.382	0.896	0.631	1.534	0.434	0.2346	5.214	0.2732	0.341	0.411	0.472
1 1/2 1.900	–	–	5S	0.065	1.770	2.461	0.375	0.497	0.463	1.274	1.067	0.1580	0.1663	0.649
	–	–	10S	0.109	1.682	2.222	0.613	0.497	0.440	2.085	0.962	0.2468	0.2599	0.634
	40	Std	40S	0.145	1.610	2.036	0.799	0.497	0.421	2.718	0.882	0.310	0.326	0.623
	80	XS	80S	0.200	1.500	1.767	1.068	0.497	0.393	3.631	0.765	0.391	0.412	0.605
	160	–	–	0.281	1.338	1.406	1.429	0.497	0.350	4.859	0.608	0.483	0.508	0.581
	–	XXS	–	0.400	1.100	0.950	1.885	0.497	0.288	6.408	0.412	0.568	0.598	0.549
	–	–	–	0.525	0.850	0.567	2.267	0.497	0.223	7.710	0.246	0.6140	0.6470	0.5200
	–	–	–	0.650	0.600	0.283	2.551	0.497	0.157	8.678	0.123	0.6340	0.6670	0.4980
2 2.375	–	–	5S	0.065	2.245	3.96	0.472	0.622	0.588	1.604	1.716	0.315	0.2652	0.817
	–	–	10S	0.109	2.157	3.65	0.776	0.622	0.565	2.638	1.582	0.499	0.420	0.802
	40	Std	40S	0.154	2.067	3.36	1.075	0.622	0.541	3.653	1.455	0.666	0.561	0.787
	80	XS	80S	0.218	1.939	2.953	1.477	0.622	0.508	5.022	1.280	0.868	0.731	0.766
	160	–	–	0.343	1.689	2.240	2.190	0.622	0.442	7.444	0.971	1.163	0.979	0.729
	–	XXS	–	0.436	1.503	1.774	2.656	0.622	0.393	9.029	0.769	1.312	1.104	0.703
	–	–	–	0.562	1.251	1.229	3.199	0.622	0.328	10.882	0.533	1.442	1.2140	0.6710
	–	–	–	0.687	1.001	0.787	3.641	0.622	0.262	12.385	0.341	1.5130	1.2740	0.6440
2 1/2 2.875	–	–	5S	0.083	2.709	5.76	0.728	0.753	0.709	2.475	2.499	0.710	0.494	0.988
	–	–	10S	0.120	2.635	5.45	1.039	0.753	0.690	3.531	2.361	0.988	0.687	0.975
	40	Std	40S	0.203	2.469	4.79	1.704	0.753	0.646	5.793	2.076	1.530	1.064	0.947
	80	XS	80S	0.276	2.323	4.24	2.254	0.753	0.608	7.661	1.837	1.925	1.339	0.924
	160	–	–	0.375	2.125	3.55	2.945	0.753	0.556	10.01	1.535	2.353	1.637	0.894
	–	XXS	–	0.552	1.771	2.464	4.03	0.753	0.464	13.70	1.067	2.872	1.998	0.844
	–	–	–	0.675	1.525	1.826	4.663	0.753	0.399	15.860	0.792	3.0890	2.1490	0.8140
	–	–	–	0.800	1.275	1.276	5.212	0.753	0.334	17.729	0.554	3.2250	2.2430	0.7860
3 3.500	–	–	5S	0.083	3.334	8.73	0.891	0.916	0.873	3.03	3.78	1.301	0.744	1.208
	–	–	10S	0.120	3.260	8.35	1.274	0.916	0.853	4.33	3.61	1.822	1.041	1.196
	40	Std	40S	0.216	3.068	7.39	2.228	0.916	0.803	7.58	3.20	3.02	1.724	1.164

(Table continued on next page)

Properties of Pipe—cont'd

nominal pipe size outside diameter, in.	schedule number* a	b	c	wall thickness, in.	inside diameter, in.	inside area, sq. in.	metal area, sq. in.	sq ft outside surface, per ft	sq ft inside surface, per ft	weight per ft, lb †	weight of water per ft, lb	moment of inertia, in.⁴	section modulus, in.³	radius gyration, in.
3	80	XS	80S	0.300	2.900	6.61	3.02	0.916	0.759	10.25	2.864	3.90	2.226	1.136
3.500	160	–	–	0.437	2.626	5.42	4.21	0.916	0.687	14.32	2.348	5.03	2.876	1.094
	–	XXS	–	0.600	2.300	4.15	5.47	0.916	0.602	18.58	1.801	5.99	3.43	1.047
	–	–	–	0.725	2.050	3.299	6.317	0.916	0.537	21.487	1.431	6.5010	3.7150	1.0140
	–	–	–	0.850	1.800	2.543	7.073	0.916	0.471	24.057	1.103	6.8530	3.9160	0.9840
3½	–	–	5S	0.083	3.834	11.55	1.021	1.047	1.004	3.47	5.01	1.960	0.980	1.385
4.000	–	–	10S	0.120	3.760	11.10	1.463	1.047	0.984	4.97	4.81	2.756	1.378	1.372
	40	Std	40S	0.226	3.548	9.89	2.680	1.047	0.929	9.11	4.28	4.79	2.394	1.337
	80	XS	80S	0.318	3.364	8.89	3.68	1.047	0.881	12.51	3.85	6.28	3.14	1.307
	–	XXS	–	0.636	2.728	5.845	6.721	1.047	0.716	22.850	2.530	9.8480	4.9240	1.2100
4	–	–	5S	0.083	4.334	14.75	1.152	1.178	1.135	3.92	6.40	2.811	1.249	1.562
4.500	–	–	10S	0.120	4.260	14.25	1.651	1.178	1.115	5.61	6.17	3.96	1.762	1.549
	–	–	–	0.188	4.124	13.357	2.547	1.178	1.082	8.560	5.800	5.8500	2.6000	1.5250
	40	Std	40S	0.237	4.026	12.73	3.17	1.178	1.054	10.79	5.51	7.23	3.21	1.510
	80	XS	80S	0.337	3.826	11.50	4.41	1.178	1.002	14.98	4.98	9.61	4.27	1.477
	120	–	–	0.437	3.626	10.33	5.58	1.178	0.949	18.96	4.48	11.65	5.18	1.445
	–	–	–	0.500	3.500	9.621	6.283	1.178	0.916	21.360	4.160	12.7710	5.6760	1.4250
	160	–	–	0.531	3.438	9.28	6.62	1.178	0.900	22.51	4.02	13.27	5.90	1.416
	–	XXS	–	0.674	3.152	7.80	8.10	1.178	0.825	27.54	3.38	15.29	6.79	1.374
	–	–	–	0.800	2.900	6.602	9.294	1.178	0.759	31.613	2.864	16.6610	7.4050	1.3380
	–	–	–	0.925	2.650	5.513	10.384	1.178	0.694	35.318	2.391	17.7130	7.8720	1.3060
5	–	–	5S	0.109	5.345	22.44	1.868	1.456	1.399	6.35	9.73	6.95	2.498	1.929
5.563	–	–	10S	0.134	5.295	22.02	2.285	1.456	1.386	7.77	9.53	8.43	3.03	1.920
	40	Std	40S	0.258	5.047	20.01	4.30	1.456	1.321	14.62	8.66	15.17	5.45	1.878
	80	XS	80S	0.375	4.813	18.19	6.11	1.456	1.260	20.78	7.89	20.68	7.43	1.839
	120	–	–	0.500	4.563	16.35	7.95	1.456	1.195	27.04	7.09	25.74	9.25	1.799
	160	–	–	0.625	4.313	14.61	9.70	1.456	1.129	32.96	6.33	30.0	10.80	1.760
	–	XXS	–	0.750	4.063	12.97	11.34	1.456	1.064	38.55	5.62	33.6	12.10	1.722
	–	–	–	0.875	3.813	11.413	12.880	1.456	0.998	43.810	4.951	36.6450	13.1750	1.6860
	–	–	–	1.000	3.563	9.966	14.328	1.456	0.933	47.734	4.232	39.1110	14.0610	1.6520
6	–	–	5S	0.109	6.407	32.2	2.231	1.734	1.677	5.37	13.98	11.85	3.58	2.304
6.525	–	–	10S	0.134	6.357	31.7	2.733	1.734	1.664	9.29	13.74	14.40	4.35	2.295
	–	–	–	0.219	6.187	30.100	4.410	1.734	1.620	15.020	13.100	22.6600	6.8400	2.2700
	40	Std	40S	0.280	6.065	28.89	5.58	1.734	1.588	18.97	12.51	28.14	8.50	2.245
	80	XS	80S	0.432	5.761	26.07	8.40	1.734	1.508	28.57	11.29	40.5	12.23	2.195
	120	–	–	0.562	5.501	23.77	10.70	1.734	1.440	36.39	10.30	49.6	14.98	2.153
	160	–	–	0.718	5.189	21.15	13.33	1.734	1.358	45.30	9.16	59.0	17.81	2.104
		XXS		0.864	4.897	18.83	15.64	1.734	1.282	53.16	8.17	66.3	20.03	2.060
	–	–	–	1.000	4.625	16.792	17.662	1.734	1.211	60.076	7.284	72.1190	21.7720	2.0200
	–	–	–	1.125	4.375	15.025	19.429	1.734	1.145	66.084	6.517	76.5970	23.1240	1.9850
8	–	–	5S	0.109	8.407	55.5	2.916	2.258	2.201	9.91	24.07	26.45	6.13	3.01
8.625	–	–	10S	0.148	8.329	54.5	3.94	2.258	2.180	13.40	23.59	35.4	8.21	3.00
	–	–	–	0.219	8.187	52.630	5.800	2.258	2.150	19.640	22.900	51.3200	11.9000	2.9700
	20	–	–	0.250	8.125	51.8	6.58	2.258	2.127	22.36	22.48	57.7	13.39	2.962
	30	–	–	0.277	8.071	51.2	7.26	2.258	2.113	24.70	22.18	63.4	14.69	2.953
	40	Std	40S	0.322	7.981	50.0	8.40	2.258	2.089	28.55	21.69	72.5	16.81	2.938
	60	–	–	0.406	7.813	47.9	10.48	2.258	2.045	35.64	20.79	88.8	20.58	2.909
	80	XS	80S	0.500	7.625	45.7	12.76	2.258	1.996	43.39	19.80	105.7	24.52	2.878
	100	–	–	0.593	7.439	43.5	14.96	2.258	1.948	50.87	18.84	121.4	28.14	2.847
	120	–	–	0.718	7.189	40.6	17.84	2.258	1.882	60.63	17.60	140.6	32.6	2.807
	140	–	–	0.812	7.001	38.5	19.93	2.258	1.833	67.76	16.69	153.8	35.7	2.777
	160	–	–	0.906	6.813	36.5	21.97	2.258	1.784	74.69	15.80	165.9	38.5	2.748
	–	–	–	1.000	6.625	34.454	23.942	2.258	1.734	81.437	14.945	177.1320	41.0740	2.7190
	–	–	–	1.125	6.375	31.903	26.494	2.258	1.669	90.114	13.838	190.6210	44.2020	2.6810
10	–	–	5S	0.134	10.482	86.3	4.52	2.815	2.744	15.15	37.4	63.7	11.85	3.75
10.750	–	–	10S	0.165	10.420	85.3	5.49	2.815	2.728	18.70	36.9	76.9	14.30	3.74
	–	–	–	0.219	10.312	83.52	7.24	2.815	2.70	24.63	36.2	100.46	18.69	3.72
	20	–	–	0.250	10.250	82.5	8.26	2.815	2.683	28.04	35.8	113.7	21.16	3.71
	30	–	–	0.307	10.136	80.7	10.07	2.815	2.654	34.24	35.0	137.5	25.57	3.69
	40	Std	40S	0.365	10.020	78.9	11.91	2.815	2.623	40.48	34.1	160.8	29.90	3.67
	60	XS	80S	0.500	9.750	74.7	16.10	2.815	2.553	54.74	32.3	212.0	39.4	3.63
	80	–	–	0.593	9.564	71.8	18.92	2.815	2.504	64.33	31.1	244.9	45.6	3.60
	100	–	–	0.718	9.314	68.1	22.63	2.815	2.438	76.93	29.5	286.2	53.2	3.56

Properties of Pipe—cont'd

nominal pipe size outside diameter, in.	schedule number*			wall thickness, in.	inside diameter, in.	inside area, sq. in.	metal area, sq. in.	sq ft outside surface, per ft	sq ft inside surface, per ft	weight per ft, lb †	weight of water per ft, lb	moment of inertia, in.⁴	section modulus, in.³	radius gyration, in.
	a	b	c											
10 10.750	120	–	–	0.843	9.064	64.5	26.24	2.815	2.373	89.20	28.0	324	60.3	3.52
	–	–	–	0.875	9.000	63.62	27.14	2.815	2.36	92.28	27.6	333.46	62.04	3.50
	140	–	–	1.000	8.750	60.1	30.6	2.815	2.291	104.13	26.1	368	68.4	3.47
	160	–	–	1.125	8.500	56.7	34.0	2.815	2.225	115.65	24.6	399	74.3	3.43
			–	1.250	8.250	53.45	37.31	2.815	2.16	126.82	23.2	428.17	79.66	3.39
			–	1.500	7.750	47.15	43.57	2.815	2.03	148.19	20.5	478.59	89.04	3.31
12 12.750	–	–	5S	0.156	12.438	121.4	6.17	3.34	3.26	20.99	52.7	122.2	19.20	4.45
	–	–	10S	0.180	12.390	120.6	7.11	3.34	3.24	24.20	52.2	140.5	22.03	4.44
	20	–	–	0.250	12.250	117.9	9.84	3.34	3.21	33.38	51.1	191.9	30.1	4.42
	30	–	–	0.330	12.090	114.8	12.88	3.34	3.17	43.77	49.7	248.5	39.0	4.39
		Std	40S	0.375	12.000	113.1	14.58	3.34	3.14	49.56	49.0	279.3	43.8	4.38
	40	–	–	0.406	11.938	111.9	15.74	3.34	3.13	53.53	48.5	300	47.1	4.37
	–	XS	80S	0.500	11.750	108.4	19.24	3.34	3.08	65.42	47.0	362	56.7	4.33
	60	–	–	0.562	11.626	106.2	21.52	3.34	3.04	73.16	46.0	401	62.8	4.31
	80	–	–	0.687	11.376	101.6	26.04	3.34	2.978	88.51	44.0	475	74.5	4.27
	–			0.750	11.250	99.40	28.27	3.34	2.94	96.2	43.1	510.7	80.1	4.25
	100	–	–	0.843	11.064	96.1	31.5	3.34	2.897	107.20	41.6	562	88.1	4.22
	–			0.875	11.000	95.00	32.64	3.34	2.88	110.9	41.1	578.5	90.7	4.21
	120	–	–	1.000	10.750	90.8	36.9	3.34	2.814	125.49	39.3	642	100.7	4.17
	140	–	–	1.125	10.500	86.6	41.1	3.34	2.749	139.68	37.5	701	109.9	4.13
	–			1.250	10.250	82.50	45.16	3.34	2.68	150.3	35.8	755.5	118.5	4.09
	160	–	–	1.312	10.126	80.5	47.1	3.34	2.651	160.27	34.9	781	122.6	4.07
14 14.000	–	–	5S	0.156	13.688	147.20	6.78	3.67	3.58	23.0	63.7	162.6	23.2	4.90
	–	–	10S	0.188	13.624	145.80	8.16	3.67	3.57	27.7	63.1	194.6	27.8	4.88
	–	–	–	0.210	13.580	144.80	9.10	3.67	3.56	30.9	62.8	216.2	30.9	4.87
	–	–	–	0.219	13.562	144.50	9.48	3.67	3.55	32.2	62.6	225.1	32.2	4.87
	10	–	–	0.250	13.500	143.1	10.80	3.67	3.53	36.71	62.1	255.4	36.5	4.86
	–	–	–	0.281	13.438	141.80	12.11	3.67	3.52	41.2	61.5	285.2	40.7	4.85
	20	–	–	0.312	13.376	140.5	13.42	3.67	3.50	45.68	60.9	314	44.9	4.84
	–	–	–	0.344	13.312	139.20	14.76	3.67	3.48	50.2	60.3	344.3	49.2	4.83
	30	Std	–	0.375	13.250	137.9	16.05	3.67	3.47	54.57	59.7	373	53.3	4.82
	40	–	–	0.437	13.126	135.3	18.62	3.67	3.44	63.37	58.7	429	61.2	4.80
	–	–	–	0.469	13.062	134.00	19.94	3.67	3.42	67.8	58.0	456.8	65.3	4.79
	–	XS	–	0.500	13.000	132.7	21.21	3.67	3.40	72.09	57.5	484	69.1	4.78
	60	–	–	0.593	12.814	129.0	24.98	3.67	3.35	84.91	55.9	562	80.3	4.74
	–	–	–	0.625	12.750	127.7	26.26	3.67	3.34	89.28	55.3	589	84.1	4.73
	80	–	–	0.750	12.500	122.7	31.2	3.67	3.27	106.13	53.2	687	98.2	4.69
	100	–	–	0.937	12.126	115.5	38.5	3.67	3.17	130.73	50.0	825	117.8	4.63
	120	–	–	1.093	11.814	109.6	44.3	3.67	3.09	150.67	47.5	930	132.8	4.58
	140	–	–	1.250	11.500	103.9	50.1	3.67	3.01	170.22	45.0	1127	146.8	4.53
	160	–	–	1.406	11.188	98.3	55.6	3.67	2.929	189.12	42.6	1017	159.6	4.48
16 16.000	–	–	5S	0.165	15.670	192.90	8.21	4.19	4.10	28	83.5	257	32.2	5.60
	–	–	10S	0.188	15.624	191.70	9.34	4.19	4.09	32	83.0	292	36.5	5.59
	10	–	–	0.250	15.500	188.7	12.37	4.19	4.06	42.05	81.8	384	48.0	5.57
	20	–	–	0.312	15.376	185.7	15.38	4.19	4.03	52.36	80.5	473	59.2	5.55
	30	Std	–	0.375	15.250	182.6	18.41	4.19	3.99	62.58	79.1	562	70.3	5.53
	40	XS	–	0.500	15.000	176.7	24.35	4.19	3.93	82.77	76.5	732	91.5	5.48
	60	–	–	0.656	14.688	169.4	31.6	4.19	3.85	107.50	73.4	933	116.6	5.43
	80	–	–	0.843	14.314	160.9	40.1	4.19	3.75	136.46	69.7	1157	144.6	5.37
	100		–	1.031	13.938	152.6	48.5	4.19	3.65	164.83	66.1	1365	170.6	5.30
	120			1.218	13.564	144.5	56.6	4.19	3.55	192.29	62.6	1556	194.5	5.24
	140	–	–	1.437	13.126	135.3	65.7	4.19	3.44	223.64	58.6	1760	220.0	5.17
	160	–	–	1.593	12.814	129.0	72.1	4.19	3.35	245.11	55.9	1894	236.7	5.12
18 18.000	–	–	5S	0.165	17.670	245.20	9.24	4.71	4.63	31	106.2	368	40.8	6.31
	–	–	10S	0.188	17.624	243.90	10.52	4.71	4.61	36	105.7	417	46.4	6.30
	10	–	–	0.250	17.500	240.5	13.94	4.71	4.58	47.39	104.3	549	61.0	6.28
	20	–	–	0.312	17.376	237.1	17.34	4.71	4.55	59.03	102.8	678	75.5	6.25
	–	Std	–	0.375	17.250	233.7	20.76	4.71	4.52	70.59	101.2	807	89.6	6.23
	30	–	–	0.437	17.126	230.4	24.11	4.71	4.48	82.06	99.9	931	103.4	6.21
	–	XS	–	0.500	17.00	227.0	27.49	4.71	4.45	93.45	98.4	1053	117.0	6.19
	40	–	–	0.562	16.876	223.7	30.8	4.71	4.42	104.75	97.0	1172	130.2	6.17
	60	–	–	0.750	16.500	213.8	40.6	4.71	4.32	138.17	92.7	1515	168.3	6.10
	80	–	–	0.937	16.126	204.2	50.2	4.71	4.22	170.75	88.5	1834	203.8	6.04
	100		–	1.156	15.688	193.3	61.2	4.71	4.11	207.96	83.7	2180	242.2	5.97

(Table continued on next page)

Properties of Pipe—cont'd

nominal pipe size outside diameter, in.	schedule number* a	b	c	wall thick-ness, in.	inside diam-eter, in.	inside area. sq. in.	metal area. sq. in.	sq ft outside surface, per ft	sq ft inside surface, per ft	weight per ft, lb †	weight of water per ft, lb	moment of inertia, in.⁴	section modulus, in.³	radius gyration, in.
18 18.000	120	–	–	1.375	15.250	182.6	71.8	4.71	3.99	244.14	79.2	2499	277.6	5.90
	140	–	–	1.562	14.876	173.8	80.7	4.71	3.89	274.23	75.3	2750	306	5.84
	160	–	–	1.781	14.438	163.7	90.7	4.71	3.78	308.51	71.0	3020	336	5.77
20 20.000	–	–	5S	0.188	19.634	302.40	11.70	5.24	5.14	40	131.0	574	57.4	7.00
	–		10S	0.218	19.564	300.60	13.55	5.24	5.12	46	130.2	663	66.3	6.99
	10	–	–	0.250	19.500	298.6	15.51	5.24	5.11	52.73	129.5	757	75.7	6.98
	20	Std	–	0.375	19.250	291.0	23.12	5.24	5.04	78.60	126.0	1114	111.4	6.94
	30	XS	–	0.500	19.000	283.5	30.6	5.24	4.97	104.13	122.8	1457	145.7	6.90
	40		–	0.593	18.814	278.0	36.2	5.24	4.93	122.91	120.4	1704	170.4	6.86
	60	–	–	0.812	18.376	265.2	48.9	5.24	4.81	166.40	115.0	2257	225.7	6.79
			–	0.875	18.250	261.6	52.6	5.24	4.78	178.73	113.4	2409	240.9	6.77
	80		–	1.031	17.938	252.7	61.4	5.24	4.70	208.87	109.4	2772	277.2	6.72
	100	–	–	1.281	17.438	238.8	75.3	5.24	4.57	256.10	103.4	3320	332	6.63
	120	–	–	1.500	17.000	227.0	87.2	5.24	4.45	296.37	98.3	3760	376	6.56
	140	–	–	1.750	16.500	213.8	100.3	5.24	4.32	341.10	92.6	4220	422	6.48
	160	–	–	1.968	16.064	202.7	111.5	5.24	4.21	379.01	87.9	4590	459	6.41
22 22.000	–	–	5S	0.188	21.624	367.3	12.88	5.76	5.66	44	159.1	766	69.7	7.71
	–		10S	0.218	21.564	365.2	14.92	5.76	5.65	51	158.2	885	80.4	7.70
	10	–	–	0.250	21.500	363.1	17.18	5.76	5.63	58	157.4	1010	91.8	7.69
	20	Std	–	0.375	21.250	354.7	25.48	5.76	5.56	87	153.7	1490	135.4	7.65
	30	XS	–	0.500	21.000	346.4	33.77	5.76	5.50	115	150.2	1953	177.5	7.61
	–	–	–	0.625	20.750	338.2	41.97	5.76	5.43	143	146.6	2400	218.2	7.56
	–	–	–	0.750	20.500	330.1	50.07	5.76	5.37	170	143.1	2829	257.2	7.52
	60	–	–	0.875	20.250	322.1	58.07	5.76	5.30	197	139.6	3245	295.0	7.47
	80	–	–	1.125	19.750	306.4	73.78	5.76	5.17	251	132.8	4029	366.3	7.39
	100	–	–	1.375	19.250	291.0	89.09	5.76	5.04	303	126.2	4758	432.6	7.31
	120	–	–	1.625	18.750	276.1	104.02	5.76	4.91	354	119.6	5432	493.8	7.23
	140	–	–	1.875	18.250	261.6	118.55	5.76	4.78	403	113.3	6054	550.3	7.15
	160	–	–	2.125	17.750	247.4	132.68	5.76	4.65	451	107.2	6626	602.4	7.07
24 24.000	10	–	–	0.250	23.500	434	18.65	6.28	6.15	63.41	188.0	1316	109.6	8.40
	20	Std	–	0.375	23.250	425	27.83	6.28	6.09	94.62	183.8	1943	161.9	8.35
	–	XS	–	0.500	23.000	415	36.9	6.28	6.02	125.49	180.1	2550	212.5	8.31
	30	–	–	0.562	22.876	411	41.4	6.28	5.99	140.80	178.1	2840	237.0	8.29
	–	–	–	0.625	22.750	406	45.9	6.28	5.96	156.03	176.2	3140	261.4	8.27
	40	–	–	0.687	22.626	402	50.3	6.28	5.92	171.17	174.3	3420	285.2	8.25
	–	–	–	0.750	22.500	398	54.8	6.28	5.89	186.24	172.4	3710	309	8.22
	–	–	5S	0.218	23.564	436.1	16.29	6.28	6.17	55	188.9	1152	96.0	8.41
	60	–	–	0.875	22.250	388.6	63.54	6.28	5.83	216	168.6	4256	354.7	8.18
	–	–	–	0.968	22.064	382	70.0	6.28	5.78	238.11	165.8	4650	388	8.15
	80	–	–	1.218	21.564	365	87.2	6.28	5.65	296.36	158.3	5670	473	8.07
	100	–	–	1.531	20.938	344	108.1	6.28	5.48	367.40	149.3	6850	571	7.96
	120	–	–	1.812	20.376	326	126.3	6.28	5.33	429.39	141.4	7830	652	7.87
	140	–	–	2.062	19.876	310	142.1	6.29	5.20	483.13	134.5	8630	719	7.79
	160	–	–	2.343	19.314	293	159.4	6.28	5.06	541.94	127.0	9460	788	7.70
26 26.000	–	–	–	0.250	25.500	510.7	19.85	6.81	6.68	67	221.4	1646	126.6	9.10
	10			0.312	25.376	505.8	25.18	6.81	6.64	86	219.2	2076	159.7	9.08
		Std		0.375	25.250	500.7	30.19	6.81	6.61	103	217.1	2478	190.6	9.06
	20	XS		0.500	25.000	490.9	40.06	6.81	6.54	136	212.8	3259	250.7	9.02
				0.625	24.750	481.1	49.82	6.81	6.48	169	208.6	4013	308.7	8.98
				0.750	24.500	471.4	59.49	6.81	6.41	202	204.4	4744	364.9	8.93
	–	–	–	0.875	24.250	461.9	69.07	6.81	6.35	235	200.2	5458	419.9	8.89
				1.000	24.000	452.4	78.54	6.81	6.28	267	196.1	6149	473.0	8.85
				1.125	23.750	443.0	87.91	6.81	6.22	299	192.1	6813	524.1	8.80
28 28.000	–	–	–	0.250	27.500	594.0	21.80	7.33	7.20	74	257.3	2098	149.8	9.81
	10			0.312	27.376	588.6	27.14	7.33	7.17	92	255.0	2601	185.8	9.79
	–	Std		0.375	27.250	583.2	32.54	7.33	7.13	111	252.6	3105	221.8	9.77
	20	XS		0.500	27.000	572.6	43.20	7.33	7.07	147	248.0	4085	291.8	9.72
	30		–	0.625	26.750	562.0	53.75	7.33	7.00	183	243.4	5038	359.8	9.68
		–		0.750	26.500	551.6	64.21	7.33	6.94	218	238.9	5964	426.0	9.64
				0.875	26.250	541.2	74.56	7.33	6.87	253	234.4	6865	490.3	9.60
			–	1.000	26.000	530.9	84.82	7.33	6.81	288	230.0	7740	552.8	9.55
			–	1.125	25.750	520.8	94.98	7.33	6.74	323	225.6	8590	613.6	9.51

Properties of Pipe—cont'd

nominal pipe size outside diameter, in.	schedule number*			wall thick-ness, in.	inside diam-eter, in.	inside area. sq. in.	metal area. sq. in.	sq ft outside surface, per ft	sq ft inside surface, per ft	weight per ft, lb †	weight of water per ft, lb	moment of inertia, in.⁴	section modulus, in.³	radius gyration, in.
	a	b	c											
			5S	0.250	29.500	683.4	23.37	7.85	7.72	79	296.3	2585	172.3	10.52
	10		10S	0.312	29.376	677.8	29.19	7.85	7.69	99	293.7	3201	213.4	10.50
		Std	–	0.375	29.250	672.0	34.90	7.85	7.66	119	291.2	3823	254.8	10.48
30	20	XS		0.500	29.000	660.5	46.34	7.85	7.59	158	286.2	5033	335.5	10.43
30.000	30		–	0.625	28.750	649.2	57.68	7.85	7.53	196	281.3	6213	414.2	10.39
	40			0.750	28.500	637.9	68.92	7.85	7.46	234	276.6	7371	491.4	10.34
				0.875	28.250	620.7	80.06	7.85	7.39	272	271.8	8494	566.2	10.30
				1.000	28.000	615.7	91.11	7.85	7.33	310	267.0	9591	639.4	10.26
				1.125	27.750	604.7	102.05	7.85	7.26	347	262.2	10653	710.2	10.22
				0.250	31.500	779.2	24.93	8.38	8.25	85	337.8	3141	196.3	11.22
	10			0.312	31.376	773.2	31.02	8.38	8.21	106	335.2	3891	243.2	11.20
		Std		0.375	31.250	766.9	37.25	8.38	8.18	127	332.5	4656	291.0	11.18
	20	XS		0.500	31.000	754.7	49.48	8.38	8.11	168	327.2	6140	383.8	11.14
32	30			0.625	30.750	742.5	61.59	8.38	8.05	209	321.9	7578	473.6	11.09
32.000	40			0.688	30.624	736.6	67.68	8.38	8.02	230	319.0	8298	518.6	11.07
				0.750	30.500	730.5	73.63	8.38	7.98	250	316.7	8990	561.9	11.05
				0.875	30.250	718.3	85.52	8.38	7.92	291	311.6	10372	648.2	11.01
				1.000	30.000	706.8	97.38	8.38	7.85	331	306.4	11680	730.0	10.95
				1.125	29.750	694.7	109.0	8.38	7.79	371	301.3	13023	814.0	10.92
				0.250	33.500	881.2	26.50	8.90	8.77	90	382.0	3773	221.9	11.93
	10		–	0.312	33.376	874.9	32.99	8.90	8.74	112	379.3	4680	275.3	11.91
		Std		0.375	33.250	867.8	39.61	8.90	8.70	135	376.2	5597	329.2	11.89
	20	XS		0.500	33.000	855.3	52.62	8.90	8.64	179	370.8	7385	434.4	11.85
34	30		–	0.625	32.750	841.9	65.53	8.90	8.57	223	365.0	9124	536.7	11.80
34.000	40		–	0.688	32.624	835.9	72.00	8.90	8.54	245	362.1	9992	587.8	11.78
			–	0.750	32.500	829.3	78.34	8.90	8.51	266	359.5	10829	637.0	11.76
				0.875	32.250	816.4	91.01	8.90	8.44	310	354.1	12501	735.4	11.72
	–			1.000	32.000	804.2	103.67	8.90	8.38	353	348.6	14114	830.2	11.67
	–	–		1.125	31.750	791.3	116.13	8.90	8.31	395	343.2	15719	924.7	11.63
	–	–	–	0.250	35.500	989.7	28.11	9.42	9.29	96	429.1	4491	249.5	12.64
	10	–		0.312	35.376	982.9	34.95	9.42	9.26	119	426.1	5565	309.1	12.62
		Std		0.375	35.250	975.8	42.01	9.42	9.23	143	423.1	6664	370.2	12.59
	20	XS		0.500	35.000	962.1	55.76	9.42	9.16	190	417.1	8785	488.1	12.55
36	30			0.625	34.750	948.3	69.50	9.42	9.10	236	411.1	10872	604.0	12.51
36.000	40			0.750	34.500	934.7	83.01	9.42	9.03	282	405.3	12898	716.5	12.46
	–	–		0.875	34.250	920.6	96.50	9.42	8.97	328	399.4	14903	827.9	12.42
	–	–		1.000	34.000	907.9	109.96	9.42	8.90	374	393.6	16851	936.2	12.38
	–			1.125	33.750	894.2	123.19	9.42	8.89	419	387.9	18763	1042.4	12.34
	–			0.250	41.500	1352.6	32.82	10.99	10.86	112	586.4	7126	339.3	14.73
		Std	–	0.375	41.250	1336.3	49.08	10.99	10.80	167	579.3	10627	506.1	14.71
	20	XS		0.500	41.000	1320.2	65.18	10.99	10.73	222	572.3	14037	668.4	14.67
42	30			0.625	40.750	1304.1	81.28	10.99	10.67	276	565.4	17373	827.3	14.62
42.000	40	–		0.750	40.500	1288.2	97.23	10.99	10.60	330	558.4	20689	985.2	14.59
	–	–		1.000	40.000	1256.6	128.81	10.99	10.47	438	544.8	27080	1289.5	14.50
	–			1.250	39.500	1225.3	160.03	10.99	10.34	544	531.2	33233	1582.5	14.41
	–			1.500	39.000	1194.5	190.85	10.99	10.21	649	517.9	39181	1865.7	14.33

Appendix B
Weights of Pipe Materials

Insulation Weight Factors

To determine the weight per foot of any piping insulation, use the pipe size and nominal insulation thickness to find the insulation weight factor F in the chart shown below. Then multiply F by the density of the insulation in pounds per cubic foot.

Example. For 4″ pipe with 4″ nominal thickness insulation, $F = .77$. If the insulation density is 12 pounds per cubic foot, then the insulation weight is $.77 \times 12 = 9.24$ lb/ft.

Nominal Pipe Size	Nominal Insulation Thickness										
	1″	1½″	2″	2½″	3″	3½″	4″	4½″	5″	5½″	6″
1	.057	.10	.16	.23	.31	.40					
1¼	.051	.12	.15	.22	.30	.39					
1½	.066	.11	.21	.29	.38	.48					
2	.080	.14	.21	.29	.37	.47	.59				
2½	.091	.19	.27	.36	.46	.58	.70	.83			
3	.10	.17	.25	.34	.44	.56	.68	.81			
3½	.15	.23	.31	.41	.54	.66	.78	..	.97		
4	.13	.21	.30	.39	.51	.63	.77	.96	1.10		
5	.15	.24	.34	.45	.58	.71	.88	1.04	1.20		
6	.17	.27	.38	.51	.64	.83	.97	1.13	1.34		
8	..	.34	.47	.66	.80	.97	1.17	1.36	1.56	1.75	
10	..	.43	.59	.75	.93	1.12	1.32	1.54	1.76	1.99	
12	..	.50	.68	.88	1.07	1.28	1.52	1.74	1.99	2.24	2.50
14	..	.51	.70	.90	1.11	1.34	1.57	1.81	2.07	2.34	2.62
16	..	.57	.78	1.01	1.24	1.49	1.74	2.01	2.29	2.58	2.88
18	..	.64	.87	1.12	1.37	1.64	1.92	2.21	2.51	2.82	3.14
20	..	.70	.96	1.23	1.50	1.79	2.09	2.40	2.73	3.06	3.40
24	..	.83	1.13	1.44	1.77	2.10	2.44	2.80	3.16	3.54	3.92

LOAD CARRYING CAPACITIES OF THREADED HOT ROLLED STEEL ROD CONFORMING TO ASTM A-36

Nominal Rod Diameter, in.	⅜	½	⅝	¾	⅞	1	1⅛	1¼	1½	1¾	2	2¼	2½	2¾	3	3¼	3½
Root Area of Thread, sq. in.	.068	.126	.202	.302	.419	.552	.693	.889	1.293	1.744	2.300	3.023	3.719	4.619	5.621	6.720	7.918
Max. Safe Load, lbs. at Rod Temp. of 650°F	610	1130	1810	2710	3770	4960	6230	8000	11630	15700	20700	27200	33500	41580	50580	60480	71280

WEIGHTS OF PIPING MATERIALS   1″ PIPE 1.313″ O.D.

	Schedule No.	40	80	160										
PIPE	Wall Designation	Std.	XS		XXS									
	Thickness—In.	.133	.179	.250	.358									
	Pipe—Lbs/Ft	**1.68**	**2.17**	**2.84**	**3.66**									
	Water—Lbs/Ft	**.37**	**.31**	**.23**	**.12**									

WELDING FITTINGS	L.R. 90° Elbow	**.3** / .3	**.4** / .3	**.6** / .3	**.7** / .3				
	S.R. 90° Elbow	**.2** / .2							
	L.R. 45° Elbow	**.2** / .2	**.3** / .2	**.4** / .2	**.4** / .2				
	Tee	**.8** / .4	**.9** / .4	**1.1** / .4	**1.3** / .4				
	Lateral	**1.7** / 1.1	**2.5** / 1.1						
	Reducer	**.3** / .2	**.4** / .2	**.4** / .2	**.5** / .2				
	Cap	**.2** / .3	**.3** / .3	**.4** / .3	**.4** / .3				

	Temperature Range °F	100–199	200–299	300–399	400–499	500–599	600–699	700–799	800–899	900–999	1000–1099	1100–1200
INSULATION	85% Magnesia Calcium Silicate — Nom. Thick., In.	1	1	1½	2	2	2½	2½	2½	3	3	3
	85% Magnesia Calcium Silicate — Lbs/Ft	**.72**	**.72**	**1.23**	**1.94**	**1.94**	**2.76**	**2.76**	**2.76**	**3.70**	**3.70**	**3.70**
	Combination — Nom. Thick., In.						2½	2½	2½	3	3	3
	Combination — Lbs/Ft						**3.30**	**3.30**	**3.30**	**4.70**	**4.70**	**4.70**
	*Asbestos Fiber— Sodium Silicate — Nom. Thick., In.	1	1	1	1	1	1½	1½	2	2	3	3
	*Asbestos Fiber— Sodium Silicate — Lbs/Ft	**.91**	**.91**	**.91**	**.91**	**.91**	**1.61**	**1.61**	**2.54**	**2.54**	**4.91**	**4.91**

	Pressure Rating psi	Cast Iron		Steel						
		125	250	150	300	400	600	900	1500	2500
FLANGES	Screwed or Slip-On	**2.5** / 1.5	**4** / 1.5	**2.5** / 1.5	**4** / 1.5	**5** / 1.5	**5** / 1.5	**12** / 1.5	**12** / 1.5	**15** / 1.5
	Welding Neck			**2.3** / 1.5	**5** / 1.5	**7** / 1.5	**7** / 1.5	**12** / 1.5	**12** / 1.5	**16** / 1.5
	Lap Joint			**2.5** / 1.5	**4** / 1.5	**5** / 1.5	**5** / 1.5	**12** / 1.5	**12** / 1.5	**15** / 1.5
	Blind	**2.5** / 1.5	**4** / 1.5	**2.5** / 1.5	**5** / 1.5	**5** / 1.5	**5** / 1.5	**12** / 1.5	**12** / 1.5	**15** / 1.5
FLANGED FITTINGS	S.R. 90° Elbow	**6** / 3.6					**15** / 3.7		**28** / 3.8	
	L.R. 90° Elbow	**8** / 3.8								
	45° Elbow	**5** / 3.2					**14** / 3.4		**26** / 3.6	
	Tee	**11** / 5.4					**20** / 5.6		**39** / 5.7	
VALVES	Flanged Bonnet Gate				**20** / 1.2		**25** / 1.5		**80** / 4.3	
	Flanged Bonnet Globe or Angle								**84** / 3.5	
	Flanged Bonnet Check									
	Pressure Seal Bonnet—Gate						**31** / 1.7	**31** / 1.7		
	Pressure Seal Bonnet—Globe									

Boldface type is weight in pounds. Lightface type beneath weight is weight factor for insulation.

Insulation thicknesses and weights are based on average conditions and do not constitute a recommendation for specific thicknesses of materials. Insulation weights are based on 85% magnesia and hydrous calcium silicate at 11 lbs/cubic foot. The listed thicknesses and weights of combination covering are the sums of the inner layer of diatomaceous earth at 21 lbs/cubic foot and the outer layer at 11 lbs/cubic foot.

Insulation weights include allowances for wire, cement, canvas, bands and paint, but not special surface finishes.

To find the weight of covering on flanges, valves or fittings, multiply the weight factor by the weight per foot of covering used on straight pipe.

Valve weights are approximate. When possible, obtain weights from the manufacturer.

Cast iron valve weights are for flanged end valves; steel weights for welding end valves.

All flanged fitting, flanged valve and flange weights include the proportional weight of bolts or studs to make up all joints.

*16 lb cu. ft. density.

1¼″ PIPE 1.660″ O.D.   WEIGHTS OF PIPING MATERIALS

		Schedule No.	40	80	160										
PIPE		Wall Designation	Std.	XS		XXS									
		Thickness—In.	.140	.191	.250	.382									
		Pipe—Lbs/Ft	**2.27**	**3.00**	**3.77**	**5.22**									
		Water—Lbs/Ft	**.65**	**.56**	**.46**	**.27**									
WELDING FITTINGS		L.R. 90° Elbow	**.6** .3	**.8** .3	**.9** .3	**1.2** .3									
		S.R. 90° Elbow	**.4** .2													
		L.R. 45° Elbow	**.4** .2	**.5** .2	**.6** .2	**.8** .2									
		Tee	**1.3** .5	**1.6** .5	**1.9** .5	**2.4** .5									
		Lateral	**2.4** 1.2	**3.9** 1.2												
		Reducer	**.5** .2	**.5** .2	**.6** .2	**.8** .2									
		Cap	**.3** .3	**.4** .3	**.5** .3	**.6** .3									

		Temperature Range °F	100–199	200–299	300–399	400–499	500–599	600–699	700–799	800–899	900–999	1000–1099	1100–1200
INSULATION	85% Magnesia Calcium Silicate	Nom. Thick., In.	1	1	1½	2	2	2½	2½	2½	3	3	3
		Lbs/Ft	**.65**	**.65**	**1.47**	**1.83**	**1.83**	**2.65**	**2.65**	**2.65**	**3.58**	**3.58**	**3.58**
	Combina-tion	Nom. Thick., In.						2½	2½	2½	3	3	3
		Lbs/Ft						**3.17**	**3.17**	**3.17**	**5.76**	**5.76**	**5.76**
	*Asbestos Fiber—Sodium Silicate	Nom. Thick., In.	1	1	1	1½	1½	1½	2	2½	2½	3	3
		Lbs/Ft	**.82**	**.82**	**.82**	**1.93**	**1.93**	**1.93**	**2.45**	**3.58**	**3.58**	**4.82**	**4.82**

		Pressure Rating psi	Cast Iron		Steel						
			125	250	150	300	400	600	900	1500	2500
FLANGES		Screwed or Slip-On	**2.5** 1.5	**5** 1.5	**3.5** 1.5	**5** 1.5	**7** 1.5	**7** 1.5	**13** 1.5	**13** 1.5	**23** 1.5
		Welding Neck			**3.5** 1.5	**7** 1.5	**8** 1.5	**8** 1.5	**13** 1.5	**13** 1.5	**25** 1.5
		Lap Joint			**3.5** 1.5	**5** 1.5	**7** 1.5	**7** 1.5	**13** 1.5	**13** 1.5	**22** 1.5
		Blind	**3.5** 1.5	**5** 1.5	**3.5** 1.5	**7** 1.5	**7** 1.5	**7** 1.5	**13** 1.5	**13** 1.5	**23** 1.5
FLANGED FITTINGS		S.R. 90° Elbow	**8** 3.6		**17** 3.7		**18** 3.8		**33** 3.9		
		L.R. 90° Elbow	**10** 3.9		**18** 3.9						
		45° Elbow	**7** 3.3		**15** 3.4		**16** 3.5		**31** 3.7		
		Tee	**13** 5.4		**23** 5.6		**28** 5.7		**49** 5.9		
VALVES		Flanged Bonnet Gate				**40** 4		**60** 4.2		**97** 4.6	
		Flanged Bonnet Globe or Angle									
		Flanged Bonnet Check				**21** 4					
		Pressure Seal Bonnet—Gate							**38** 1.8	**38** 1.1	
		Pressure Seal Bonnet—Globe									

Boldface type is weight in pounds. Lightface type beneath weight is weight factor for insulation.

Insulation thicknesses and weights are based on average conditions and do not constitute a recommendation for specific thicknesses of materials. Insulation weights are based on 85% magnesia and hydrous calcium silicate at 11 lbs/cubic foot. The listed thicknesses and weights of combination covering are the sums of the inner layer of diatomaceous earth at 21 lbs/cubic foot and the outer layer at 11 lbs/cubic foot.

Insulation weights include allowances for wire, cement, canvas, bands and paint, but not special surface finishes.

To find the weight of covering on flanges, valves or fittings, multiply the weight factor by the weight per foot of covering used on straight pipe.

Valve weights are approximate. When possible, obtain weights from the manufacturer.

Cast iron valve weights are for flanged end valves; steel weights for welding end valves.

All flanged fitting, flanged valve and flange weights include the proportional weight of bolts or studs to make up all joints.

*16 lb cu. ft. density.

WEIGHTS OF PIPING MATERIALS   1.900″ O.D. 1½″ PIPE

	Schedule No.	40	80	160									
PIPE	Wall Designation	Std.	XS		XXS								
	Thickness—In.	.145	.200	.281	.400								
	Pipe—Lbs/Ft	**2.72**	**3.63**	**4.86**	**6.41**								
	Water—Lbs/Ft	**.88**	**.77**	**.61**	**.41**								

WELDING FITTINGS	L.R. 90° Elbow	**.8** .4	**1.1** .4	**1.4** .4	**1.8** .4
	S.R. 90° Elbow	**.6** .3	**.7** .3		
	L.R. 45° Elbow	**.5** .2	**.7** .2	**.8** .2	**1** .2
	Tee	**2** .6	**2.5** .6	**3.1** .6	**3.7** .6
	Lateral	**3.3** 1.3	**5.4** 1.3		
	Reducer	**.6** .2	**.7** .2	**.9** .2	**1.2** .2
	Cap	**.4** .3	**.5** .3	**.7** .3	**.7** .3

			Temperature Range °F	100–199	200–299	300–399	400–499	500–599	600–699	700–799	800–899	900–999	1000–1099	1100–1200
INSULATION	85% Magnesia Calcium Silicate		Nom. Thick., In.	1	1	1½	2	2	2½	2½	2½	3	3	3
			Lbs/Ft	**.84**	**.84**	**1.35**	**2.52**	**2.52**	**3.47**	**3.47**	**3.47**	**4.52**	**4.52**	**4.52**
	Combina-tion		Nom. Thick., In.						2½	2½	2½	3	3	3
			Lbs/Ft						**4.20**	**4.20**	**4.20**	**5.62**	**5.62**	**5.62**
	*Asbestos Fiber— Sodium Silicate		Nom. Thick., In.	1	1	1	1½	1½	2	2	2½	2½	3	3
			Lbs/Ft	**1.07**	**1.07**	**1.07**	**1.85**	**1.85**	**3.50**	**3.50**	**4.76**	**4.76**	**6.16**	**6.16**

		Pressure Rating psi	Cast Iron		Steel						
			125	250	150	300	400	600	900	1500	2500
FLANGES	Screwed or Slip-On		**3.5** 1.5	**7** 1.5	**3.5** 1.5	**8** 1.5	**9** 1.5	**9** 1.5	**19** 1.5	**19** 1.5	**31** 1.5
	Welding Neck				**4** 1.5	**9** 1.5	**12** 1.5	**12** 1.5	**19** 1.5	**19** 1.5	**34** 1.5
	Lap Joint				**3.5** 1.5	**8** 1.5	**9** 1.5	**9** 1.5	**19** 1.5	**19** 1.5	**31** 1.5
	Blind		**3.5** 1.5	**7** 1.5	**3.5** 1.5	**9** 1.5	**10** 1.5	**10** 1.5	**19** 1.5	**19** 1.5	**31** 1.5
FLANGED FITTINGS	S.R. 90° Elbow		**10** 3.7		**12** 3.7	**23** 3.8		**26** 3.9		**46** 4		
	L.R. 90° Elbow		**12** 4		**13** 4	**24** 4						
	45° Elbow		**9** 3.4		**11** 3.4	**21** 3.5		**23** 3.5		**39** 3.7		
	Tee		**17** 5.6		**20** 5.6	**30** 5.7		**37** 5.8		**70** 6		
VALVES	Flanged Bonnet Gate		**27** 6.8			**55** 4.2		**70** 4.5		**125** 5		
	Flanged Bonnet Globe or Angle					**40** 4.2		**45** 4.2		**170** 5		
	Flanged Bonnet Check				**30** 4.1	**35** 4.1		**40** 4.2		**110** 4.5		
	Pressure Seal Bonnet—Gate								**42** 1.9	**42** 1.2		
	Pressure Seal Bonnet—Globe											

Boldface type is weight in pounds. Lightface type beneath weight is weight factor for insulation.

Insulation thicknesses and weights are based on average conditions and do not constitute a recommendation for specific thicknesses of materials. Insulation weights are based on 85% magnesia and hydrous calcium silicate at 11 lbs/cubic foot. The listed thicknesses and weights of combination covering are the sums of the inner layer of diatomaceous earth at 21 lbs/cubic foot and the outer layer at 11 lbs/cubic foot.

Insulation weights include allowances for wire, cement, canvas, bands and paint, but not special surface finishes.

To find the weight of covering on flanges, valves or fittings, multiply the weight factor by the weight per foot of covering used on straight pipe.

Valve weights are approximate. When possible, obtain weights from the manufacturer.

Cast iron valve weights are for flanged end valves; steel weights for welding end valves.

All flanged fitting, flanged valve and flange weights include the proportional weight of bolts or studs to make up all joints.

*16 lb cu. ft. density.

2" PIPE 2.375" O.D.   WEIGHTS OF PIPING MATERIALS

PIPE	Schedule No.	40	80	160	
	Wall Designation	Std.	XS		XXS
	Thickness—In.	.154	.218	.343	.436
	Pipe—Lbs/Ft	**3.65**	**5.02**	**7.44**	**9.03**
	Water—Lbs/Ft	**1.46**	**1.28**	**.97**	**.77**

WELDING FITTINGS					
L.R. 90° Elbow	**1.5** .5	**2** .5	**2.9** .5	**3.5** .5
S.R. 90° Elbow	**1** .3	**1.3** .3		
L.R. 45° Elbow	**.8** .2	**1.1** .2	**1.6** .2	**1.8** .2
Tee	**3** .6	**3.7** .6	**5** .6	**5.7** .6
Lateral	**5** 1.4	**7.8** 1.4		
Reducer	**.9** .3	**1.2** .3	**1.6** .3	**1.9** .3
Cap	**.5** .4	**.7** .4	**1.2** .4	**1.2** .4

INSULATION

Temperature Range °F		100–199	200–299	300–399	400–499	500–599	600–699	700–799	800–899	900–999	1000–1099	1100–1200
85% Magnesia Calcium Silicate	Nom. Thick., In.	1	1	1½	2	2	2½	2½	3	3	3	3½
	Lbs/Ft	**1.01**	**1.01**	**1.71**	**2.53**	**2.53**	**3.48**	**3.48**	**4.42**	**4.42**	**4.42**	**5.59**
Combination	Nom. Thick., In.						2½	2½	3	3	3	3½
	Lbs/Ft						**4.28**	**4.28**	**5.93**	**5.93**	**5.93**	**7.80**
*Asbestos Fiber—Sodium Silicate	Nom. Thick., In.	1	1	1	1½	1½	2	2	2½	2½	3	3
	Lbs/Ft	**1.26**	**1.26**	**1.26**	**2.20**	**2.20**	**3.32**	**3.32**	**4.57**	**4.57**	**5.99**	**5.99**

	Pressure Rating psi	Cast Iron		Steel						
		125	250	150	300	400	600	900	1500	2500
FLANGES	Screwed or Slip-On	**6** 1.5	**9** 1.5	**6** 1.5	**9** 1.5	**11** 1.5	**11** 1.5	**32** 1.5	**32** 1.5	**48** 1.5
	Welding Neck			**6** 1.5	**10** 1.5	**13** 1.5	**13** 1.5	**31** 1.5	**31** 1.5	**48** 1.5
	Lap Joint			**6** 1.5	**9** 1.5	**12** 1.5	**12** 1.5	**32** 1.5	**32** 1.5	**48** 1.5
	Blind	**6** 1.5	**10** 1.5	**4.8** 1.5	**10** 1.5	**12** 1.5	**12** 1.5	**31** 1.5	**31** 1.5	**49** 1.5
FLANGED FITTINGS	S.R. 90° Elbow	**16** 3.8	**24** 3.8	**19** 3.8	**29** 3.8		**35** 4		**83** 4.2	
	L.R. 90° Elbow	**18** 4.1	**27** 4.1	**22** 4.1	**31** 4.1					
	45° Elbow	**14** 3.4	**22** 3.5	**16** 3.4	**24** 3.5		**33** 3.7		**73** 3.9	
	Tee	**23** 5.7	**37** 5.7	**27** 5.7	**41** 5.7		**52** 6		**129** 6.3	
VALVES	Flanged Bonnet Gate	**37** 6.9	**52** 7.1	**40** 4	**65** 4.2		**80** 4.5		**190** 5	
	Flanged Bonnet Globe or Angle	**30** 7	**64** 7.3	**30** 3.8	**45** 4		**85** 4.5		**235** 5.5	
	Flanged Bonnet Check	**26** 7	**51** 7.3	**35** 3.8	**40** 4		**60** 4.2		**300** 5.8	
	Pressure Seal Bonnet—Gate								**150** 2.5	
	Pressure Seal Bonnet—Globe								**165** 3	

Boldface type is weight in pounds. Lightface type is weight factor for insulation.

Insulation thicknesses and weights are based on average conditions and do not constitute a recommendation for specific thicknesses of materials. Insulation weights are based on 85% magnesia and hydrous calcium silicate at 11 lbs/cubic foot. The listed thicknesses and weights of combination covering are the sums of the inner layer of diatomaceous earth at 21 lbs/cubic foot and the outer layer at 11 lbs/cubic foot.

Insulation weights include allowances for wire, cement, canvas, bands and paint, but not special surface finishes.

To find the weight of covering on flanges, valves or fittings, multiply the weight factor by the weight per foot of covering used on straight pipe.

Valve weights are approximate. When possible, obtain weights from the manufacturer.

Cast iron valve weights are for flanged end valves; steel weights are for welding end valves.

All flanged fitting, flanged valve and flange weights include the proportional weight of bolts or studs to make up all joints.

*16 lb cu. ft. density.

WEIGHTS OF PIPING MATERIALS   2.875″ O.D. 2½″ PIPE

PIPE					
Schedule No.	40	80	160	
Wall Designation	Std.	XS		XXS
Thickness—In.	.203	.276	.375	.552
Pipe—Lbs/Ft	**5.79**	**7.66**	**10.01**	**13.70**
Water—Lbs/Ft	**2.08**	**1.84**	**1.54**	**1.07**

WELDING FITTINGS

L.R. 90° Elbow	**2.9** .6	**3.8** .6	**4.9** .6	**6.5** .6
S.R. 90° Elbow	**1.9** .4	**2.5** .4		
L.R. 45° Elbow	**1.6** .3	**2.1** .3	**2.7** .3	**3.5** .3
Tree	**5.2** .8	**6.4** .8	**7.9** .8	**9.9** .8
Lateral	**9.2** 1.5	**14** 1.5		
Reducer	**1.6** .3	**2.1** .3	**2.7** .3	**3.4** .3
Cap	**.8** .4	**1** .4	**2** .4	**2.1** .4

INSULATION

Temperature Range °F	100–199	200–299	300–399	400–499	500–599	600–699	700–799	800–899	900–999	1000–1099	1100–1200
85% Magnesia Calcium Silicate — Nom. Thick., In.	1	1	1½	2	2	2½	2½	3	3	3½	3½
Lbs/Ft	**1.14**	**1.14**	**2.29**	**3.23**	**3.23**	**4.28**	**4.28**	**5.46**	**5.46**	**6.86**	**6.86**
Combination — Nom. Thick., In.						2½	2½	3	3	3½	3½
Lbs/Ft						**5.20**	**5.20**	**7.36**	**7.36**	**9.58**	**9.58**
*Asbestos Fiber—Sodium Silicate — Nom. Thick., In.	1	1	1	1½	1½	2	2	2½	2½	3	3
Lbs/Ft	**1.44**	**1.44**	**1.44**	**3.09**	**3.09**	**4.34**	**4.34**	**5.75**	**5.75**	**7.34**	**7.34**

FLANGES

Pressure Rating psi	Cast Iron		Steel						
	125	250	150	300	400	600	900	1500	2500
Screwed or Slip-On	**8** 1.5	**14** 1.5	**9** 1.5	**14** 1.5	**17** 1.5	**17** 1.5	**46** 1.5	**46** 1.5	**69** 1.5
Welding Neck			**9** 1.5	**14** 1.5	**20** 1.5	**20** 1.5	**46** 1.5	**46** 1.5	**66** 1.5
Lap Joint			**9** 1.5	**14** 1.5	**18** 1.5	**18** 1.5	**45** 1.5	**45** 1.5	**67** 1.5
Blind	**8** 1.5	**15** 1.5	**9** 1.5	**14** 1.5	**19** 1.5	**19** 1.5	**45** 1.5	**45** 1.5	**70** 1.5

FLANGED FITTINGS

	125	250	150	300	400	600	900	1500	2500
S.R. 90° Elbow	**21** 3.8	**36** 3.9	**27** 3.8	**42** 3.9		**50** 4.1		**114** 4.4	
L.R. 90° Elbow	**25** 4.2	**40** 4.2	**30** 4.2	**47** 4.2					
45° Elbow	**19** 3.5	**34** 3.6	**22** 3.5	**35** 3.6		**46** 3.8		**99** 3.9	
Tee	**32** 5.7	**55** 5.8	**42** 5.7	**61** 5.9		**77** 6.2		**169** 6.6	

VALVES

	125	250	150	300	400	600	900	1500	2500
Flanged Bonnet Gate	**50** 7	**82** 7.1	**60** 4	**100** 4.2		**105** 4.6		**275** 5.2	
Flanged Bonnet Globe or Angle	**43** 7.1	**87** 7.4	**50** 4	**70** 4.1		**120** 4.6		**325** 5.5	
Flanged Bonnet Check	**36** 7.1	**71** 7.4	**40** 4	**50** 4		**105** 4.6		**320** 5.5	
Pressure Seal Bonnet—Gate								**215** 2.5	
Pressure Seal Bonnet—Globe								**230** 2.8	

Boldface type is weight in pounds. Lightface type beneath weight is weight factor for insulation.

Insulation thicknesses and weights are based on average conditions and do not constitute a recommendation for specific thicknesses of materials. Insulation weights are based on 85% magnesia and hydrous calcium silicate at 11 lbs/cubic foot. The listed thicknesses and weights of combination covering are the sums of the inner layer of diatomaceous earth at 21 lbs/cubic foot and the outer layer at 11 lbs/cubic foot.

Insulation weights include allowances for wire, cement, canvas, bands and paint, but not special surface finishes.

To find the weight of covering on flanges, valves or fittings, multiply the weight factor by the weight per foot of covering used on straight pipe.

Valve weights are approximate. When possible, obtain weights from the manufacturer.

Cast iron valve weights are for flanged end valves; steel weights for welding end valves.

All flanged fitting, flanged valve and flange weights include the proportional weight of bolts or studs to make up all joints.

*16 lb cu. ft. density.

3″ PIPE 3.500″ O.D.   WEIGHTS OF PIPING MATERIALS

PIPE	Schedule No.	40	80	160								
	Wall Designation	Std.	XS		XXS							
	Thickness—In.	.216	.300	.438	.600							
	Pipe—Lbs/Ft	**7.58**	**10.25**	**14.32**	**18.58**							
	Water—Lbs/Ft	**3.20**	**2.86**	**2.35**	**1.80**							

WELDING FITTINGS												
L.R. 90° Elbow	**4.6** .8	**6.1** .8	**8.4** .8	**10.7** .8								
S.R. 90° Elbow	**3** .5	**4** .5										
L.R. 45° Elbow	**2.4** .3	**3.2** .3	**4.4** .3	**5.4** .3								
Tee	**7.4** .8	**9.5** .8	**12.2** .8	**14.8** .8								
Lateral	**13** 1.8	**19** 1.8										
Reducer	**2.2** .3	**2.9** .3	**3.7** .3	**4.7** .3								
Cap	**1.4** .5	**1.8** .5	**3.5** .5	**3.7** .5								

INSULATION		Temperature Range °F	100–199	200–299	300–399	400–499	500–599	600–699	700–799	800–899	900–999	1000–1099	1100–1200
85% Magnesia Calcium Silicate	Nom. Thick., In.		1	1	1½	2	2	2½	3	3	3	3½	3½
	Lbs/Ft		**1.25**	**1.25**	**2.08**	**3.01**	**3.01**	**4.07**	**5.24**	**5.24**	**5.24**	**6.65**	**6.65**
Combination	Nom. Thick., In.							2½	3	3	3	3½	3½
	Lbs/Ft							**5.07**	**6.94**	**6.94**	**6.94**	**9.17**	**9.17**
*Asbestos Fiber— Sodium Silicate	Nom. Thick., In.		1	1	1	1½	1½	2	2	3	3	3½	3½
	Lbs/Ft		**1.61**	**1.61**	**1.61**	**2.74**	**2.74**	**3.98**	**3.98**	**6.99**	**6.99**	**8.99**	**8.99**

	Pressure Rating psi	Cast Iron		Steel						
		125	250	150	300	400	600	900	1500	2500
FLANGES	Screwed or Slip-On	**9** 1.5	**17** 1.5	**9** 1.5	**17** 1.5	**20** 1.5	**20** 1.5	**37** 1.5	**61** 1.5	**102** 1.5
	Welding Neck			**11** 1.5	**19** 1.5	**27** 1.5	**27** 1.5	**38** 1.5	**61** 1.5	**113** 1.5
	Lap Joint			**9** 1.5	**17** 1.5	**19** 1.5	**19** 1.5	**36** 1.5	**60** 1.5	**99** 1.5
	Blind	**10** 1.5	**19** 1.5	**10** 1.5	**20** 1.5	**24** 1.5	**24** 1.5	**38** 1.5	**61** 1.5	**105** 1.5
FLANGED FITTINGS	S.R. 90° Elbow	**26** 3.9	**46** 4	**32** 3.9	**53** 4		**67** 4.1	**98** 4.3	**150** 4.6	
	L.R. 90° Elbow	**30** 4.3	**50** 4.3	**40** 4.3	**63** 4.3					
	45° Elbow	**22** 3.5	**41** 3.6	**28** 3.5	**46** 3.6		**60** 3.8	**93** 3.9	**135** 4	
	Tee	**39** 5.9	**67** 6	**52** 5.9	**81** 6		**102** 6.2	**151** 6.5	**238** 6.9	
VALVES	Flanged Bonnet Gate	**66** 7	**112** 7.4	**70** 4	**125** 4.4		**155** 4.8	**260** 5	**410** 5.5	
	Flanged Bonnet Globe or Angle	**56** 7.2	**121** 7.6	**60** 4.3	**95** 4.5		**155** 4.8	**225** 5	**495** 5.5	
	Flanged Bonnet Check	**46** 7.2	**100** 7.6	**60** 4.3	**70** 4.4		**120** 4.8	**150** 4.9	**440** 5.8	
	Pressure Seal Bonnet—Gate							**208** 3	**235** 3.2	
	Pressure Seal Bonnet—Globe							**135** 2.5	**180** 3	

Boldface type is weight in pounds. Lightface type beneath weight is weight factor for insulation.

Insulation thicknesses and weights are based on average conditions and do not constitute a recommendation for specific thicknesses of materials. Insulation weights are based on 85% magnesia and hydrous calcium silicate at 11 lbs/cubic foot. The listed thicknesses and weights of combination covering are the sums of the inner layer of diatomaceous earth at 21 lbs/cubic foot and the outer layer at 11 lbs/cubic foot.

Insulation weights include allowances for wire, cement, canvas, bands and paint, but not special surface finishes.

To find the weight of covering on flanges, valves or fittings, multiply the weight factor by the weight per foot of covering used on straight pipe.

Valve weights are approximate. When possible, obtain weights from the manufacturer.

Cast iron valve weights are for flanged end valves; steel weights are for welding end valves.

All flanged fitting, flanged valve and flange weights include the proportional weight of bolts or studs to make up all joints.

*16 lb cu. ft. density.

WEIGHTS OF PIPING MATERIALS   4.000″ O.D. 3½″ PIPE

PIPE		Schedule No.	40	80											
		Wall Designation	Std.	XS	XXS										
		Thickness—In.	.226	.318	.636										
		Pipe—Lbs/Ft	9.11	12.51	22.85										
		Water—Lbs/Ft	4.28	3.85	2.53										

WELDING FITTINGS															
	L.R. 90° Elbow	6.4 / .9	8.7 / .9	15.4 / .9											
	S.R. 90° Elbow	4.3 / .6	5.8 / .6												
	L.R. 45° Elbow	3.3 / .4	4.4 / .4	7.5 / .4											
	Tee	9.9 / .9	12.6 / .9	20 / .9											
	Lateral	17 / 1.8	26 / 1.8												
	Reducer	3.1 / .3	4.1 / .3	6.9 / .3											
	Cap	2.1 / .6	2.8 / .6	5.5 / .6											

INSULATION		Temperature Range °F	100–199	200–299	300–399	400–499	500–599	600–699	700–799	800–899	900–999	1000–1099	1100–1200
	85% Magnesia Calcium Silicate	Nom. Thick., In.	1	1	1½	2	2½	2½	3	3	3½	3½	3½
		Lbs/Ft	1.83	1.83	2.77	3.71	4.88	4.88	6.39	6.39	7.80	7.80	7.80
	Combination	Nom. Thick., In.						2½	3	3	3½	3½	3½
		Lbs/Ft						6.49	8.71	8.71	10.8	10.8	10.8
	*Asbestos Fiber—Sodium Silicate	Nom. Thick., In.	1	1	1	1½	1½	2	2	3	3	3½	3½
		Lbs/Ft	2.41	2.41	2.41	3.65	3.65	5.07	5.07	8.66	8.66	10.62	10.62

		Pressure Rating psi	Cast Iron		Steel						
			125	250	150	300	400	600	900	1500	2500
FLANGES	Screwed or Slip-On		13 / 1.5	21 / 1.5	13 / 1.5	21 / 1.5	27 / 1.5	27 / 1.5			
	Welding Neck				14 / 1.5	22 / 1.5	32 / 1.5	32 / 1.5			
	Lap Joint				13 / 1.5	21 / 1.5	26 / 1.5	26 / 1.5			
	Blind		14 / 1.5	23 / 1.5	15 / 1.5	25 / 1.5	35 / 1.5	35 / 1.5			
FLANGED FITTINGS	S.R. 90° Elbow		35 / 4	56 / 4.1	49 / 4			82 / 4.3			
	L.R. 90° Elbow		40 / 4.4	62 / 4.4	54 / 4.4						
	45° Elbow		31 / 3.6	51 / 3.7	39 / 3.6			75 / 3.9			
	Tee		54 / 6	86 / 6.2	70 / 6			133 / 6.4			
VALVES	Flanged Bonnet Gate		82 / 7.1	143 / 7.5	90 / 4.1	155 / 4.5		180 / 4.8	360 / 5	510 / 5.5	
	Flanged Bonnet Globe or Angle		74 / 7.3	137 / 7.7				160 / 4.7			
	Flanged Bonnet Check		71 / 7.3	125 / 7.7				125 / 4.7			
	Pressure Seal Bonnet—Gate							140 / 2.5	295 / 2.8	380 / 3	
	Pressure Seal Bonnet—Globe										

Boldface type is weight in pounds. Lightface type beneath weight is weight factor for insulation.

Insulation thicknesses and weights are based on average conditions and do not constitute a recommendation for specific thicknesses of materials. Insulation weights are based on 85% magnesia and hydrous calcium silicate at 11 lbs/cubic foot. The listed thicknesses and weights of combination covering are the sums of the inner layer of diatomaceous earth at 21 lbs/cubic foot and the outer layer at 11 lbs/cubic foot.

Insulation weights include allowances for wire, cement, canvas, bands and paint, but not special surface finishes.

To find the weight of covering on flanges, valves or fittings, multiply the weight factor by the weight per foot of covering used on straight pipe.

Valve weights are approximate. When possible, obtain weights from the manufacturer.

Cast iron valve weights are for flanged end valves; steel weights for welding end valves.

All flanged fitting, flanged valve and flange weights include the proportional weight of bolts or studs to make up all joints.

*16 lb cu. ft. density.

4″ PIPE 4.500″ O.D.   WEIGHTS OF PIPING MATERIALS

PIPE

	40	80	120	160	XXS
Schedule No.	40	80	120	160	
Wall Designation	Std.	XS			XXS
Thickness—In.	.237	.337	.438	.531	.674
Pipe—Lbs/Ft	**10.79**	**14.98**	**18.96**	**22.51**	**27.54**
Water—Lbs/Ft	**5.51**	**4.98**	**4.48**	**4.02**	**3.38**

WELDING FITTINGS

(Boldface = weight in lbs; lightface beneath = weight factor)

	40	80	120	160	XXS
L.R. 90° Elbow	8.7 / 1	11.9 / 1		17.6 / 1	21 / 1
S.R. 90° Elbow	5.8 / .7	7.9 / .7			
L.R. 45° Elbow	4.3 / .4	5.9 / .4		8.5 / .4	10.1 / .4
Tee	12.6 / 1	16.4 / 1		23 / 1	27 / 1
Lateral	21 / 2.1	33 / 2.1			
Reducer	3.6 / .3	4.9 / .3		6.6 / .3	8.2 / .3
Cap	2.6 / .6	3.4 / .6		6.5 / .6	6.7 / .6

INSULATION

Temperature Range °F	100–199	200–299	300–399	400–499	500–599	600–699	700–799	800–899	900–999	1000–1099	1100–1200
85% Magnesia Calcium Silicate — Nom. Thick., In.	1	1	1½	2	2½	2½	3	3	3½	3½	4
85% Magnesia Calcium Silicate — Lbs/Ft	1.62	1.62	2.55	3.61	4.66	4.66	6.07	6.07	7.48	7.48	9.10
Combination — Nom. Thick., In.						2½	3	3	3½	3½	3½
Combination — Lbs/Ft						6.07	8.30	8.30	10.6	10.6	10.6
*Asbestos Fiber—Sodium Silicate — Nom. Thick., In.	1	1	1	1½	1½	2	2	3	3	3½	3½
*Asbestos Fiber—Sodium Silicate — Lbs/Ft	2.04	2.04	2.04	3.28	3.28	4.70	4.70	8.29	8.29	10.25	10.25

FLANGES / FLANGED FITTINGS / VALVES

(Boldface = weight in lbs; lightface beneath = weight factor)

Pressure Rating psi	Cast Iron 125	Cast Iron 250	Steel 150	Steel 300	Steel 400	Steel 600	Steel 900	Steel 1500	Steel 2500
FLANGES									
Screwed or Slip-On	16 / 1.5	26 / 1.5	15 / 1.5	26 / 1.5	32 / 1.5	43 / 1.5	66 / 1.5	90 / 1.5	158 / 1.5
Welding Neck			17 / 1.5	29 / 1.5	41 / 1.5	48 / 1.5	64 / 1.5	90 / 1.5	177 / 1.5
Lap Joint			15 / 1.5	26 / 1.5	31 / 1.5	42 / 1.5	64 / 1.5	92 / 1.5	153 / 1.5
Blind	18 / 1.5	29 / 1.5	19 / 1.5	31 / 1.5	39 / 1.5	47 / 1.5	67 / 1.5	90 / 1.5	164 / 1.5
FLANGED FITTINGS									
S.R. 90° Elbow	45 / 4.1	72 / 4.2	59 / 4.1	85 / 4.2	99 / 4.3	128 / 4.4	185 / 4.5	254 / 4.8	
L.R. 90° Elbow	52 / 4.5	79 / 4.5	72 / 4.5	98 / 4.5					
45° Elbow	40 / 3.7	65 / 3.8	51 / 3.7	78 / 3.8	82 / 3.9	119 / 4	170 / 4.1	214 / 4.2	
Tee	70 / 6.1	109 / 6.3	86 / 6.1	121 / 6.3	153 / 6.4	187 / 6.6	262 / 6.8	386 / 7.2	
VALVES									
Flanged Bonnet Gate	109 / 7.2	188 / 7.5	100 / 4.2	175 / 4.5	195 / 5	255 / 5.1	455 / 5.4	735 / 6	
Flanged Bonnet Globe or Angle	97 / 7.4	177 / 7.8	95 / 4.3	145 / 4.8	215 / 5	230 / 5.1	415 / 5.5	800 / 6	
Flanged Bonnet Check	80 / 7.4	146 / 7.8	80 / 4.3	105 / 4.5	160 / 4.8	195 / 5	320 / 5.6	780 / 6	
Pressure Seal Bonnet—Gate						215 / 2.8	380 / 3	520 / 4	
Pressure Seal Bonnet—Globe							240 / 2.7	290 / 3	

Boldface type is weight in pounds. Lightface type beneath weight is weight factor for insulation.

Insulation thicknesses and weights are based on average conditions and do not constitute a recommendation for specific thicknesses of materials. Insulation weights are based on 85% magnesia and hydrous calcium silicate at 11 lbs/cubic foot. The listed thicknesses and weights of combination covering are the sums of the inner layer of diatomaceous earth at 21 lbs/cubic foot and the outer layer at 11 lbs/cubic foot.

Insulation weights include allowances for wire, cement, canvas, bands and paint, but not special surface finishes.

To find the weight of covering on flanges, valves or fittings, multiply the weight factor by the weight per foot of covering used on straight pipe.

Valve weights are approximate. When possible, obtain weights from the manufacturer.

Cast iron valve weights are for flanged end valves; steel weights are for welding end valves.

All flanged fitting, flanged valve and flange weights include the proportional weight of bolts or studs to make up all joints.

*16 lb cu. ft. density.

WEIGHTS OF PIPING MATERIALS   5.563″ O.D. 5″ PIPE

PIPE

Schedule No.	40	80	120	160	
Wall Designation	Std.	XS			XXS
Thickness—In.	.258	.375	.500	.625	.750
Pipe—Lbs/Ft	**14.62**	**20.78**	**27.04**	**32.96**	**38.55**
Water—Lbs/Ft	**8.66**	**7.89**	**7.09**	**6.33**	**5.62**

WELDING FITTINGS (boldface = weight lbs, lightface = weight factor)

Fitting	40	80	120	160	XXS
L.R. 90° Elbow	**14.7** / 1.3	**21** / 1.3		**32** / 1.3	**37** / 1.3
S.R. 90° Elbow	**9.8** / .8	**13.7** / .8			
L.R. 45° Elbow	**7.3** / .5	**10.2** / .5		**15.6** / .5	**17.7** / .5
Tee	**19.8** / 1.2	**26** / 1.2		**39** / 1.2	**43** / 1.2
Lateral	**31** / 2.5	**50** / 2.5			
Reducer	**6** / .4	**8.3** / .4		**12.4** / .4	**14.2** / .4
Cap	**4.2** / .7	**5.7** / .7		**11** / .7	**11** / .7

INSULATION

Temperature Range °F	100–199	200–299	300–399	400–499	500–599	600–699	700–799	800–899	900–999	1000–1099	1100–1200
*Asbestos Fiber—Sodium Silicate — Nom. Thick., In.	1	1½	1½	2	2½	2½	3	3½	3½	4	4
Lbs/Ft	**1.86**	**2.92**	**2.92**	**4.08**	**5.38**	**5.38**	**6.90**	**8.41**	**8.41**	**10.4**	**10.4**
Combination — Nom. Thick., In.						2½	3	3½	3½	4	4
Lbs/Ft						**7.01**	**9.30**	**11.8**	**11.8**	**14.9**	**14.9**
85% Magnesia Calcium Silicate — Nom. Thick., In.	1	1	1	1½	1½	2½	2½	3	3	4	4
Lbs/Ft	**2.34**	**2.34**	**2.34**	**3.76**	**3.76**	**7.35**	**7.35**	**9.31**	**9.31**	**14.37**	**14.37**

FLANGES

Pressure Rating psi	Cast Iron 125	Cast Iron 250	Steel 150	300	400	600	900	1500	2500
Screwed or Slip-On	**20** / 1.5	**32** / 1.5	**18** / 1.5	**32** / 1.5	**37** / 1.5	**73** / 1.5	**100** / 1.5	**162** / 1.5	**259** / 1.5
Welding Neck			**22** / 1.5	**36** / 1.5	**49** / 1.5	**78** / 1.5	**103** / 1.5	**162** / 1.5	**293** / 1.5
Lap Joint			**18** / 1.5	**32** / 1.5	**35** / 1.5	**71** / 1.5	**98** / 1.5	**168** / 1.5	**253** / 1.5
Blind	**23** / 1.5	**37** / 1.5	**23** / 1.5	**39** / 1.5	**50** / 1.5	**78** / 1.5	**104** / 1.5	**172** / 1.5	**272** / 1.5

FLANGED FITTINGS

Fitting	125	250	150	300	400	600	900	1500
S.R. 90° Elbow	**58** / 4.3	**94** / 4.3	**80** / 4.3	**113** / 4.3	**123** / 4.5	**205** / 4.7	**268** / 4.8	**435** / 5.2
L.R. 90° Elbow	**68** / 4.7	**105** / 4.7	**91** / 4.7	**128** / 4.7				
45° Elbow	**51** / 3.8	**83** / 3.8	**66** / 3.8	**98** / 3.8	**123** / 4	**180** / 4.2	**239** / 4.3	**350** / 4.5
Tee	**90** / 6.4	**145** / 6.5	**119** / 6.4	**172** / 6.4	**179** / 6.8	**304** / 7	**415** / 7.2	**665** / 7.8

VALVES

Valve	125	250	150	300	400	600	900	1500
Flanged Bonnet Gate	**138** / 7.3	**264** / 7.9	**150** / 4.3	**265** / 4.9	**310** / 5.3	**455** / 5.5	**615** / 6	**1340** / 7
Flanged Bonnet Globe or Angle	**138** / 7.6	**247** / 8	**155** / 4.3	**215** / 5	**355** / 5.2	**515** / 5.8	**555** / 5.8	**950** / 6
Flanged Bonnet Check	**118** / 7.6	**210** / 8	**110** / 4.3	**165** / 5	**185** / 5	**350** / 5.8	**560** / 6	**1150** / 7
Pressure Seal Bonnet—Gate						**350** / 3.1	**520** / 3.8	**865** / 4.5
Pressure Seal Bonnet—Globe							**280** / 4	**450** / 4.5

Boldface type is weight in pounds. Lightface type beneath weight is weight factor for insulation.

Insulation thicknesses and weights are based on average conditions and do not constitute a recommendation for specific thicknesses of materials. Insulation weights are based on 85% magnesia and hydrous calcium silicate at 11 lbs/cubic foot. The listed thicknesses and weights of combination covering are the sums of the inner layer of diatomaceous earth at 21 lbs/cubic foot and the outer layer at 11 lbs/cubic foot.

Insulation weights include allowances for wire, cement, canvas, bands and paint, but not special surface finishes.

To find the weight of covering on flanges, valves or fittings, multiply the weight factor by the weight per foot of covering used on straight pipe.

Valve weights are approximate. When possible, obtain weights from the manufacturer.

Cast iron valve weights are for flanged end valves; steel weights for welding end valves.

All flanged fitting, flanged valve and flange weights include the proportional weight of bolts or studs to make up all joints.

*16 lb cu. ft. density.

6" PIPE 6.625" O.D.   WEIGHTS OF PIPING MATERIALS

PIPE

Schedule No.	40	80	120	160		
Wall Designation	Std.	XS			XXS	
Thickness—In.	.280	.432	.562	.718	.864	
Pipe—Lbs/Ft	**18.97**	**28.57**	**36.39**	**45.3**	**53.2**	
Water—Lbs/Ft	**12.51**	**11.29**	**10.30**	**9.2**	**8.2**	

WELDING FITTINGS

	40	80	120	160	
L.R. 90° Elbow	**23** 1.5	**34** 1.5		**53** 1.5	**62** 1.5
S.R. 90° Elbow	**15.2** 1	**23** 1			
L.R. 45° Elbow	**11.3** .6	**16.7** .6		**26** .6	**30** .6
Tee	**29.3** 1.4	**42** 1.4		**60** 1.4	**68** 1.4
Lateral	**42** 2.9	**79** 2.9			
Reducer	**8.7** .5	**12.6** .5		**18.8** .5	**21** .5
Cap	**6.4** .9	**9.2** .9		**17.5** .9	**17.5** .9

INSULATION

Temperature Range °F		100–199	200–299	300–399	400–499	500–599	600–699	700–799	800–899	900–999	1000–1099	1100–1200
85% Magnesia Calcium Silicate	Nom. Thick., In.	1	1½	2	2	2½	3	3	3½	3½	4	4
	Lbs/Ft	**2.11**	**3.28**	**4.57**	**4.57**	**6.09**	**7.60**	**7.60**	**9.82**	**98.2**	**11.5**	**11.4**
Combination	Nom. Thick., In.						3	3	3½	3½	4	4
	Lbs/Ft						**10.3**	**10.3**	**13.4**	**13.4**	**16.6**	**16.6**
*Asbestos Fiber—Sodium Silicate	Nom. Thick., In.	1	1	1	1½	1½	2½	2½	3½	3½	4	4
	Lbs/Ft	**2.57**	**2.57**	**2.57**	**4.18**	**4.18**	**8.10**	**8.10**	**13.31**	**13.31**	**15.85**	**15.85**

FLANGES / FLANGED FITTINGS / VALVES

Pressure Rating psi	Cast Iron		Steel						
	125	250	150	300	400	600	900	1500	2500
FLANGES									
Screwed or Slip-On	**25** 1.5	**42** 1.5	**22** 1.5	**45** 1.5	**54** 1.5	**95** 1.5	**128** 1.5	**202** 1.5	**396** 1.5
Welding Neck			**27** 1.5	**48** 1.5	**67** 1.5	**96** 1.5	**130** 1.5	**202** 1.5	**451** 1.5
Lap Joint			**22** 1.5	**45** 1.5	**52** 1.5	**93** 1.5	**125** 1.5	**208** 1.5	**387** 1.5
Blind	**28** 1.5	**51** 1.5	**29** 1.5	**56** 1.5	**71** 1.5	**101** 1.5	**133** 1.5	**197** 1.5	**418** 1.5
FLANGED FITTINGS									
S.R. 90° Elbow	**74** 4.3	**125** 4.4	**90** 4.3	**147** 4.4	**184** 4.6	**275** 4.8	**375** 5	**566** 5.3	
L.R. 90° Elbow	**91** 4.9	**145** 4.9	**126** 4.9	**182** 4.9					
45° Elbow	**66** 3.8	**115** 3.9	**82** 3.8	**132** 3.9	**149** 4.1	**240** 4.3	**320** 4.3	**476** 4.6	
Tee	**114** 6.5	**195** 6.6	**149** 6.5	**217** 6.6	**279** 6.9	**400** 7.2	**565** 7.5	**839** 8	
VALVES									
Flanged Bonnet Gate	**172** 7.3	**359** 8	**190** 4.3	**360** 5	**435** 5.5	**620** 5.8	**835** 6	**1595** 7	
Flanged Bonnet Globe or Angle	**184** 7.8	**345** 8.2	**185** 4.4	**275** 5	**415** 5.3	**645** 5.8	**765** 6	**1800** 7	
Flanged Bonnet Check	**154** 7.8	**286** 8.2	**150** 4.8	**200** 5	**360** 5.4	**445** 6	**800** 6.4	**1630** 7	
Pressure Seal Bonnet—Gate							**580** 3.5	**750** 4	**1215** 5
Pressure Seal Bonnet—Globe								**730** 4	**780** 5

Boldface type is weight in pounds. Lightface type beneath weight is weight factor for insulation.

Insulation thicknesses and weights are based on average conditions and do not constitute a recommendation for specific thicknesses of materials. Insulation weights are based on 85% magnesia and hydrous calcium silicate at 11 lbs/cubic foot. The listed thicknesses and weights of combination covering are the sums of the inner layer of diatomaceous earth at 21 lbs/cubic foot and the outer layer at 11 lbs/cubic foot.

Insulation weights include allowances for wire, cement, canvas, bands and paint, but not special surface finishes.

To find the weight of covering on flanges, valves or fittings, multiply the weight factor by the weight per foot of covering used on straight pipe.

Valve weights are approximate. When possible, obtain weights from the manufacturer.

Cast iron valve weights are for flanged end valves; steel weights are for welding end valves.

All flanged fitting, flanged valve and flange weights include the proportional weight of bolts or studs to make up all joints.

*16 lb cu. ft. density.

WEIGHTS OF PIPING MATERIALS   8.625″ O.D. 8″ PIPE

PIPE	Schedule No.	20	30	40	60	80	100	120	140		160		
	Wall Designation			Std.		XS					XXS		
	Thickness—In.	.250	.277	.322	.406	.500	.593	.718	.812	.875	.906		
	Pipe—Lbs/Ft	22.36	24.70	28.55	35.64	43.4	50.9	60.6	67.8	72.4	74.7		
	Water—Lbs/Ft	22.48	22.18	21.69	20.79	19.8	18.8	17.6	16.7	16.1	15.8		

WELDING FITTINGS

		40	80		160	
L.R. 90° Elbow		**46** 2	**69** 2		**114** 2	**117** 2
S.R. 90° Elbow		**31** 1.3	**46** 1.3			
L.R. 45° Elbow		**23** .8	**34** .8		**55** .8	**56** .8
Tee		**54** 1.8	**76** 1.8		**118** 1.8	**120** 1.8
Lateral		**76** 3.8	**140** 3.8			
Reducer		**13.9** .5	**20** .5		**32** .5	**33** .5
Cap		**11.3** 1	**16.3** 1		**31** 1	**32** 1

INSULATION

	Temperature Range °F	100–199	200–299	300–399	400–499	500–599	600–699	700–799	800–899	900–999	1000–1099	1100–1200
85% Magnesia Calcium Silicate	Nom. Thick., In.	$1\frac{1}{2}$	$1\frac{1}{2}$	2	2	$2\frac{1}{2}$	3	$3\frac{1}{2}$	$3\frac{1}{2}$	4	4	$4\frac{1}{2}$
	Lbs/Ft	4.13	4.13	5.64	5.64	7.85	9.48	11.5	11.5	13.8	13.8	16.0
Combination	Nom. Thick., In.						3	$3\frac{1}{2}$	$3\frac{1}{2}$	4	4	$4\frac{1}{2}$
	Lbs/Ft						12.9	16.2	16.2	20.4	20.4	23.8
*Asbestos Fiber— Sodium Silicate	Nom. Thick., In.	$1\frac{1}{2}$	$1\frac{1}{2}$	$1\frac{1}{2}$	$1\frac{1}{2}$	$1\frac{1}{2}$	$2\frac{1}{2}$	$2\frac{1}{2}$	$3\frac{1}{2}$	$3\frac{1}{2}$	$4\frac{1}{2}$	$4\frac{1}{2}$
	Lbs/Ft	5.38	5.38	5.38	5.38	5.38	10.60	10.60	15.85	15.85	20.85	20.85

FLANGES

	Pressure Rating psi	Cast Iron		Steel						
		125	250	150	300	400	600	900	1500	2500
Screwed or Slip-On		**34** 1.5	**64** 1.5	**33** 1.5	**67** 1.5	**82** 1.5	**135** 1.5	**207** 1.5	**319** 1.5	**601** 1.5
Welding Neck				**42** 1.5	**76** 1.5	**104** 1.5	**137** 1.5	**222** 1.5	**334** 1.5	**692** 1.5
Lap Joint				**33** 1.5	**67** 1.5	**79** 1.5	**132** 1.5	**223** 1.5	**347** 1.5	**587** 1.5
Blind		**45** 1.5	**83** 1.5	**48** 1.5	**90** 1.5	**115** 1.5	**159** 1.5	**232** 1.5	**363** 1.5	**649** 1.5

FLANGED FITTINGS

		125	250	150	300	400	600	900	1500	2500
S.R. 90° Elbow		**117** 4.5	**201** 4.7	**157** 4.5	**238** 4.7	**310** 5	**435** 5.2	**639** 5.4	**995** 5.7	
L.R. 90° Elbow		**152** 5.3	**236** 5.3	**202** 5.3	**283** 5.3					
45° Elbow		**101** 3.9	**171** 4	**127** 3.9	**203** 4	**215** 4.1	**360** 4.4	**507** 4.5	**870** 4.8	
Tee		**175** 6.8	**304** 7.1	**230** 6.8	**337** 7.1	**445** 7.5	**610** 7.8	**978** 8.1	**1465** 8.6	

VALVES

		125	250	150	300	400	600	900	1500	2500
Flanged Bonnet Gate		**251** 7.5	**583** 8.1	**305** 4.5	**505** 5.1	**730** 6	**960** 6.3	**1180** 6.6	**2740** 7	
Flanged Bonnet Globe or Angle		**317** 8.4	**554** 8.6	**475** 5.4	**505** 5.5	**610** 5.9	**1130** 6.3	**1160** 6.3	**2865** 7	
Flanged Bonnet Check		**302** 8.4	**454** 8.6	**235** 5.2	**310** 5.3	**475** 5.6	**725** 6	**1140** 6.4	**2075** 7	
Pressure Seal Bonnet—Gate							**925** 4.5	**1185** 4.7	**2345** 5.5	
Pressure Seal Bonnet—Globe								**1550** 4	**1680** 5	

Boldface type is weight in pounds. Lightface type beneath weight is weight factor for insulation.

Insulation thicknesses and weights are based on average conditions and do not constitute a recommendation for specific thicknesses of materials. Insulation weights are based on 85% magnesia and hydrous calcium silicate at 11 lbs/cubic foot. The listed thicknesses and weights of combination covering are the sums of the inner layer of diatomaceous earth at 21 lbs/cubic foot and the outer layer at 11 lbs/cubic foot.

Insulation weights include allowances for wire, cement, canvas, bands and paint, but not special surface finishes.

To find the weight of covering on flanges, valves or fittings, multiply the weight factor by the weight per foot of covering used on straight pipe.

Valve weights are approximate. When possible, obtain weights from the manufacturer.

Cast iron valve weights are for flanged end valves; steel weights are for welding end valves.

All flanged fitting, flanged valve and flange weights include the proportional weight of bolts or studs to make up all joints.

*16 lb cu. ft. density.

10" PIPE 10.750" O.D. WEIGHTS OF PIPING MATERIALS

PIPE	Schedule No.	20	30	40	60	80	100	120	140	160			
	Wall Designation			Std.	XS								
	Thickness—In.	.250	.307	.365	.500	.593	.718	.843	1.000	1.125			
	Pipe—Lbs/Ft	28.04	34.24	40.5	54.7	64.3	76.9	89.2	104.1	115.7			
	Water—Lbs/Ft	33.77	34.98	34.1	32.3	31.1	29.5	28.0	26.1	24.6			

WELDING FITTINGS

	20	30	40	60	80	100	120	140	160			
L.R. 90° Elbow			82 / 2.5	109 / 2.5					226 / 2.5			
S.R. 90° Elbow			54 / 1.7	73 / 1.7								
L.R. 45° Elbow			40 / 1	54 / 1					109 / 1			
Tee			91 / 2.1	118 / 2.1					222 / 2.1			
Lateral			124 / 4.4	202 / 4.4								
Reducer			23 / .6	31 / .6					58 / .6			
Cap			20 / 1.3	26 / 1.3					54 / 1.3			

INSULATION

	Temperature Range °F	100–199	200–299	300–399	400–499	500–599	600–699	700–799	800–899	900–999	1000–1099	1100–1200
85% Magnesia Calcium Silicate	Nom. Thick., In.	1½	1½	2	2½	2½	3	3½	3½	4	4	4½
	Lbs/Ft	5.20	5.20	7.07	8.93	8.93	11.0	13.2	13.2	15.5	15.5	18.1
Combination	Nom. Thick., In.						3	31/2	31/2	4	4	41/2
	Lbs/Ft						15.4	19.3	19.3	23	23	27.2
*Asbestos Fiber—Sodium Silicate	Nom. Thick., In.	1½	1½	1½	1½	1½	2½	2½	3½	3½	4½	4½
	Lbs/Ft	6.77	6.77	6.77	6.77	6.77	12.03	12.03	18.39	18.39	25.21	25.21

FLANGES

	Pressure Rating psi	Cast Iron		Steel						
		125	250	150	300	400	600	900	1500	2500
Screwed or Slip-On		53	97	51	100	117	213	293	528	1148
		1.5	1.5	1.5	1.5	1.5	1.5	1.5	1.5	1.5
Welding Neck				60	110	152	225	316	546	1291
				1.5	1.5	1.5	1.5	1.5	1.5	1.5
Lap Joint				51	110	138	231	325	577	1120
				1.5	1.5	1.5	1.5	1.5	1.5	1.5
Blind		71	136	78	146	181	267	338	599	1248
		1.5	1.5	1.5	1.5	1.5	1.5	1.5	1.5	1.5

FLANGED FITTINGS

	125	250	150	300	400	600	900	1500	2500
S.R. 90° Elbow	190	323	240	343	462	747	995		
	4.8	4.9	4.8	4.9	5.2	5.6	5.8		
L.R. 90° Elbow	245	383	290	438					
	5.8	5.8	5.8	5.8					
45° Elbow	160	273	185	288	332	572	732		
	4.1	4.2	4.1	4.2	4.3	4.6	4.7		
Tee	293	479	353	527	578	1007	1417		
	7.2	7.4	7.2	7.4	7.8	8.4	8.7		

VALVES

	125	250	150	300	400	600	900	1500	2500
Flanged Bonnet Gate	471	899	455	750	1035	1575	2140	3690	
	7.7	8.3	4.5	5	6	6.9	7.1	8	
Flanged Bonnet Globe or Angle	541	943	485	855	1070	1500	2500	4160	
	9.1	9.1	4.5	5.5	6	6.3	6.8	8	
Flanged Bonnet Check	453	751	370	485	605	1030	1350	2280	
	9.1	9.1	6	6.1	6.3	6.8	7	7.5	
Pressure Seal Bonnet—Gate							1450	1860	3150
							4.9	5.5	6
Pressure Seal Bonnet—Globe								1800	1910
								5	6

Boldface type is weight in pounds. Lightface type beneath weight is weight factor for insulation.

Insulation thicknesses and weights are based on average conditions and do not constitute a recommendation for specific thicknesses of materials. Insulation weights are based on 85% magnesia and hydrous calcium silicate at 11 lbs/cubic foot. The listed thicknesses and weights of combination covering are the sums of the inner layer of diatomaceous earth at 21 lbs/cubic foot and the outer layer at 11 lbs/cubic foot.

Insulation weights include allowances for wire, cement, canvas, bands and paint, but not special surface finishes.

To find the weight of covering on flanges, valves or fittings, multiply the weight factor by the weight per foot of covering used on straight pipe.

Valve weights are approximate. When possible, obtain weights from the manufacturer.

Cast iron valve weights are for flanged end valves; steel weights for welding end valves.

All flanged fitting, flanged valve and flange weights include the proportional weight of bolts or studs to make up all joints.

*16 lb cu. ft. density.

WEIGHTS OF PIPING MATERIALS   12.750″ O.D. 12″ PIPE

PIPE	Schedule No.	20	30		40		60	80	100	120	140	160	
	Wall Designation			Std.		XS							
	Thickness—In.	.250	.330	.375	.406	.500	.562	.687	.843	1.000	1.125	1.312	
	Pipe—Lbs/Ft	**33.38**	**43.8**	**49.6**	**53.5**	**65.4**	**73.2**	**88.5**	**107.2**	**125.5**	**139.7**	**160.3**	
	Water—Lbs/Ft	**51.10**	**49.7**	**49.0**	**48.5**	**47.0**	**46.0**	**44.0**	**41.6**	**39.3**	**37.5**	**34.9**	

WELDING FITTINGS	L.R. 90° Elbow			**119** 3		**157** 3						**375** 3	
	S.R. 90° Elbow			**80** 2		**104** 2							
	L.R. 45° Elbow			**60** 1.3		**78** 1.3						**181** 1.3	
	Tee			**132** 2.5		**167** 2.5						**360** 2.5	
	Lateral			**180** 5.4		**273** 5.4							
	Reducer			**33** .7		**44** .7						**94** .7	
	Cap			**30** 1.5		**38** 1.5						**89** 1.5	

		100–199	200–299	300–399	400–499	500–599	600–699	700–799	800–899	900–999	1000–1099	1100–1200
INSULATION	Temperature Range °F											
85% Magnesia Calcium Silicate	Nom. Thick., In.	1½	1½	2	2½	3	3	3½	4	4	4½	4½
	Lbs/Ft	**6.04**	**6.04**	**8.13**	**10.5**	**12.7**	**12.7**	**15.1**	**17.9**	**17.9**	**20.4**	**20.4**
Combination	Nom. Thick., In.						3	3½	4	4	4½	4½
	Lbs/Ft						**17.7**	**21.9**	**26.7**	**26.7**	**31.1**	**31.1**
*Asbestos Fiber—Sodium Silicate	Nom. Thick., In.	1½	1½	1½	1½	1½	2½	2½	4	4	5	5
	Lbs/Ft	**5.22**	**5.22**	**5.22**	**5.22**	**5.22**	**14.20**	**14.20**	**24.64**	**24.64**	**32.40**	**32.40**

		Cast Iron		Steel						
	Pressure Rating psi	125	250	150	300	400	600	900	1500	2500
FLANGES	Screwed or Slip-On	**71** 1.5	**137** 1.5	**72** 1.5	**140** 1.5	**164** 1.5	**261** 1.5	**388** 1.5	**820** 1.5	**1611** 1.5
	Welding Neck			**88** 1.5	**163** 1.5	**212** 1.5	**272** 1.5	**434** 1.5	**843** 1.5	**1919** 1.5
	Lap Joint			**72** 1.5	**164** 1.5	**187** 1.5	**286** 1.5	**433** 1.5	**902** 1.5	**1573** 1.5
	Blind	**96** 1.5	**177** 1.5	**118** 1.5	**209** 1.5	**261** 1.5	**341** 1.5	**475** 1.5	**928** 1.5	**1775** 1.5
FLANGED FITTINGS	S.R. 90° Elbow	**265** 5	**453** 5.2	**345** 5	**509** 5.2	**669** 5.5	**815** 5.8	**1474** 6.2		
	L.R. 90° Elbow	**375** 6.2	**553** 6.2	**485** 6.2	**624** 6.2			**1598** 6.2		
	45° Elbow	**235** 4.3	**383** 4.3	**282** 4.3	**414** 4.3	**469** 4.5	**705** 4.7	**1124** 4.8		
	Tee	**403** 7.5	**684** 7.8	**513** 7.5	**754** 7.8	**943** 8.3	**1361** 8.7	**1928** 9.3		
VALVES	Flanged Bonnet Gate	**687** 7.8	**1298** 8.5	**635** 4	**1015** 5	**1420** 5.5	**2155** 7	**2770** 7.2	**4650** 8	
	Flanged Bonnet Globe or Angle	**808** 9.4	**1200** 9.5	**710** 5	**1410** 5.5					
	Flanged Bonnet Check	**674** 9.4	**1160** 9.5	**560** 6	**720** 6.5		**1410** 7.2	**2600** 8	**3370** 8	
	Pressure Seal Bonnet—Gate						**1975** 5.5	**2560** 6	**4515** 7	
	Pressure Seal Bonnet—Globe									

Boldface type is weight in pounds. Lightface type is weight factor for insulation.

Insulation thicknesses and weights are based on average conditions and do not constitute a recommendation for specific thicknesses of materials. Insulation weights are based on 85% magnesia and hydrous calcium silicate at 11 lbs/cubic foot. The listed thicknesses and weights of combination covering are the sums of the inner layer of diatomaceous earth at 21 lbs/cubic foot and the outer layer at 11 lbs/cubic foot.

Insulation weights include allowances for wire, cement, canvas, bands and paint, but not special surface finishes.

To find the weight of covering on flanges, valves or fittings, multiply the weight factor by the weight per foot of covering used on straight pipe.

Valve weights are approximate. When possible, obtain weights from the manufacturer.

Cast iron valve weights are for flanged end valves; steel weights for welding end valves.

All flanged fitting, flanged valve and flange weights include the proportional weight of bolts or studs to make up all joints.

*16 lb cu. ft. density.

14" PIPE 14" O.D.   WEIGHTS OF PIPING MATERIALS

		Schedule No.	10	20	30	40		60	80	100	120	140	160
PIPE		Wall Designation			Std.		XS						
		Thickness—In.	.250	.312	.375	.438	.500	.593	.750	.937	1.093	1.250	1.406
		Pipe—Lbs/Ft	**36.71**	**45.7**	**54.6**	**63.4**	**72.1**	**84.9**	**106.1**	**130.7**	**150.7**	**170.2**	**189.1**
		Water—Lbs/Ft	62.06	60.92	59.7	58.7	57.5	55.9	53.2	50.0	47.5	45.0	42.6

		10	20	30	40	XS	60
WELDING FITTINGS	L.R. 90° Elbow			**154** 3.5		**202** 3.5	
	S.R. 90° Elbow			**102** 2.3		**135** 2.3	
	L.R. 45° Elbow			**77** 1.5		**100** 1.5	
	Tee			**159** 2.8		**203** 2.8	
	Lateral			**218** 5.8		**340** 5.8	
	Reducer			**63** 1.1		**83** 1.1	
	Cap			**35** 1.7		**46** 1.7	

		Temperature Range °F	100–199	200–299	300–399	400–499	500–599	600–699	700–799	800–899	900–999	1000–1099	1100–1200
INSULATION	85% Magnesia Calcium Silicate	Nom Thick., In.	1½	1½	2	2½	3	3	3½	4	4	4½	4½
		Lbs/Ft	**6.16**	**6.16**	**8.38**	**10.7**	**13.1**	**13.1**	**15.8**	**18.5**	**18.5**	**21.3**	**21.3**
	Combination	Nom. Thick., In.						3	3½	4	4	4½	4½
		Lbs/Ft						**18.2**	**22.8**	**27.5**	**27.5**	**32.4**	**32.4**
	*Asbestos Fiber—Sodium Silicate	Nom. Thick., In.	1½	1½	1½	2	2	3	3	4	4	5	5
		Lbs/Ft	**7.90**	**7.90**	**7.90**	**11.18**	**11.18**	**18.00**	**18.00**	**25.42**	**25.42**	**33.53**	**33.53**

		Pressure Rating psi	Cast Iron		Steel						
			125	250	150	300	400	600	900	1500	2500
FLANGES	Screwed or Slip-On		**93** 1.5	**184** 1.5	**96** 1.5	**195** 1.5	**235** 1.5	**318** 1.5	**460** 1.5	**1016** 1.5	
	Welding Neck				**113** 1.5	**217** 1.5	**277** 1.5	**406** 1.5	**642** 1.5	**1241** 1.5	
	Lap Joint				**110** 1.5	**220** 1.5	**254** 1.5	**349** 1.5	**477** 1.5	**1076** 1.5	
	Blind		**126** 1.5	**239** 1.5	**142** 1.5	**267** 1.5	**354** 1.5	**437** 1.5	**574** 1.5		
FLANGED FITTINGS	S.R. 90° Elbow		**372** 5.3	**617** 5.5	**497** 5.3	**632** 5.5	**664** 5.7	**918** 5.9	**1549** 6.4		
	L.R. 90° Elbow		**492** 6.6	**767** 6.6	**622** 6.6	**772** 6.6					
	45° Elbow		**292** 4.3	**497** 4.4	**377** 4.3	**587** 4.4	**638** 4.6	**883** 4.8	**1246** 4.9		
	Tee		**563** 8	**956** 8.4	**683** 8	**968** 8.3	**1131** 8.6	**1652** 8.9	**2318** 9.6		
VALVES	Flanged Bonnet Gate		**921** 7.9	**1762** 8.8	**905** 4.9	**1525** 6	**1920** 6.3	**2960** 7	**4170** 8	**6425** 8.8	
	Flanged Bonnet Globe or Angle		**1171** 9.9								
	Flanged Bonnet Check		**885** 9.9		**1010** 5	**1155** 5.2					
	Pressure Seal Bonnet—Gate							**2620** 6	**3475** 6.5	**6380** 7.5	
	Pressure Seal Bonnet—Globe										

Boldface type is weight in pounds. Lightface type beneath weight is weight factor for insulation.

Insulation thicknesses and weights are based on average conditions and do not constitute a recommendation for specific thicknesses of materials. Insulation weights are based on 85% magnesia and hydrous calcium silicate at 11 lbs/cubic foot. The listed thicknesses and weights of combination covering are the sums of the inner layer of diatomaceous earth at 21 lbs/cubic foot and the outer layer at 11 lbs/cubic foot.

Insulation weights include allowances for wire, cement, canvas, bands and paint, but not special surface finishes.

To find the weight of covering on flanges, valves or fittings, multiply the weight factor by the weight per foot of covering used on straight pipe.

Valve weights are approximate. When possible, obtain weights from the manufacturer.

Cast iron valve weights are for flanged end valves; steel weights for welding end valves.

All flanged fitting, flanged valve and flange weights include the proportional weight of bolts or studs to make up all joints.

*16 lb cu. ft. density

WEIGHTS OF PIPING MATERIALS 16″ O.D. 16″ PIPE

	Schedule No.	10	20	30	40	60	80	100	120	140	160		
PIPE	Wall Designation			Std.	XS								
	Thickness—In.	.250	.312	.375	.500	.656	.843	1.031	1.218	1.438	1.593		
	Pipe—Lbs/Ft	**42.1**	**52.4**	**62.6**	**82.8**	**107.5**	**136.5**	**164.8**	**192.3**	**223.6**	**245.1**		
	Water—Lbs/Ft	**81.8**	**80.5**	**79.1**	**76.5**	**73.4**	**69.7**	**66.1**	**62.6**	**58.6**	**55.9**		

WELDING FITTINGS	L.R. 90° Elbow			**201** 4	**265** 4	
	S.R. 90° Elbow			**135** 2.5	**177** 2.5	
	L.R. 45° Elbow			**100** 1.7	**132** 1.7	
	Tee			**202** 3.2	**257** 3.2	
	Lateral			**275** 6.7	**433** 6.7	
	Reducer			**78** 1.2	**102** 1.2	
	Cap			**44** 1.8	**58** 1.8	

	Temperature Range °F	100–199	200–299	300–399	400–499	500–599	600–699	700–799	800–899	900–999	1000–1099	1100–1200
INSULATION	85% Magnesia Calcium Silicate — Nom. Thick., In.	$1\frac{1}{2}$	$1\frac{1}{2}$	2	$2\frac{1}{2}$	3	3	$3\frac{1}{2}$	4	4	$4\frac{1}{2}$	$4\frac{1}{2}$
	Lbs/Ft	**6.90**	**6.90**	**9.33**	**12.0**	**14.6**	**14.6**	**17.5**	**20.5**	**20.5**	**23.6**	**23.6**
	Combination — Nom. Thick., In.						3	$3\frac{1}{2}$	4	4	$4\frac{1}{2}$	$4\frac{1}{2}$
	Lbs/Ft						**20.3**	**25.2**	**30.7**	**30.7**	**36.0**	**36.0**
	*Asbestos Fiber—Sodium Silicate — Nom. Thick., In.	$1\frac{1}{2}$	$1\frac{1}{2}$	$1\frac{1}{2}$	$1\frac{1}{2}$	$1\frac{1}{2}$	$2\frac{1}{2}$	$2\frac{1}{2}$	$3\frac{1}{2}$	$3\frac{1}{2}$	$4\frac{1}{2}$	$4\frac{1}{2}$
	Lbs/Ft	**9.26**	**9.26**	**9.26**	**9.26**	**9.26**	**16.35**	**16.35**	**24.11**	**24.11**	**32.57**	**32.57**

	Pressure Rating psi	Cast Iron		Steel						
		125	250	150	300	400	600	900	1500	2500
FLANGES	Screwed or Slip-On	**120** 1.5	**233** 1.5	**108** 1.5	**262** 1.5	**310** 1.5	**442** 1.5	**559** 1.5	**1297** 1.5	
	Welding Neck			**142** 1.5	**288** 1.5	**351** 1.5	**577** 1.5	**785** 1.5	**1597** 1.5	
	Lap Joint			**143** 1.5	**282** 1.5	**337** 1.5	**476** 1.5	**588** 1.5	**1372** 1.5	
	Blind	**175** 1.5	**308** 1.5	**185** 1.5	**349** 1.5	**455** 1.5	**603** 1.5	**719** 1.5		
FLANGED FITTINGS	S.R. 90° Elbow	**501** 5.5	**826** 5.8	**656** 5.5	**958** 5.8	**1014** 6	**1402** 6.3	**1886** 6.7		
	L.R. 90° Elbow	**701** 7	**1036** 7	**781** 7	**1058** 7					
	45° Elbow	**391** 4.3	**696** 4.6	**481** 4.3	**708** 4.6	**839** 4.7	**1212** 5	**1586** 5		
	Tee	**746** 8.3	**1263** 8.7	**961** 8.3	**1404** 8.6	**1671** 9	**2128** 9.4	**3054** 10		
VALVES	Flanged Bonnet Gate	**1254** 8	**2321** 9	**1190** 5	**2015** 7	**2300** 7.2	**3675** 7.9	**4950** 8.2	**7875** 9	
	Flanged Bonnet Globe or Angle									
	Flanged Bonnet Check	**1166** 10.5			**1225** 6					
	Pressure Seal Bonnet—Gate							**3230** 7		**8130** 8
	Pressure Seal Bonnet—Globe									

Boldface type is weight in pounds. Lightface type beneath weight is weight factor for insulation.

Insulation thicknesses and weights are based on average conditions and do not constitute a recommendation for specific thicknesses of materials. Insulation weights are based on 85% magnesia and hydrous calcium silicate at 11 lbs/cubic foot. The listed thicknesses and weights of combination covering are the sums of the inner layer of diatomaceous earth at 21 lbs/cubic foot and the outer layer at 11 lbs/cubic foot.

Insulation weights include allowances for wire, cement, canvas, bands and paint, but not special surface finishes.

To find the weight of covering on flanges, valves or fittings, multiply the weight factor by the weight per foot of covering used on straight pipe.

Valve weights are approximate. When possible, obtain weights from the manufacturer.

Cast iron valve weights are for flanged end valves; steel weights for welding end valves.

All flanged fitting, flanged valve and flange weights include the proportional weight of bolts or studs to make up all joints.

*16 lb cu. ft. density.

18″ PIPE 18″ O.D.   WEIGHTS OF PIPING MATERIALS

PIPE

Schedule No.	10	20		30		40	60	80	100	120	140	160
Wall Designation			Std.		XS							
Thickness—In.	.250	.312	.375	.438	.500	.562	.750	.937	1.156	1.375	1.562	1.781
Pipe—Lbs/Ft	47.4	59.0	70.6	82.1	93.5	104.8	138.2	170.8	208.0	244.1	274.2	303.5
Water—Lbs/Ft	104.3	102.8	101.2	99.9	98.4	97	92.7	88.5	83.7	79.2	75.3	71.0

WELDING FITTINGS

Fitting	Std (.375)	XS (.500)
L.R. 90° Elbow	256 / 4.5	338 / 4.5
S.R. 90° Elbow	171 / 2.8	225 / 2.8
L.R. 45° Elbow	128 / 1.9	168 / 1.9
Tee	258 / 3.6	328 / 3.6
Lateral	326 / 7.5	526 / 7.5
Reducer	94 / 1.3	123 / 1.3
Cap	57 / 2.1	75 / 2.1

INSULATION

Temperature Range °F	100–199	200–299	300–399	400–499	500–599	600–699	700–799	800–899	900–999	1000–1099	1100–1200
85% Magnesia Calcium Silicate — Nom. Thick., In.	1½	1½	2	2½	3	3	3½	4	4	4½	4½
85% Magnesia Calcium Silicate — Lbs/Ft	7.73	7.73	10.4	13.3	16.3	16.3	19.3	22.6	22.6	25.9	25.9
Combination — Nom. Thick., In.						3	3½	4	4	4½	4½
Combination — Lbs/Ft						22.7	28.0	33.8	33.8	39.5	39.5
*Asbestos Fiber—Sodium Silicate — Nom. Thick., In.	1½	1½	1½	2	2	3	3	4	4	5	5
*Asbestos Fiber—Sodium Silicate — Lbs/Ft	9.93	9.93	9.93	13.72	13.72	21.84	21.84	31.22	31.22	40.77	40.77

FLANGES, FLANGED FITTINGS, VALVES

Pressure Rating psi	Cast Iron 125	250	Steel 150	300	400	600	900	1500	2500
FLANGES									
Screwed or Slip-On	140 / 1.5		140 / 1.5	331 / 1.5	380 / 1.5	573 / 1.5	797 / 1.5	1694 / 1.5	
Welding Neck			160 / 1.5	355 / 1.5	430 / 1.5	652 / 1.5	1074 / 1.5	2069 / 1.5	
Lap Joint			166 / 1.5	355 / 1.5	415 / 1.5	566 / 1.5	820 / 1.5	1769 / 1.5	
Blind	210 / 1.5	396 / 1.5	229 / 1.5	440 / 1.5	572 / 1.5	762 / 1.5	1030 / 1.5		
FLANGED FITTINGS									
S.R. 90° Elbow	621 / 5.8	1060 / 6	711 / 5.8	1126 / 6	1340 / 6.2	1793 / 6.6	2817 / 7		
L.R. 90° Elbow	881 / 7.4	1350 / 7.4	941 / 7.4	1426 / 7.4					
45° Elbow	461 / 4.4	870 / 4.7	521 / 4.4	901 / 4.7	1040 / 4.8	1543 / 5	2252 / 5.2		
Tee	921 / 8.6	1625 / 9	1010 / 8.6	1602 / 9	1909 / 9.3	2690 / 9.9	4327 / 10.5		
VALVES									
Flanged Bonnet Gate	1629 / 8.2	2578 / 9.3	1510 / 6	2505 / 6.5	3765 / 7	4460 / 7.8	6675 / 8.5		
Flanged Bonnet Globe or Angle									
Flanged Bonnet Check	1371 / 10.5								
Pressure Seal Bonnet—Gate						3100 / 5.5	3400 / 5.6	4200 / 6	
Pressure Seal Bonnet—Globe									

Boldface type is weight in pounds. Lightface type is weight factor for insulation.

Insulation thicknesses and weights are based on average conditions and do not constitute a recommendation for specific thicknesses of materials. Insulation weights are based on 85% magnesia and hydrous calcium silicate at 11 lbs/cubic foot. The listed thicknesses and weights of combination covering are the sums of the inner layer of diatomaceous earth at 21 lbs/cubic foot and the outer layer at 11 lbs/cubic foot.

Insulation weights include allowances for wire, cement, canvas, bands and paint, but not special surface finishes.

To find the weight of covering on flanges, valves or fittings, multiply the weight factor by the weight per foot of covering used on straight pipe.

Valve weights are approximate. When possible, obtain weights from the manufacturer.

Cast iron valve weights are for flanged end valves; steel weights for welding end valves.

All flanged fitting, flanged valve and flange weights include the proportional weight of bolts or studs to make up all joints.

*16 lb cu. ft. density.

WEIGHTS OF PIPING MATERIALS   20″ O.D. 20″ PIPE

PIPE	Schedule No.	10	20	30	40	60	80	100	120	140	160		
	Wall Designation		Std.	XS									
	Thickness—In.	.250	.375	.500	.593	.812	1.031	1.281	1.500	1.750	1.968		
	Pipe—Lbs/Ft	52.7	78.6	104.1	122.9	166.4	208.9	256.1	296.4	341.1	379.0		
	Water—Lbs/Ft	129.5	126.0	122.8	120.4	115.0	109.4	103.4	98.3	92.6	87.9		

WELDING FITTINGS

Fitting	Sch 20	Sch 30
L.R. 90° Elbow	317 / 5	419 / 5
S.R. 90° Elbow	212 / 3.4	278 / 3.4
L.R. 45° Elbow	158 / 2.1	208 / 2.1
Tee	321 / 4	407 / 4
Lateral	396 / 8.3	628 / 8.3
Reducer	142 / 1.7	186 / 1.7
Cap	72 / 2.3	94 / 2.3

INSULATION

		Temperature Range °F	100–199	200–299	300–399	400–499	500–599	600–699	700–799	800–899	900–999	1000–1099	1100–1200
85% Magnesia Calcium Silicate		Nom. Thick., In.	1½	1½	2	2½	3	3	3½	4	4	4½	4½
		Lbs/Ft	8.45	8.45	11.6	14.6	17.7	17.7	21.1	24.6	24.6	28.1	28.1
Combination		Nom. Thick., In.						3	3½	4	4	4½	4½
		Lbs/Ft						24.7	30.7	37.0	37.0	43.1	43.1
*Asbestos Fiber— Sodium Silicate		Nom. Thick., In.	1½	1½	1½	2	2	3	3	4	4	5	5
		Lbs/Ft	10.96	10.96	10.96	14.86	14.86	24.24	24.24	33.79	33.79	44.03	44.03

FLANGES / FLANGED FITTINGS / VALVES

	Pressure Rating psi	Cast Iron		Steel						
		125	250	150	300	400	600	900	1500	2500
FLANGES	Screwed or Slip-On	176 / 1.5		181 / 1.5	378 / 1.5	468 / 1.5	733 / 1.5	972 / 1.5	2114 / 1.5	
	Welding Neck			196 / 1.5	431 / 1.5	535 / 1.5	811 / 1.5	1344 / 1.5	2614 / 1.5	
	Lap Joint			211 / 1.5	428 / 1.5	510 / 1.5	725 / 1.5	1048 / 1.5	2189 / 1.5	
	Blind	276 / 1.5	487 / 1.5	298 / 1.5	545 / 1.5	711 / 1.5	976 / 1.5	1287 / 1.5		
FLANGED FITTINGS	S.R. 90° Elbow	792 / 6	1315 / 6.3	922 / 6	1375 / 6.3	1680 / 6.5	2314 / 6.9	3610 / 7.3		
	L.R. 90° Elbow	1132 / 7.8	1725 / 7.8	1352 / 7.8	1705 / 7.8					
	45° Elbow	592 / 4.6	1055 / 4.8	652 / 4.6	1105 / 4.8	1330 / 4.9	1917 / 5.2	2848 / 5.4		
	Tee	1178 / 9	2022 / 9.5	1378 / 9	1908 / 9.5	2370 / 9.7	3463 / 10.1	5520 / 11		
VALVES	Flanged Bonnet Gate	1934 / 8.3	3823 / 9.5	1855 / 6	3370 / 7	5700 / 8	5755 / 8			
	Flanged Bonnet Globe or Angle									
	Flanged Bonnet Check	1772 / 11								
	Pressure Seal Bonnet—Gate									
	Pressure Seal Bonnet—Globe									

Boldface type is weight in pounds. Lightface type beneath weight is weight factor for insulation.

Insulation thicknesses and weights are based on average conditions and do not constitute a recommendation for specific thicknesses of materials. Insulation weights are based on 85% magnesia and hydrous calcium silicate at 11 lbs/cubic foot. The listed thicknesses and weights of combination covering are the sums of the inner layer of diatomaceous earth at 21 lbs/cubic foot and the outer layer at 11 lbs/cubic foot.

Insulation weights include allowances for wire, cement, canvas, bands and paint, but not special surface finishes.

To find the weight of covering on flanges, valves or fittings, multiply the weight factor by the weight per foot of covering used on straight pipe.

Valve weights are approximate. When possible, obtain weights from the manufacturer.

Cast iron valve weights are for flanged end valves; steel weights are for welding end valves.

All flanged fitting, flanged valve and flange weights include the proportional weight of bolts or studs to make up all joints.

*16 lb cu. ft. density.

24″ PIPE 24″ O.D.   WEIGHTS OF PIPING MATERIALS

PIPE

Schedule No.	10	20		30	40	60	80	100	120	140	160
Wall Designation		Std.	XS								
Thickness—In.	.250	.375	.500	.562	.687	.968	1.218	1.531	1.812	2.062	2.343
Pipe—Lbs/Ft	63.4	94.6	125.5	140.8	171.2	238.1	296.4	367.4	429.4	483.1	541.9
Water—Lbs/Ft	188.0	183.8	180.1	178.1	174.3	165.8	158.3	149.3	141.4	134.5	127.0

WELDING FITTINGS

Fitting		
L.R. 90° Elbow	458 / 6	606 / 6
S.R. 90° Elbow	305 / 3.7	404 / 3.7
L.R. 45° Elbow	229 / 2.5	302 / 2.5
Tee	445 / 4.9	563 / 4.9
Lateral	544 / 10	882 / 10
Reducer	167 / 1.7	220 / 1.7
Cap	102 / 2.8	134 / 2.8

INSULATION

Temperature Range °F	100–199	200–299	300–399	400–499	500–599	600–699	700–799	800–899	900–999	1000–1099	1100–1200
85% Magnesia Calcium Silicate — Nom. Thick., In.	1½	1½	2	2½	3	3	3½	4	4	4½	4½
Lbs/Ft	10.0	10.0	13.4	17.0	21.0	21.0	24.8	28.7	28.7	32.9	32.9
Combination — Nom. Thick., In.						3	3½	4	4	4½	4½
Lbs/Ft						29.2	36.0	43.1	43.1	50.6	50.6
*Asbestos Fiber—Sodium Silicate — Nom. Thick., In.	1½	1½	1½	2	2	3	3	4½	4½	5	5
Lbs/Ft	13.55	13.55	13.55	18.44	18.44	28.38	28.38	45.06	45.06	50.97	50.97

FLANGES

Pressure Rating psi	Cast Iron		Steel						
	125	250	150	300	400	600	900	1500	2500
Screwed or Slip-On	255 / 1.5		245 / 1.5	577 / 1.5	676 / 1.5	1056 / 1.5	1823 / 1.5	3378 / 1.5	
Welding Neck			295 / 1.5	632 / 1.5	777 / 1.5	1157 / 1.5	2450 / 1.5	4153 / 1.5	
Lap Joint			295 / 1.5	617 / 1.5	752 / 1.5	1046 / 1.5	2002 / 1.5	3478 / 1.5	
Blind	405 / 1.5	757 / 1.5	446 / 1.5	841 / 1.5	1073 / 1.5	1355 / 1.5	2442 / 1.5		

FLANGED FITTINGS

	125	250	150	300	400	600	900
S.R. 90° Elbow	1231 / 6.7	2014 / 6.8	1671 / 6.7	2174 / 6.8	2474 / 7.1	3506 / 7.6	6155 / 8.1
L.R. 90° Elbow	1711 / 8.7	2644 / 8.7	1821 / 8.7	2874 / 8.7			
45° Elbow	871 / 4.8	1604 / 5	1121 / 4.8	1634 / 5	1974 / 5.1	2831 / 5.5	5124 / 6
Tee	1836 / 10	3061 / 10.2	2276 / 10	3161 / 10.2	3811 / 10.6	5184 / 11.4	9387 / 12.1

VALVES

	125	250	150	300	400	600
Flanged Bonnet Gate	3062 / 8.5	6484 / 9.8	2500 / 5	4675 / 7	6995 / 8.7	8020 / 9.5
Flanged Bonnet Globe or Angle						
Flanged Bonnet Check	2956 / 12					
Pressure Seal Bonnet—Gate						
Pressure Seal Bonnet—Globe						

Boldface type is weight in pounds. Lightface type beneath weight is weight factor for insulation.

Insulation thicknesses and weights are based on average conditions and do not constitute a recommendation for specific thicknesses of materials. Insulation weights are based on 85% magnesia and hydrous calcium silicate at 11 lbs/cubic foot. The listed thicknesses and weights of combination covering are the sums of the inner layer of diatomaceous earth at 21 lbs/cubic foot and the outer layer at 11 lbs/cubic foot.

Insulation weights include allowances for wire, cement, canvas, bands and paint, but not special surface finishes.

To find the weight of covering on flanges, valves or fittings, multiply the weight factor by the weight per foot of covering used on straight pipe.

Valve weights are approximate. When possible, obtain weights from the manufacturer.

Cast iron valve weights are for flanged end valves; steel weights for welding end valves.

All flanged fitting, flanged valve and flange weights include the proportional weight of bolts or studs to make up all joints.

*16 lb cu. ft. density.

WEIGHTS OF PIPING MATERIALS   26″ O.D. 26″ PIPE

PIPE													
	Schedule No.	10		20									
	Wall Designation		Std.	XS									
	Thickness—In.	.312	.375	.500									
	Pipe—Lbs/Ft	85.7	102.6	136.2									
	Water—Lbs/Ft	219.2	216.8	212.5									

WELDING FITTINGS

L.R. 90° Elbow	602 8.5	713 8.5	
S.R. 90° Elbow	359 5	474 5	
L.R. 45° Elbow	269 3.5	355 3.5	
Tee	634 6.8	794 6.8	
Lateral			
Reducer	200 2.5	272 2.5	
Cap	110 4.3	145 4.3	

INSULATION

Temperature Range °F		100–199	200–299	300–399	400–499	500–599	600–699	700–799	800–899	900–999	1000–1099	1100–1200
85% Magnesia Calcium Silicate	Nom. Thick., In.	1½	1½	2	2½	3	3½	4	4½	5	5	6
	Lbs/Ft	10.4	10.4	14.1	18.0	21.9	26.0	30.2	34.6	39.1	39.1	48.4
Combination	Nom. Thick., In.						3½	4½	5½	6	6½	7
	Lbs/Ft						37.0	51.9	67.8	76.0	84.5	93.2
*Asbestos Fiber Sodium Silicate	Nom. Thick., In.	2	2	2	2	2	3	3	4½	4½	5	5
	Lbs/Ft	18.93	18.93	18.93	18.93	18.93	29.87	29.87	47.60	47.60	53.85	53.85

FLANGES / FLANGED FITTINGS / VALVES

	Pressure Rating psi	Cast Iron		Steel						
		125	250	150	300	400	600	900	1500	2500
Screwed or Slip-On				250 1.5	570 1.5	650 1.5	950 1.5	1525 1.5		
Welding Neck				300 1.5	670 1.5	750 1.5	1025 1.5	1575 1.5		
Lap Joint										
Blind				525 1.5	1050 1.5	1125 1.5	1525 1.5	2200 1.5		
S.R. 90° Elbow										
L.R. 90° Elbow										
45° Elbow										
Tee										
Flanged Bonnet Gate										
Flanged Bonnet Globe or Angle										
Flanged Bonnet Check										
Pressure Seal Bonnet—Gate										
Pressure Seal Bonnet—Globe										

Boldface type is weight in pounds. Lightface type beneath weight is weight factor for insulation.

Insulation thicknesses and weights are based on average conditions and do not constitute a recommendation for specific thicknesses of materials. Insulation weights are based on 85% magnesia and hydrous calcium silicate at 11 lbs/cubic foot. The listed thicknesses and weights of combination covering are the sums of the inner layer of diatomaceous earth at 21 lbs/cubic foot and the outer layer at 11 lbs/cubic foot.

Insulation weights include allowances for wire, cement, canvas, bands and paint, but not special surface finishes.

To find the weight of covering on flanges, valves or fittings, multiply the weight factor by the weight per foot of covering used on straight pipe.

Valve weights are approximate. When possible, obtain weights from the manufacturer.

Cast iron valve weights are for flanged end valves; steel weights for welding end valves.

All flanged fitting, flanged valve and flange weights include the proportional weight of bolts or studs to make up all joints.

*16 lb cu. ft. density.

28" PIPE 28" O.D.   WEIGHTS OF PIPING MATERIALS

PIPE					
Schedule No.	10		20	30	
Wall Designation		Std.	XS		
Thickness—In.	.312	.375	.500	.625	
Pipe—Lbs/Ft	92.4	110.6	146.9	182.7	
Water—Lbs/Ft	255.0	252.7	248.1	243.6	

WELDING FITTINGS

L.R. 90° Elbow	626 / 9	829 / 9
S.R. 90° Elbow	415 / 5.4	551 / 5.4
L.R. 45° Elbow	312 / 3.6	413 / 3.6
Tee	729 / 7	910 / 7
Lateral		
Reducer	210 / 2.7	290 / 2.7
Cap	120 / 4.5	160 / 4.5

INSULATION

Temperature Range °F	100–199	200–299	300–399	400–499	500–599	600–699	700–799	800–899	900–999	1000–1099	1100–1200
85% Magnesia Calcium Silicate — Nom. Thick., In.	1½	1½	2	2½	3	3½	4	4½	5	5	6
85% Magnesia Calcium Silicate — Lbs/Ft	11.2	11.2	15.1	19.2	23.4	27.8	32.3	36.9	41.6	41.6	51.4
Combination — Nom. Thick., In.						3½	4½	5½	6	6½	7
Combination — Lbs/Ft						39.5	55.4	72.2	80.9	89.8	99.0
*Asbestos Fiber—Sodium Silicate — Nom. Thick., In.	2	2	2	2	2	3	3	4½	4½	5	5
*Asbestos Fiber—Sodium Silicate — Lbs/Ft	20.26	20.26	20.26	20.26	20.26	31.90	31.90	52.51	52.51	59.17	59.17

FLANGES

Pressure Rating psi	Cast Iron		Steel						
	125	250	150	300	400	600	900	1500	2500
Screwed or Slip-On			285 / 1.5	720 / 1.5	780 / 1.5	1075 / 1.5	1800 / 1.5		
Welding Neck			315 / 1.5	810 / 1.5	880 / 1.5	1175 / 1.5	1850 / 1.5		
Lap Joint									
Blind			620 / 1.5	1275 / 1.5	1425 / 1.5	1750 / 1.5	2575 / 1.5		

FLANGED FITTINGS

S.R. 90° Elbow	
L.R. 90° Elbow	
45° Elbow	
Tee	

VALVES

Flanged Bonnet Gate	
Flanged Bonnet Globe or Angle	
Flanged Bonnet Check	
Pressure Seal Bonnet—Gate	
Pressure Seal Bonnet—Globe	

Boldface type is weight in pounds. Lightface type beneath weight is weight factor for insulation.

Insulation thicknesses and weights are based on average conditions and do not constitute a recommendation for specific thicknesses of materials. Insulation weights are based on 85% magnesia and hydrous calcium silicate at 11 lbs/cubic foot. The listed thicknesses and weights of combination covering are the sums of the inner layer of diatomaceous earth at 21 lbs/cubic foot and the outer layer at 11 lbs/cubic foot.

Insulation weights include allowances for wire, cement, canvas, bands and paint, but not special surface finishes.

To find the weight of covering on flanges, valves or fittings, multiply the weight factor by the weight per foot of covering used on straight pipe.

Valve weights are approximate. When possible, obtain weights from the manufacturer.

Cast iron valve weights are for flanged end valves; steel weights for welding end valves.

All flanged fitting, flanged valve and flange weights include the proportional weight of bolts or studs to make up all joints.

*16 lb cu. ft. density.

WEIGHTS OF PIPING MATERIALS 30″ O.D. 30″ PIPE

PIPE	Schedule No.	10			20		30							
	Wall Designation		Std.		XS									
	Thickness—In.	.312	.375	.438	.500	.562	.625							
	Pipe—Lbs/Ft	98.9	118.7	138.0	157.6	176.8	196.1							
	Water—Lbs/Ft	293.5	291.0	288.4	286.0	283.6	281.1							

WELDING FITTINGS

	Std.	XS
L.R. 90° Elbow	775 / 10	953 / 10
S.R. 90° Elbow	470 / 5.9	644 / 5.9
L.R. 45° Elbow	358 / 3.9	475 / 3.9
Tee	855 / 7.8	1065 / 7.8
Lateral		
Reducer	220 / 3.9	315 / 3.9
Cap	125 / 4.8	175 / 4.8

INSULATION

		100–199	200–299	300–399	400–499	500–599	600–699	700–799	800–899	900–999	1000–1099	1100–1200
Temperature Range °F												
85% Magnesia Calcium Silicate	Nom. Thick., In.	1½	1½	2	2½	3	3½	4	4½	5	5	6
	Lbs/Ft	11.9	11.9	16.1	20.5	25.0	29.5	34.3	39.1	44.1	44.1	54.4
Combination	Nom. Thick., In.						3½	4½	5½	6	6½	7
	Lbs/Ft						42.1	58.9	76.5	85.7	95.1	104.7
*Asbestos Fiber—Sodium Silicate	Nom. Thick., In.	2½	2½	2½	2½	2½	3	3	4½	4½	5	5
	Lbs/Ft	27.17	27.17	27.17	27.17	27.17	33.43	33.43	55.18	55.18	62.18	62.18

	Pressure Rating psi	Cast Iron		Steel						
		125	250	150	300	400	600	900	1500	2500
FLANGES	Screwed or Slip-On			315 / 1.5	810 / 1.5	900 / 1.5	1175 / 1.5	2075 / 1.5		
	Welding Neck			360 / 1.5	930 / 1.5	1000 / 1.5	1300 / 1.5	2150 / 1.5		
	Lap Joint									
	Blind			720 / 1.5	1500 / 1.5	1675 / 1.5	2000 / 1.5	3025 / 1.5		
FLANGED FITTINGS	S.R. 90° Elbow									
	L.R. 90° Elbow									
	45° Elbow									
	Tee									
VALVES	Flanged Bonnet Gate									
	Flanged Bonnet Globe or Angle									
	Flanged Bonnet Check									
	Pressure Seal Bonnet—Gate									
	Pressure Seal Bonnet—Globe									

Boldface type is weight in pounds. Lightface type beneath weight is weight factor for insulation.

Insulation thicknesses and weights are based on average conditions and do not constitute a recommendation for specific thicknesses of materials. Insulation weights are based on 85% magnesia and hydrous calcium silicate at 11 lbs/cubic foot. The listed thicknesses and weights of combination covering are the sums of the inner layer of diatomaceous earth at 21 lbs/cubic foot and the outer layer at 11 lbs/cubic foot.

Insulation weights include allowances for wire, cement, canvas, bands and paint, but not special surface finishes.

To find the weight of covering on flanges, valves or fittings, multiply the weight factor by the weight per foot of covering used on straight pipe.

Valve weights are approximate. When possible, obtain weights from the manufacturer.

Cast iron valve weights are for flanged end valves; steel weights for welding end valves.

All flanged fitting, flanged valve and flange weights include the proportional weight of bolts or studs to make up all joints.

*16 lb cu. ft. density.

32″ PIPE 32″ O.D.   **WEIGHTS OF PIPING MATERIALS**

		Schedule No.	10		20	30	40							
PIPE		Wall Designation		Std.	XS									
		Thickness—In.	.312	.375	.500	.625	.688							
		Pipe—Lbs/Ft	105.8	126.7	168.2	209.4	229.9							
		Water—Lbs/Ft	335.0	332.3	327.0	321.8	319.2							
WELDING FITTINGS		L.R. 90° Elbow		**818** 10.5	**1090** 10.5									
		S.R. 90° Elbow		**546** 6.3	**722** 6.3									
		L.R. 45° Elbow		**408** 4.2	**541** 4.2									
		Tee		**991** 8.4	**1230** 8.4									
		Lateral												
		Reducer		**255** 3.1	**335** 3.1									
		Cap		**145** 5.2	**190** 5.2									

		Temperature Range °F	100–199	200–299	300–399	400–499	500–599	600–699	700–799	800–899	900–999	1000–1099	1100–1200
INSULATION	85% Magnesia Calcium Silicate	Nom. Thick., In.	1½	1½	2	2½	3	3½	4	4½	5	5	6
		Lbs/Ft	12.7	12.7	17.1	21.7	26.5	31.3	36.3	41.4	46.6	46.6	57.5
	Combination	Nom. Thick., In.						3½	4½	5½	6	6½	7
		Lbs/Ft						44.7	62.3	80.9	90.5	100.4	110.5
	*Asbestos Fiber— Sodium Silicate	Nom. Thick., In.	3	3	3	3	3	3	3	4½	4½	5	5
		Lbs/Ft	41.50	41.50	41.50	41.50	41.50	41.50	41.50	57.27	57.27	64.49	64.49

		Pressure Rating psi	Cast Iron		Steel						
			125	250	150	300	400	600	900	1500	2500
FLANGES		Screwed or Slip-On			**395** 1.5	**890** 1.5	**1025** 1.5	**1375** 1.5	**2500** 1.5		
		Welding Neck			**435** 1.5	**1025** 1.5	**1150** 1.5	**1500** 1.5	**2575** 1.5		
		Lap Joint									
		Blind			**870** 1.5	**1775** 1.5	**1975** 1.5	**2300** 1.5	**3650** 1.5		
FLANGED FITTINGS		S.R. 90° Elbow									
		L.R. 90° Elbow									
		45° Elbow									
		Tee									
VALVES		Flanged Bonnet Gate									
		Flanged Bonnet Globe or Angle									
		Flanged Bonnet Check									
		Pressure Seal Bonnet—Gate									
		Pressure Seal Bonnet—Globe									

Boldface type is weight in pounds. Lightface type beneath weight is weight factor for insulation.

Insulation thicknesses and weights are based on average conditions and do not constitute a recommendation for specific thicknesses of materials. Insulation weights are based on 85% magnesia and hydrous calcium silicate at 11 lbs/cubic foot. The listed thicknesses and weights of combination covering are the sums of the inner layer of diatomaceous earth at 21 lbs/cubic foot and the outer layer at 11 lbs/cubic foot.

Insulation weights include allowances for wire, cement, canvas, bands and paint, but not special surface finishes.

To find the weight of covering on flanges, valves or fittings, multiply the weight factor by the weight per foot of covering used on straight pipe.

Valve weights are approximate. When possible, obtain weights from the manufacturer.

Cast iron valve weights are for flanged end valves; steel weights for welding end valves.

All flanged fitting, flanged valve and flange weights include the proportional weight of bolts or studs to make up all joints.

*16 lb cu. ft. density.

WEIGHTS OF PIPING MATERIALS   34″ O.D. 34″ PIPE

PIPE	Schedule No.	10		20	30	40								
	Wall Designation		Std.	XS										
	Thickness—In.	.312	.375	.500	.625	.688								
	Pipe—Lbs/Ft	112.4	134.7	178.9	222.8	244.6								
	Water—Lbs/Ft	379.1	376.0	370.3	365.0	362.2								

WELDING FITTINGS	L.R. 90° Elbow	926 11	1230 11										
	S.R. 90° Elbow	617 5.5	817 5.5										
	L.R. 45° Elbow	463 4.4	615 4.4										
	Tee	1136 8.9	1420 8.9										
	Lateral												
	Reducer	270 3.3	355 3.3										
	Cap	160 5.6	210 5.6										

		100–199	200–299	300–399	400–499	500–599	600–699	700–799	800–899	900–999	1000–1099	1100–1200
INSULATION	85% Magnesia Calcium Silicate	Nom. Thick., In.	1½	1½	2	2½	3	3½	4	4½	5	5	6
		Lbs/Ft	13.4	13.4	18.2	23.0	28.0	33.1	38.3	43.7	49.1	49.1	60.5
	Combination	Nom. Thick., In.						3½	4½	5½	6	6½	7
		Lbs/Ft						47.2	65.8	85.3	95.4	105.7	116.3
	*Asbestos Fiber— Sodium Silicate	Nom. Thick., In.	3	3	3	3	3	3	3	4½	4½	5	5
		Lbs/Ft		38.74	38.74	38.74	38.74	38.74	38.74	60.50	60.50	68.04	68.04

	Pressure Rating psi	Cast Iron		Steel							
		125	250	150	300	400	600	900	1500	2500
FLANGES	Screwed or Slip-On			420 1.5	1075 1.5	1150 1.5	1500 1.5	2950 1.5		
	Welding Neck			465 1.5	1200 1.5	1300 1.5	1650 1.5	3025 1.5		
	Lap Joint									
	Blind			990 1.5	2025 1.5	2250 1.5	2575 1.5	4275 1.5		
FLANGED FITTINGS	S.R. 90° Elbow									
	L.R. 90° Elbow									
	45° Elbow									
	Tee									
VALVES	Flanged Bonnet Gate									
	Flanged Bonnet Globe or Angle									
	Flanged Bonnet Check									
	Pressure Seal Bonnet—Gate									
	Pressure Seal Bonnet—Globe									

Boldface type is weight in pounds. Lightface type beneath weight is weight factor for insulation.

Insulation thicknesses and weights are based on average conditions and do not constitute a recommendation for specific thicknesses of materials. Insulation weights are based on 85% magnesia and hydrous calcium silicate at 11 lbs/cubic foot. The listed thicknesses and weights of combination covering are the sums of the inner layer of diatomaceous earth at 21 lbs/cubic foot and the outer layer at 11 lbs/cubic foot.

Insulation weights include allowances for wire, cement, canvas, bands and paint, but not special surface finishes.

To find the weight of covering on flanges, valves or fittings, multiply the weight factor by the weight per foot of covering used on straight pipe.

Valve weights are approximate. When possible, obtain weights from the manufacturer.

Cast iron valve weights are for flanged end valves; steel weights are for welding end valves.

All flanged fitting, flanged valve and flange weights include the proportional weight of bolts or studs to make up all joints.

*16 lb cu. ft. density.

36″ PIPE 36″ O.D.   WEIGHTS OF PIPING MATERIALS

PIPE	Schedule No.	10		20	30	40
	Wall Designation		Std.	XS		
	Thickness—In.	.312	.375	.500	.625	.750
	Pipe—Lbs/Ft	**119.1**	**142.7**	**189.6**	**236.1**	**282.4**
	Water—Lbs/Ft	**425.9**	**422.6**	**416.6**	**411.0**	**405.1**

WELDING FITTINGS

	Std. (.375)	XS (.500)
L.R. 90° Elbow	**1040** / 12	**1380** / 12
S.R. 90° Elbow	**692** / 5	**913** / 5
L.R. 45° Elbow	**518** / 4.8	**686** / 4.8
Tee	**1294** / 9.5	**1610** / 9.5
Lateral		
Reducer	**340** / 3.6	**360** / 3.6
Cap	**175** / 6	**235** / 6

INSULATION

Temperature Range °F	100–199	200–299	300–399	400–499	500–599	600–699	700–799	800–899	900–999	1000–1099	1100–1200
85% Magnesia Calcium Silicate — Nom. Thick., In.	1½	1½	2	2½	3	3½	4	4½	5	5	6
85% Magnesia Calcium Silicate — Lbs/Ft	**14.2**	**14.2**	**19.2**	**24.2**	**29.5**	**34.8**	**40.3**	**45.9**	**51.7**	**51.7**	**63.5**
Combination — Nom. Thick., In.						3½	4½	5½	6	6½	7
Combination — Lbs/Ft						**49.8**	**69.3**	**89.7**	**100.2**	**111.0**	**122.0**
*Asbestos Fiber—Sodium Silicate — Nom. Thick., In.	3	3	3	3	3	3	3	4½	4½	5	5
*Asbestos Fiber—Sodium Silicate — Lbs/Ft	**40.84**	**40.84**	**40.84**	**40.84**	**40.84**	**40.84**	**40.84**	**55.83**	**55.83**	**71.48**	**71.48**

FLANGES

Pressure Rating psi	Cast Iron		Steel						
	125	250	150	300	400	600	900	1500	2500
Screwed or Slip-On			**480** / 1.5	**1200** / 1.5	**1325** / 1.5	**1600** / 1.5	**3350** / 1.5		
Welding Neck			**520** / 1.5	**1300** / 1.5	**1475** / 1.5	**1750** / 1.5	**3450** / 1.5		
Lap Joint									
Blind			**1125** / 1.5	**2275** / 1.5	**2525** / 1.5	**2950** / 1.5	**4900** / 1.5		

FLANGED FITTINGS

	125	250	150	300	400	600	900	1500	2500
S.R. 90° Elbow									
L.R. 90° Elbow									
45° Elbow									
Tee									

VALVES

	125	250	150	300	400	600	900	1500	2500
Flanged Bonnet Gate									
Flanged Bonnet Globe or Angle									
Flanged Bonnet Check									
Pressure Seal Bonnet—Gate									
Pressure Seal Bonnet—Globe									

Boldface type is weight in pounds. Lightface type beneath weight is weight factor for insulation.

Insulation thicknesses and weights are based on average conditions and do not constitute a recommendation for specific thicknesses of materials. Insulation weights are based on 85% magnesia and hydrous calcium silicate at 11 lbs/cubic foot. The listed thicknesses and weights of combination covering are the sums of the inner layer of diatomaceous earth at 21 lbs/cubic foot and the outer layer at 11 lbs/cubic foot.

Insulation weights include allowances for wire, cement, canvas, bands and paint, but not special surface finishes.

To find the weight of covering on flanges, valves or fittings, multiply the weight factor by the weight per foot of covering used on straight pipe.

Valve weights are approximate. When possible, obtain weights from the manufacturer.

Cast iron valve weights are for flanged end valves; steel weights for welding end valves.

All flanged fitting, flanged valve and flange weights include the proportional weight of bolts or studs to make up all joints.

*16 lb cu. ft. density.

WEIGHTS OF PIPING MATERIALS   42″ O.D. 42″ PIPE

PIPE	Schedule No.												
	Wall Designation	Std.	XS										
	Thickness—In.	.375	.500										
	Pipe—Lbs/Ft	**166.7**	**221.6**										
	Water—Lbs/Ft	**578.7**	**571.7**										

WELDING FITTINGS													
	L.R. 90° Elbow	**1420** 15	**1880** 15										
	S.R. 90° Elbow	**1079** 9	**1430** 9										
	L.R. 45° Elbow	**707** 6	**937** 6										
	Tee												
	Lateral												
	Reducer	**290** 4.5	**385** 4.5										
	Cap	**230** 7.5	**300** 7.5										

INSULATION	Temperature Range °F	100– 199	200– 299	300– 399	400– 499	500– 599	600– 699	700– 799	800– 899	900– 999	1000– 1099	1100– 1200
85% Magnesia Calcium Silicate	Nom. Thick., In.	$1\frac{1}{2}$	$1\frac{1}{2}$	2	$2\frac{1}{2}$	3	$3\frac{1}{2}$	4	$4\frac{1}{2}$	5	5	6
	Lbs/Ft	**16.5**	**16.5**	**22.2**	**28.0**	**34.0**	**40.1**	**46.4**	**52.7**	**59.2**	**59.2**	**72.6**
Combination	Nom. Thick., In.						$3\frac{1}{2}$	$4\frac{1}{2}$	$5\frac{1}{2}$	6	$6\frac{1}{2}$	7
	Lbs/Ft						**57.4**	**79.7**	**102.8**	**114.8**	**126.9**	**139.3**
*Asbestos Fiber— Sodium Silicate	Nom. Thick., In.	3	3	3	3	3	3	3	$4\frac{1}{2}$	$4\frac{1}{2}$	5	5
	Lbs/Ft	**47.06**	**47.06**	**47.06**	**47.06**	**47.06**	**47.06**	**47.06**	**72.92**	**72.92**	**83.22**	**83.22**

FLANGES	Pressure Rating psi	Cast Iron		Steel						
		125	250	150	300	400	600	900	1500	2500
	Screwed or Slip-On			**680** 1.5	**1610** 1.5	**1759** 1.5	**2320** 1.5			
	Welding Neck			**750** 1.5	**1739** 1.5	**1879** 1.5	**2414** 1.5			
	Lap Joint									
	Blind			**1625** 1.5	**3164** 1.5	**3576** 1.5	**4419** 1.5			

FLANGED FITTINGS										
	S.R. 90° Elbow									
	L.R. 90° Elbow									
	45° Elbow									
	Tee									

VALVES										
	Flanged Bonnet Gate									
	Flanged Bonnet Globe or Angle									
	Flanged Bonnet Check									
	Pressure Seal Bonnet—Gate									
	Pressure Seal Bonnet—Globe									

Boldface type is weight in pounds. Lightface type beneath weight is weight factor for insulation.

Insulation thicknesses and weights are based on average conditions and do not constitute a recommendation for specific thicknesses of materials. Insulation weights are based on 85% magnesia and hydrous calcium silicate at 11 lbs/cubic foot. The listed thicknesses and weights of combination covering are the sums of the inner layer of diatomaceous earth at 21 lbs/cubic foot and the outer layer at 11 lbs/cubic foot.

Insulation weights include allowances for wire, cement, canvas, bands and paint, but not special surface finishes.

To find the weight of covering on flanges, valves or fittings, multiply the weight factor by the weight per foot of covering used on straight pipe.

Valve weights are approximate. When possible, obtain weights from the manufacturer.

Cast iron valve weights are for flanged end valves; steel weights for welding end valves.

All flanged fitting, flanged valve and flange weights include the proportional weight of bolts or studs to make up all joints.

*16 lb cu. ft. density.

Appendix C
Formulas for Pipe, Internal Fluid, and Insulation Weights

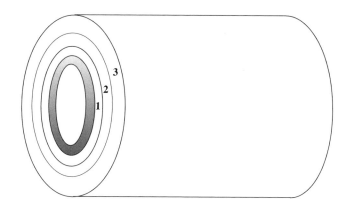

W_p = weight per unit length of pipe (shaded area), lb_m/ft
D_i = inside diameter of pipe, in.
D_o = outside diameter of pipe, in.
D_1 = outside diameter of insulation 1, in.
D_2 = outside diameter of insulation 2, in.
D_3 = outside diameter of insulation 3, in.
T_p = thickness of pipe, in.
T_1 = thickness of insulation 1, in.
T_2 = thickness of insulation 2, in.
T_3 = thickness of insulation 3, in.
ρ_p = specific weight of pipe material, lb_m/ft^3
ρ_1 = specific weight of insulation material 1, lb_m/ft^3

543

ρ_2 = specific weight of insulation material 2, lb_m/ft^3
ρ_3 = specific weight of insulation material 3, lb_m/ft^3
W_1 = insulation 1 weight per unit length, lb_m/ft
W_2 = insulation 2 weight per unit length, lb_m/ft
W_3 = insulation 3 weight per unit length, lb_m/ft
W_T = total weight of pipe, insulation, and internals per unit length, lb_m/ft
γ = specific gravity of process fluid

Weight of the pipe per unit length is as follows:

$$W_p = 0.0218\rho_p T_p(D_i + T_p) = 0.0218\rho_p T_p(D_o - T_p)$$

Weight of the internal fluid is as follows:

$$W_f = 0.340\gamma(D_o - 2T_p)^2$$

The insulation weights are calculated as follows:

$$D_1 = D_o + 2T_1$$

$$D_2 = D_1 + 2T_2$$

$$D_3 = D_2 + 2T_3$$

$$W_1 = 0.0218\rho_1 T_1(D_p + T_1)$$

$$W_2 = 0.0218\rho_2 T_2(D_1 + T_2)$$

$$W_3 = 0.0218\rho_3 T_3(D_2 + T_3)$$

$$W_T = W_1 + W_2 + W_3$$

General formula for the insulation weights is

$$W_{Tins} = \sum_{i=1}^{n} W_i = W_1 + W_2 + \cdots + W_i + \cdots + W_n$$

The total weight of the piping is

$$W_T = W_p + W_f + W_{Tins}$$

Example 1: A 12″ ϕ standard weight (STD WGT) pipe contains water and has no insulation. What is the weight of the pipe per unit length?

The weight of steel is 0.283 lb_m/ft^3 = 489.02 lb_m/ft
The thickness of a 12″ ϕ STD WGT pipe is 0.375 in.
The weight of the pipe is W_p = 0.0218(489.02)(0.375)(12.75 − 0.375)
= 49.47 lb/ft
The weight of the water is W_f = 0.340(1.0)(12.75 − 0.375)²
= 48.96 lb/ft
The total weight of the pipe is W_T = 49.47 + 48.96 = <u>98.43 lb/ft</u>

Example 2: A 36″ ϕ STD WGT gas line has two layers of insulation. The pipe is 0.375 in. nominal wall, the first layer of insulation is 4.92 in. thick with a specific weight of 13.5 lb/ft^3, the second layer of insulation is 1.58 in. thick with a specific weight of 7.49 lb/ft^3. What is the weight of the pipe per unit length?
In spreadsheet form the solution is as follows:

Rules Sheet

Rules

$$\rho_p = \text{pmatl} \cdot 1728$$
$$W_p = 0.0218 \cdot \rho_p \cdot T_p \cdot (D_o \cdot T_p)$$
$$W_f = 0.340 \cdot \gamma \cdot (D_o \cdot 2 \cdot T_p)^2$$
$$W_1 = 0.0218 \cdot \rho_1 \cdot T_1 \cdot (D_o + T_1)$$
$$D_1 = D_o + 2 \cdot T_1$$
$$W_2 = 0.0218 \cdot \rho_2 \cdot T_2 \cdot (D_1 + T_2)$$
$$D_2 = D_1 + 2 \cdot T_2$$
$$W_3 = 0.0218 \cdot \rho_3 \cdot T_3 \cdot (D_2 + T_3)$$
$$W_T = W_p + W_f + W_1 + W_2 + W_3$$

Variables Sheet

Input	Name	Output	Unit	Comment
.283	pmatl			Weight of pipe per unit length, lbm/ln^3
	W_p	142.420599		Weight of pipe per unit length, lbm/ft
	ρ_p	489.024		Specific weight of pipe material, lbm/ft^3
.375	T_p		in	Nominal wall thickness of pipe
36	D_o		in	Outside diameter of pipe
	W_f	0		Weight of internal fluid per unit length, lbm/ft

0	γ			Specific gravity of process fluid
	W_1	59.250360		First insulation layer weight per unit length, lbm/ft
13.5	ρ_1			Specific weight of first layer of insulation per unit length, lbm/ft^3
4.92	T_1		in	Thickness of first layer of insulation
	W_2	12.233675		Second insulation layer weight per unit length, lbm/ft
7.49	ρ_2			Specific weight of second layer of insulation per unit length, lbm/ft^3
1.58	T_2		in	Thickness of second layer of insulation
	D_1	45.84	in	Outside diameter of first layer of insulation
	W_3	0		Third insulation layer weight per unit length, lbm/ft
0	ρ_3			Specific weight of third layer of insulation per unit length, lbm/ft^3
0	T_3		in	Thickness of third layer of insulation
	D_2	49	in	Outside diameter of second layer of insulation
	W_t	213.904634		Total weight of pipe per unit length, lbm/ft

Thus the total weight of the 36″ ϕ pipe with two layers of insulation is 213.9 lb$_m$/ft.

Derivations of Formulas

Weight per unit length of pipe:

Let A = cross-sectional area of pipe wall
D = inside diameter of pipe
T = nominal thickness of pipe wall

$$A = \frac{\pi}{4}\left[(D + 2T)^2 - D^2\right]$$

$$A = \pi T(D + T) \qquad\qquad\qquad \text{Eq. C-1}$$

Let W = weight per unit length

$$W = \rho A \qquad\qquad\qquad\qquad\qquad \text{Eq. C-2}$$

where ρ = specific weight of pipe material, lb$_m$/ft^3

Substituting Eq. C-1 into Eq. C-2, we have

$$W = \pi\rho\left(\frac{lb_m}{ft^3}\right)T(D + T)\,in.^2\left(\frac{1\,ft^2}{144\,in.^2}\right)$$

$$W = 0.0218\rho T(D + T)\frac{lb_m}{ft}$$

Or

$$W = 0.0218\rho T(D_o - T)\frac{lb_m}{ft}$$

Eq. C-3

where D_o = pipe outside diameter

Equation C-3 can be used similarly for the insulation weights.

The formula for the weight of the liquid in the pipe is as follows:
Using the specific weight of water, 62.4 lb_m/ft^3, the weight of the water per unit length inside the pipe is

$$W_{water} = 62.4\frac{lb_m}{ft^3}\frac{\pi(D_o - 2T)^2}{4}\,in.^2\left(\frac{1\,ft^2}{144\,in.^2}\right)$$

with D_o = outside diameter of the pipe and T = nominal pipe wall thickness

$$W_{water} = 0.340(D_o - 2T)^2\frac{lb_m}{ft}$$

Eq. C-4

For any liquid with a specific gravity of γ, Eq. C-4 becomes

$$W_{liquid} = 0.340\gamma(D_o - 2T)^2\frac{lb_m}{ft}$$

Eq. C-5

In Metric SI Units

The weight of the pipe (kg/m) is

$$W_p = \frac{\pi\rho T(D_o - T)}{1,000,000}\frac{kg}{m}$$

Eq. C-6

The weight of the water inside the pipe is

The specific weight of water is 62.4 $\text{lb}_\text{m}/\text{ft}^3$ = 999.543 kg/m³

$$W_{water} = (999.543)\frac{\text{kg}}{\text{m}^3}\frac{\pi(D_o - 2T)^2}{4}\text{mm}^2\left(\frac{1\,\text{m}}{1000\,\text{mm}}\right)^2$$

$$\boxed{W_{water} = 0.000785(D_o - 2T)^2\frac{\text{kg}}{\text{m}}}$$ Eq. C-7

The weight of any liquid in the pipe is as follows:

$$\boxed{W_{water} = 0.000785\gamma(D_o - 2T)^2\frac{\text{kg}}{\text{m}}}$$ Eq. C-8

Example 3: Consider the 12" ϕ STD WGT pipe in Example 1, but use metric SI units. Thus we have the following:

The specific weight of steel = 0.283 lb/in.³ = 7833.37 kg/m³

Using Eq. C-6,

D_o = 12.75 in. = 323.85 mm, T = 0.375 in. = 9.525 mm

$$W_p = \frac{\pi(7833.37)\frac{\text{kg}}{\text{m}^3}(9.525)\,\text{mm}\,(323.85 - 9.525)\,\text{mm}^2}{1,000,000\frac{\text{mm}^3}{\text{m}^3}}$$

W_p = 73.67 kg/m

Check:

$$W_p = 49.47\frac{\text{lb}_\text{m}}{\text{ft}}\left(\frac{3.281\,\text{ft}}{\text{m}}\right)\left(\frac{\text{kg}}{2.2046\,\text{lb}_\text{m}}\right) = 73.62\frac{\text{kg}}{\text{m}}$$

Using Eq. C-7 the weight of the water is

$$W_{water} = 0.000785\left[323.85 - 2(9.525)\right]^2\frac{\text{kg}}{\text{m}}$$

$W_{water} = 72.93$ kg/m

Check:

$$W_{water} = 48.96 \frac{lb_m}{ft} \left(\frac{3.281 \text{ ft}}{m} \right) \left(\frac{kg}{2.2046 \text{ lb}_m} \right)$$

$W_{water} = 72.86$ kg/m

The total weight of the pipe with water $= 73.62$ kg/m $+ 72.93$ kg/m $= 146.55$ kg/m.

Or $W_T = 146.55$ kg/m $= 98.47$ lb$_m$/ft.

The small errors are due to round-off in the units.

Index

551